The Ecology of
Desert Communities

Desert Ecology Series

The Ecology of Desert Communities

Edited by
Gary A. Polis

The University of Arizona Press   Tucson

The University of Arizona Press
www.uapress.arizona.edu

Printed in the United States of America
21 20 19 18 17 16   7 6 5 4 3 2

ISBN-13: 978-0-8165-1186-0 (cloth)
ISBN-13: 978-0-8165-3539-2 (Century Collection paper)

Library of Congress Cataloging-in-Publication Data
The Ecology of desert communities / edited by Gary A. Polis.
    p.   cm. — (Desert ecology series)
    Volume is an outgrowth of the Intercol meetings held in Syracuse in 1986
    Includes index.
    ISBN 0-8165-1186-1 (cl : acid free paper)
    1. Desert ecology. 2. Anthropods—Ecology. I. Polis, Gary A., 1946– .
II. International Congress of Ecology (4th : 1986 : Syracuse, N.Y.) III. Series.
QH541.5.D4E28   1991
574.5'2652—dc20   90-20183
                                                        CIP

British Library Cataloguing-in-Publication Data are available.

♾ This paper meets the requirements of ANSI/NISO Z39.48-1992
(Permanence of Paper).

Contents

Preface

Deserts are one of the major terrestrial ecosystems on this planet. For this reason, as well as their relative lack of complexity and the apparently stressful conditions imposed upon desert biota, scientists have long conducted research in arid lands. Early research was primarily descriptive and autecological. During the last 20 years, a "second generation" of ecologists has refocused research to include more experimental and holistic analyses of the patterns and processes that structure desert communities. Consequently, we have progressed considerably in our understanding how abiotic and biotic factors interact to influence the distribution and abundance of individual species and groups of species. (This is not to say that this research has not generated debate over the relative contribution of these factors.)

The purpose of this book is to present the relatively new ideas and syntheses of this beta generation of desert biologists. This is the first work centering on the structure of desert communities since the early 1970s when the International Biological Program reports on deserts and the two volumes of *Desert Biology* (edited by G. W. Brown) were published. From that time to the present, several excellent books have been written about deserts, for example, Gideon Louw and Mary Seely's *Ecology of Desert Organisms* and Clifford Crawford's *Biology of Desert Invertebrates*. However, these books only tangentially discussed desert community structure. I, along with many of my colleagues, perceive that we are now in a position *to begin* to discern how desert communities are structured. There is now a need and the time is right for a book that will synthesize the new ideas on desert communities. This consensus motivated the writing of this book.

In organizing the book, I established guidelines to create continuity between chapters and facilitate comparisons between different groups of organisms. All authors were asked to produce a synthetic overview and objective presentation of what we know about their groups of organisms. However, in addition to a simple review, I stressed that it is appropriate and important that each author should have the latitude to present his or her idiosyncratic view of what is important in desert community ecology. Further, the authors were asked to suggest directions for future research. Consequently, although this book reviews our knowledge, it is designed to look forward and, I hope, to indicate fruitful topics for research over the next several decades. Thus, hypotheses are presented and evaluated, disagreements are expressed, tentative conclusions are proffered, and new research is suggested.

Although this book embodies most of what we know about desert community ecology, holes exist. Foremost, a chapter viewing desert communities from a historical and paleogeographical perspective would be quite useful. Moreover, there is no discussion on the ecology of perennial plants. I solicited chapters for both subjects; however, various circumstances intervened and neither chapter was written. Further, a chapter on desert riparian communities would be interesting and valuable. Readers should also note that 5 of the 13 chapters focus on arthropods. In part, this reflects my own subjective bias and view of deserts. However, the book does represent arthropods in proportion to their importance in desert communities: they play key roles in and above the soil, as decomposers, as herbivores, as granivores, and as predators. Their biomass and species diversity is much greater than that of all other desert animals combined.

In closing, reading the chapters again leaves me with two overall impressions about the book and the community ecology of deserts. First, we now have a very large amount of information. Scientists have studied deserts for years and their work has been fruitful and interesting. We are now beginning to perceive the real issues and frame the critical questions. Second, as much as we know, we have just scratched the surface. Tremendous disagreement and large lacunae in our knowledge exist even for such well-known groups as ants, birds, and rodents. Our ignorance is even more vast for other groups. So much needs to be done. I view our stage of understanding of (desert) community ecology as equivalent to the gastrula stage in the development of a metazoan embryo: growth and development will continue for a long time before we discern the outline of a mature body of knowledge. This is an exciting insight. The present generation of community and desert ecologists can make great contributions at this stage. I hope this book will contribute to this process.

Acknowledgments

The idea for this book came from the participants at a symposium on desert ecology organized by Claude Grenot and me for the Intercol meetings at Syracuse in 1986. They included Cliff Crawford, Claude Grenot, Ken Nagy, Bob Petruszka, Mary Seely, Vaughn Shoemaker, and Roland Vernet. Barbara Beatty, acquiring editor for the University of Arizona Press, present at these meetings, encouraged the idea to germinate and eventually to grow into this book. Many people read and commented on all or part of the book. In particular, James Brown and Eric Pianka each carefully read the entire book and offered many valuable criticisms and suggestions. The ecology group at Vanderbilt also made many important suggestions. My special thanks to Tracey Wadsworth and Phomma Phothimat for their help with proofreading and compiling the index. Mary Seely and all the wonderful people at the Desert Ecological Research Unit of Namibia provided tremendous hospitality that greatly facilitated many stages in the preparation of this book. The Natural Science Committee and the University Research Council of Vanderbilt have supplied me with funds during the preparation of this book.

Finally, I must acknowledge the many people who have been instrumental in my growth as an ecologist and a desert biologist. My parents first introduced me to nature and to deserts; Larry Pomeroy and Ken Sculteure seemed to always be with me when I first discovered all the joys of deserts; Roger Farley, Marie Turner, and Bill Mayhew taught me about deserts; my students and co-workers have continued to let me learn and wonder about what it all means; and Sharon Lee Polis (almost) always encourages me to "take off" to study the faraway deserts. To everyone above, a heartfelt thanks.

Gary A. Polis

The Ecology of
Desert Communities

Desert Communities: An Overview of Patterns and Processes

1

Gary A. Polis

Deserts are one of the major terrestrial habitats on this planet, occupying one-quarter to one-third of the world's land surface (McGinnies, Goldman, and Paylore 1968; Crawford 1981). They are stressful environments where annual water loss by evapotranspiration may exceed that gained by precipitation. More importantly, desert rainfall is the most unpredictable and sporadic on the planet (Crawford 1981; MacMahon 1981). Further, temperate deserts experience air temperatures that annually exceed 40–50°C. Such conditions translate into rates of primary productivity that are one to three orders of magnitude below those of other habitats (Louw and Seely 1982; Ludwig 1987; Seely this volume). Standing crops of desert plants and animals— functions of precipitation—are characteristically low. The lowest standing biomass ever recorded in a terrestrial ecosystem occurred in the Namib Desert after a prolonged dry period (Seely and Louw 1980). Water availability not only governs productivity but also limits the distribution and abundance of many species and thus restricts membership in desert communities.

The limitations imposed by low and unpredictable water supplies, high temperatures, and low productivity have fostered the suggestion that desert communities are primarily structured by abiotic factors, specifically water availability (Noy-Meir 1973, 1974, 1980, 1985; Brown, Davidson, and Reichman 1979). A further characterization firmly entrenched in the literature is that deserts are relatively simple ecosystems marked by low diversity and the absence of strong biological controls and feedback (Noy-Meir 1974, 1980; Seely and Louw 1980; Wallwork 1982; Whitford 1986). However, the contents of this book argue strongly that deserts are not the simple systems that they first appear. Each chapter focuses on a particular taxonomic group by first presenting patterns of diversity, abundance, and distribution, then analyzing processes that may produce these patterns. The overall document shows that deserts are relatively complex and biologically rich places. The surprising diversity of some taxa, strong spatial and temporal heterogeneity, and a varied array of biotic interactions are some of the factors that promote complexity.

In this chapter, I present an overview of the patterns of diversity and heterogeneity and inspect important processes that may shape desert communities. I focus on themes and questions posed throughout the book: What factors limit the distribution and abundance of member species? What factors establish species diversity in a guild or community? To what extent is community structure influenced by abiotic conditions, low productivity, history, and environmental variation? Are the dynamics of individual species and communities largely controlled by the physical environment or do species interactions shape community structure? How do species interact? Are interactions constant in space and time?

Some terms should be clarified. A community is a group of species that occurs at the same place and time; these species usually interact with one another trophically or through resource use. Some authors use the word community to describe all species in a particular place (e.g., the Namib dune community), whereas others apply it to taxonomic subsets (e.g., the reptile community). "Community structure" refers to several attributes: species diversity, the abundance of member species, their distribution in space and time, and their use of important resources (e.g., water, food).

Patterns

Diversity and Abundance

Deserts support a surprisingly diverse fauna and flora including almost every taxon of terrestrial plant and animal. Even some taxa normally considered to be mesic (algae, mosses, ferns, isopods, and a few species of salamanders and amphisbaenians) occur in deserts. Analysis of species diversity in particular locales (Table 1.1) demonstrates that individual desert communities support hundreds or thousands of species (see also Polis, Polis and Yamashita, Seely this volume).

It would be interesting to compare these data to determine where deserts stand relative to other habitats. Unfortunately, I do not have sufficient data to make this comparison. MacMahon (1981) argues that biotic diversity in deserts is comparable to (or even exceeds) diversity in grasslands and temperate forests. However, Noy-Meir (1985) uses limited evidence to suggest that, in general, the number of plant and animal species declines with increasing aridity. He notes that a positive correlation exists between species richness and rainfall throughout North America for ants, grasshoppers, and mammals (but a negative correlation for reptiles). Within North American and the Namibian deserts, ant and rodent diversity decreases with increasing aridity (MacKay this volume; Marsh 1986; Morton and Davidson 1988). However, diversity of ants in Australia tends to increase with aridity (Morton and Davidson 1988). Wiens (this volume) notes that avian species richness increases with increasing precipitation in Australian deserts but decreases in North American deserts.

Table 1.1 Recorded Species Diversity in Deserts.

	Study Site													
	RSR	CV	DC	NTS₁	NTS₂	SD	AD	CD	ND₁	ND₂	JO₁	JO₂	NEG	MD
Vascular plants	125[a] / 86	174	>600	38	27	250	75		20					250
Arachnids	>100	55	97	41	60	>65			>35	133			>130	
Acarines	>29	>30		8	17				10		3	0		
Isopods	1	1	3	1	1	1			0	0	0	0	1	
Insects	1,231[b] / 4,397	>1,000	>2,540	84[c]	85[c]	138[d]			>90		>300	>300		155[d]
Tenebrionids	109	14	34	26	26		23		33	200				
Ants	27	16	59	17	17	25		23->50	2	36	22	15		25
Myriapoda	2	1	>2	3	7						1	1		
Amphibians	1	0	2	0	0	12	1	5	0	8	6	1	1	14
Reptiles	20	20	34	20	16	43	14	36	9	96	17	20	49	29
Lizards	11	11	20	11	7	19[e] / 8.8	9	17	7		8	9		15
Birds	25[f] / 150	56[f] / 97	84[f] / 123	84	55	57	15[f] / 50	47[f] / 121	5	>79	14	15	57[f] / 80	61
Mammals	21	18	32	23	20	64	16	30	9	>79	16	19	49	3
Rodents	6	8	10	11	11			13	11	2			12	15

STUDY SITES AND SOURCES: RSR = Repetek Sandy Desert Biosphere Reserve (E. Kara Kum, Turkmen Republic, U.S.S.R.; Walter & Box 1983; Reserve pamphlet; V. Fet pers. comm. 1988). CV = Coachella Valley, sandy habitats (Riverside Co., Calif.; Polis this volume); DC = Deep Canyon Reserve (Riverside Co., Calif.; Polis this volume) (diversities of vertebrates are at the station; all other diversities refer to the entire reserve); NTS₁ = Nevada Test Site. *Larrea-Franseria* habitat; NTS₂ = Nevada Test Site. *Grayia-Lycium* habitat (Allred, Beck, & Jorgensen 1963); SD = Sonoran Desert, International Biological Program at Avra Valley (Pima Co., Ariz.; Orians & Solbrig 1977; arachnid diversity from Chew 1961, Polis unpublished; average lizard diversity from Pianka 1986); AD = Algodones Dunes (Sonoran Desert, Imperial Co., Calif.; Bury & Luckenbach 1983); CD = Chihuahuan Desert, Bolson de Mapimi (Durango, Mexico; Barbault & Halffter 1981; ant diversity from MacKay this volume); ND₁ = Namib Desert, Sand Dunes at Gobabeb; ND₂ = bajada site; W. Whitford pers. comm. 1989); NEG = Negev Desert, Sede Boqer, Israel (Y. Lubin pers. comm. 1989); MD = Monte Desert International Biological Program at Bolson de Pipanaco (Orians & Solbrig 1977).

[a]Vascular species (above); non-vascular species (below).

[b]Identified (above); estimated (below).

[c]Insects include only Orthoptera, five families of Coleoptera, and ants.

[d]Insects include only Orthoptera, bees, and ants.

[e]Total number (above); average for each site (below).

[f]Only nesting species (above); all birds (below).

The diversity of scorpions in Australia decreases with increasing aridity; however, scorpion diversity in North America increases with increasing aridity for the continent as a whole but not within deserts (Polis 1989). Apparently the diversity of rodents, reptiles, some insects (e.g., tenebrionids and bombyliids), solpugids, and scorpions is higher in deserts than most other habitats (Reichman, Vitt, Polis, Polis and Yamashita this volume). The diversity of birds and amphibians is undoubtedly lower (Wiens, Woodward, and Mitchell this volume). I suspect that the diversity of other mammals and insects, spiders, and vascular plants is about the same as in other temperate habitats but lower than the diversity of these taxa in the tropics. For example, the number of either summer or winter annual plants in deserts averages 48.5 species (range = 5–187 species; Inouye this volume).

Furthermore, some taxa exhibit extraordinary densities and population biomass (e.g., species of desert ants, termites, scorpions, and isopods) (see Tables 7.3 and 7.4). The density and population biomass of these groups are probably higher in the desert than in most non-arid habitats (MacKay, Crawford, Polis and Yamashita this volume). The density and biomass of most other groups is likely lower, with the possible exceptions of rodents and lizards in some deserts.

Regardless of the exact relative values of these parameters, it is apparent that deserts are home to many species, some of which are quite abundant. These data should dispel the notion that deserts are depauperate places populated by only a few highly adapted species capable of existing under physically harsh conditions.

Trophic Structure and Food Webs

The trophic relations among desert species are quite complex (Polis this volume). This is a product of the opportunistic and catholic behavior of desert consumers (Noy-Meir 1974; Orians et al. 1977; Seely and Louw 1980; Brown 1986; Polis in press; Wiens, Polis, Polis and Yamashita, Vitt this volume). Few desert consumers exhibit specialized diets. The majority are either resource generalists (consuming several species within a similar taxonomic group—e.g., within plants) and/or trophic generalists (eating foods from several different resources—e.g., plants, detritus, arthropods). For example, desert granivores regularly supplement seeds with arthropods, not only for protein but to balance water budgets (MacKay, Polis, Wiens this volume). Extensive omnivory occurs simply because predators eat prey types regardless of the prey's feeding history. Thus, arthropodovores eat arthropods that are herbivores, detritivores, parasitoids, predators, and predators of predators (Polis in press, this volume).

Such great omnivory produces a high degree of connectivity in desert food webs (Polis in press, this volume). Most species (even from different microhabitats and times) are linked directly or indirectly via trophic interactions. At first inspection, several distinct energy channels or subwebs appear in deserts, for example, diurnal versus nocturnal or surface versus subsurface species. However, extensive crossover exists among potential compartments: "different channel omnivores" link compartments when they feed on prey from different subwebs (Polis this volume). For example, detritus and plants are eaten both by species that live above and below the surface. Many desert arthropods feed in the soil as larvae but export energy when they become surface-dwelling adults. Predators eat prey types rather than specializing on particular energy channels or trophic levels.

Omnivory, decompartmentalization, age-structured populations, and heterogeneity in space and time are the four factors that produce very complex food webs in deserts (Polis in press, this volume). It is unclear if desert food webs are more complex than those from other habitats; these same factors are nearly universally distributed and should promote complexity in all ecosystems.

Spatial and Temporal Heterogeneity

All communities vary markedly through time and space. Spatial and temporal differences occur on many different scales and significantly influence community structures (Wiens et al. 1986; Brown 1987; Giller and Gee 1987; Kotler and Brown 1988). Attempting to account for such differences is one of the most challenging but most important tasks of community ecology. In particular, deserts show dramatic spatial variability on scales ranging from square centimeters to square kilometers (Noy-Meir 1973; Brown 1987; Inouye, MacKay, Vitt, Wiens, Zak and Freckman this volume). Important changes also occur temporally, on daily, seasonal, yearly, or historical bases. Such spatial and temporal heterogeneity is manifested in a hierarchy which ultimately creates differences in community structure: environmental heterogeneity sequentially influences first the distribution and abundance of particular taxa, then species composition and trophic relationships, and finally food web structure and community energetics. Many of the authors in this book argue that heterogeneity is one of the more important factors influencing desert communities (Polis, Wiens, Wisdom, Woodward and Mitchell, Zak and Freckman this volume), What is the nature of such heterogeneity and how does it affect the structure of desert communities?

I will document the great heterogeneity in deserts by illustrating spatial and temporal patterns that probably are ubiquitous. Spatially, we see differences on several scales. Desert soils vary in their mineral composition, nutrient reserve, organic content, and capacity to hold water (Noy-Meir 1973, 1981; Crawford, Seely, Zak and Freckman this volume). These differences may be due to erosion and drainage patterns, substrate composition, wind, and biological influences (e.g., termite and ant colonies enrich the areas around their nests; MacKay this volume). Soil nutrients are extremely patchy, found in concentrations around shrubs and in the topsoil layer (Noy-Meir 1985 and included references). Such edaphic factors drastically affect the distribution and abundance of plants (Noy-Meir 1973; Inouye this volume), productivity (Noy-Meir 1981; Ludwig 1986), and the composition and abundance of primary consumers (detritivores, herbivores) and their predators (Noy-Meir 1985; Morton and James 1988; Polis and Yamashita, Seely, Zak and Freckman this volume).

The abundance, distribution, and diversity of detritivores and other soil biota vary greatly as a function of such factors as soil depth and moisture, substrate stability, and spatial distribution of detritus (Seely and Louw 1980; Whitford 1986; Crawford, Seely, Zak and Freckman this volume). Microbial populations are 200 times more abundant in stabilized versus unstabilized sand (Venkateswarlu and Rao 1981). Substantially different species combinations and food webs exist in adjacent desert soils (e.g., buried versus surface litter, under different shrub species, at different distances from shrub canopies) (Zak and Freckman this volume). For example, wind deposits detritus into discrete and rich concentrations on sand dunes (Seely this volume); detritivore abundance and diversity peak within such local accumulations.

The distribution and abundance of herbivores and granivores are correlated closely with spatial differences among desert plants. Seed reserves in the soil vary greatly in abundance among patches (e.g., by over two orders of magnitude; Reichman, Inouye this volume). Such differences translate into differences in the distribution and abundance of granivores (Kotler and Brown 1988; Reichman, MacKay, Wiens this volume). Wisdom (this volume) emphasizes that the community ecology of desert herbivorous insects is strongly influenced by the heterogeneous distribution of their host plants. Not only do plant communities vary in their local species composition and abundance (Inouye this volume) but individual plants of the same species vary in their palatability to herbivores (Wisdom this volume).

Heterogeneous edaphic factors also directly influence the distribution and abundance of animals, independent of their feeding on plants. For example, the biogeographic and microhabitat distribution of the desert iguana (*Dipsosaurus dorsalis*) is determined by soil moisture (Muth 1980). Eggs of these lizards incubate properly in soils only within certain levels of moisture. Eggs desiccate in dry soils and rot in wet soils. Similar limitations are undoubtedly widespread among desert species, for example, grasshoppers (Andrewartha and Birch 1954). Many desert species are adapted to particular substrates or must burrow in soils with specific characteristics (Polis and Yamashita this volume). For example, congeneric African scorpions segregate spatially according to small differences in soil hardness (Lamoral 1978). Lithophilic species of scorpions, lizards, and mammals often are restricted to rocky areas that provide shelter from predators.

Spatial heterogeneity occurs on larger scales. Patterns of precipitation and water infiltration into soils are quite patchy in deserts. Rains due to thunderstorms (rather than to large fronts) may soak one area yet not wet areas just a few kilometers away. Runoff patterns after rains contribute to a mosaic of patches differing in soil moisture and subsequent productivity (Noy-Meir 1981). Productivity varies greatly as a function of topography and runoff, for example, from small gullies to large, arid river valleys (Ludwig 1986). Topographical differences translate into highly heterogeneous desert habitats including some local "hot spots" of abundance and diversity (from well-watered perennial shrubs to riparian communities). These areas provide different degrees of food and shelter for desert animals. However, alluvial areas are not optimal for all animals; flooding can cause high mortality to burrowing animals and in many cases limits the abundance of taxa that would otherwise flourish there (e.g., Bradley 1986). Areas in which water stands likewise vary in productivity according to substances contained in the runoff. For example, artificial ponds in Arizona are rich in nutrients and support a great variety and abundance of desert plants and animals after they dry (Pomeroy 1981). Conversely, depressions that collect saline or alkaline runoff eventually support practically no life as they dry into salt pans.

Precipitation and water availability often vary along geographic gradients or clines (McGinnies, Goldman, and Paylore 1968). Clines may exist on several scales. For example, Polis (this volume) illustrates marked changes in precipitation, productivity, species composition, and feeding ecology at two sites 65 km apart within the Coachella Valley (Colorado Desert, California). In the Namib, rainfall increases but water from fog condensation decreases as one proceeds east from the Atlantic coast (Seely this volume). In the southwestern deserts of the United States, patterns of precipitation gradually change from predominantly winter rains in the west (Mojave Desert) to predominantly summer rains 1,000 km to the east (Sonoran and Chihuahuan deserts). (Predictability also changes along both clines: In the Namib, rain is relatively unpredictable, whereas fog is relatively predictable; in the southwest U.S., winter rains are more sporadic than the regular summer thunderstorms.) Productivity, biomass, species composition, activity patterns, and plant phenology show regular and predictable changes along aridity clines. The intensity of biological interactions also varies along these clines (see Processes, below).

It is important to note that spatial heterogeneity of some types is less in deserts than forest habitats. In particular, foliage height diversity in deserts must be among the lowest of any habitat (perhaps surpassing only grasslands and tundra). Consequently, whatever effects such architecture exerts on community characteristics (e.g., increasing bird diversity by providing more spatial niches) must be less in deserts than forests.

Deserts also are heterogeneous through time. The most important changes occur in precipitation and temperature (Zak and Freckman this volume). Annual variation in precipitation is an inverse function of the amount of precipitation; consequently, patterns of precipitation in deserts are among the most unpredictable on this planet (MacMahon 1981). Annual rainfall may vary by one or two orders of magnitude in deserts (e.g., Southern California: 34–301 mm; Polis this volume; the Namib: 2.2–134 mm; Seely and Louw 1980). Productivity mirrors precipitation during these periods (Seely and Louw 1980; Noy-Meir 1981; and references included in each). For example, the standing biomass in the Namib after a 13-year dry period increased from 600 to 900 percent (plants and animals respectively) after heavy rains (Seely and Louw 1980). Yearly net primary productivity (above-ground) varied by about 20-fold between years in the Chihuahuan Desert (Ludwig 1986). The great variation in the timing of rainfall is a major stochastic influence on productivity (Ludwig 1986, 1987) and community structure (see below).

Longer-term changes in precipitation (decades to centuries to millions of years) also occur. For example, American rain shadow deserts were produced from more mesic areas during relatively recent orogenous events that built the Andes and the Sierra Nevada (Axelrod 1979). Large parts of the southwestern deserts of North America were under great lakes as little as 10,000–12,000 years ago (Benson and Thompson 1987). Temporal heterogeneity at this geological scale could exert strong historical influences (see Processes, below). Desert biota probably experience relatively rapid changes in patterns of distribution and abundance during wet-dry cycles.

The above discussion illustrates the nature of temporal and spatial heterogeneity in deserts. Although heterogeneity characterizes all communities, its effects are particularly exaggerated in deserts where variation is relatively great and extreme variation causes some areas to receive no water for long and irregular periods. As a result, it influences many key processes and is central to our understanding of deserts.

Probably the most apparent effect of heterogeneity is the temporally unpredictable "feast or famine" nature of primary productivity and food availability in deserts (MacMahon 1981; Polis, Vitt, Wiens this volume). Under "bad" conditions, precipitation is low or non-existent and plants grow little if at all. In "good" periods of adequate-to-heavy rains, relatively luxuriant plant growth occurs. Dramatic changes in productivity stimulated Noy-Meir (1973, 1974) to propose his "pulse-reserve hypothesis" as a paradigm for arid areas in general. Noy-Meir argues that plants and animals grow and establish reserves (e.g., seeds, tubers, tissue, eggs) during good times; these reserves maintain the population or individual during interim dry periods.

Several basic features of desert organisms may have evolved in response to unpredictable productivity; for example, divergent life history strategies (Polis and Farley 1980; Louw and Seely 1982; Morton and James 1988; Crawford, Polis and Yamashita, Vitt, Wiens this volume) and opportunistic feeding habits (Noy-Meir 1974; Brown 1986; Polis this volume). Differences in life history and trophic opportunism exert great influence on such biological interactions as competition and predation (Polis and Yamashita, Vitt, Wiens this volume). For example, opportunistic animals exhibit quick functional responses to prey eruptions but probably cannot tightly regulate particular prey species. Crawford (1986) notes that, in general, spatial and temporal patchiness of nutritional reserves in deserts combine with stochastic arrival of moisture to limit the accuracy of what can be predicted about foraging and the impact of feeding guilds.

Spatial patchiness, temporal variation, and aridity clines produce a dynamic, non-equilibrium community. Productivity, species composition, patterns of abundance, and species interactions are not constant in space or time. Such variation exerts several consequences at the community level. First, it likely inhibits a universal outcome of any particular abiotic or biotic process. "Hide and seek" dynamics in heterogeneous environments allow for local extinctions followed by recolonization (Taylor 1988). Extinctions may be caused by physical disturbances, competitive exclusion, or mortality from predators or pathogens. For example, areas of relatively high productivity (e.g., runoff areas) serve as refugia in times of severe drought (Noy-Meir 1981). Heterogeneity allows persistence and coexistence among species that are engaged in otherwise deterministic interactions (Caswell 1978; Taylor 1988; Polis and Yamashita this volume). Inferior competitors or prey can escape elimination by being distributed in periods or places that are enemy-free. Thus, heterogeneity spreads the risk of extinction (denBoer 1968) and increases population persistence by decreasing overall susceptibility to various mortality factors. These processes promote biotic diversity.

Second, the quantity and quality of resources vary greatly between patches in heterogeneous environments. "Hot spots" of relatively high primary productivity and biological activity apparently characterize deserts. Plant production in these areas travels up the food web to affect secondary productivity and the distribution and abundance of heterotrophs (Noy-Meir 1985). For example, some desert shrubs support a high diversity of herbivorous insects (Wisdom this volume). Predators are attracted to these areas: 26 spider species occur on creosote *(Larrea divaricata)* and more than 25 species on saltbush *(Atriplex canescens)* (Polis and Yamashita this volume). Patchiness between good and bad microhabitats produces variation in feeding success, growth rates, and reproduction of individual consumers (Polis and Yamashita this volume).

Third, heterogeneity creates a mix of good and bad periods and habitats. Resource limitation and competition may occur only during periods of "ecological crunches" (Wiens 1977, this volume). Periods between crunches are marked by little or no competition and by relaxed selection. Heterogeneity in production between (micro-)habitats may partition a species's population into individuals that live in "source" and "sink" habitats (Pulliam 1988). Source habitats produce a net surplus of individuals. Sink habitats are suboptimal areas where populations are not self-sustaining but exist only because of migration from source habitats. Such a system might occur commonly in heterogeneous deserts.

Finally, heterogeneity should slow the speed of evolution and the rate that species coevolve in a community (predators and prey; competitors, mutualists). The strength of selection is not constant and gene flow may disrupt locally adaptive changes in gene frequency. Since interacting populations only contact intermittently or at local points of sympatry, we expect to see species interactions that are relatively less tightly coevolved. Effects at the level of the community should be manifested in the organization of exploiter-victim systems or guilds of potential competitors that are looser relative to those in more homogeneous environments. Woodward and Mitchell (this volume) discuss how heterogeneity (patch structure) has influenced evolution of communities of desert anurans.

Processes

One purpose of this book is to evaluate the contribution of different processes to the distribution and abundance of individual species and (consequently) on community structure. Many different factors have been advanced as explanations for the structure of communities (see Strong et al. 1984; Diamond and Case 1986; Gee and Giller 1987). These factors can be subtended under seven general hypotheses that are not mutually exclusive: (1) productivity, (2) autecology, (3) history and chance, (4) resource limitation and exploitation competition, (5) interference competition, (6) predators, parasites, and pathogens, and (7) disturbance. Each is examined by authors in this book. The last four posit that an interactive community is formed by species that are linked to one another either trophically, via competition, or by other biological interactions (e.g., mutualism). The second argues that species groupings are an assemblage of largely non-interactive units, and the third downplays the importance of interactions. The first two and the last hypotheses explicitly recognize that abiotic factors affect community structure.

In this section I present an overview of the importance of these processes in terms of their potential effect on desert communities. It is no surprise that data are lacking to evaluate fully the relative importance of any one process or the relative effects of different processes even on well-studied groups such as ants, reptiles, rodents, and birds (MacKay, Vitt, Reichman, Wiens this volume). Consequently, I approach this section by addressing the conditions necessary for each process to operate, by showing the potential importance of each, and by indicating which process may be expected to operate in particular taxa. Not surprisingly, I conclude that each process contributes to producing the observed community patterns in deserts, with the exact importance of any one process varying from desert to desert and group to group.

Hypothesis 1: Productivity

Deserts are often described as "harsh" environments largely controlled by extreme abiotic factors, the most stressful and limiting being water. Noy-Meir (1973, 1974, 1980, 1985) proposed that an understanding of the limits imposed by water would be sufficient to represent much of the dynamics of entire desert communities. The basic rationale is that water limits primary productivity, and primary productivity limits energy available to consumers and thus controls secondary productivity. Primary productivity is tightly correlated with precipitation (Davidson 1977; Noy-Meir 1981, 1985; Ludwig 1986, 1987). Deserts receive less water than any other habitat and consequently are the least productive habitats on earth (Louw and Seely 1982; Ludwig 1987; Seely this volume). However, this water-centered view must be modified to include other factors that also limit primary productivity in deserts: nitrogen, phosphates, the often-slow rate of decomposition of litter, the exhaustion of nutrients before they can be replaced (Hadley 1980; Inouye, Wisdom this volume), and many of the processes discussed below.

Secondary productivity (actually consumer biomass) shows a high positive correlation with primary productivity and thus (indirectly) with precipitation (see earlier in this chapter and Orians and Solbrig 1977; Davidson 1977; Noy-Meir 1981, 1985; MacKay and MacKay 1984; Pianka 1986; Marsh 1986; Brown 1987; MacKay, Vitt, Wisdom this volume). However, consumers are limited by more than a deficiency in the total energy available to them. Secondary production appears to be limited by food quality (water and protein content), low levels of available water, extreme temperatures, and reduced spatial heterogeneity (Andrewartha and Birch 1954; Hadley 1980; Seely and Louw 1980; Crawford 1981; Louw and Seely 1982; see also processes below). These factors especially are noted by Wisdom (this volume) to limit the distribution and abundance of desert herbivorous insects.

Productivity influences other aspects of communities. The diversity and abundance of several taxa are apparently correlated with precipitation: some groups are more abundant or diverse with increasing aridity; others, less so (see earlier). Diversity of many groups of consumers also is correlated with spatial diversity of vegetation, a function of primary productivity (e.g., foliage height diversity; see Noy-Meir 1981, Pianka 1986, Wisdom this volume; but see Morton and James 1988, Pianka 1989, Vitt, Wiens this volume for conflicting evidence on Australian lizards and desert birds). Simple environments offer fewer spatial niches and a limited number of nesting or foraging sites and refugia (shelters) from enemies and extreme physical conditions.

As importantly, the strength of different interactions may vary with water availability and productivity. Noy-Meir (1980) envisions a cline running from extremely arid to more mesic conditions. At the arid end, few biological feedbacks (via competition or predation) exist and assemblages of species are abiotically controlled; at the mesic end, feedbacks are relatively more important. This hypothesis has not been tested rigorously. However, Seely's analysis (this volume) of an extremely arid desert (the Namib) shows that biotic processes are relatively unimportant to the structure of this community as compared with abiotic factors. This Namib dune ecosystem apparently stands at the pole where physical factors predominate (parts of the Atacama and Iranian deserts are similarly extreme). At the opposite pole would be deserts with greater rainfall and influenced relatively less by physical factors (e.g., the Sonoran Desert).

The intensity of predation apparently varies along aridity clines. For example, Péfaur (1981) found that the proportion of predators decreased from low elevation, xeric sites on the coastal desert of Peru to more mesic and productive sites at higher elevations. Seely and Louw's (1980) data from the Namib report analogous findings: the ratio of carnivores to detritivores and omnivores decreased from 1:1.2 to 1:7.4 from a dry to a wet period. These data represent species (Péfaur 1981) and biomass (Seely and Louw 1980) of predator; it is unclear if predation intensity similarly increases with increasing aridity (see Polis and Yamashita this volume).

Analogous clines in the presence and intensity of exploitation competition may or may not occur. Desert species may be resource-stressed mostly at the low end of a productivity cline. However, it is not deductively clear where competition should occur. On the one hand, productivity is so low at the arid end that resources should become limited and competition could occur. Alternatively, the abundance of potential competitors may be too low in these areas to allow resource-mediated interactions.

Resource use and niche overlap among potential competitors may vary with changes in productivity. Seely and Louw (1980) found that overlap increases with productivity: increased precipitation resulted in more homogeneously distributed resources and a subsequent increase in spatial niche breadth by both plant and consumer species. Whether overlap always increases with productivity, and whether increased overlap during good periods causes more competition are moot questions (Schoener 1982; Fowler 1986; Polis 1988b; Polis and Yamashita, Wiens this volume).

Thus, physical factors (low or variable water availability and extreme temperatures, or both) exert great effects on the biological community by directly limiting productivity, recycling rates, nutrient availability, and even spatial heterogeneity. These limitations are then expressed in the diversity of plants and animals, in relatively low secondary productivity, and in an apparent decrease in the importance of biological interactions. It is upon this background that subsequent processes operate to produce the community structure of deserts.

Hypothesis 2: Autecology

The second hypothesis regarding community structure proposes that communities are noninteractive assemblages of species whose membership is established almost entirely by the autecology of individual species (Noy-Meir 1980; Strong 1983; Seely, Vitt, Wiens this volume). Those species possessing features (e.g., response to physical factors, foraging biology) that allow them to be present in a particular time and place form such assemblages. Absent are populations unable to respond to the unpredictable vagaries of the desert environment or those that do not possess the requisite ecophysiological or behavioral adaptions to withstand the extremes of heat and water stress (Noy-Meir 1974; Crawford 1981; Louw and Seely 1982; Morton and James 1988). Community structure is viewed as little more than the sum of the individual biologies of member species (Strong 1983). Species may adapt vertically to species above and below them in the food chain (i.e., their predators and prey). However horizontal, synecological processes such as competition contribute relatively little to species existence or community structure. Note that individual responses can produce community-level patterns of niche separation similar to those hypothetically produced by biological interactions. For example, resources (food, space) will be partitioned among species if species specialize in order to use that resource efficiently. Such autecological tracking by consumers could produce trophic differences within a guild similar to those produced by competition (e.g., Bloom 1981).

Deserts are one of the few habitats where the autecological hypothesis has been suggested to be a major force producing community structure (Andrewartha and Birch 1954; Noy-Meir 1973, 1974, 1980; Seely, Wiens this volume). This view may originate in the fact that autecology determines, in large part, not only membership in desert communities but also patterns of distribution, abundance, and diversity by promoting differential success among taxa. First, certain taxa are well represented in deserts because they possess a suite of characteristics that is particularly well-adapted to the abiotic extremes of the desert (e.g., scorpions; Polis and Yamashita this volume) or to unique traits of certain deserts (e.g., Australian lizards feeding on termites that decompose *Spinifex* grasses; Morton and James 1988; but see Pianka 1989). Other taxa, less well-adapted, are less successful (e.g., amphibians; Woodward and Mitchell this volume). Second, life history responses to perturbations (e.g., dry periods, precipitation) partially set population abundance: some species closely track changes in productivity, others are relatively insensitive (Polis and Farley 1980; Crawford 1981; Louw and Seely 1982; Crawford, Polis and Yamashita this volume). Third, autecology establishes limits to distribution in space and time via (ecophysiological and behavioral) adaptions that restrict individuals to a subset of all possible times and places.

The autecological hypothesis forms an important foundation to our understanding of community structure of deserts. Wiens (this volume; see especially Figure 10.1) demonstrates how autecological responses integrate with other processes to shape communities. Even if not stated explicitly, autecology contributes in various degrees to the community ecology of each group in this book. For example, the distribution and abundance of desert shrubs may be established primarily by propagule distribution and species-specific responses to topography, edaphic factors, and soil moisture; other factors (herbivory, granivory, allelochemistry, competition for water, "nurse plants") fine tune these patterns. A prima facie argument for the autecological hypothesis (and against competitive hypotheses) among perennial plants is the lack of good evidence that biologically driven succession occurs among desert plant communities (see Seely this volume and Fowler 1986 for arguments on this subject). Autecology is the key factor explaining patterns of some desert arthropods (e.g., herbivores and detritivores; Crawford, Wisdom, Zak and Freckman this volume); although still important, it is relatively less so among arthropods that interact competitively (predators, ants, and termites; MacKay, Polis and Yamashita this volume). Similar differences occur among vertebrates (e.g., Vitt this volume). Wiens (this volume) hypothesizes that autecology is the major factor determining the distribution and abundance of desert birds. Conversely, biological interactions appear to be more important to desert rodents, although autecology may be the key to understanding coexistence of some aridland African ungulates (see Reichman this volume).

The autecological hypothesis may be particularly useful in explaining much of the community structure in extreme desert environments. Seely (this volume) finds little evidence that biological interactions are important in the Namib dune system. Populations there are generally at such low densities that interactions are infrequent. Although a few interactions are important (e.g., among sparassid spiders), even relatively abundant species (e.g., tenebrionids) appear to exert little influence on each other. The distribution and abundance of Namib species are established largely by the autecological responses of resident species. However, as deserts become less extreme with increasing precipitation (e.g., toward the east in the Namib), productivity increases, densities increase, and the importance of biological interactions also likely increases. Note that important biological interactions presented in the other chapters in this book were studied, for the most part, in deserts that are much less extreme than the Namib.

Although the autecological hypothesis is the most parsimonious explanation of community structure, it is obvious that other factors modify basic autecological patterns. For example, competitive interactions may constrain species distribution to a smaller subset, the realized niche.

Hypothesis 3: History and Chance

The third hypothesis proposes that the specific assemblage of species in a community (and hence much of community structure) is largely a product of the unique past events experienced by that community. Geological, climatic, and biogeographic history combine with the vagaries of dispersal, colonization, and extinction to produce the particular taxonomic set that occurs in one place at one time (see Chesson and

Case 1986). History is expected to influence all communities if nothing more than by offering different species pools with different biologies. For example, Australian deserts exhibit the highest diversities of ants, lizards, and burrowing spiders; solpugids are totally absent. Pleistocene extinctions in North American deserts have left a fauna relatively depauperate of large mammals but relatively rich in small mammals (Martin and Klein 1984; Webb 1984, 1987). The presence and diversity of many taxa are influenced by their Pleistocene distribution (e.g., see Brown 1987 for Great Basin Desert rodents). The paleobiogeography of scorpions has produced faunas in Australia and Africa characterized by a high proportion of fossorial species that forage from their burrows; in North America, a taxon absent on the other two continents (the Vaejovidae) has radiated such that most American species forage primarily on the surface.

Such paleohistorical factors can exert tremendous influence on community organization (Martin and Klein 1984; Janzen 1986; Webb 1984, 1987). Historical and biogeographic events may explain why mammals are a relatively unimportant part of the granivore communities in South America and Australia, but quite important in North America and Israel (Morton 1985). Pleistocene extinction of large mammals apparently has radically changed the selective regime for Chihuahuan and other American desert vegetation (Janzen 1986). Janzen presents evidence that such extinctions cascade down to directly and indirectly affect the abundance and diversity of the entire community, including herbivorous and granivorous insects and vertebrates and even predaceous vertebrates.

At a different scale, Vitt (this volume) shows that historical distributions of *Urosaurus* lizards may be a more powerful explanation than interspecific competition for their present distribution patterns. Niche differences may reflect autecological adaptations to different environments in which these lizards lived in the past when they were allopatric.

However, in theory, "history's thumbprint" may be masked or erased by interactive biological factors that act to organize the structure of communities into similar patterns. Specifically, competition has been theorized to produce such phenomena as limiting similarity in niche overlap, size ratios, and other assemblage rules within communities as well as convergence of these parameters across communities. In particular, convergence of community characteristics should be manifest in simple environments such as deserts. The degree that deserts converge is one measure of the relative influence of historical versus biotic factors in community structure (Orians and Solbrig 1977; Orians and Paine 1983). In fact deserts have been analyzed repeatedly at the community level to test for ecological equivalence and convergence (e.g., the International Biological Program on deserts; Orians and Solbrig 1977.) Such studies show that great differences are observed between independently evolved faunas of desert lizards, ants, birds, and rodents occurring in areas with similar climatic, vegetative, and topographic features (Brown, Davidson, and Reichman, 1979; Morton 1985; Pianka 1986; Morton and Davidson 1988; Morton and James 1988; MacKay, Reichman, Vitt, Wiens this volume). Pianka found that desert lizard communities in apparently similar habitats in Australian, North American, and African deserts varied significantly in the number of species (18–42, 4–11, 12–18 species, respectively), density, foraging tactics, diet, and spatial and temporal patterns (see Morton and James 1988 for a hypothesis why Australian deserts support such a diverse lizard fauna; see also Pianka 1989.) The diversity and density of desert granivores and the patterns and intensity of seed use vary greatly between continents (see Orians and Paine 1983; Morton 1985; Riechman this volume).

This evidence suggests that history is a very important factor that often overrides factors promoting convergence. Convergence at the community level apparently does not occur, because of marked taxonomic differences in biogeographic species pools, differential radiation of existing taxa, and the different ages of deserts. Age is particularly important. Community features (e.g., coevolved biologies, the degree that speciation and radiation have occurred) in recently formed deserts may be quite different from those of much older deserts. This may be one reason why community structure is so different in the comparatively young (4,000–12,000[?]-year-old; Van Devender and Spaulding 1979; Van Devender, Thompson, and Betancourt 1987) North and South American deserts as compared with the multimillion-year-old African and Australian deserts. However, it is important to note that age and other historical differences cannot fully explain community structure. For example, although many taxa (see above) are more diverse in Australian deserts than other equivalent deserts, Australian aridland scorpions are the least diverse of any comparable area (Polis and Yamashita this volume).

Hypothesis 4: Resource Limitation and Exploitation Competition

The fourth hypothesis is that exploitation competition (EC) for limited resources is a key interaction structuring guilds and communities. Contested resources for animals are usually food and occasionally space, and usually water and space for plants. EC is the negative effect exerted by one individual or population on a second due to its differential ability to harvest a resource. In theory, EC selects for species to diverge ecologically from one another in the use of resources. Such divergence decreases overlap, thereby dividing resources and allowing coexistence of potential competitors.

The significance (and indeed the very existence) of EC in natural communities is a hotly debated topic (Schoener 1982; Strong et al. 1984; Diamond and Case 1986; Gee and Giller 1987; the entire issue of *American Naturalist* [1983] 122[5]). Conclusive examples of its existence in nature are notably rare. Because food (and water) is in short supply in deserts, at least seasonally, one may expect that EC is particularly important in arid areas. In fact, EC has been proposed as a major process structuring communities of desert granivores (both ants and rodents; see MacKay, Reichman this volume), birds (Cody 1974), reptiles (Pianka 1986), amphibians (Woodward and Mitchell this volume), and plants (Inouye this volume). However, its significance among plants (Fowler 1986; Inouye this volume), birds (Wiens this volume), reptiles (Vitt this volume), and amphibians (Woodward and Mitchell this volume) is uncertain. EC is considered relatively unimportant among arthropod detritivores (Crawford this volume), herbivores (Otte and Joern 1977; Wisdom this volume), or predators (Polis and Yamashita this volume), among soil biota (Zak and Freckman this volume), or in the Namib sand desert (Seely this volume).

It is not surprising that the importance of EC varies from group to group or that a disagreement exists over its role. Different taxa exploit resources in different ways, and various types of resources are distributed differently in space and time. For example, 90–95 percent of the seed reserves in the desert are depleted by actively searching seed predators (Noy-Meir 1985; Inouye this volume); thus we may expect and do see EC among granivores (MacKay, Reichman this volume). Conversely, only 2–10 percent of leaf and stem plant biomass is consumed by herbivores (Noy-Meir 1985). EC among desert herbivores is probably unimportant because they apparently cannot respond quickly enough to the large and rapid increases in available plant biomass after precipitation (Wisdom this volume; see also Wiens this volume).

However, part of the problem fueling the debate over EC involves the types of evidence marshalled in its support. For example, EC theoretically can produce resource partitioning, niche segregation, and character displacement. This causal relationship has been turned around so that the existence of these patterns is used as evidence that EC operates. As is obvious, correlations do not imply causality. Several distinct processes likewise can produce similar patterns (e.g., autecology, interference, predation). Alternatively, the existence of food limitation is used as evidence. However, two conditions are necessary and sufficient to demonstrate EC in nature (Polis and McCormick 1987): (1) food levels must affect growth, survival, or reproduction ("food limitation condition"); and (2) lowered performance or abundance of species A must be mediated through a change in resource availability clearly due to the presence or absence of species B. This "resource depletion condition" would occur if harvesting by species B sufficiently decreased resource levels, thus suppressing the success of A.

Several lines of evidence suggest that the vast majority of desert organisms are limited by resource availability, at least seasonally (water for plants, food for animals). Water availability, primary production, plant biomass, and seed densities change dramatically through time (see above). The abundance of arthropod prey fluctuates one to two orders of magnitude during the year (Polis, Polis and Yamashita, Seely, Wisdom this volume). Analyses of feeding rates, growth rates, body size, or reproduction invariably demonstrate that these parameters are functions of food availability (Polis 1988b). As shown by several authors in this book, individuals do better in good years or in good microhabitats.

Thus food frequently limits consumer success in deserts. However, this does not imply competition for energy, as two distinct factors can cause food limitation. First, harvesting may deplete the supply of food. In this case, EC occurs. Alternatively, inherent limitations in the foraging biology of the consumer may not allow sufficient capture or ingestion of food. Consumers may not have enough time to forage, may not be efficient under all concentrations of prey, or may not be able to handle enough prey to provide the energy necessary for maximum growth or reproduction. Foraging time limitations may be exerted by constraints that are either abiotic (e.g., high temperatures) or biotic (e.g., predator avoidance; see below). For example, prey may be at such low densities (e.g., in extreme deserts or during bad years) and so widely distributed that foraging time per se may be the factor limiting energy acquisition (Seely and Louw 1980).

Such inherent limitations were described by Andrewartha and Birch (1954) as "relative food shortage" (see also Andrewartha and Browning 1961; White 1978; Polis and McCormick 1987). I speculate that, although difficult to assess, relative food shortage produces food limitation in deserts as frequently if not more so than does resource depletion via EC. It is a more parsimonious explanation of food limitation than that of EC: consumers simply experience difficulties in finding and capturing food. Analogous arguments can be made for relative water shortage as an alternate explanation to EC for water among desert plants.

So what is the importance of EC in deserts? It seems to be a major factor in some groups whose foods are modular (seeds, animal prey) and potentially depleted by harvesting, for example, lizards and certain granivores. It may also be important in desert plants when soil water is the limiting resource (Fowler 1986; Inouye this volume). However, its importance to other groups of similar consumers (birds and predaceous arthropods) is debatable at best (Polis and Yamashita, Wiens this volume). It is interesting that standard correlates of EC (resource partitioning and niche shifts) are generally discussed only by researchers working on ants, rodents, and lizards; other authors largely ignore these themes. This represents a marked shift away from analyses of community structure during the last three decades, which were dominated by EC-generated concepts.

Hypothesis 5: Interference Competition

Interference (e.g., territoriality, aggression, allelopathy) is a process whereby the actions of one entity (individual, species, or age class) limit the access of a second entity to resources. Interference causes community patterns similar to those caused by exploitation competition (Polis 1988a; Polis, Myers, and Holt 1989). Dominant entities occupy the most productive times and places whereas subordinates temporally and spatially avoid more dominant species. Such avoidance can occur in both ecological and evolutionary time scales. Thus interference decreases overlap in resource use, decreases "limiting similarities," and increases resource partitioning (Polis 1988a; Polis, Myers, and Holt 1989). These changes allow coexistence. In theory, the monopolization of resources by interfering dominants can relax resource limitation due to exploitation.

Interference in various forms is sometimes important among desert perennials, some soil biota, social insects, arthropod predators and herbivores, amphibian larvae, reptiles, birds, and mammals (see appropriate chapters). For example, allelopathy is one mechanism whereby desert shrubs inhibit the success of nearby plants, thus decreasing the possibility of exploitation competition for water (Fowler 1986). Ants are notorious for fighting, for killing workers and alates, and for intraguild predation among species (MacKay this volume). Such interference produces a more regular distribution of colonies and more exclusive use of food within their territories. The distribution and abundance of predaceous arachnids and insects in the deserts frequently are established by interference (e.g., territoriality, aggression, intraguild predation; Polis and Yamashita this volume). Territoriality, aggression, or even intraguild predation (Polis, Myers, and Holt 1989) occur in various degrees and combinations among most of the vertebrates in desert communities (see appropriate chapters).

Hypothesis 6: Predators, Parasites, and Pathogens

Predators, parasites, and pathogens lower success or produce mortality when they feed on host or prey. Although recent work has shown that parasites and pathogens can exert great impact on populations and communities (May 1983), research on population- and community-level effects of these enemies is much less well-developed than that on predation. For this reason, the authors in this book restrict themselves almost exclusively to predators in their presentations of the effect of enemies on desert communities (but see Wisdom this volume for the effect of parasites on herbivorous insects). However, it is likely that parasites universally afflict desert species (Polis this volume), whereas pathogens are relatively less important (due to the aridity and the great spacing between individuals in deserts). Whatever, all types of enemies should produce direct and indirect effects (see below) generally similar to those caused by predation.

Predation is expected to be important in communities simply because many secondary consumers feed on live animals and most species are eaten by consumers. Strong evidence shows that predation significantly influences the structure of many terrestrial and aquatic communities (e.g., Connell 1975; Kerfoot and Sih 1987). These studies show that predation exerts both direct and indirect effects on prey communities. Direct effects due to mortality decrease prey populations and may directly reduce community diversity by eliminating prey species. Indirect effects take several forms. One possibility is that changes in the dynamics of one prey population may affect other populations (e.g., competitors; see keystone predator and intermediate disturbance hypotheses below under Hypothesis 7: Disturbance). Another is that prey may avoid predators by refuging in enemy-free space, time, or both (Holt 1984; Jeffries and Lawton 1984). Such avoidance can increase competition when prey, crowded into refuges, deplete local resources. Avoidance also can establish niche patterns similar to those caused by competition (e.g., large guild members, which are invulnerable to predators, co-occur with predators, whereas small species segregate into predator-free areas; similar species occupy different enemy-free space; see Otte and Joern 1977 as applied to desert grasshoppers and Brown et al. 1988, Kotler and Brown 1988 as applied to desert rodents). A third possible indirect effect is "apparent competition" (Holt 1984), which produces patterns similar to those predicted for competition. For example, the introduction of A (a potential competitor of C) can increase the population of predator B sufficiently that B now eats a higher proportion of species C; thus, the introduction of A decreases the population of C, apparently by competition.

A priori evidence suggests that predation should be important in deserts. Predators form a high proportion of all consumers (Crawford, Polis and Yamashita, Polis this volume) and food is limiting for most of these species. Further, the fact that desert organisms possess many well-known anti-predator traits argues that predation is an important selective factor in evolution. (Adaptations include venoms, crypticity, morphological and behavioral features, nocturnality, and foraging strategies; Polis and Yamashita, Reichman this volume). Nevertheless, the contribution of predation to the structure of desert communities is unclear (Seely 1985). Although all authors acknowledge its potential role, few studies demonstrate population or community effects. Perhaps predation itself is a relatively unimportant ecological process, or (more likely) the community-level effects of predation have not been well researched. We do know that it is a key factor capable of reducing densities of seeds in the soil bank (Inouye, MacKay, Reichman this volume), detritivorous beetles (Crawford this volume; but see Seely 1985), spiders and scorpions (Polis and Yamashita this volume), amphibian larvae (Woodward and Mitchell this volume), and rodents (Reichman this volume). Predation also affects spatial or temporal patterns, or both, in some desert rodents, lizards, amphibians, spiders, scorpions, and ants (see appropriate chapters). Predation may be particularly important among soil and litter biota. Apparently, energy flow in the soil subweb is regulated by mites that prey on nematodes that feed, in turn, on microorganisms (Zak and Freckman, Polis this volume).

Hypothesis 7: Disturbance

Theoretically, in simple environments certain processes, if left unchecked, will proceed deterministically until species are eliminated, at least locally. However, other factors can disturb or interrupt this process, allowing coexistence and thus increasing diversity (Caswell 1978; Connell 1978). For example, physical disturbance (e.g., drought, storms) may promote coexistence by not allowing sufficient time for biotic processes (competition and predation) to proceed to the point where species are lost from a community. Alternatively, predators may differentially exploit competitively superior species and therefore allow coexistence by preventing exclusion of inferior species ("keystone predation"; Paine 1969). Note, however, that if predation and physical disturbance restrict the abundance of certain species (e.g., competitive subordinates or vulnerable prey) to a greater degree than their competitors or enemies, these processes will accelerate species elimination (Lubchenco 1978).

Intense predation or severe physical stress also can reduce diversity simply by eliminating all but the least vulnerable species. Conversely, very weak predation or mild disturbance can decrease diversity by allowing deterministic processes to proceed until species are eliminated. Intermediate levels of predation or disturbance, however, have been shown to increase diversity by providing a balance between these two sources of mortality. This entire process (the "intermediate disturbance hypothesis") successfully describes changes in diversity in terrestrial (plant) and marine communities (Lubchenco 1978; Sousa 1984) as a function of the intensity of predation or disturbance.

The heterogeneous and sometimes stressful nature of the desert suggests that intermittent disturbance may be an important process influencing community structure. If selective factors change through time and space, no one species will be always at an advantage. Predation, competition, or even mortality due to abiotic factors may never proceed to the point of species elimination because temporal changes create new selective environments. Thus heterogeneity and periodic physical stress in deserts should promote species coexistence and increased diversity.

In particular, intermittent disturbance may exert great effects on the interplay between desert consumers and their resources. If periods of severe mortality (e.g., ecological crunches due to dry periods) cause bottlenecks in the population of consumers, consumers may not be able to significantly affect their resources, especially during periodic flushes of productivity. The fact that only 2–10 percent of plant biomass is eaten before entering the detrital system indicates that herbivore populations are too low to process seasonal plant production (see also Seely and Louw 1980). Analogously, the periodic accumulation of detritus suggests that detritivores are unable to eat detritus at a rate equal to its input (Crawford, Seely, Zak and Freckman this volume). Populations of predators take much longer to recover from crashes compared to prey populations ("Volterra's principle"). Thus frequent disturbance in deserts will continually benefit prey because low predator populations are unable to increase at a sufficient rate to check increasing prey populations. It is important to note that the impact of periodic disturbance on desert communities is a function of the life history of the consumer; it will be more important in systems with "opportunistic" consumers whose populations fluctuate seasonally in response to food as compared to systems with "long-lived" consumers whose populations are less tightly linked to short-term changes in resources (Polis and Farley 1980; Crawford 1981; Louw and Seely 1982; Polis and Yamashita this volume).

Periodic ecological crunches may also decrease the importance of competition for resources (Wiens 1977, this volume). Potential competitors may influence each other only when their densities are high relative to resource levels. During periods between crunches, the populations of competitors may be too low relative to resource levels to produce resource depletion and subsequent exploitation competition. Crunches also affect the intensity of interference (Polis and Yamashita, Wiens this volume). In general, the probability of contact and interference increases as the number of individuals in competing populations increases and as food stress increases.

Pluralism

Are deserts thus simple systems so overwhelmed by the physical environment that biological processes are seemingly unimportant? Or are they places where such biological interactions as competition and predation play a paramount role in shaping community structure? Stated differently, are the dynamics and structure of desert communities "environment-driven" or "organism-driven"? To varying degrees, each chapter in this book addresses both of these polarized views. It is thought-provoking that different groups of organisms apparently are affected so differently by abiotic and biotic factors. Undoubtedly the particular view of community structure espoused in each chapter reflects reality but also to some extent mirrors the biases of the author. However, it is encouraging (or is it discouraging?) to note that the book, as a unit, demonstrates convincingly that no one factor is of paramount importance for all taxa and in all systems. The rich and probably quite accurate view emerges that a pluralism of factors interact and combine to produce the patterns observed in any community. Several chapters in this book are explicitly pluralistic in their assessment of processes that shape communities. This more realistic and reasonable view is beginning to replace the sometimes extreme polarization that previously characterized community ecology (see, for example, the integrative approaches to desert communities in Brown 1987; Kotler and Brown 1988; Morton and James 1988).

This book manifests some of the maturity that is slowly developing in ecology. Unfortunately, this maturity is accompanied by a loss of security: the world is no longer a simple and deducibly comprehensible place. We now cannot rely on one or two processes (e.g., exploitation competition or abiotic stress) as being central to understanding whole communities. Simple answers in a complex world are no longer acceptable or satisfying. In this way the book represents a microcosm of the important issues currently debated among community ecologists.

Conclusions

More than anything, this book conveys a sense of the great complexity in deserts. Complexity is due to diversity, age structure, trophic characteristics, spatial and temporal heterogeneity, historical effects, and a variety of phenomena that act intermittently to shape community structure. Community- and species-level responses to biotic and abiotic factors occur against a background of spatial and temporal heterogeneity. Such non-equilibrium community dynamics are probably widespread in heterogeneous environments (Chesson and Case 1986) such as the desert. These factors complicate research and frustrate modeling efforts designed to reduce community dynamics to simpler terms (Noy-Meir 1981). Nevertheless, a profound understanding of desert communities may lie in accepting that the variance of a particular process may be more important than its mean value. It is possible that our most realistic understanding of (desert) communities will come only when we embrace the variation and complexity of the system (denBoer 1968). This reality will not vanish if ignored. However, I do not feel that this insight should provide an excuse for us to abandon the hope of understanding with the facile (and, in the extreme, nihilistic) maxim, "it's complex." Deserts are complex but not incomprehensible. However, the cost of comprehension cannot include the sacrifice of reality.

Future Research

The authors were asked to delineate areas that each considered important to advancing our understanding of deserts. Their replies make obvious that there is no shortage of important questions and topics in need of research. Many authors point to the fact that our knowledge is quite varied. Most work has been concentrated on only a certain few taxa (e.g., heteromyid rodents) and in certain places (North American deserts, the Namib). This creates a two-edged challenge. We are now beginning to appreciate the complexity and richness of intensively studied systems. More work needs to be devoted to further developing this knowledge. However, we also need to determine if what we have learned from these systems is unique or has general application. So, what is new? The authors highlight the need for more extensive and more intensive research.

A second common theme is the need to determine the relative contribution of various factors influencing the structure of desert communities. In particular, autecological and historical factors must be integrated with the effects of interactive processes. Even more challenging, the import of these factors must be organized on a mosaic created by the most heterogeneous habitat on this planet.

These are herculean tasks. It is up to the individual empiricist to analyze his or her system to determine how various factors coalesce at that time and place to produce the observed community structure. This is an involved process that will take years (if not decades or centuries) to achieve. Discouraging as this insight may be, this book demonstrates that we continue to understand more and more about deserts.

We will always learn more as long as people love to study the desert. Researchers travel to deserts not only to appease intellectual curiosity but to nourish aesthetic and emotional needs. The joy of working in deserts, the awe when the sun nears the horizon, and the surprise and wonder at so much life in these arid places continue to attract scientists. John Steinbeck (from *The Log from the Sea of Cortez*) embraces many of these feelings: "At night in this waterless air the stars come down just out of reach of your fingers. In such a place lived the hermits of the early church piercing to infinity with uninhibited minds. The great concept of oneness and of majestic order seems always to be born in the desert."

Acknowledgments

Mary Seely, William MacKay, Cliff Crawford, Jim Reichman, Jim Brown, Eric Pianka, Chris Myers, Tsunemi Yamashita, David McCauley, and many of the Vanderbilt graduate seminar group made very helpful comments to improve this chapter. Saul Frommer, Walt Whitford, Yael Lubin, Mary Seely, Martin Muma, and Victor Fet provided data for Table 1.1.

Bibliography

Allred, D., D. Beck, and C. Jorgensen. 1963. *Biotic communities of the Nevada Test Site.* Brigham Young University Science Bulletin 2: 1–50.

Andrewartha, H., and L. C. Birch. 1954. *The Distribution and Abundance of Animals.* Chicago: University of Chicago Press.

Andrewartha, H., and T. Browning. 1961. An analysis of the idea of "resources" in animal ecology. *Journal of Theoretical Biology* 8: 83–97.

Axelrod, D. 1979. Age and origin of Sonoran Desert vegetation. *Occasional Papers of the California Academy of Sciences* 12: 1–74.

Barbault, R., and G. Halffter. 1981. *Ecology of the Chihuahuan Desert*. Mexico City, Mexico: Instituto de Ecología, 167 pp.

Benson, L., and R. Thompson. 1987. The physical record of lakes in the Great Basin. In W. Ruddiman and H. Wright (eds.), *North America and Adjacent Oceans During the Last Deglaciation*. Boulder, Colo.: Geological Society of America.

Bloom, S. 1981. Specialization and noncompetitive resource partitioning among sponge-eating dorid nudibranchs. *Oecologia* 49: 305–315.

Bradley, R. 1986. The relationship between population density of *Paruroctonus utahensis* (Scorpionida: Vaejovidae) and characteristics of its habitat. *Journal of Arid Environments* 11: 165–171.

Brown, J. H. 1986. The role of vertebrates in desert ecosystems. Pp. 51–71 in W. Whitford (ed.), *Pattern and Process in Desert Ecosystems*. Albuquerque, N.Mex.: University of New Mexico Press. 139 pp.

———. 1987. Variation in desert rodent guilds: patterns, processes, and scales. Pp. 185–204 in J. Gee and P. Giller (eds.), *Organization of Communities, Past and Present*. London: Blackwell Scientific Publications, 576 pp.

Brown, J. H., D. Davidson, and O. J. Reichman. 1979. Granivory in desert ecosystems. *Annual Review of Ecology and Systematics* 10: 210–227.

Brown, J. S., B. Kotler, R. Smith, and W. Wirtz. 1988. The effects of owl predation on the foraging behavior of heteromyid rodents. *Oecologia* 76: 408–415.

Bury, C., and R. Luckenbach. 1983. The Algodones Dunes: A case study. In R. Webb and H. Wilshire (eds.), *Environmental Effects of Off-Road Vehicles: Impact and Management in Arid Regions*. New York: Springer-Verlag, 534 pp.

Caswell, H. 1978. Predator-mediated coexistence: a non-equilibrium model. *American Naturalist* 112: 127–154.

Chesson, P., and T. Case. 1986. Overview: Nonequilibrium community theories: chance, variability, history and coexistence. Pp. 229–239 in J. Diamond and T. Case (eds.), *Community Ecology*. New York: Harper and Row.

Chew, R. 1961. Ecology of the spiders of a desert community. *Journal of the New York Entomological Society* 69: 5–41.

Cody, M. 1974. *Competition and the Structure of Bird Communities*. Princeton, N.J.: Princeton University Press.

Connell, J. 1975. Some mechanisms producing structure in natural communities: a model and evidence from field experiments. Pp. 460–490 in M. Cody and J. Diamond (eds.), *Ecology and Evolution of Communities*. Cambridge, Mass.: Belknap.

———. 1978. Diversity in tropical rainforests and coral reefs. *Science* 199: 1302–1309.

Crawford, C. S. 1981. *Biology of Desert Invertebrates*. New York: Springer-Verlag.

———. 1986. The role of invertebrates in desert ecosystems. Pp. 73–92 in W. Whitford (ed.), *Pattern and Process in Desert Ecosystems*. Albuquerque, N.Mex.: University of New Mexico Press, 139 pp.

Davidson, D. 1977. Species diversity and community organization in desert seed eating ants. *Ecology* 58: 711–724.

denBoer, P. J. 1968. Spreading of risk and stabilization of animal numbers. *Acta Biotheoretica* 18: 165–194.

Diamond, J., and T. Case. 1986. *Community Ecology*. New York: Harper and Row.

Fowler, N. 1986. The role of competition in plant communities in arid and semiarid regions. *Annual Review of Ecology and Systematics* 17: 89–110.

Gee, J., and P. Giller. 1987. *Organization of Communities, Past and Present*. London: Blackwell Scientific Publications, 576 pp.

Giller, P., and J. Gee. 1987. The analysis of community organization: the influence of equilibrium, scale and terminology. Pp. 519–542 in J. Gee and P. Giller (eds.), *Organization of Communities, Past and Present*. London: Blackwell Scientific Publications, 576 pp.

Hadley, N. 1980. Productivity of desert ecosystems. Section B in *Handbook of Nutrition*. West Palm Beach, Fla.: Chemical Rubber Company Press.

Holm, E., and C. Scholtz. 1980. Structure and pattern of the Namib Desert dune ecosystem at Gobabeb. *Madoqua* 12: 3–39.

Holt, R. 1984. Spatial heterogeneity, indirect interactions, and the coexistence of prey species. *American Naturalist* 124: 377–406.

Janzen, D. 1986. Chihuahuan desert nopaleras: defaunated big mammal vegetation. *Annual Review of Ecology and Systematics* 17: 595–636.

Jeffries, M., and J. Lawton. 1984. Enemy free space and the structure of ecological communities. *Biological Journal of the Linnean Society* 23: 269–286.

Kerfoot, W., and A. Sih. 1987. *Predation: Direct and Indirect Impacts on Aquatic Communities*. Hanover, N.H.: University Press of New England.

Kotler, B., and J. S. Brown. 1988. Environmental heterogeneity and coexistence of desert rodents. *Annual Review of Ecology and Systematics* 19: 281–308.

Kzivolchatsky, V. 1985. *Insects of Repetek*. Ylym: Askhabad, 71 pp. (in Russian).

Lamoral, B. 1978. Soil hardness, an important and limiting factor in burrowing scorpions of the genus *Opisthophthalmus* C. L. Koch, 1837 (Scorpionidae, Scorpionida). *Symposium of the Zoological Society of London* 42: 171–181.

Louw, G., and M. K. Seely. 1982. *Ecology of Desert Organisms*. London: Longman.

Lubchenco, J. 1978. Plant species diversity in a marine intertidal community: importance of herbivore food preference and algal competitive abilities. *American Naturalist* 112: 23–39.

Ludwig, J. 1986. Primary production variability in desert ecosystems. Pp. 5–17 in W. Whitford (ed.), *Pattern and Process in Desert Ecosystems*. Albuquerque, N.Mex.: University of New Mexico Press, 139 pp.

———. 1987. Primary productivity in arid lands: myths and realities. *Journal of Arid Environments* 13: 1–7.

McGinnies, W., B. Goldman, and P. Paylore. 1968. *Deserts of the World*. Tucson, Ariz.: University of Arizona Press, 788 pp.

MacKay, W., and E. MacKay. 1984. Why do harvester ants store seeds in their nests? *Sociobiology* 9: 31–47.

MacMahon, J. A. 1981. Introduction. Pp. 263–269 in D. Goodall and R. Perry (eds.), *Arid-Land Ecosystems: Structure, Functioning and Management*, vol 2. Cambridge: Cambridge University Press.

Marsh, A. 1986. Ant species richness along a climatic gradient in the Namib Desert. *Journal of Arid Environments* 11: 235–241.

Martin, P., and R. Klein. 1984. *Quaternary Extinctions: A Prehistoric Revolution*. Tucson, Ariz.: University of Arizona Press.

May, R. 1983. Parasitic infections as regulators of animal populations. *American Scientist* 71: 36–45.

Morton, S. 1985. Granivory in arid regions: comparison of Australia with North and South America. *Ecology* 66: 1859–1866.

Morton, S., and D. Davidson. 1988. Comparative structure of harvester ant communities in arid Australia and North America. *Ecological Monographs* 58: 19–38.

Morton, S., and C. James. 1988. The diversity and abundance of lizards in arid Australia: a new hypothesis. *American Naturalist* 132: 237–256.

Muth, A. 1980. Physiological ecology of desert iguana (*Dipsosaurus dorsalis*) eggs: temperature and water relations. *Ecology* 61: 1335–1343.

Noy-Meir, I. 1973. Desert ecosystems: environment and producers. *Annual Review of Ecology and Systematics* 4: 25–41.

———. 1974. Desert ecosystems: higher trophic levels. *Annual Review of Ecology and Systematics* 5: 195–214.

———. 1980. Structure and function of desert ecosystems. *Israel Journal of Botany* 28: 1–19.

———. 1981. Spatial effects in modeling of arid ecosystems. Pp. 411–432 in D. Goodall and R. Perry (eds.), *Arid-Land Ecosystems: Structure, Functioning and Management*, vol. 2. Cambridge: Cambridge University Press.

———. 1985. Desert ecosystem structure and function. Pp. 93–103 in M. Evenari (ed.), *Hot Deserts and Arid Shrublands*. Amsterdam: Elsevier Science Publishers.

Orians, G., R. Cates, M. Mares, A. Moldenke, J. Neff, D. Rhoades, M. Rosenzweig, B. Simpson, J. Schultz, and C. Tomoff. 1977. Resource utilization systems. Pp. 164–224 in G. Orians and O. Solbrig (eds.), *Convergent Evolution in Warm Deserts*. Stroudsburg, Pa.: Dowden, Hutchinson & Ross.

Orians, G., and R. Paine. 1983. Convergent evolution at the community level. Pp. 431–458 in D. Futuyma and M. Slatkin (eds.), *Coevolution*. Sunderland, Mass.: Sinauer Associates.

Orians, G., and O. Solbrig. 1977. Degree of convergence of ecosystem characteristics. Pp. 226–255 in G. Orians and O. Solbrig (eds.), *Convergent Evolution in Warm Deserts*. Stroudsburg, Pa.: Dowden, Hutchinson & Ross.

Otte, D., and A. Joern. 1977. On feeding patterns in desert grasshoppers and the evolution of specialized diets. *Proceedings of the Academy of Natural Sciences of Philadelphia* 128: 89–129.

Paine, R. 1969. A note on trophic complexity and community stability. *American Naturalist* 103: 91–93.

Péfaur J. 1981. Composition and phenology of epigeic animal communities of the Lomas of southern Peru. *Journal of Arid Environments* 4: 1–42.

Pianka, E. 1986. *Ecology and Natural History of Desert Lizards*. Princeton, N.J.: Princeton University Press, 208 pp.

———. 1989. Desert lizard diversity: additional comments and some data. *American Naturalist* 134: 344–364.

Polis, G. A. 1988a. Exploitation competition and the evolution of interference, cannibalism and intraguild predation in age/size structured populations. In L. Perrson and B. Ebenmann (eds.), *Size Structured Populations: Ecology and Evolution*. New York: Springer-Verlag.

———. 1988b. Foraging and evolutionary responses of desert scorpions to harsh environmental periods of food stress. *Journal of Arid Environments* 14: 123–134.

———. 1989. Ecology. Chapter 6 in G. A. Polis (ed.), *Biology of Scorpions*. Stanford, Calif.: Stanford University Press.

———. In press. Complex trophic interactions in deserts: an empirical assessment of food web theory. *American Naturalist*.

Polis, G. A., and R. Farley. 1980. Population biology of a desert scorpion: survivorship, microhabitat and the evolution of life history strategy. *Ecology* 61: 620–629.

Polis, G. A., and S. J. McCormick. 1987. Intraguild predation and competition among desert scorpions. *Ecology* 68: 332–343.

Polis, G. A., C. A. Myers, and R. Holt. 1989. The ecology and evolution of intraguild predation: potential competitors that eat each other. *Annual Review of Ecology and Systematics* 20: 297–330.

Pomeroy, L. V. 1981. Developmental polymorphism in the tadpoles of the spadefoot toad, *Scaphiopus multiplicatus*. Ph.D. diss., University of California at Riverside.

Pulliam, R. 1988. Sources, sinks and population regulation. *American Naturalist* 132: 652–661.

Schoener, T. W. 1982. The controversy over interspecific competition. *American Scientist* 70: 586–595.

Seely, M. K. 1985. Predation and environment as selective forces in the Namib Desert. Pp. 161–165 in E. Vrba (ed.), *Species and Speciation*, Transvaal Museum Monograph No. 4. Pretoria: Transvaal Museum.

Seely, M. K., and M. Griffin. 1986. Animals of the Namib Desert: Interactions with their physical environment. *Revue de Zoologie Africaine*: 100: 47–61.

Seely, M. K., and G. Louw. 1980. First approximation of the effect of rainfall on the ecology and energetics of a Namib Desert dune ecosystem. *Journal of Arid Environments* 3: 25–54.

Sousa, W. 1984. The role of disturbance in natural communities. *Annual Review of Ecology and Systematics* 15: 5–92.

Strong, D. 1983. Natural variability and the manifold mechanisms of ecological communities. *American Naturalist* 122: 636–660.

Strong, D., D. Simberloff, L. Abele, and A. Thistle. 1984. *Ecological Communities: Conceptual Issues and the Evidence*. Princeton, N.J.: Princeton University Press.

Taylor, A. 1988. Large-scale spatial structure and population dynamics in arthropod predator-prey systems. *Annales Zoologici Fennici* 25: 63–74.

Van Devender, T., and W. Spaulding. 1979. Development of vegetation and climate in the Southwestern deserts. *Science* 204: 701–710.

Van Devender, T., R. Thompson, and J. Betancourt. 1987. Vegetation history of the deserts of southwestern North America. In W. Ruddiman and H. Wright (eds.), *North America and Adjacent Oceans During the Last Deglaciation*. Boulder, Colo.: Geological Society of America.

Venkateswarlu, B., and A. Rao. 1981. Distribution of microorganisms in stabilised and unstabilised sand dunes of the Indian desert. *Journal of Arid Environments* 4: 203–207.

Wallwork, J. 1982. *Desert soil fauna*. New York: Praeger Publishers, 296 pp.

Walter, H., and E. Box. 1983. The Karakum Desert, an example of a well-studied eu-biome. Pp. 105–159 in N. West (ed.), *Ecosystems of the World*. Vol. 5: *Temperate Deserts and Semi-deserts*. Amsterdam: Elsevier.

Webb, S. 1984. Ten million years of mammal extinctions in North America. In P. Martin and R. Klein (eds.), *Quaternary Extinctions*. Tucson, Ariz.: University of Arizona Press.

———. 1987. Community patterns in extinct terrestrial vertebrates. Pp. 421–468 in J. Gee and P. Giller (eds.), *Organization of Communities, Past and Present*. London: Blackwell Scientific Publications, 576 pp.

White, T. 1978. The importance of relative shortage of food in animal ecology. *Oecologia* 33: 71–86.

Whitford, W. G. 1986. Decomposition and nutrient cycling in deserts. Chapter 5 in W. G. Whitford (ed.), *Pattern and process in desert ecosystems*. Albuquerque, N. Mex.: University of New Mexico Press.

Wiens, J. A. 1977. On competition in variable environments. *American Scientist* 65: 590–597.

Wiens, J. A., J. Addicott, T. Case, and J. Diamond. 1986. Overview: the importance of spatial and temporal scale in ecological investigations. Pp. 145–153 in J. Diamond and T. Case (eds.), *Community Ecology*. New York: Harper and Row.

Population Biology of Desert Annual Plants

2

Richard S. Inouye

Annual plants are usually an inconspicuous part of the desert flora, hidden in the soil as seeds or growing just large enough to produce a few flowers and seeds before dying. In some years, however, the proper combination of temperature, quantity of rainfall, timing of rainfall, and light stimulates widespread germination and rapid growth of these plants. When this happens, annuals are for a time the most abundant and showiest plants growing in the desert, exploding into a living carpet of flowers.

While desert annuals may grow and bloom for only a fraction of the year, seeds of these plants are present all year. Because of their constant availability, and because a seed is a concentrated package of calories and protein, seeds of desert annuals form the base of a diverse food chain. They are eaten by ants, rodents, and birds, which in turn are eaten by a host of secondary consumers including snakes, lizards, birds, and mammalian predators.

There is no question that abiotic factors such as limited and variable rainfall, a short growing season, and extremes of temperature have played important roles in the evolution of desert annuals. However, an important message of this chapter is that biotic interactions have also been important influences on the evolution of desert annual plants and continue to influence density, diversity, and species composition of desert annual communities today.

Phenology, Diversity, and Abundance

Desert annuals typically germinate after a rain, grow vegetatively, flower, set seed, and die. This apparently simple, and predictable, life cycle is complicated by the variability that is characteristic of many desert climates. The time between germination and death can vary from weeks to months, depending on patterns of temperature and rainfall in a given year. Plant size at maturity is highly variable. If water is readily available individual plants can achieve the stature of small shrubs and produce many thousands of seeds. In years with sparse rainfall annual plants may produce only one or a few flowers before they die, if they survive to flower at all. There are large year-to-year differences in the density, diversity, and productivity of annual plants due to variability in rainfall and temperature. Table 2.1 summarizes productivity of desert annuals for studies with data from more than 1 year. The range in productivity between relatively wet and dry years can exceed an order of magnitude.

Populations of desert annuals survive successive years with little or no rain because seeds of these plants can remain dormant in the soil for long periods. Not all seeds will germinate at once, even in a good year. In the event that a cohort of seeds that has germinated does not survive to flower and set seed, seeds that remained dormant in the soil can survive to germinate and reproduce in future years. Seed densities of more than 220,000 per m^2 in desert soils have been reported (Table 2.2). The dynamics and movement of these seeds remain relatively little-studied, primarily due to the difficulty of following the movements of individual seeds.

In years with abundant rain, both the density and diversity of desert annuals can be very high (Tables 2.3, 2.4). Plant densities of nearly 1,000 per m^2 have been reported, while as many as 187 species of annuals have been found in 1 ha, and more than 60 species in a single m^2 (Shmida and Ellner 1984). Four factors probably contribute to the maintenance of this high diversity. The first is the limited amount of precipitation that characterizes deserts. Because productivity is limited by rainfall, deserts lack a closed canopy of perennial vegetation. When water is available open spaces can be colonized by annual plants. In contrast with many more productive habitats, annuals in deserts do not only persist as fugitive species growing in disturbed areas for a brief time before being competitively displaced by perennials.

The second factor affecting annual plant diversity is seasonality in precipitation. Some deserts have distinct summer and winter rainy seasons. In these areas the vast majority of annual plants grow only during one of the two growing seasons. A few species may grow during both seasons; these species may provide an important link between what are otherwise distinct annual communities.

The third factor that promotes a high diversity of desert annuals is variation in the availability of water in space and in time. Relatively small differences in surface topography can create significant differences in soil moisture by channeling surface flow away from some areas and concentrating it in others. Spatial variation in clay, organic matter, and litter can amplify small-scale variation in moisture availability. Year-to-year variation in precipitation may cause species to germinate in varying proportions because of differences in germination requirements. Unpredictability of rainfall can favor an annual growth habit because seeds are better than plants at surviving extended droughts. This is consistent with the observation that the proportion of annuals in five desert habitats in the U.S. increased with the coefficient of variation of total annual rainfall (Schaffer and Gadgil 1975).

Competitive relationships within the annual community may also be influenced by variation in weather. Temperature and rainfall patterns that favor one species are

Table 2.1 Biomass (kg/ha) of desert annuals.

	Peak Biomass		
	1964	1965	1966
Mojave Desert sites			
(Beatley 1969)			
Rock Valley, #3	3.5	0.2	136.8
Jackass Flats, #5	85.7	47.9	616.2
Jackass Flats, #15	26.6	3.6	292.2

	Peak Biomass		
	1964	1965	1966
Great Basin Desert sites			
(Beatley 1969)			
East, #30	3.4	26.7	0.6
East, #58	8.5	3.5	92.4
Forty-Mile Canyon, #63	89.2	185.6	55.8

	Mid-Season Biomass 1968–1969
Sonoran Desert Cave Creek, AZ (Halvorson & Patten 1975)	26.0–76.0

	Season Productivity	
	1972–73	1973–74
Sonoran Desert Cave Creek, AZ (Patten 1978)	952.0	94.0

	End-of-Season Above-Ground Biomass 1976–1977
Sonoran Desert Silverbell, AZ (Inouye, Byers, & Brown 1980)	58.0

	End-of-Season Above-Ground Biomass	
	1972–73	1973–74
Negev Desert Sede Zin (Loria & Noy-Meir 1979)	3.0	136.0

	October Biomass	
	1981	1982
Chihuahuan Desert Jornada Site, NM (Gutierrez & Whitford 1987)	43.2	1.7

not likely to be repeated in successive years. Year-to-year variation in weather may shift the competitive balance within the annual community and allow a greater number of species to persist.

Finally, predation contributes to the maintenance of high diversity in desert annual communities. Density-dependent, frequency-dependent, and size-dependent predation all can reduce the extent to which competitively dominant annual species can displace poorer competitors. These four factors will be discussed in more detail in the sections that follow.

Germination

Seed dormancy and germination behavior have been the focus of a large body of work on desert annual plants, from the first papers published on desert annuals by Went in the 1940s until the present day. One of the most widely drawn conclusions of empirical studies of desert annuals is that germination behavior plays a key role in their persistence and dynamics (Went 1948, 1949; Went and Westergaard 1949; Juhren, Went, and Philipps 1956; Koller 1957; Tevis 1958a,b; Guttermann 1972; Mott 1972, 1974; Beatley 1974; Bowers 1987). Field studies invariably record

Table 2.2 Seed densities (seeds per square meter).

Location and Source	Average Seed Density[a]	Maximum Seed Density[b]
Mojave Desert Rock Valley, NV (Childs & Goodall 1973)	4,270	. . .
Mojave Desert Rock Valley, NV (Nelson & Chew 1977)	8,000–187,000[c]	. . .
Sonoran Desert Silverbell, AZ (Reichman 1984)	5,600	63,800 (76.6 per 12 cm^2)
Sonoran Desert Santa Rita Mountains, AZ (Price & Reichman 1987)	7,000	160,000 (256 per 16 cm^2)
Chihuahuan Desert Portal, AZ (Huntly & Chesson unpublished)	75,639	229,254 (1536 per 67 cm^2)
Great Basin Desert Kemmerer, WY (Parmenter & MacMahon 1983)	3,600	. . .

[a]Density averaged over a site.
[b]Highest density of seeds reported for a single sample at that site. Actual area and number of seeds in parethenses.
[c]Variation across habitats and years.

Table 2.3 Average densities of annual plants.

Location and Source	Plants (per m^2)
Sonoran Desert Palm Springs, CA (Tevis 1958a)	792
Chihuahuan Desert Jornada Validation Site (Whitson 1975)	360
Mojave Desert Nevada Test Site (Beatley 1969)	975
Mojave Desert Joshua Tree Nat. Mon., CA (Tevis 1958b)	791
Negev Desert Avdat Research Station (Evenari & Gutterman 1976)	5.9– 96.1[a]
Sonoran Desert Cave Creek, AZ (Halvorson & Patten 1975)	551–1,877[b]
Negev Desert Sede Zin (Loria & Noy-Meir 1979–80)	212; 244[c]

NOTE: The actual area sampled varied in these studies, making
direct quantitative comparisons difficult.
[a]Eight years data.
[b]Four sites in one year.
[c]Two years data.

emergence, rather than germination; however, the term germination usually is used instead.

Two environmental factors have been identified as crucial in determining the germination response of seeds to a rainfall event. The first of these, not surprisingly, is the amount of precipitation. Several studies have suggested that the minimal amount of rainfall necessary to stimulate germination in desert annuals is about 25 mm; less rainfall results in very little germination, whereas more rainfall, up to a certain level, stimulates greater numbers of seeds to germinate (Went 1948, 1949; Tevis 1958a,b; Beatley 1974; but see Gutierrez and Whitford 1987). The minimum amount of precipitation required to trigger germination may differ for summer and winter annuals. Mott (1972, 1974) reported that summer annuals in Western Australia required at least 25 mm of rain to germinate, while winter annuals in the same desert would germinate after 15 mm. Differences in the quantity of rainfall required to trigger seasonal floras are probably related to the rate at which soils dry during summer and winter. Rainfall earlier in the same year is likely to influence soil moisture, so the effect of a particular rain may depend in part on rainfall earlier in the year.

Table 2.4   Species richness of desert annual plants.

Location and Source	Season	No. Annual Species
Baja California (Aschmann 1967)	Winter	114
Mojave Desert Rock Valley, NV (Beatley 1967)	Winter	58
Mojave Desert Rock Valley, NV (Bowers 1987)	Winter	56
Mojave Desert Joshua Tree Nat. Mon., CA (Juhren, Went, & Philipps 1956)	Winter Summer	40 9
Sonoran Desert Silverbell, AZ (Thames 1974, 1975)	Winter Summer	46 11
Sonoran Desert Sonora & Arizona (Shreve & Wiggins 1964)	Winter	100
Sonoran Desert Cave Creek, AZ (Halvorson & Patten 1975)	Winter	38
Negev Desert (Loria & Noy-Meir 1979–80)	Winter	62
Judean Desert (Danin 1978)	Winter	38
Palestine (Negbi 1968)	Summer	5
Dead Sea Valley (Danin 1976)	Winter	15
Har Gilo Israel (Shmida & Ellner 1984)	. . .	187

Temperature is the second environmental factor shown to influence germination of desert annuals. This response is particularly important where there are two distinct rainy seasons that differ in their average temperatures. The Sonoran Desert, for example, has distinct summer and winter peaks in precipitation which result from storms moving west from the Gulf of Mexico and east from the Pacific coast, respectively. Because of this seasonality in precipitation, the Sonoran Desert has two annual floras that are almost entirely distinct. Winter annuals germinate from November to March in response to rain and cool temperatures; summer annuals germinate in July and August at much higher temperatures associated with convective thunderstorms. Relatively specific temperature requirements for germination assure that seeds do not germinate in response to rainfall at the wrong time of year. Similar patterns have been reported for annual plants in Australia (Mott 1972, 1974).

Differences in temperature and moisture requirements for germination may also play a role in maintaining high diversity of desert annuals within winter and summer floras. Differences in germination requirements, combined with variability in the timing of precipitation and a long-lived seed bank, could result in the coexistence of many annual species because different years favor different species (e.g., Chesson and Huntly 1988). Data for several species suggest that, even within a seasonal flora, species differ in temperature and moisture thresholds to trigger germination (Koller 1957; McDonough 1963; Huntly and Chesson unpublished data). In a 6-year study of annuals in the northern Mojave Desert, Bowers (1987) found that there was a high degree of between-year variability in annual species composition, and that species composition was correlated with the pattern of precipitation. In separate studies of winter annuals in the Negev Desert, Loria and Noy-Meir (1979–80) and Evenari and Gutterman (1976) found substantial variation in density and species composition both between years and among sites.

The interaction between temperature and rainfall has been further demonstrated by work on winter annuals in the Negev Desert. Evenari and Gutterman (1976) observed relatively little germination when mean daily evaporation (MDE) was greater than 5–5.7 mm, even in response to increased rainfall. When MDE was less than 5.7 mm they recorded germination of winter annuals after as little as 4.9 mm of rain. Two years later, when MDE was 6.9 mm, they found no germination after 13.5 mm of rain.

Biotic factors also can influence germination of desert annuals. Juhren et al. (1956) suggested that there might be inhibition of germination at high seed densities, and Inouye (1980) found that germination in response to a second winter storm was influenced by the density of seedlings that emerged in response to an earlier storm. Data supporting this inhibition hypothesis were obtained from unmanipulated plots in the Chihuahuan Desert and from experimental plots in the Sonoran Desert, where removal of seedlings as they emerged resulted in a greater total number of seedlings emerging (Inouye 1980). This inhibition probably occurs because seedlings that emerge where there is a high density of established plants are likely to face significant competition. Data from the Sonoran and Chihuahuan deserts indicate that growth and survival can be negatively correlated with annual plant density (Inouye, Byers, and Brown 1980; Inouye 1980, 1982).

Differences in phenology and growth that are functions of seed size probably play a role in the germination inhibition observed at high seedling density. In the Sonoran Desert some of the relatively large-seeded annuals (e.g., *Erodium cicutarium*, *E. texanum*, *Lotus humistratus*) germinate in response to early winter rains in November or December. Many of these plants establish a rosette and then remain quiescent during a dry period before further rains in January or February. With further rains these plants grow rapidly, having an established root system with which to take up water and nutrients, and because of their initial size advantage inhibitory effects are likely to be magnified. Even if they germinate at the same time, the inhibitory effect of seedlings on subsequent germination is also related to seed size. A factorial experiment in which large- and small-seeded annuals were thinned separately and together as they emerged demonstrated that large-seeded species significantly reduced subsequent germination of small-seeded annuals, whereas removing small-seeded annuals had no impact on subsequent germination of large-seeded annuals (Inouye 1982).

The germination response of plants in unpredictable environments, deserts in particular, has been the focus of several theoretical studies that have addressed how plants can best "program" seeds to germinate in order to maximize long-term reproduction. Dormancy, delaying germination in time, has the benefit of reducing the risk that all of a plant's progeny will germinate at the same time and die before reproducing because of a lack of subsequent rainfall. The two costs associated with dormancy are the continued risk of mortality as a seed and the cost of delayed reproduction. Cohen (1966) and MacArthur (1972) developed simple models to predict the fraction of seeds that should germinate in any given year. Their models suggest that the fraction of seeds germinating in any year should increase as the probability that there will be enough subsequent rain to flower and set seed increases. Where the probability of subsequent rainfall is low, the proportion of seeds that should germinate in any year is also low.

Wilcott (1973) developed a more detailed model that included seed predation and an age-structured seed bank. Wilcott assigned different probabilities of predation to seeds in the year they were produced and in subsequent years, after they were part of the soil seed bank. His simulations suggest that the probability of predation in the first year can have an important effect on optimal germination strategies.

Venable and Lawlor (1980) discussed the differences between Cohen's and MacArthur's models, and presented a model that considered the interaction between dormancy and dispersal as ways of dispersing in time and in space. Dispersal in space or in time are separate ways for a plant to hedge its bets, or reduce the risk that all of its seeds will germinate where and when they will be unable to survive to reproduce. Their model predicted that, as the fraction of seeds dispersing in space increases, the fraction of seeds that are dormant (the fraction dispersing in time) should decrease. This should be most pronounced where the proportion of seeds that disperses is relatively low and where the probability of being in a good environment is low. In other words, increased bet hedging in one way reduces the benefit of spreading the risk in other ways. Data for species with dimorphic seeds were consistent with these predictions; seeds that were more widely dispersed almost always had less dormancy than seeds that were less widely dispersed in space.

One way that plants might use germination behavior to spread their risk is to have relatively narrow conditions under which seeds will germinate. Seeds that are dispersed into different microsites are likely to experience different conditions of temperature, light, and moisture. If the range of conditions required by each seed for germination is very narrow it is unlikely that all the seeds will germinate at one time. In this way a plant could take advantage of environmental variability at the microsite level to survive temporal environmental unpredictability.

If conditions at the time of germination are correlated with subsequent growing conditions, environments are not truly unpredictable. Although the year-to-year sequence of conditions may effectively be random, information may be available in any particular year that allows plants to "predict" subsequent conditions in that year with some accuracy. The benefits of bet hedging are reduced when plants are able to use cues of this sort, and plants can avoid the cost of delaying reproduction if reliable cues indicate that conditions will remain favorable long enough to reproduce. The relationship between germination and mean daily evaporation reported by Evenari and Gutterman (1976) may be an example of seeds using multiple cues to predict future environmental conditions. Freas and Kemp (1983) suggested that greater reliability of summer rainfall contributed to the higher germination fraction of summer annuals compared with winter annuals in the Chihuahuan Desert. This observation is consistent with the hypothesis that greater environmental predictability should favor reduced dormancy.

Some of the implications of dormancy with respect to the genetics of a population were considered in a 15-year field study of a Mojave Desert annual, *Linanthus parryae*, by Epling et al. (1960). This study reported that the proportions of blue- and white-flowered morphs of *L. parryae* were relatively constant through time, despite large differences in plant density and seed production among years. The authors concluded that the ability of seeds to remain dormant for at least 7–10 years, combined with differences in the germination requirements of seeds, produces a population of plants that arise from seeds produced in a number of different years. Under these conditions, the effective population size, or the size of the breeding group, is actually much larger than the number of plants that grow and reproduce in any given year, and the mating system is similar to that of long-lived perennial plants. The high degree of constancy in color morphs over the length of the study suggests that there is relatively little genetic drift in these populations, and that flower color, or traits that are linked to flower color, is under strong selective pressure.

Published data for survival of desert annuals from germination to maturity are remarkably similar for studies in the Mojave and Sonoran deserts and Western Australia (Table 2.5). This similarity is probably due to the phenotypic plasticity of desert annuals as well as their use of cues such as precipitation and temperature to trigger germination. Loria and Noy-Meir (1979–80) reported much lower survival for one year in the Negev Desert in which there was very little rainfall (Table 2.5). This high level of mortality illustrates the potential cost of germinating in a year when there is not sufficient rain to mature and reproduce.

Distribution in Space

Studies have shown both positive and negative associations of annual plants with desert shrubs. Went (1942) reported positive associations of several annual species with several shrub species. Muller (1953) reported a positive association of annuals with *Ambrosia dumosa*, and a negative association of annuals with *Encelia farinosa*. Muller and Muller (1956) and Halvorson and Patten (1975) reported differences in

Table 2.5 Survival estimates (germination to reproduction) for desert annual plants.

Location and Source	Season	Percent Survival
Sonoran Desert Silverbell, AZ (Inouye, Byers, & Brown, 1980)	Winter	42
Negev Desert Avdat Research Station (Evanari & Gutterman 1976)	Winter	43–67[a]
Mojave Desert Joshua Tree Nat. Mon., CA (Juhren, Went, & Philipps 1956)	Winter	46
Negev Desert Sede Zin (Loria & Noy-Meir 1979–80)	Winter	11; 72[b]
Mojave Desert Joshua Tree Nat. Mon., CA (Tevis 1958a)	Winter	50
Mojave Desert Nevada Test Site (Beatley 1967)	Winter	38
Western Australia Mileura Station (Mott & McComb 1974)	Winter	37; 46; 57[c]
Western Australia Mileura Station (Mott & McComb 1974)	Summer	41

[a]Eight years data.
[b]Two years data.
[c]Three years data.

the species composition of annuals under shrubs and in open areas, and Adams, Strain, and Adams (1970) reported no annuals under several perennial species.

Several factors probably contribute to these differences in the distribution of annuals with respect to shrubs. Shrubs may have direct effects on annuals by providing shade, thereby reducing water stress, and by acting as a wind shadow and trapping seeds and organic matter. Muller (1953) suggested that the positive association of annuals with *Ambrosia dumosa* was due to higher soil organic content at the base of the shrub. The branched stems of *A. dumosa* are much more likely to trap organic matter than the single stem of *E. farinosa*. The importance of shade, and reduced water stress, probably varies between years depending on the amount and timing of rainfall. Comparing the distribution of annuals with respect to shrubs in relatively wet and dry years might provide some indication of the role of shade in producing these patterns. It has also been suggested that shrubs may have direct allelopathic effects on annuals. Muller (1953) and Muller and Muller (1956) showed the potential for shrub extracts to inhibit growth of annual plants in the lab, but there has been no clear demonstration of this kind of effect in the field.

The distribution of annuals with respect to shrubs can also be influenced by consumers. Some seed-eating rodents (e.g., pocket mice) forage primarily under shrubs which provide cover from some of their predators (e.g., Kotler 1984; Price, Waser, and Bass 1984). Other rodents (e.g., bipedal kangaroo rats) forage primarily in open areas between shrubs. Differences in the ways that these rodents harvest seeds may influence the species composition of annuals found under shrubs and in open areas. There are other ways that seed-eating rodents influence the spatial distribution of desert annuals. Large mounds produced by some kangaroo rats (e.g., *Dipodomys spectabilis*) can influence the abundance, species composition, and growth of annual plants. It is not unusual to see higher densities of relatively large annuals on these mounds (personal observation). Such differences in annual density and species composition may be due to the direct effects of rodents on the seed bank, and they also may be influenced by changes in soil moisture or nutrient levels that are caused by rodent activity. Ants may have similar direct and indirect effects on the distribution of desert annuals (see Dispersal, below).

Small-scale topographic variability probably contributes to the high degree of spatial variation in annual density and species composition that has been reported (e.g., Evenari and Gutterman 1976; Loria and Noy-Meir 1979–80). Small differences in the height of the soil surface can channel more moisture to certain areas, with the result that germination and growth responses for different species may vary significantly among closely adjacent plots.

Surface flow, or sheet flooding, caused by heavy rainfall can redistribute seeds and organic matter. Mott and McComb (1974) reported that winter annuals in Western Australia were positively associated with small mounds that rose about 6 cm above adjacent depressions. Germination of seeds planted in depressions and on mounds was not significantly different, and survival of winter annual seedlings was similar in depressions and on mounds. Mott and McComb suggested that sheet flooding deposited seeds on mounds as water slowed to move around obstructions, rather than in depressions, where water flow rates were higher. They also found that soil depth and organic matter were higher on mounds than in depressions. They suggested that these differences were responsible for the higher moisture content of mound soil and the greater size of annual plants on mounds. Wind currents act in much the same way as sheet flooding, redistributing seeds and creating dense clumps of seeds. Reichman (1984) reported extremely high seed densities associated with objects that slowed wind speed, causing seeds to collect in wind shadows.

Growth

The physically stringent desert environment has focused attention on physiological adaptations of desert plants. This is true of annuals and perennials, although it has been suggested that winter annuals, because they may grow for extended periods under relatively mesic conditions, are not really desert-adapted plants (Went 1949; Shreve 1951; Juhren, Went, and Philipps 1956). Mulroy and Rundel (1977) reviewed some of the adaptations of annual plants to growth in the desert. One of the most striking patterns they reported was a correlation between photosynthetic pathway and the time of year that a species grows. Most annuals growing during the cooler winter months used the C_3 photosynthetic pathway, while most annuals that grew during the summer used the C_4 pathway. Similar data were reported for annual plants in the Chihuahuan Desert by Kemp (1983). Other adaptations that were discussed by Mulroy and Rundel included changes in growth form with temperature (change from a rosette to a more upright growth form as temperatures increase), and high phenotypic plasticity.

The relatively short growing season apparently has favored rapid growth. Solar tracking and high nitrogen concentrations in tissues contribute to high growth rates (e.g., Forseth and Ehleringer 1982a,b, 1983; Werk et al. 1983; Ehleringer 1983). Winter annuals have some of the highest photosynthetic rates ever measured (Mooney et al. 1976; Ehleringer, Mooney, and Berry 1979). In cool winter months annuals may be limited more by low temperatures than by soil moisture (Wallace and Szarek 1981; Toft and Pearcy 1982). At high spring and summer temperatures water can rapidly become limiting (Tevis 1958a,b; Beatley 1974; Pavlik 1980; Bowers 1987; Gutierrez and Whitford 1987).

In addition to temperature and soil moisture, soil nitrogen also may limit growth of desert annuals. Nitrogen addition experiments have produced significant increases in annual biomass (Wallace et al. 1978; Romney, Wallace, and Hunter 1978; Williams and Bell 1981). Gutierrez and Whitford (1987) suggested that the lack of response of annuals in the Chihuahuan Desert in the second of two years of watering was due to nitrogen limitation. Soil nitrogen is distributed non-randomly with respect to desert shrubs (Parker et al. 1982); this is probably one of several factors that influences the commonly observed higher abundance of annuals near shrubs (e.g., Halvorson and Patten 1975; Patten 1978). Differential responses of annual species to nitrogen addition (Williams and Bell 1981; see also Parker et al. 1982) suggest that differences in soil nutrient requirements may be another factor that contributes to the high diversity of desert annual plants (e.g., Tilman 1982).

Allocation

Allocation refers to the way that biomass or nutrients are distributed among the various parts of a plant. The way that a plant allocates resources has important consequences for the fitness of that plant. Carbon or nitrogen can be used to make more leaves, thereby increasing the leaf surface area available to capture sunlight. Alternatively, the same resources can be used to make roots, enabling the plant to extract water and nutrients from a larger volume of soil. Flowering creates yet another demand for potentially limiting resources. While resources that are used to build leaves or roots increase the size of a plant, resources allocated to reproduction make a much smaller contribution to plant growth. Because of this tradeoff, a plant that delays flowering may grow larger, but where the growing season is limited such a plant will have less time in which to flower and set seed. A plant that begins flowering sooner will have more time in which to reproduce, but it will pay a price in reduced growth.

Annual plants are often used to model optimal allocation strategies because it is relatively easy to quantify their fitness. In the absence of a surviving adult, fitness at the end of the growing season can be estimated by the number of seeds that an annual plant has produced. Several theoretical and empirical studies have addressed the question of allocation patterns in desert annuals. Optimization arguments have been used to predict patterns of allocation to vegetative versus reproductive structures (Cohen 1971; Paltridge and Denholm 1974; Denholm 1975; Vincent and Pulliam 1980; King and Roughgarden 1982a); some of these models may be particularly relevant to desert annuals (e.g., Schaffer, Inouye, and Whittam 1982, King and Roughgarden 1982b). Where the growing season is of a fixed length, allocation models generally predict a "bang-bang" strategy, in which all of a plant's photosynthate is allocated to making a larger plant until a switch point is reached, at which time all photosynthate is shunted to producing reproductive structures and seeds. Where the length of the growing season is not fixed, optimization models can predict graded allocation patterns, in which photosynthate is directed to more than one function at a time (King and Roughgarden 1982b). Because of the high variability in timing and amount of rainfall, this modification seems particularly appropriate for desert annuals.

Empirical data suggest that bang-bang allocation patterns are not the rule, and that unpredictability in the length of the growing season has selected for graded allocation patterns (Dina and Klikoff 1974; Bell, Hiatt, and Niles 1979; Clark and Burk 1980; Williams and Bell 1981). Desert annuals usually initiate flowering early enough that most plants are able to produce at least a few seeds, but maintain the ability to reinitiate vegetative growth in response to more rain. For example, *Schismus arabicus* seedlings in the Negev Desert produce a single flowering spike very quickly in dry years, setting about 10 seeds. If soil and moisture conditions remain favorable they can continue to flower for a prolonged period, setting up to 100 seeds (Loria and Noy-Meir 1979–80).

A critical allocation "decision" for desert annuals is related to the tradeoffs between seed size and seed number (e.g., Salisbury 1942; Harper, Lovell, and Moore 1970; Werner and Platt 1976; Loria and Noy-Meir 1979–80; Ellner and Shmida 1981; Silvertown 1981; Inouye 1982; Gross 1984; Stanton 1984, 1985; Venable and Brown 1988). Using a given amount of energy and nutrients, small seeds can be produced in greater numbers, are more readily dispersed, and may more easily fall into sites that are protected from predators or that are suitable for germination. The primary advantage of large seed size is that it provides a greater reserve with which a seedling can establish and begin growth, a reserve that typically confers a competitive advantage over plants emerging from smaller seeds (e.g., Black 1956, 1957). Differences in seed size are often correlated with differences in dormancy (see Germination, above).

Several theoretical studies have examined the implications of seed size for dispersal or for dormancy (Templeton and Levin 1979; Venable and Lawlor 1980; Levin, Cohen, and Hastings 1984; Brown and Venable 1986). Venable and Brown (1988) considered all three parameters simultaneously. They argued that seed size, dispersal, and dormancy are all bet-hedging devices that reduce the year-to-year variance in fitness. Increased seed size buffers a plant against environmental deterioration by increasing the likelihood of successful establishment. Dispersal in space and in time both reduce the risk that all of a plant's progeny will die if local conditions deteriorate the following year. They also argued that these three parameters reduce the impact of competition from other annuals, and from sibs in particular. Their model demonstrated that seed size, dispersal, and dormancy are tightly coupled. Changes in any one of the parameters select for changes in the other two parameters as well. Their model also suggests that the three parameters are partially substitutable. Conclusions of their model appear to be greatly influenced by their assumption that competitive interactions are not important, hence a constraint on seed size that has been shown to be important in the Sonoran and Chihuahuan deserts is not included in their model.

Dispersal

Soil samples from particular microhabitats (e.g., small depressions or wind shadows behind rocks or shrubs) and the use of artificial seed traps have demonstrated that seeds of desert annuals are redistributed across the desert floor throughout the year by wind and water (e.g., Went 1942; Nelson and Chew 1977; Reichman 1984). Seed densities are highly variable both in space and in time. Reichman (1984) reported 78-fold variation in seed density in 4-cm^2 patches and 28-fold variation in density through time in the Sonoran Desert. Price and Reichman (1987) reported significant spatial and temporal variability in seed density and seed mass for 16-cm^2 samples in the Sonoran Desert. The average number of seeds per sample in various habitats ranged from fewer than 1 to more than 20, with a maximum of 256 seeds (160,000/m^2). Huntly and Chesson (unpublished) found 17-fold variation among 67-cm^2 samples in the Chihuahuan Desert. This kind of variation makes characterization of seed densities and the dynamics of the entire annual plant population extremely difficult.

Perennial plants influence the distribution of seeds both directly and indirectly. Airborne material, including seeds, can fall out of the airstream when wind speed slows behind shrubs and other obstacles. Several studies have reported higher seed densities under shrubs than in open areas. In the Mojave Desert, Nelson and Chew (1977) found nearly 10 times more seeds under shrubs (2,800–7,600/m^2) than between shrubs (120–580/m^2). In the Sonoran Desert Reichman (1984) reported about twice as many seeds under shrubs (10,000–15,000/m^2) as between shrubs (4,000–5,000/m^2), and Price and Reichman (1987) found more seeds under large bushes (8,262/m^2) than in large open areas (5,625/m^2). Working against this accumulation of seeds, shrubs provide cover for seed-eating rodents that can greatly reduce seed densities and may well change the relative abundance of annual plants in different microhabitats.

Although some seeds are passively dispersed about the desert floor, it is not clear how important this part of the seed bank is to the dynamics of the plant population. In April of 1978, I harvested annuals on 0.5 by 0.5-m plots in the Chihuahuan

Desert near Portal, Arizona. Above-ground parts of plants were collected near the end of the growing season before most seeds had dispersed. In November 1978, fewer seedlings emerged on harvested plots than on control plots (harvested plots: $\bar{X}$ = 105.5; control plots: $\bar{X}$ = 189.6; t' = 5.81, $p < .01$)(Inouye 1982). The difference in seedling density on harvested and control plots was greatest at high seedling densities; where November seedling density averaged fewer than 100 plants/m^2 there was no significant difference in density on harvested and control plots.

Two factors are likely to have contributed to the difference in seedling density on harvested and control plots. First, dead standing parts of annual plants may help to trap dispersing seeds. Second, seedlings may germinate primarily from seeds that were produced on the same plot. More detailed experiments will be necessary to see just how large a contribution seeds that disperse across the desert floor make to future seed production. The most important role that these seeds play may be that they form the relatively dense clumps of seeds that are used by some seed-eating rodents (e.g., Price 1978; Price and Reichman 1987).

Interestingly, the study of *Linanthus parryae* by Epling, Lewis, and Ball (1960) concluded that dispersal by wind and water was relatively unimportant, and perhaps maladaptive, in the dynamics of this Mojave Desert species. Their conclusion was based on the observation that local frequencies of two color morphs of this species remained constant. Local patches of one color that appeared where the other color was predominant tended to be short-lived, suggesting that individuals that dispersed into the "wrong" areas were not successful.

In contrast to the plots at Portal, where plant density was signficantly lower on plots where plants were harvested during seed production, densities of *L. parryae* did not decrease even after plants were harvested prior to seed set for 5 years. There is probably greater variability in precipitation at the Mojave Desert site than at the Portal site (*L. parryae* was extremely rare or absent in several years during the study). If dormancy increases with environmental variability, as predicted by Cohen (1966, 1967) and MacArthur (1972) and as recently demonstrated by Philippi and Seger (1989), then a larger proportion of the plants in the Mojave Desert may germinate from seeds that have been in the soil for more than 1 year. Such an increase in dormancy would reduce the short-term effect of harvesting plants because harvesting would have little or no effect on seeds in the soil at the start of the experiment.

Adaptations for long-range dispersal are relatively rare among desert plants. In contrast to the commonly accepted hypothesis that this is the direct result of selection to remain in what was a good microsite, Ellner and Shmida (1981) suggested that the absence of dispersal mechanisms is the result of selection against investment in dispersal structures. They reason that where there is greater unpredictability in time than in space, selection should favor investment in mechanisms that contribute to dispersal in time at the expense of those that contribute to dispersal in space. By allocating resources to produce more seeds, rather than investing in structures to aid in dispersal in space, a plant can increase the probability that at least one seed will germinate and reproduce.

Some desert annuals do invest energy to increase dispersal by producing elaiosomes, special structures that aid in dispersal by attracting ants. Seeds of these species may be dispersed greater distances and with greater reliability than seeds lacking elaiosomes. Because they are moved away from the parent plant the risk of predation by rodents may be decreased (e.g., O'Dowd and Hay 1980). Seeds that are carried back to the nest by ants are often discarded in refuse piles, together with other organic material. Several authors have noted that plants growing in such refuse piles grow larger than plants growing away from refuse piles (e.g., Golley and Gentry 1964; Wight and Nichols 1966). Rissing (1986) reported positive associations between harvester ant colonies and densities of two annual species that lack elaiosomes, *Schismus arabicus* and *Plantago insularis*. Harvester ants eat seeds of both species; however, some seeds apparently end up in refuse piles, where they germinate and grow. Seed production of both species was siginficantly greater for plants growing in refuse piles than for control plants growing away from ant colonies. Greater growth or fecundity of plants growing on refuse piles is surprising because plant densities may be significantly higher near colonies; however, increased nutrients and organic matter may more than compensate for increased plant density.

Biotic Interactions

Although desert annuals provide dramatic examples of adaptations to unpredictable and stressful abiotic environments, populations of desert annuals are strongly influenced by biotic interactions as well. Competition and predation play significant roles in determining the species composition and abundance of communities of desert annuals. Experimental studies of desert granivores and annual plants have provided some of the best demonstrations of the importance of indirect interactions in natural communities (e.g., Inouye 1982; Davidson, Samson, and Inouye 1985; Brown et al. 1986).

Because of environmental variability, and the resulting variation in the abundance of annual plants (as seeds and as growing plants), the relative importance of biotic interactions does vary in time and in space. Competition for resources (e.g., nitrogen) is less likely to be important when relatively few seeds germinate. Similarly, the intensity of competition for nitrogen may vary with the abundance of annuals in the previous year (Gutierrez and Whitford 1987). The impact of seed predators probably varies with the density of annual plants and with predator density. Seed-eating birds that forage in flocks may have a very large impact on the annual community in one year and little or no effect the next year if they do not return to the same location. These sources of variability, the high degree of phenotypic plasticity of desert annuals, and the high density of seedlings that can emerge after extended periods of low seed production have led some people to believe that biotic interactions such as competition and predation do not influence the dynamics or abundance of desert annual plants. Quite to the contrary, it is now apparent that these biotic factors play very important roles in determining the characteristics of desert annual communities.

It is possible to estimate seed production and seed consumption and conclude that granivores do not have a significant impact on the annual community because they remove only a small percentage of the seeds that are present (e.g., Tevis 1958c). The weakness of this argument is that seed-eating ants, rodents, and birds are not indiscriminate predators. Any selectively-foraging predator can alter the composition of its prey community, even if it takes only a fraction of all of the prey that are present. That this is true for desert annual communities has been clearly demonstrated by experimentally removing groups of granivores and following changes in the annual plant community (Inouye, Byers, and Brown 1980; Inouye 1982; Davidson, Samson, and Inouye 1985; Brown et al. 1986). Both density and species composition of annual plants changed significantly in three separate sets of granivore exclosures, two in the Sonoran Desert (Inouye, Byers, and Brown 1980; Brown et al. 1986) and one in the Chihuahuan Desert (Davidson, Samson, and Inouye 1985).

Seed size influences the risk of predation by different groups of granivores. Seed-eating rodents consume more of the relatively large seeds (>1 mg), while seed-eating ants collect more of the relatively small seeds (<1 mg). Where rodents have been removed, densities of large-seeded annuals, such as *Erodium cicutarium, E. texanum*, and *Lotus humistratus*, increased by as much as three orders of magnitude (Brown et al. 1986). By influencing the more numerically abundant, smaller-seeded annuals, ants can significantly increase the evenness in species numbers (Inouye, Byers, and Brown 1980).

Experiments with artificial seed clumps have demonstrated the efficiency of rodents and ants as seed predators, as well as showing differences in the types of seed clumps that these two groups of granivores are able to exploit (Reichman 1979; Parmenter, MacMahon, and Van der Wall 1984). Reichman (1979) put out seeds that were clumped or dispersed and buried or on the surface. In 24 hours rodents found all the groups of seeds, both on and below the soil surface, and took more than 80 percent of the individual seeds. Ants found about 85 percent of the surface groups and took about 45 percent of the seeds in those groups.

Grazing by cattle also can significantly affect diversity of annual plants. In a study at Organ Pipe Cactus National Monument, Waser and Price (1981) found significantly lower diversity of annuals on grazed plots; however, they did not find any consistent effect of cattle on abundance of annuals as measured by cover. Such grazing effects are common in other environments (e.g., Harper 1977; Crawley 1983).

Experimental manipulations of plant density clearly demonstrate significant direct effects of competition on growth rates and reproduction of desert annual plants. In one of the earliest experimental studies of a desert annual, Klikoff (1966) manipulated density of *Plantago insularis* under different moisture regimes and found significant reductions in productivity and survival at high plant density. Inouye, Byers, and Brown (1980) found significant increases in biomass and fecundity of several species on plots where densities were reduced experimentally by about two-thirds.

Inhibition of germination by established plants is also a competitive effect. Where high seedling density reduces further germination (Inouye 1980), the short-term impact of competition is much less apparent than it would be if dormancy were not an option. Plants can avoid potential reductions in survival or fecundity associated with high plant densities by remaining dormant, but by doing so they pay a cost in delayed reproduction. Delayed germination may have greater costs if nutrients are incorporated into organic matter during one year and made unavailable to plants germinating the following year. These kinds of costs are not likely to be measured in the course of a 1- or 2-year study, but they can contribute to significant changes in species composition over time (Inouye 1982).

Interactions between seasonal floras also may be common, but they are relatively hard to detect. Shimshi (1971) suggested that winter annuals may inhibit summer annuals, and Davidson, Samson, and Inouye (1985) reported that winter annuals in the Chihuahuan Desert, near Portal, Arizona, affected summer annuals by means of their effect on *Eriogonum abertianum*, a winter-germinating, summer-flowering species. High densities of winter annuals inhibited *E. abertianum*, which, when it was present at high density, in turn inhibited summer annuals. Negbi (1968) suggested that winter annuals in Palestine may inhibit summer annuals by extracting moisture from the soil that would otherwise be available for summer-active species. Seasonal floras might also influence each other by incorporating nutrients in seeds and in organic matter, thereby reducing the level of nutrients that are available the following season. A third way that summer and winter floras probably influence each other is through seed predators. Together these relatively distinct groups of annuals may support a higher density of granivores than would be present if either the winter or summer flora were absent.

Seed size plays an important role in determining competitive relationships in desert annuals, just as it influences the risk of predation by different groups of granivores. Seed size affects initial growth rate and competitive ability of many plant species (e.g., Black 1956, 1957; Stanton 1984), and experiments in the Sonoran and Chihuahuan deserts indicate that this generalization holds true in these deserts as well. As discussed earlier, the reduction in subsequent germination caused by established seedlings is a function of seed size. Although it has not been demonstrated, differences in initial growth rates probably also translate into superior competitive abilities for large-seeded relative to small-seeded annuals during periods of vegetative growth. Data for *Erodium cicutarium* plants grown in the lab indicate that differences in seed size can translate into significant differences in plant size. In the absence of competition, plants grown from larger seeds maintained a significant size advantage for more than 6 weeks (unpublished data).

Differences in competitive ability associated with seed size, taken together with differences in predation by ants and rodents on large- and small-seeded annuals, are responsible for some of the significant indirect effects between granivorous ants and rodents. Although the short-term effect of removing either group of granivores is an increase in density or activity of the other group (e.g., Brown et al. 1986), the asymmetry in competition between large- and small-seeded species can make the net long-term effect of rodents on ants positive rather than negative. Where rodents are removed, density of large-seeded annuals increases, and densities of small-seeded annuals ultimately decrease because of increased competition from large-seeded annuals (Inouye 1982). Thus, by keeping densities of large-seeded annuals lower, rodents effectively maintain higher densities of the small-seeded annuals that are most commonly consumed by ants.

Mares and Rosenzweig (1978) suggested that indirect mutualisms of this sort might be responsible for differences in the granivore fauna between the Sonoran Desert in Arizona and the Monte Desert in Argentina. They argued that the loss of predators specializing on one end of a seed size spectrum would allow plants to evolve toward that end of the spectrum and thus escape predators that had specialized on the opposite end of the seed size spectrum. Data from the granivore exclosures in the Sonoran and Chihuahuan deserts suggests that the asymmetry in competition between large- and small-seeded annuals makes the indirect effects between ants and rodents asymmetric as well; thus, the net interaction between ants and rodents is not mutually beneficial. Whereas rodents benefit ants by keeping densities of large-seeded annuals low, a reduction in density of small-seeded annuals by ants does not have a similar effect on the density of large-seeded annuals.

A large-seeded annual, *E. cicutarium*, is the key link in another surprising interaction between two predators, granivorous rodents and *Synchytrium papillatum*, a fungal parasite of *E. cicutarium*. The infection rate of a host-specific fungus increased on plots in the Sonoran Desert where rodents were removed, indicating that these two taxonomically unrelated predators, granivorous rodents and a fungus, probably compete for their shared prey species even though they have their greatest impact on different life stages of the plant (Inouye 1981).

Another unexpected and interesting indirect link involving desert annuals was demonstrated by Parker et al. (1982). Experimental removal of subterranean termites resulted in increased soil nitrogen near the soil surface and in changes in the abundance of several annual species. Responses of annuals were species specific; although total biomass did not differ between treatments, the relative abundance of several species did vary significantly between termite removal and control plots. These data suggest that heterogeneity in soil nutrients may play an important role in the coexistence of desert annuals.

A final example of the way that biotic interactions continue to shape the biology of desert annual plant populations comes from the long-term experimental studies of granivores and plants in the Chihuahuan Desert near Portal, Arizona. Treatments for these 50 by 50-m plots included removal of all rodents (Munger and Brown 1981; Davidson, Samson, and Inouye 1985). Excluding rodents resulted in significant increases in the density of *Erodium cicutarium*, a large-seeded annual plant (Brown et al. 1986).

In the face of reduced seed predation, one advantage of producing a larger number of seeds is reduced. If the probability that a seed will be eaten is high, a plant can increase the likelihood that at least one seed will survive to reproduce by making a larger number of seeds. Given finite resources, an increase in the number of seeds produced will require a reduction in average seed size. Where predation is reduced, this advantage of producing a greater number of smaller seeds is also reduced.

Where annual plant density is high, for example inside a rodent exclosure, the potential for competition within the annual community also is high. Because seed size can be positively correlated with competitive ability (e.g., Black 1956, 1957), the principal advantage of large seed size also is likely to increase as plant density and competition increase.

Based on these arguments, I predicted that there would be selection for increased seed size where removal of rodents had resulted in high density of large-seeded annuals. To test this prediction I compared seeds from *Erodium cicutarium* plants growing in four rodent exclosures with seeds from an unfenced area away from the exclosures. These seeds were stored in a desiccator at room temperature and weighed individually.

Data for average seed mass are shown in Table 2.6. In exclosure 7, where the density of *E. cicutarium* increased soonest after the experiment was begun, average seed mass was significantly greater than in the unfenced area. Differences between the other three exclosures and the unfenced area were not significant; however, differences in these means were in the direction predicted. These data strongly suggest that selective pressures operating on seed size changed as a result of excluding rodents. The direct effect of excluding rodents was a reduction in predation pressure. An important indirect effect of removing rodents was an increase in competition within the annual plant community. Together, these shifts in selective pressures produced significant changes in seed size.

These data are striking because they suggest that a significant evolutionary change took place over a relatively brief period in which the biotic environment was altered experimentally. The rodent exclosures were built in 1977, 11 years before these seeds were collected. *E. cicutarium* densities did not immediately increase in all of the rodent exclosures (Table 2.6); competition within the annual plant community may not have increased substantially immediately after the exclosures were built.

Nearly all of these biotic interactions involving desert annuals have been uncovered only because of relatively long-term manipulative studies. The importance of this kind of study, especially in a habitat where a major controlling factor, precipitation, is patchy both in space and in time, cannot be overstated. More long-term, detailed studies will be necessary to further unravel the factors that control the dynamics of this diverse group of plants. For example, testing indications that nitrogen availability may influence abundance and species composition of annuals will no doubt require a multi-year field study, particularly if interactions between resource competition among annuals and predation are considered.

Future Studies

Many important questions about the physiology and population biology of desert annuals remain unanswered. Although both herbivory and granivory have been shown to influence annual diversity, it is unlikely that the high diversities reported for desert annuals can be explained entirely by predation. Theoretical studies suggest that various forms of environmental heterogeneity may play an important role in allowing diverse assemblages of species to persist. Tests of the importance of spatial and temporal heterogeneity may shed new light on the mechanisms that allow so many annual species to coexist. The roles that temporal heterogeneity in precipitation and spatial heterogeneity in soil properties play in the coexistence of annuals in the Chihuahuan Desert are currently being tested by N. Huntly and P. Chesson.

Nutrient relationships of desert annuals are also worth exploring. Most studies of desert annuals have assumed that water was the primary limiting resource; however,

Table 2.6 *Erodium cicutarium* seed mass as a function of plant density.

| | Mean Seed Mass | | | Plant Density | |
	mg/seed	Significant difference?	s.d.	Year[a]	Plants/m²[b]
Location					
Rodent exclosure 7	1.297[c]	yes	0.120	1979	47.8
Rodent exclosure 10	1.138	no	0.120	1984	5.1
Rodent exclosure 16	1.155	no	0.129	1984	9.6
Rodent exclosure 23	1.183	no	0.113	[d]	0.8
Unfenced area	1.116	no	0.179		[e]

NOTE: Plant density data courtesy of J. H. Brown.
[a]Year in which average *E. cicutarium* first exceeded 10 plants/m².
[b]Average *E. cicutarium* density 1978–87.
[c]Significantly greater than mean seed mass in other exclosures or unfenced areas ($p < .05$).
[d]*E. cicutarium* density had not increased above 10 plants/m² by 1987.
[e]*E. cicutarium* density was not measured.

there are indications that soil nutrients may play an important role in the biology of these plants. If plants are superior competitors at different ratios of limiting resources (e.g., Tilman 1982), then spatial heterogeneity in soil resources may contribute to the coexistence of annual species. There may also be temporal heterogeneity in available levels of soil resources that influences the abundance of annuals on a year-to-year basis. An abundant crop of annuals in one year may reduce the level of available nutrients the following year, thereby inhibiting growth in the second of two succeeding years with abundant rainfall (e.g., Ettershank et al. 1978; Parker et al. 1982).

There is very little quantitative data on dispersal of desert annuals. What is the distribution of dispersal distances, and how does it vary for seeds of different shapes and sizes? How successful are seeds that disperse different distances? Are dispersing seeds more likely to successfully germinate and reproduce, or do they suffer a higher rate of mortality than seeds that remain stationary in the soil? Janzen (1972) has suggested that in tropical forests, due to density-dependent predation, dispersal away from a parent tree may reduce the probability that a seed will be eaten. Seeds in the desert that are moved by wind or water may be more likely to become part of a dense clump of seeds, with the result that they are more likely to be found and eaten by a seed predator (e.g., Reichman 1979).

Seed dispersal is difficult to quantify in virtually all systems. The small size of desert annual seeds exacerbates the difficulties of this kind of study. The ability to label seeds with different types of radionuclides may help to address this general question. Primack and Levy (1988) labeled annual plants and were able to identify seeds and seedlings produced by the labeled plants by testing plant samples in the lab for gamma emissions. The ability to label and identify individual seeds in the field would be even more useful.

Desert annuals are a good group of organisms with which to address certain evolutionary questions. Differences between deserts provide natural experiments with which to make comparative studies. Differences in precipitation patterns and in the variability of rainfall provide opportunities to test theories about bet hedging (Philippi and Seger 1989). Life history traits, including phenological and allocation patterns, can be compared both within and between species growing in areas with different rainfall patterns (e.g., Philippi and Seger 1989); there is still more to learn about how climatic variability has influenced the evolution of these traits. Although the seed bank can serve as an important stabilizing force in the evolution of desert annuals (Epling, Lewis, and Ball 1960), their short life cycle still makes possible relatively rapid evolutionary changes. Data presented here on changes in seed size of *Erodium cicutarium* from Portal, Arizona, suggest that evolutionary changes can occur in response to experimental manipulations on a time scale that we can observe.

Finally, interactions within the annual plant community and interactions between desert annuals and other groups of organisms should provide some of the most exciting answers, and questions, in the future. For example, in some areas perennials have significant direct and indirect effects on the density and distribution of annuals. Do annuals reduce the level of nutrients available to perennials, thereby reducing perennial growth and reproduction? If high densities of annuals reduce nutrient availability to perennials, then seed-eating rodents may benefit the shrubs that provide their cover by reducing densities of desert annuals.

Reproductive biology has been studied for very few desert annuals, and little is known about mutualistic relationships that may exist between these plants and their pollinators. It seems likely that most desert annuals would have the potential for self-fertilization, rather than requiring the services of pollinators that might or might not be present.

Pathogens are a relatively unstudied group that may have significant effects on the population dynamics of desert annuals, as is suggested by the interaction between granivorous rodents, *Erodium cicutarium*, and *Synchytrium papillatum* (Inouye 1981). We have some good evidence about how granivores in the southwest United States influence the annual community, but relatively little is known about interactions between annuals and other taxonomic or functional groups. Termites are an important group in many deserts. Work by Parker et al. (1982) suggests that there may be important links between termites and desert annuals that are the result of nutrient dynamics in desert soils. Soil algae are another group that may be linked with desert annuals by nutrient dynamics. Data on nutrient dynamics both within the annual community and between annuals and other groups should provide us with a more mechanistic understanding of these potentially important interactions.

Acknowledgments

This chapter was improved by comments from N. Huntly, G. Polis, and J. H. Brown, and by financial support from the Department of Biological Sciences at Idaho State University. Measurement of seed sizes in the rodent exclosures at Portal, Arizona, was graciously permitted by J. H. Brown, and was supported by a grant from the Idaho State University Faculty Research Committee.

Bibliography

Adams, S., B. R. Strain, and M. S. Adams. 1970. Water-repellent soils and annual plant cover in a desert community of southeast California. *Ecology* 51: 696–700.

Aschmann, H. 1967. *The Central Desert of Baja California: Demography and Ecology.* Riverside, Calif.: Manessier: 282 pp.

Beatley, J. C. 1967. Survival of winter annuals in the northern Mojave Desert. *Ecology* 48: 745–750.

———. 1969. Biomass of desert winter annual plant populations in southern Nevada. *Oikos* 20: 261–263.

———. 1974. Phenological events and their environmental triggers in Mojave Desert ecosystems. *Ecology* 55: 856–863.

Bell, K. T., H. D. Hiatt, and W. E. Niles. 1979. Seasonal changes in biomass allocation in eight winter annuals of the Mojave Desert. *Journal of Ecology* 67: 781–787.

Black, J. N. 1956. The influence of seed size and depth of sowing on preemergence and early vegetative growth of subterranean clover (*Trifolium subterraneum* L.) in relationship to size of seed. *Australian Journal of Agricultural Research* 7: 98–109.

———. 1957. Seed size as a factor in the growth of subterranean clover (*Trifolium subterraneum* L.) under spaced and sward conditions. *Australian Journal of Agricultural Research* 8: 1–14.

Bowers, M. A. 1987. Precipitation and the relative abundances of desert winter annuals: a 6-year study in the northern Mohave Desert. *Journal of Arid Environments* 12: 141–149.

Brown, J. H., D. W. Davidson, J. C. Munger, and R. S. Inouye. 1986. Experimental community ecology: The desert granivore system. In J. Diamond and T. J. Case (eds.), *Community Ecology.* New York: Harper and Row.

Brown, J. S., and D. L. Venable. 1986. Evolutionary ecology of seed-bank annuals in temporally varying environments. *American Naturalist* 127: 31–47.

Chesson, P., and N. Huntly. 1988. Community consequences of life-history traits in a variable environment. *Annales Zoologici Fennici* 25: 5–16.

Childs, S., and D. W. Goodall. 1973. *Seed Reserves of Desert Soils.* U.S. International Biological Program Desert Biome Research Memo. 73–5. Logan, Utah: Utah State University: 23 pp.

Clark, D. D., and J. H. Burk. 1980. Resource allocation patterns of two California-Sonoran Desert ephemerals. *Oecologia* 46: 86–91.

Cohen, D. 1966. Optimizing reproduction in a randomly varying environment. *Journal of Theoretical Biology* 12: 119–129.

———. 1967. Optimizing reproduction in a randomly varying environment, when a correlation may exist between the condition at the time a choice has to be made and the subsequent outcome. *Journal of Theoretical Biology* 16: 1–14.

———. 1971. Maximizing final yield when growth is limited by time or by limiting resources. *Journal of Theoretical Biology* 33: 299–307.

Crawley, M. J. 1983. *Herbivory: The Dynamics of Animal-Plant Interactions.* Oxford: Blackwell Scientific Publications.

Danin, A. 1976. Plant species diversity under desert conditions. I. Annual species diversity in the Dead Sea Valley. *Oecologia* 22: 251–259.

———. 1978. Species diversity of semishrub xerohalophyte communities in the Judean desert of Israel. *Israel Journal of Botany* 27: 66–76.

Davidson, D. W., D. A. Samson, and R. S. Inouye. 1985. Granivory in the Chihuahuan Desert: interactions within and between trophic levels. *Ecology* 66: 486–502.

Denholm, J. V. 1975. Necessary condition for maximum yield in a senescing two-phase plant. *Journal of Theoretical Biology* 52: 251–254.

Dina, S. J., and L. G. Klikoff. 1974. Carbohydrate cycle of *Plantago insularis* var. *fastigiata*, a winter annual from the Sonoran Desert. *Botanical Gazette* 135: 13–18.

Ehleringer, J. 1983. Ecophysiology of *Amaranthus palmeri*, a Sonoran Desert summer annual. *Oecologia* 57: 107–112.

Ehleringer, J., H. A. Mooney, and J. A. Berry. 1979. Photosynthesis and microclimate of *Cammissonia claviformis*, a desert winter annual. *Ecology* 60: 280–286.

Ellner, S., and A. Shmida. 1981. Why are adaptations for long-range seed dispersal rare in desert plants? *Oecologia* 51: 133–144.

Epling, C., H. Lewis, and F. Ball. 1960. The breeding group and seed storage: a study in population dynamics. *Evolution* 14: 238–255.

Ettershank, G., J. Ettershank, M. Bryant, and W. G. Whitford. 1978. Effects of nitrogen fertilization on primary production in a Chihuahuan Desert ecosystem. *Journal of Arid Environments* 1: 135–139.

Evenari, M., and Y. Gutterman. 1976. Observations on the secondary succession of three plant communities in the Negev Desert, Israel. I. *Artemisietum herbae albae*. In R. Jacques (ed.), *Etudes de Biologie Vegetale*. Paris: CNRS.

Forseth, I. N., and J. R. Ehleringer. 1982a. Ecophysiology of two solar tracking desert winter annuals. I. Photosynthetic acclimation to growth temperature. *Australian Journal of Plant Physiology* 9: 321–332.

――――. 1982b. Ecophysiology of two solar tracking desert winter annuals. II. Leaf movements, water relations, and microclimate. *Oecologia* 54: 41–49.

――――. 1983. Ecophysiology of two solar tracking desert winter annuals. III. Gas exchange responses to light, CO_2 and VPD in relation to long-term drought. *Oecologia* 57: 344–351.

Freas, K. E., and P. R. Kemp. 1983. Some relationships between environmental reliability and seed dormancy in desert annual plants. *Journal of Ecology* 71: 211–217.

Golley, F. B., and J. B. Gentry. 1964. Bioenergetics of the southern harvester ant *Pogonomyrmex badius*. *Ecology* 45: 217–225.

Gross, K. L. 1984. Effects of seed size and growth form on seedling establishment of six monocarpic perennial plants. *Journal of Ecology* 72: 369–387.

Gutierrez, J. R., and W. G. Whitford. 1987. Responses of Chihuahuan Desert herbaceous annuals to rainfall augmentation. *Journal of Arid Environments* 12: 127–139.

Guttermann, Y. 1972. Delayed seed dispersal and rapid germination as survival mechanisms of the desert plant *Blepharis persica* (Burm.) Kuntze. *Oecologia* 10: 145–149.

Halvorson, W. L., and D. T. Patten. 1975. Productivity and flowering of winter ephemerals in relation to Sonoran Desert shrubs. *American Midland Naturalist* 93: 311–319.

Harper, J. L. 1977. *Population Biology of Plants*. New York: Academic Press.

Harper, J. L., P. H. Lovell, and K. G. Moore. 1970. The shapes and sizes of seeds. *Annual Review of Ecology and Systematics* 1: 327–356.

Inouye, R. S. 1980. Density-dependent germination response by seeds of desert annuals. *Oecologia* 46: 235–238.

――――. 1981. Interactions among unrelated species: granivorous rodents, a parasitic fungus, and a shared prey species. *Oecologia* 49: 425–427.

――――. 1982. Population biology of desert annual plants. Ph.D. dissertation, University of Arizona, Tucson.

Inouye, R. S., G. S. Byers, and J. H. Brown. 1980. Effects of predation and competition on survivorship, fecundity, and community structure of desert annuals. *Ecology* 61: 1344–1351.

Janzen, D. H. 1972. Escape in space by *Sterculia apetala* seeds from the bug *Dysdercus fasciatus* in a Costa Rican deciduous forest. *Ecology* 53: 350–361.

Juhren, M., F. W. Went, and E. Philipps. 1956. Ecology of desert plants. IV. Combined field and laboratory work on germination of annuals in the Joshua Tree National Monument, California. *Ecology* 37: 318–330.

Kemp, P. R. 1983. Phenological patterns of Chihuahuan desert plants in relation to the timing of water availability. *Journal of Ecology* 71: 427–436.

King, D., and J. Roughgarden. 1982a. Multiple switches between vegetative and reproductive growth in annual plants. *Theoretical Population Biology* 21: 194–204.

———. 1982b. Graded allocation between vegetative and reproductive growth for annual plants in growing seasons of random length. *Theoretical Population Biology* 22: 1–16.

Klikoff, L. G. 1966. Competitive response to moisture stress of a winter annual of the Sonoran Desert. *American Midland Naturalist* 75: 383–391.

Koller, D. 1957. Germination-regulating mechanisms in some desert seeds. IV. *Atriplex dimorphostegia* Kar. et Kir. *Ecology* 38: 1–13.

Kotler, B. P. 1984. Risk of predation and the structure of desert rodent communities. *Ecology* 65: 689–701.

Levin, S., D. Cohen, and A. Hastings. 1984. Dispersal strategies in patchy environments. *Theoretical Population Biology* 26: 165–191.

Loria, M., and I. Noy-Meir. 1979–80. Dynamics of some annual populations in a desert loess plain. *Israel Journal of Botany* 28: 211–225.

MacArthur, R. H. 1972. *Geographical Ecology*. New York: Harper and Row.

McDonough, W. T. 1963. Interspecific associations among desert plants. *American Midland Naturalist* 70: 291–299.

Mares, M. A., and M. L. Rosenzweig. 1978. Granivory in North and South American deserts: rodents, birds, and ants. *Ecology* 59: 235–241.

Mooney, H. A., J. Ehleringer, and J. A. Berry. 1976. High photosynthetic capacity of a winter annual in Death Valley. *Science* 194: 322–324.

Mott, J. J. 1972. Germination studies on some annual species from an arid region of western Australia. *Journal of Ecology* 60: 293–304.

———. 1974. Factors affecting seed germination of three annual species from an arid region of western Australia. *Journal of Ecology* 62: 699–709.

Mott, J. J., and A. J. McComb. 1974. Patterns in annual vegetation and soil microrelief in an arid region of western Australia. *Journal of Ecology* 62: 115–126.

Muller, C. H. 1953. The association of desert annuals with shrubs. *American Journal of Botany* 40: 53–60.

Muller, W. H., and C. H. Muller. 1956. Association patterns involving desert plants that contain toxic products. *American Journal of Botany* 43: 354–361.

Mulroy, T. W., and P. W. Rundel. 1977. Annual plants: adaptations to desert environments. *BioScience* 27: 109–114.

Munger, J. C., and J. H. Brown. 1981. Competition in desert rodents: an experiment with semipermeable exclosures. *Science* 211: 510–512.

Negbi, M. 1968. The status of summer annuals in Palestine. *Israel Journal of Botany* 17: 217–221.

Nelson, J. F., and R. M. Chew. 1977. Factors affecting seed reserves in the soil of a Mojave Desert ecosystem, Rock Valley, Nye County, Nevada. *American Midland Naturalist* 97: 300–320.

O'Dowd, D. J., and M. E. Hay. 1980. Mutualism between ants and a desert ephemeral: seed escape from rodents. *Ecology* 61: 531–540.

Paltridge, G. W., and J. V. Denholm. 1974. Plant yield and the switch from vegetative to reproductive growth. *Journal of Theoretical Biology* 44: 23–34.

Parker, L. W., H. G. Fowler, G. Ettershank, and W. G. Whitford. 1982. The effects of subterranean termite removal on desert soil nitrogen and ephemeral flora. *Journal of Arid Environments* 5: 53–59.

Parmenter, R. R., and J. A. MacMahon. 1983. Factors determining the abundance and distribution of rodents in a shrub-steppe ecosystem: the role of shrubs. *Oecologia* 59: 145–156.

Parmenter, R. R., J. A. MacMahon, and S. B. Van der Wall. 1984. The measurement of granivory by desert rodents, birds and ants: a comparison of an energetics approach and a seed-dish technique. *Journal of Arid Environments* 7: 75–92.

Patten, D. T. 1978. Productivity and production efficiency of an upper Sonoran Desert ephemeral community. *American Journal of Botany* 65: 891–895.

Pavlik, B. M. 1980. Patterns of water potential and photosynthesis of desert sand dune plants, Eureka Valley, California. *Oecologia* 46: 147–154.

Philippi, T., and J. Seger. 1989. Hedging one's evolutionary bets, revisited. *Trends in Ecology and Evolution* 4: 41–44.

Price, M. V. 1978. Seed dispersion preferences of coexisting desert rodent species. *Journal of Mammalogy* 59: 624–626.

Price, M. V., and O. J. Reichman. 1987. Distribution of seeds in Sonoran Desert soils: implications for heteromyid rodent foraging. *Ecology* 68: 1797–1811.

Price, M. V., N. W. Waser, and T. A. Bass. 1984. Effects of moonlight on microhabitat use by desert rodents. *Journal of Mammalogy* 65: 353–356.

Primack, R. B., and C. K. Levy. 1988. A method to label seeds and seedlings using radionuclides. *Ecology* 69: 796–800.

Reichman, O. J. 1979. Desert granivore foraging and its impact on seed densities and distributions. *Ecology* 60: 1085–1092.

———. 1984. Spatial and temporal variation of seed distributions in Sonoran Desert soils. *Journal of Biogeography* 11: 1–11.

Rissing, S. W. 1986. Indirect effects of granivory by harvester ants: plant species composition and reproductive increase near ant nests. *Oecologia* 68: 231–234.

Romney, E. M., A. Wallace, and R. B. Hunter. 1978. Plant responses to nitrogen fertilization in the northern Mojave Desert and its relationship to water manipulation. Pp. 237–243 in N. E. West and J. Skujins (eds.), *Nitrogen in Desert Ecosystems*. U.S. International Biological Program Synthesis Series 9. Stroudsburg, Pa.: Dowden, Hutchinson and Ross.

Salisbury, E. 1942. *The Reproductive Capacity of Plants*. London: G. Bell.

Schaffer, W. M., and M. D. Gadgil. 1975. Selection for optimal life histories in plants. In M. L. Cody and J. M. Diamond (eds.), *Ecology and Evolution of Communities*. Cambridge, Mass.: Belknap Press.

Schaffer, W. M., R. S. Inouye, and T. S. Whittam. 1982. Energy allocation by an annual plant when the effects of seasonality on growth and reproduction are decoupled. *American Naturalist* 120: 787–815.

Shimshi, D. 1971. Population dynamics of *Aellenia hierochuntica* (Bornm.) Aellen. *Israel Journal of Botany* 20: 44–47.

Shmida, A., and S. Ellner. 1984. Coexistence of plant species with similar niches. *Vegetatio* 58: 29–55.

Shreve, F. 1951. *Vegetation of the Sonoran Desert*. Carnegie Institute of Washington Publication 591.

Shreve, F., and I. L. Wiggins. 1964. *Vegetation and Flora of the Sonoran Desert*. Stanford, Calif.: Stanford University Press, 1740 pp.

Silvertown, J. W. 1981. Seed size, lifespan, and germination date as co-adapted features of plant life history. *American Naturalist* 118: 860–864.

Stanton, M. L. 1984. Seed variation in wild radish: effect of seed size on components of seedling and adult fitness. *Ecology* 65: 1105–1112.

―――. 1985. Seed size and emergence time within a stand of wild radish (*Raphanus raphanistrum* L.): the establishment of a fitness hierarchy. *Oecologia* 67: 524–531.

Templeton, A. R., and D. A. Levin. 1979. Evolutionary consequences of seed pools. *American Naturalist* 114: 232–249.

Tevis, L. 1958a. Germination and growth of ephemerals induced by sprinkling a sandy desert. *Ecology* 39: 681–688.

―――. 1958b. A population of desert ephemerals germinated by less than one inch of rain. *Ecology* 39: 688–695.

―――. 1958c. Interrelations between the harvester ant *Veromessor pergandei* (Mayr) and some desert ephemerals. *Ecology* 39: 695–704.

Thames, J. L. 1974. *Tucson Basin Validation Site Report*, U.S. International Biological Program Desert Biome Research Memo 74-3. Logan, Utah: Utah State University, 33 pp.

―――. 1975. *Tucson Basin Validation Site Report*, U.S. International Biological Program Desert Biome Research Memo 75-3. Logan, Utah: Utah State University, 26 pp.

Tilman, D. 1982. *Resource Competition and Community Structure*. Princeton, N.J.: Princeton University Press.

Toft, N. L., and R. W. Pearcy. 1982. Gas exchange characteristics and temperature relations of two desert annuals: a comparison of winter-active and a summer-active species. *Oecologia* 55: 170–177.

Venable, D. L., and J. S. Brown. 1988. The selective interactions of dispersal, dormancy, and seed size as adaptations for reducing risk in variable environments. *American Naturalist* 131: 360–384.

Venable, D. L., and L. Lawlor. 1980. Delayed germination and dispersal in desert annuals: escape in space and time. *Oecologia* 46: 272–282.

Vincent, T. L., and H. R. Pulliam. 1980. Evolution of life history strategies for an asexual annual plant model. *Theoretical Population Biology* 17: 215–231.

Wallace, A., E. M. Romney, G. E. Kleinkopf, and S. M. Soufi. 1978. Uptake of mineral forms of nitrogen by desert plants. Pp. 130–151 in N. E. West and J. J. Skujins (eds.), *Nitrogen in Desert Ecosystems*. Stroudsburg, Pa.: Dowden, Hutchinson, and Ross.

Wallace, C. S., and S. R. Szarek. 1981. Ecophysiological studies of Sonoran Desert plants. VII. Photosynthetic gas exchange of winter ephemerals from sun and shade environments. *Oecologia* 51: 57–61.

Waser, N. M., and M. V. Price. 1981. Effects of grazing on diversity of annual plants in the Sonoran Desert. *Oecologia* 50: 407–411.

Went, F. W. 1942. The dependence of certain annual plants on shrubs in a Southern California desert. *Bulletin of the Torrey Botanical Club* 69: 100–114.

―――. 1948. Ecology of desert plants. I. Observations on germination in the Joshua Tree National Monument, California. *Ecology* 29: 242–253.

―――. 1949. Ecology of desert plants. II. The effect of rain and temperature on germination and growth. *Ecology* 30: 1–13.

Went, F. W., and M. Westergaard. 1949. Ecology of desert plants. III. Development of plants in the Death Valley National Monument, California. *Ecology* 30: 26–38.

Werk, K. S., J. Ehleringer, I. N. Forseth, and C. S. Cook. 1983. Photosynthetic characteristics of Sonoran Desert winter annuals. *Oecologia* 59: 101–105.

Werner, P. A., and W. J. Platt. 1976. Ecological relationships of co-occurring goldenrods (*Solidago:* Compositae). *American Naturalist* 110: 959–971.

Whitson, P. D. 1975. Pp. 117–126 in *Factors Contributing to Production and Distribution of Chihuahuan Desert Annuals.* U.S. International Biological Program Desert Biome Research Memo 75-11. Logan, Utah: Utah State University.

Wight, J. R., and J. T. Nichols. 1966. Effects of harvester ants on production of saltbush community. *Journal of Range Management* 19: 68–71.

Wilcott, J. C. 1973. A seed demography model for finding optimal strategies for desert annuals. Ph.D. dissertation, Utah State University, Logan, 137 pp.

Williams, R. B., and K. L. Bell. 1981. Nitrogen allocation in Mojave Desert winter annuals. *Oecologia* 48: 145–150.

Soil Communities in Deserts: Microarthropods and Nematodes

3

John C. Zak and
Diana W. Freckman

In desert ecosystems, as with all terrestrial systems, the rates of decomposition and subsequent mineralization are important determinants and regulators of primary production. Even when moisture levels are adequate, low mineralization rates, particularly for nitrogen, can significantly limit plant growth. When two wet seasons occur in succession in the northern Chihuahuan Desert, shrub and annual plant growth is usually reduced the second year (Parker, Santos et al. 1984) even though moisture may be optimum for plant growth. Parker, Santos et al. (1984) hypothesized that low nitrogen mineralization rates during the second wet season limited plant growth during this period. Subsequent field studies by Gutierrez and Whitford (1987) found that species composition, life span, and productivity of spring and summer annual plant assemblages in the Chihuahuan Desert were determined by the interaction between soil nitrogen and water availability. Fisher et al. (1988) reported that the growth of *Larrea tridentata* (D.C.) Cov (creosotebush), the dominant shrub in the Chihuahuan Desert, was limited by the combined effects of soil moisture and nitrogen availability. In a series of supplemental watering experiments in which field plots of creosotebush received either frequent rain events (6 mm/wk) or infrequent precipitation (25 mm once a month) with or without the addition of 10 g of nitrogen/m^2, Fisher et al. (1988) noted that frequent small rain events were more effective in promoting plant growth than large, infrequent ones. Fisher et al. (1987) proposed that the small rain events increased the turnover of soil nitrogen, thus increasing mineralization rates.

Decomposition rates in arid systems are controlled not only through abiotic constraints such as moisture and temperature, but also by the effects of moisture and temperature on the activity of the microflora and microfauna and their interactions (see Whitford and Freckman 1988; Zak and Whitford 1988 for a review). The importance of microfloral-microfaunal interactions to decomposition dynamics has previously been addressed (Coleman et al. 1984). In attempting to examine the patterns of microarthropod and nematode diversity, species composition, and community structure within and across deserts, it is necessary first to understand the roles these organisms play in decomposition dynamics, and second to determine how patterns of microarthropod and nematode diversity and structure are influenced by the quantity and quality of the soil organic matter and by the abiotic characteristics of each macro- and microhabitat.

On the soil surface, plant litter can accumulate directly under shrubs, in rodent excavations, or in arroyos following rain events (Whitford 1986). The pattern of litter input into the surface detrital pool differs among desert ecosystems and is determined by many factors, including precipitation patterns, animal activity, and wind speed.

The input of litter from perennials is seasonal and directly related to precipitation patterns (e.g., West 1979; Crawford and Gosz 1986). While many shrubs in the North American deserts are either winter deciduous (e.g., *Flourensia cernua*, *Chilopsis linearis*) or drought deciduous (*Fouquieria splendens*, *Cercidium* spp.) the timing of litter input from the same plant species can vary between deserts. In the Chihuahuan Desert, peak leaf fall of *L. tridentata* occurs from October through November, while in the Mojave, where rainfall occurs primarily in the winter, litter fall takes place in March and April (Strojan et al. 1979). With perennial grasses, the timing of litter input may depend upon fragmentation by wind and animal activity. The highest rate of litter input from perennial grasses in the Chihuahuan Desert occurs primarily during January through April, which is the driest and the windiest period. The interactions between precipitation and litter input patterns and temperature constraints on activity may regulate microfaunal abundances, species occurrence, and trophic relationships.

The three main types of belowground litter are surface litter that becomes buried as a result of wind or water action and rodent activity, decomposing plant roots, and the root region of living plants. These microhabitats are centers of intense microbial and microfaunal activity (e.g., Santos et al. 1981; Santos and Whitford 1981; Parker, Santos et al. 1984; Parker, Freckman et al. 1984; Whitford, Stinnett, and Anderson 1988). By ameliorating the adverse abiotic conditions associated with surface litter, the seasonal activity of the microfauna in belowground organic matter can be increased, resulting in a higher rate of decomposition and mineralization (Santos, Phillips, and Whitford 1981). The root region in particular represents an important center of microbial-microfaunal activity that significantly affects plant growth through increased mineralization of nitrogen (e.g., Coleman et al. 1984; Clarholm 1985).

From this base of information, we will examine the general spatial and temporal patterns of species abundance, richness, diversity, and trophic relationships of soil microarthropods and nematodes associated with surface litter, buried litter, and the root region of plants in desert systems. For comparison purposes we will restrict our discussion to data collected from the four major North American deserts (Figure

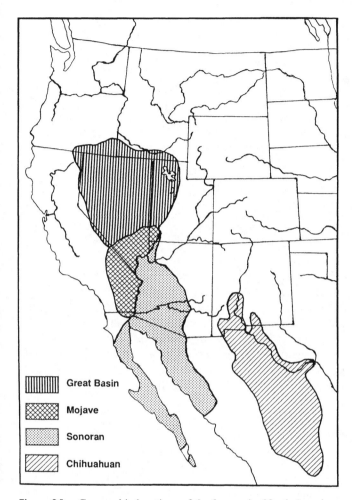

Figure 3.1. Geographic locations of the four major North American deserts.

3.1). Our major emphasis will be to elucidate those patterns of activity, species composition, and trophic relationships that may account for the influence of microarthropods and nematodes in the decomposition of organic matter.

Site Descriptions

The geographic locations of the four North American deserts (Chihuahuan, Great Basin, Mojave, and Sonoran) are presented in Figure 3.1. (For details see MacMahon 1979 and MacMahon and Wagner 1985.) Although, as a consequence of geography and topography, each desert is characterized by within-site variability in rainfall

Table 3.1 Precipitation pattern, elevation, and temperature ranges for the four major
North American deserts.

Desert	Total Precipitation (mm)	No. of Days Precipitation	Pattern of Precipitation	Elevation Range (m)	Average Minimum-Maximum Temperatures (°C)
Chihuahuan	223–244	33–55	summer	915–1,675	7.2–27.8
Great Basin	131–408	49–91	winter–early spring	0–4,354	−7.5–25.0
Mojave	35–130	18–26	winter–early spring	−90–1,525	7.2–30.0
Sonoran	trace–320	15–23	summer and winter	−83–2,755	10.0–30.6

SOURCE: Data from MacMahon (1979); McKell (1985).

amounts and pattern, generalizations in yearly precipitation pattern can be made
(Table 3.1). The Great Basin and Mojave deserts are characterized by winter-spring
rains, while summer-winter rains are typical in the Sonoran. The majority of the
precipitation in the Chihuahuan Desert falls during the summer from convectional
storms. The number of precipitation days varies considerably between the deserts,
with the Great Basin having the highest number of precipitation days and the Mojave
and Sonoran the least. Differences in precipitation patterns, total amounts of moisture received, and temperatures during the year, along with soil edaphic characteristics, are major determinants of microfaunal species abundance, diversity, and trophic
relationships in each of the deserts.

Microarthropods

Surface Litter

Surface litter in arid habitats is subjected to extremes in temperature and moisture
availability throughout the year. Thus, one would predict that microfaunal activity in
this material should closely track seasonal moisture and temperature patterns (Noy-
Meir 1973, 1974; Whitford et al. 1981; Whitford et al. 1983). Correlated to increased
microfaunal activity should be an increase in litter decomposition rates. However,
Whitford, Repass et al. (1982) found no effects of rainfall on surface litter disappearance, and suggested that the litter microfauna have evolved adaptations that allow
these organisms to process surface litter independent of rainfall.

Although actual decomposition rates in deserts may be independent of rainfall
(Zak and Whitford 1988), microarthropod density, reproduction, and activity are
linked to "predictable" rainfall patterns (Wallwork et al. 1986). A distinct seasonal
pattern in species abundance for microarthropods in juniper (*Juniperus monosperma*)
litter from a Mojave Desert site (Figure 3.2) was noted by Wallwork (1972). Total
microarthropod numbers were highest during December and January, the period of
maximum rainfall in this area, with a second and smaller peak occurring in April

CHIHUAHUAN DESERT

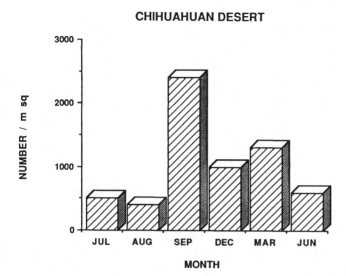

MOJAVE DESERT

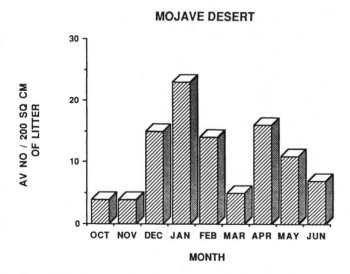

Figure 3.2. Seasonal patterns in microarthropod occurrences in surface litter from the Chihuahuan (data from Silva, MacKay, & Whitford 1985) and the Mojave (data from Wallwork 1972) deserts.

and May during the spring rains. The majority of the microfauna from the Mojave site belonged to the orders Astigmata, Cryptostigmata, and Prostigmata. *Joshuella striata*, *Haplochthonius variabilis*, and *Eremaeus magniporus* (cryptostigmatids) comprised 50 percent of the total number of mites collected, *Glycyphagus* spp. (astigmatid) 33 percent, and *Spinibdella cronini* and *Speleorchestes* spp. (predatory prostigmatids) about 10 percent. Wallwork (1972) determined that the observed seasonal increases in microarthropod densities were due to recruitment of juveniles into the various populations. Juvenile densities were greatest during December and January for *Joshuella striata*, and during April and May for *H. variabilis*. *Spinibdella cronini* exhibited two recruitment peaks, which coincided with juvenile emergence of *J. striata* in January and *H. variabilis* in April and May.

In the Chihuahuan Desert, microarthropod densities in surface litter of *Erioneuron pulchellum* (fluffgrass) were highest during September and declined during the winter months (Silva, MacKay, and Whitford 1985) (Figure 3.2). The observed pattern in microarthropod densities corresponded to long-term precipitation patterns for the Chihuahuan Desert. Wallwork (1972) had previously suggested that litter microfauna closely "track" moisture inputs and should respond to the environment in an opportunistic, "*r*-selected" manner. In a later study, however, additional moisture inputs to creosotebush litter in the Chihuahuan Desert, either at the rate of 6 mm per week or 25 mm once a month, did not significantly increase total microarthropod numbers compared to the unwatered control (Wallwork, Kamill, and Whitford 1984). The lack of a significant change in total microarthropod numbers in the water-supplemented litter of Wallwork, Kamill, and Whitford (1984) was not a result of compensating changes in individual species numbers. Wallwork, Kamill, and Whitford (1984) did observe that watering intensified the seasonality of the breeding response in three dominant cryptostigmatid mites from the site: *Jornadia larreae*, *Passalozetes californicus*, and *P. neomexicanus*. The two *Passolozetes* species and *J. larreae* produced eggs primarily from August through October. This reproductive pattern corresponds to the period of maximum rainfall in the Chihuahuan Desert. A fourth species, *Joshuella striata*, exhibited two peaks in egg production that coincided with the natural rainfall pattern in the Chihuahuan Desert (August through October) and winter rains in the Mojave Desert (December through February) (Wallwork 1972). Wallwork, Kamill, and Whitford (1985) indicated that Chihuahuan Desert mites could be classified primarily as seasonal breeders. That is, they evolved a reproductive pattern that is synchronized to the long-term "predictable" season of precipitation and do not respond to favorable moisture conditions outside their optimal reproductive window. The seasonal breeding and activity patterns exhibited by litter microarthropods in desert systems have evolved in response to the optimum conditions for microbial growth and favorable microclimate that occur at predictable periods of the year (Wallwork, Kamill, and Whitford 1985).

Although reproduction and population growth of soil mites in the Chihuahuan Desert were related to moisture patterns, Loring, Weems, and Whitford (1988) found that population densities of epigeic collembola associated with different plant communities along a Chihuahuan Desert watershed were unaffected by rainfall patterns.

Table 3.2   Effects of water and shade on species numbers, diversity ($\bar{H}$), and trophic relationships of microarthropods associated with surface litter in the Chihuahuan Desert.

	Treatment			
	Control	Water	Shade	Water & Shade
No. of taxa	12	13	10	12
No. of indiv./taxon ($\bar{x} \pm$ s.d.)	2.9 ± 4.8	2.5 ± 4.3	8.7 ± 10.4	38.6 ± 61.2
$\bar{H}$	1.62	1.70	1.62	1.59

Trophic Group	Mean Number of Microarthropods[a]			
Nematode predator	6.6 (19.1)	3.9 (12.2)	12.7 (14.8)	10.4 (2.2)
General predator	1.6 (4.6)	2.2 (6.9)	2.7 (3.2)	85.4 (18.4)
Fungivores & detritivores	26.4 (76.3)	25.8 (80.9)	70.2 (82.0)	367.4 (79.3)

SOURCE: Data from MacKay et al. (1986).
[a]Number in parentheses = % of the total number of microarthropods for that treatment that were assigned to the specific trophic group.

Using the inverse Simpson index (N_2) (Peet 1974) to determine species diversity of collembola along the watershed and estimating variance and confidence intervals for N_2 by a Jacknife procedure (Routledge 1984), Loring, Weems, and Whitford (1988) found that diversity estimates were significantly correlated with long-term temperature patterns. Collembolan abundance was greatest during the warm-wet parts of the year, indicating that collembolan activity, like that of the oribatid mites, may be in response to increased microbial activity at this time. Loring, Weems, and Whitford (1988) indicated that temporal and spatial patterns in soil nitrogen could contribute to the observed collembolan densities along the watershed.

The effects of available moisture and soil temperatures on litter microarthropod activity in a mixed herbaceous bajada area of the Jornada NSF, Long Term Ecological Research (LTER) site in the Chihuahuan Desert were examined by MacKay et al. (1986). A sprinkler system delivered 12 mm of water at irregular intervals throughout the summer. Soil temperature was modified by suspending nylon netting above the soil surface. As had been found by Wallwork et al. (1986), the additional water did not change trophic relationships, numbers of microarthropod taxa, or the mean number of individuals per taxon (Table 3.2). Fungivores and detritivores constituted the largest group in both the control and added water-only plots. We calculated Shannon-Weaver diversity indices ($\bar{H}$) and species abundance relationships from the data presented by MacKay et al. (1986) and found that higher moisture levels through the summer increased diversity ($\bar{H}$) (Table 3.2) and that the species abundance distributions of the litter microarthropod assemblages were altered as compared with the control (Figure 3.3). Species abundance distribution for the control was best described by an exponential function, while the species abundance distribution from the watered-only plots was best fitted by a logarithmic curve as determined by their r^2 values. These data indicate that while the number of taxa and mean number of individuals per taxon were unaffected by watering, the distribution of individuals

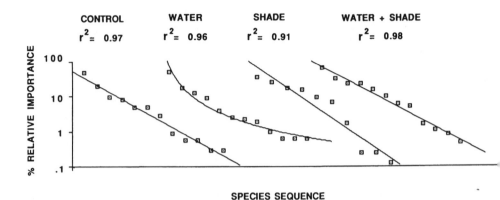

Figure 3.3. Effects of water and shade on species abundance distributions of microarthropods in surface litter from the Chihuahuan Desert (data from MacKay et al. 1986).

among the microarthropod species was altered by the increased moisture. In general terms, species abundance distributions convey information on the structure of the observed assemblages and provide a theoretical basis for determining the roles of various abiotic or biotic parameters responsible for the observed distribution (May 1975, 1981).

Lowering the soil temperature through shading resulted in the loss of two taxa, although overall numbers of individuals per taxon increased substantially in the McKay et al. (1986) study (Table 3.2). Smith, Smith, and Patten (1987) also found increased densities of mites in litter under *Ambrosia* and creosotebush in the Sonoran Desert after shading. The diversity ($\overline{H}$) of the microarthropod assemblages was unaffected by shading alone, as were the trophic relationships (Table 3.2) and species abundance distributions (Figure 3.3). Lower soil temperatures may allow more microarthropods to be active in the litter, though cooler soil temperatures did not alter either trophic relationships or the structure of the assemblage. All microarthropods appear to benefit from lower soil temperatures, while additional moisture may actually stimulate some taxa and suppress others, which alters the species abundance distribution.

Analyses of the data in MacKay et al. (1986) indicated that when water and shade were combined in one treatment, taxon numbers and species abundance distributions of the litter microarthropod assemblage were not altered compared to the control plots (Table 3.2). However, the number of individuals per taxon increased 10-fold with water and shade treatment. The resulting increase in microarthropod densities suggests a positive synergistic effect of water and shade on microarthropod populations. Fungivores-detritivores were the dominant functional microarthropod group in the water and shade treatment. However, general microarthropod predators were more abundant in the water and shade treatment than in the control, where nematophagous microarthropods were dominant.

Buried Organic Matter

Surface litter in deserts becomes buried through the action of sheet flow during intense rain, by windblown sand, and by rodent activity. Dead roots of annual and perennial plants also provide habitat for microarthropod feeding and reproduction. Once belowground, organic matter is protected from the harsh abiotic constraints associated with surface litter. Therefore, temporal patterns in microarthropod densities should not fluctuate as much as for surface litter. Similarly, temporal differences in microarthropod abundances between deserts should be less for buried litter than for surface organic matter because seasonal moisture patterns restrict surface activity.

Using litter from a common Chihuahuan Desert annual, *Lepidium lasiocarpum*, and selected biocides to remove either fungi or microarthropods, Parker, Santos et al. (1984) found that the removal of the microfauna resulted in a significant decrease in buried litter decomposition rates. Analyzed by stepwise regression models, the carbon dynamics of *Lepidium* litter decomposition appeared to result from the action of various microarthropod groups. Santos and Whitford (1981) previously showed that exclusion of microarthropods from buried *Larrea tridentata* litter in the Chihuahuan Desert also resulted in a marked decline in the decomposition rate of this material compared with the control treatment. Collembola and psocopterans, counted as detritivores-fungivores by Parker, Santos et al. (1984), were primarily responsible for the majority of the mass loss from the buried *Lepidium* litter. Microarthropod activity in the buried litter increased as decomposition progressed; Tydeidae and Anoetidae were the major colonizers of the material during the early stages of decomposition (Table 3.3). Tydeid numbers increased for 32 days following burial and declined during the later stages of decomposition. As decomposition progressed, other microarthropod groups colonized the litter, with tarsonemid mites becoming numerically dominant by the 56th day following burial. However, by 50 percent mass loss (day 96), mesostigmatid mites and collembola were as prevalent in the litter as tarsonemid mites. Parker, Santos et al. (1984) suggested that, by grazing on fungi, tarsonemid mites were important regulators of microbial activity in decomposing organic matter. D. Walter (personal communication) has observed that tydeids occurring belowground preferentially feed on fungi while those species occurring aboveground are predators.

The work of Parker, Santos et al. (1984) in the Chihuahuan Desert corroborated the results of Santos and Whitford (1981), who showed that a specific microarthropod colonization sequence occurs on buried litter that is related to the decomposition stage of the material. Santos and Whitford (1981) found that tydeid and, occasionally, paratydeid mites were the only microarthropods associated with buried litter until 30 percent of mass was lost. Heterostigmatid mite populations were usually not present until 30 to 40 percent of the organic matter decomposed. Collembola and predatory gamasina occurred in the litter only when mass loss was greater than 40 percent. Santos and Whitford (1981) suggested that the observed sequence of microarthropod colonization could be accounted for by trophic relationships. Tydeid mites were considered to be nematode predators consuming rhabditid nematodes in the litter. Rhabditid nematodes feed on yeasts and bacteria (Freckman 1988). Once the material was sufficiently colonized by fungi, heterostigmatid mites occurred. After fungivorous mites obtained high densities, predatory taxa (gamasine and pro-stigmatid mites) increased in the litter, which caused a significant decline in hetero-stigmatid mite numbers.

Santos et al. (1984) compared the decomposition of buried litter of creosotebush in the four hot desert sites in North America and found that from March through October litter decomposition was significantly less in the Mojave compared with the Colorado (a subcomponent of the Sonoran; see MacMahon and Wagner 1985), Chihuahuan, or Sonoran deserts. The last three deserts were not significantly differ-ent as a group. However, all four sites exhibited decomposition rates greater than 35 percent mass loss during March through October. During the first 3 months of the study, decomposition rates were significantly higher in the Sonoran Desert, attribut-able to the higher rainfall in this desert during the early summer.

Although litter decomposition rates of buried litter did not vary greatly between the four desert sites, there were significant differences in the species composition of the microarthropod assemblages. Tydeid mites were associated with litter during the first 3 months of decomposition in all sites, but were most abundant in the Chihua-huan Desert (Table 3.4). However, after 6 months of decomposition, they were not associated with the buried litter from any location. Tarsonemid mites were found only in the Chihuahuan Desert and constituted the largest numerical group obtained

Table 3.3 Mean microarthropod densities from buried *Lepidium lasiocarpum* litter (number/100 cm^2 litter bags) in the Chihuahuan Desert.

	Days Following Burial			
Taxon	10	32	56	96
Astigmata				
Acaridae	0	2	0	0
Anoetidae	19	0	0	0
Collembola				
Isotoma sp.	0	0	0	8
Lepidocyrtus sp.	0	0	0	2
Onychiurus sp.	0	0	0	22
Mesostigmata	0	29	16	56
Prostigmata				
Cunaxidae	0	0	0	1
Pyemotidae	0	0	0	2
Tarsonemidae	0	76	546	35
Tydeidae	18	93	24	1
% mass loss	10	25	35	50

SOURCE: Data from Parker, Santos et al. (1984).

Table 3.4 Mean number per litter bag of selected microarthropods found in buried *Larrea tridentata* leaf litter from four North American deserts.

Sampling Period	Desert	Taxon					
		Coll.	Tyd.	Tars.	Gam.	Psocop.	Meso.
June 5	Chihuahuan	0	157	7,345	461	0	0
	Colorado	0	3	0	0	12	0
	Mojave	0	8	0	0.1	126	0
	Sonoran	0	17	0	0.02	153	0
August 7	Chihuahuan	0	15	94	18	0	625
	Colorado	0	10	0	16	7	7
	Mojave	521	21	0	21	0.05	37
	Sonoran	0	144	0	0	0	0
October 2	Chihuahuan	10	0	367	5	87	0
	Colorado	0	0	0	0	163	0
	Mojave	0	0	0	0	6	0
	Sonoran	0	0	0	0.3	32	0

SOURCE: Data from Santos et al. (1984).
[a]Coll. = Collembola; Tyd. = Tydeidae; Tars. = Tarsonemidae; Gam. = Gamasina;
Psocop. = Psocoptera; Meso. = Mesostigmata.

during the 6-month study by Santos et al. (1984). Isotomid and sminthurid collembola were dominant in the Mojave Desert only in August. Psocoptera were abundant in buried litter from the Colorado, Mojave, and Sonoran sites, but were absent in the Chihuahuan Desert until the 6-month sampling period. The occurrence of psocopterans during the early stages of decomposition in the Colorado, Mojave, and Sonoran sites indicated that the microarthropod colonization sequence previously described by Santos and Whitford (1981) applied only to buried organic matter in the Chihuahuan Desert. Recent work by Lundquist (personal communication) has revealed that tarsonemoid mites do occur in all of the four North American deserts. Discrepancies between the work of Santos et al. (1984) and Lundquist probably result from differences in microarthropod extraction procedures and efficiencies, sampling periods, and yearly patterns of species occurrences.

As with surface litter, buried litter decomposition rates appear to be independent of rainfall and microarthropod densities. Santos et al. (1984) indicated that above some threshold value for soil moisture availability, the activity of the soil microfauna is not affected by additional moisture. Obviously, some moisture is necessary for activity to begin. However, unlike surface litter, which can be subjected to repeated wetting and drying events, buried litter can be ameliorated by the soil environment and may remain moist for longer periods of time once wetted. The potentially stable moisture conditions associated with buried litter increase the time soil microflora and microfauna are active during periods of favorable temperatures, irrespective of rainfall. Santos and Whitford (1981) suggested that through the activity of predatory microarthropods, the decomposition of buried organic matter is partially uncoupled from environmental constraints. Santos and Whitford (1981) speculated that, without grazing of fungi and nematodes by microarthropods, microfloral activity of buried litter would be regulated strictly by abiotic constraints.

Litter and Soil Beneath Perennial Plants

The litter and associated soil beneath perennial forbs and shrubs in arid habitats constitute important centers of microbial and microfaunal activity (Wallwork 1982; Freckman and Mankau 1986). The distribution, quantity, and quality of the organic matter beneath perennial plants, and the degree of disturbance during rain events, all interact to affect spatial and temporal patterns of microarthropods in the litter and adjacent soil.

The previous discussions concerning abundance relationships and species composition of microarthropod assemblages in litter focused on those studies that examined either only surface or buried litter. In this section, we will examine those investigations that combined litter and soil beneath perennial plants. Because litter and adjacent soil are subjected to diurnal fluctuations in temperature and moisture (Wallwork 1982), sampling of litter and soil should provide a more accurate and detailed picture of microfaunal population densities and species occurrences than surface litter or buried litter alone. Whitford et al. (1981) suggested that diurnal migration of microarthropods from mineral soil into litter occurs as litter moisture increases during early morning. One of the few studies that experimentally examined microarthropod migration into litter (MacKay, Silva, and Whitford 1987) observed that only certain taxa (e.g., *Cunaxa* spp., *Eupodes* sp., *Spinibdella cronini*, and *Speleorchestes* spp.) migrated into creosotebush litter (Chihuahuan Desert) during a 24-hour period. *Jornadia larreae*, *Liposcelis* sp., *Tarsonemus* sp., and collembola appeared to enter a cryptobiotic state during the day and became active when litter temperature and moisture conditions were more favorable. The importance of diurnal migration and cryptobiosis in determining the species composition and structure of the surface litter microarthropod assemblages in various deserts has not been examined. The extent of diurnal migration and cryptobiosis among microarthropods should represent a major research focus.

In the northern Mojave, Franco, Edney, and McBrayer (1979) found that cryptostigmatid and tydeid mites exhibited the highest densities over most of the year in soil and litter under shrubs (Table 3.5). While total numbers of microarthropods were greatest in the fall and winter when soil moisture levels were the highest, we determined that diversity ($\overline{H}$) was actually greater in April and July (Table 3.5). This discrepancy is due to an overall lower abundance of the dominant microarthropod groups at this time, which results in a more equitable distribution of individuals among the taxa and a higher diversity index. When we compared the species composition of the litter microarthropod assemblages between sampling times, the highest degree of similarity was found between January and July samples (Table 3.6). In general, the spring sample (April) was least similar to samples from the other collection dates; the assemblages from January and October and July and October had a high degree of similarity. Because a major rainfall event occurred in the Mojave

Table 3.5 Seasonal patterns in density (number/m^3 and diversity ($\bar{H}$) of soil microarthropods in the northern Mojave Desert.

Taxon	Jan.	April	July	Oct.
		Season		
Astigmata				
Acaroidea	31	186	22	3
Cryptostigmata	978	751	642	2,786
Mesostigmata	71	74	45	57
Prostigmata				
Pachygnathoidea	50	233	92	601
Tydeoidea	213	214	190	1,156
Bdelloidea	24	131	123	171
Caeculoidea	1	4	3	9
Raphignathoidea	4	53	105	206
Anystoidea & Erythraeoidea	74	31	12	13
Thrombidioidea	14	12	4	22
Other Prostigmata	138	665	123	192
Collembola	843	12	175	629
Total	2,441	2,366	1,536	5,845
$\bar{H}$	1.54	1.84	1.86	1.57

SOURCE: Data from Franco, Edney, & McBrayer (1979).

during July (Franco, Edney, and McBrayer 1979), it seems probable that higher soil moisture levels at this time and in January and October accounted for the greater degree of community similarity between these sampling times. As previously stated, additional water may not affect microarthropod densities in the litter (e.g., Wallwork, Kamill, and Whitford 1984) if litter and soil moisture are at or near optimum levels for microarthropod activity. Franco, Edney, and McBrayer (1979) observed, however, that soil moisture levels (0–5 cm) in the Mojave were less than 2 percent for the June sampling period, suggesting that following low soil moisture, major rainfall events (25 mm for July in the Mojave) may affect litter microarthropod species composition, if soil temperatures are not low. The low similarity between the April sample and the other dates resulted from a significant increase in the density of Acaroidea and a category labeled "other prostigmata." These microarthropods may be more adapted to the xeric conditions that occurred during April.

Using information from Wallwork (1982), Parker, Santos et al. (1984), Walter and Ikonen (in press), and D. Walter (personal communication), the species obtained by Franco, Edney, and McBrayer (1979) for the Mojave Desert were placed into trophic groups to examine seasonal changes in trophic relationships. Fungivores and detritivores were the largest trophic group of the Mojave Desert soil litter microarthropods (Figure 3.4). The proportion of the total number of microarthropods in this group was highest in January. Densities of general predators were low in January and increased to peak density in the summer. The decline in the fungivore-detritivore trophic group could be from increased predation during the spring and summer when predator densities were highest. Santos and Whitford (1981) also reported predatory mites controlling population densities of microarthropods (fungivores-detritivores) associated with buried litter in the Chihuahuan Desert. Densities of nematophagous microarthropods in the Mojave Desert were relatively constant throughout most of the year until the October sample when they constituted 20 percent of the total number of microarthropods. The increase in nematophages at the October sample may reflect an increase in nematodes as a result of higher soil moisture levels and more favorable soil temperatures than in January.

Although microarthropod species densities in the Mojave Desert changed during the year, we determined that species abundance distributions of these assemblages at each sampling time were similar (Figure 3.5). All distributions were best described by exponential functions and had r^2 values between 0.92 and 0.98. The similarity of the species abundance distributions during the year suggests that the structure of these microarthropod assemblages is determined by trophic relationships and resultant food webs which are modified by abiotic constraints of the habitat. Certain taxa may become seasonally dominant, but the structure of the microarthropod assemblage remains unchanged.

Within desert ecosystems, the heterogeneity in the quantity and quality of the litter may contribute to increased diversity and complexity of the microarthropod fauna (Wallwork 1982). However, Franco, Edney, and McBrayer (1979) found no effect of shrub type on average arthropod densities beneath four dominant plant species (*Ambrosia dumosa, Krameria parvifolia, Larrea tridentata*, and *Lycium andersonii*) in the northern Mojave Desert. A positive relationship has been noticed between organic matter content and microarthropod abundances along a moisture gradient in the Chihuahuan Desert (Santos, Depree, and Whitford 1978; Wallwork, Kamill, and Whitford 1985). Although microarthropod densities tended to increase with greater soil organic matter content, reanalyzing data from Wallwork, Kamill, and Whitford (1985) we found no significant positive correlation (Spearman Rank) between total numbers of microarthropod species and soil organic matter content. Steinberger and Whitford (1984) also found no significant correlation between microarthropod abundances and organic matter content of the soil along a Chihuahuan Desert watershed. However, the range for organic matter content of their samples

Table 3.6 Similarity in taxonomic composition of microarthropod assemblages under shrubs in the northern Mojave Desert.

Sampling Period	Similarity Index[a]		
	April	July	Oct.
Jan.	0.70	0.93	0.85
Apr.		0.69	0.61
July			0.82

SOURCE: Data from Franco, Edney, & McBrayer (1979).
[a]Similarity index = $2W/(a + b)$ (Ludwig and Reynolds 1988).

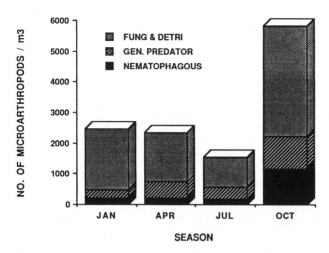

Figure 3.4. Seasonal patterns in trophic relationships of microarthropods from soil and litter beneath shrubs in the northern Mojave Desert (data from Franco, Edney, & McBrayer 1979) (Fung. & Detri. = Fungivore-Detritivore; Gen. Predator = General Predator).

was small (2.8–7.4 percent), thus not providing a sufficient range to critically test if high soil organic matter content does result in increased microarthropod densities. Species abundances may indeed increase in desert soils as organic matter content increases, based on results from Wallwork, Kamill, and Whitford (1985), but competitive and predatory interactions between microarthropod taxa appear to limit species richness in the high organic matter soils. Zak et al. (n.d.) have found that microarthropod abundances in the litter layer of packrat middens in the Chihuahuan Desert were comparable to densities found in litter from eastern deciduous forests. Additional research is needed to definitively address the question of microarthropod abundances in relation to soil organic matter content.

Responses of individual taxa to changes in organic matter content have been reported. Abundances of prostigmatid versus cryptostigmatid mites under various shrubs along an arroyo system in the northern Chihuahuan Desert were related to organic matter content (Kamill, Steinberger, and Whitford 1985). Wallwork, Kamill, and Whitford (1985) observed that prostigmatid mites tended to be numerically dominant in soils with up to 30 percent organic matter content, after which cryptostigmatids became dominant. However, when we plotted data from Wallwork, Kamill, and Whitford (1985), we found that cryptostigmatid mite densities were higher than those of prostigmatid mites at both very low and high levels of soil organic matter (Figure 3.6). This relationship could be aberrant in that Wallwork, Kamill, and Whitford (1985) did not obtain soil organic matter data in the range of 10–25 percent. Prostigmatid mite densities were equal to or greater than cryptostigmatid densities only within a very narrow range of the observed soil organic matter levels. We determined that species richness of prostigmatid versus cryptostigmatid mites in the

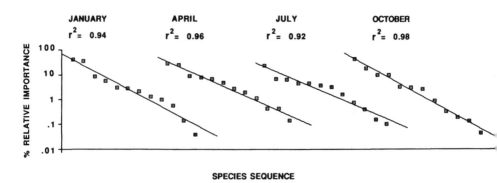

Figure 3.5. Seasonal patterns in species abundance distributions of soil and litter microarthropods from the northern Mojave Desert (data from Franco, Edney, & McBrayer 1979).

Chihuahuan Desert sites examined by Wallwork, Kamill, and Whitford (1985) also did not follow any overall trend when plotted against percent soil organic matter (Figure 3.7). However, at both of the extreme levels of soil organic matter, species numbers of prostigmatids were higher than those of cryptostigmatids, the reverse of the pattern observed for abundances. The complex relationship between densities and species richness of prostigmatid mites versus cryptostigmatids obviously results from the synergistic effects of several factors which themselves are influenced by soil organic matter content (e.g., soil moisture, microbial activity).

Wallwork (1982) and Kamill, Steinberger, and Whitford (1985) have suggested that while the relationship between cryptostigmatid and prostigmatid mites may be due to interference competition, trophic relationships of these mites were a more important factor determining their densities in soils of varying organic matter content. However, when the ratio of number of prostigmatid to cryptostigmatid species was plotted against the ratio of abundances (from Wallwork, Kamill, and Whitford 1985), we found the species ratio (species of prostigmatid/cryptostigmatid) was greater when cryptostigmatid abundances were higher than those of prostigmatids (Figure 3.8). As prostigmatid abundances increased, numbers of prostigmatid species declined and leveled off, indicating that competition or habit homogeneity may be prime factors structuring these assemblages. If high densities of cryptostigmatid mites are associated with high soil organic matter content, the greater structural heterogeneity of these sites could allow for increased species packing of prostigmatid mites.

Root Region

Results from microcosm studies have shown that the root region represents a center of intense microfloral-microfaunal interactions (Anderson, Coleman, and Cole 1981; Coleman et al. 1984; Clarholm 1985). Grazing of bacteria and fungi in the rhizosphere and along the root surface by nematodes and microarthropods can

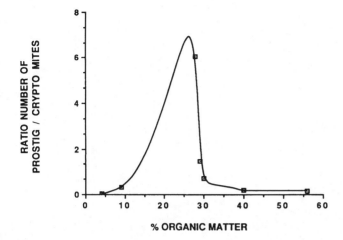

Figure 3.6. Relationship between prostigmatid and cryptostigmatid mite abundances and soil organic matter in the Chihuahuan Desert (data from Wallwork, Kamill, & Whitford 1985).

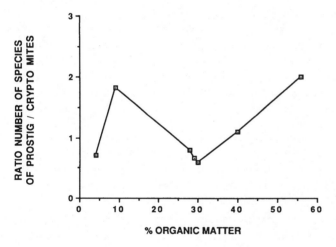

Figure 3.7. Ratio of the number of species of prostigmatid to cryptostigmatid mites in the Chihuahuan Desert as influenced by soil organic matter (data from Wallwork, Kamill, & Whitford 1985).

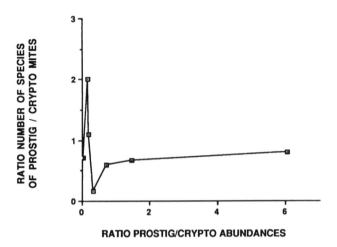

Figure 3.8. The relationship between prostigmatid and
cryptostigmatid species abundances and number of species for
microarthropod assemblages under shrubs in the Chihuahuan Desert
(data from Wallwork, Kamill, & Whitford 1985).

increase nitrogen mineralization and thus influence plant growth. Ingham et al.
(1985) indicated that microfloral grazing by microfauna may be particularly crucial
in semi-arid habitats when plant demand exceeds mineralization rates by the microflora alone. Since plant growth and microfloral and microfaunal activity in deserts
are highly regulated by long-term rainfall patterns (e.g., Whitford and Freckman
1988), trophic relationships among the soil biota during predictable periods of optimum temperature and moisture become major regulators of decomposition and
mineralization in the root region.

Contrary to the patterns of microarthropod species richness and abundances observed in surface and buried litter, Zak, Freckman, and Loring (n.d.) found that in
the Chihuahuan Desert numbers of microarthropod species and abundances (Figures
3.9 and 3.10) per root system of *Erioneuron pulchellum* (fluffgrass) were highest in
January (winter) and declined by April (spring), with no major changes in species
numbers occurring during the rainy season. Soil moisture was highest in January
when evapotranspiration was the lowest and declined during the spring and summer
(Figure 3.9). Microarthropod abundances in the root region increased during October, probably in response to increased soil moisture and lower soil temperatures.
The densities of microarthropods in the root regions of desert plants indicate a
regulation by the interaction between soil moisture and temperature, irrespective of
the status of the plant.

The largest trophic group of microarthropods for any sampling period was the
fungivores-detritivores, comprising approximately 60 percent of the microarthropods at each sampling period (Figure 3.10). Termite predators were abundant only
during the January sample. Abundances of nematophagous microarthropods were
greatest at the October sample; however, on a percentage basis, these predators

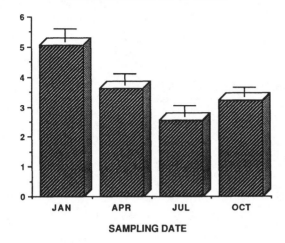

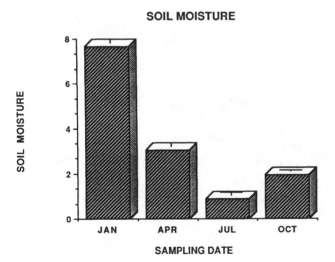

Figure 3.9. Numbers of microarthropod species and soil moisture ($\bar{x} \times \pm$ s.e.) within the root region of fluffgrass from the northern Chihuahuan Desert (data from Zak, Freckman, & Loring n.d.).

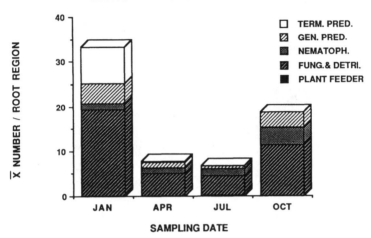

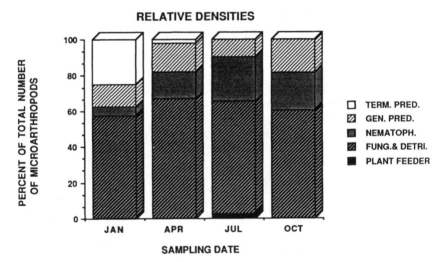

Figure 3.10.   Trophic relationships of microarthropods within the root region of fluffgrass growing in the Chihuahuan Desert (data from Zak, Freckman, & Loring n.d.) (Term. Pred. = Termite Predator; Gen. Pred. = General Predator; Nematoph. = Nematophagous; Fung. & Detri. = Fungivore-Detritivore).

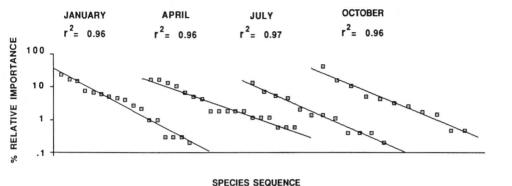

Figure 3.11. Temporal changes in species abundance patterns of root region microarthropods of fluffgrass from the northern Chihuahuan Desert (data from Zak, Freckman, & Loring n.d.).

constituted 10 to 15 percent of the microarthropods from April onward. General predators also were more abundant in October as compared with the other sampling periods; however, their proportional abundance fluctuated during the year, perhaps in response to prey densities.

Species abundance distributions of the root region microarthropod assemblages were best described by exponential functions (r^2 around 0.96) and did not change over the growing season (Figure 3.11). This form of distribution we have shown to occur for microarthropod assemblages enumerated from surface litter and buried soil organic matter and may be typical of microarthropod assemblages from semi-arid and arid habitats. The lack of change in these distributions indicates that while species abundances and trophic groups are changing through the season, the distribution of individuals within a given set of taxa is constant, even though there may be a different subset of taxa from the species pool occurring in the root at each sampling time. Another point indicated by the species abundance distributions is that during periods of low abundances (e.g., July), all taxa are equally reduced.

Nematodes

Nematodes, microscopic roundworms, are major microfaunal components of all belowground systems. Soil nematodes are all heterotrophic and depend on organic food sources. Because of difficulties in identifying species of free-living nematodes (Freckman 1982; Freckman and Baldwin in press), and to show linkages with other soil biota, nematodes are classified to at least the trophic level. Based on morphology and literature accounts of feeding activity, they are placed into four trophic groups: bacterivores, fungivores, omnivore-predators, and plant feeders. The bacterivores, fungivores, and omnivore-predators indirectly influence plant production and organic matter decomposition via their consumption, regulation, and movement of the

Table 3.7 Number of nematodes (corrected for
extraction efficiency), and nematode biomass from
four North American deserts.

	Number/ 500 cm^3	Biomass (g/m^2)
Great Basin		
Agropyron	2,266	0.21
Atriplex	2,482	0.24
Artemisia	4,299	0.42
Halogeton	2,599	0.25
Chrysothamnus	1,584	0.15
Bassia	2,012	0.19
Mojave	1,553	0.15
Sonoran		
Larrea	1,381	0.08
Chihuahuan	839	0.05

SOURCE: Data from Freckman, Sher, & Mankau (1974); Freckman,
Mankau, & Sher (1975).

microflora, while the plant feeders directly consume cytoplasm of roots and foliage
(Freckman and Caswell 1985; Freckman 1988; Freckman and Baldwin in press).

Nematodes require a film of moisture around soil particles for their movement
and activity (Wallace 1959; Demeure et al. 1979). Without water, the nematodes
slowly become inactive. This inactivation, which is reversible, is known as anhydrobiosis, meaning life without water, and is a means of survival for nematodes of
all trophic groups (Freckman and Womersley 1983; Freckman 1986). The ability to
survive in the soil of hot, dry deserts is a fascinating part of soil ecology. Are, for
example, nematodes inactive and anhydrobiotic during most of the year in deserts
and therefore uncoupled from processes of decomposition and herbivory? Is
nematode distribution limited to "islands" of plants where soil moisture is greatest?

The study of the distribution and function of nematodes in desert soils of the U.S.
was initiated as part of the International Biological Program (Freckman, Sher, and
Mankau 1974; Freckman, Mankau, and Sher 1975). Four desert sites were compared
for species diversity, abundance, and trophic groups. They were the Great Basin
Desert at Curlew Valley, Utah, the Mojave Desert at Rock Valley, Nevada, the Sonoran Desert at the Silverbell site, Tucson, Arizona, and the Chihuahuan Desert at
the Jornada site, Las Cruces, New Mexico. Unfortunately, the sampling design
varied between deserts. During 1973–1974, data were obtained for the Chihuahuan
Desert from 20 spaced samples in a square hectare grid. At the other three deserts,
spatial distribution relative to the dominant plants was obtained by trenching to 30
cm and removing 0–10 cm, 11–20 cm, and 21–30 cm soil samples at three distances
(plant base, edge of plant canopy, and three shrub radii) from the plant. The results
(Table 3.7) of this preliminary comparison indicated that the three warm deserts
(Chihuahuan, Sonoran, and Mojave) had a similar nematode abundance and trophic

structure. The lower densities at the Chihuahuan Desert probably resulted from a spatially different sampling pattern (grid versus plant), with samples taken in many bare ground spaces. However, this desert had a trophic structure similar to the other warm desert sites (50 percent microbivores, 30 percent omnivore-predators, 9 percent fungivores, 6 percent plant feeders, and the remainder unidentifiable).

Nematode abundances from the Sonoran, Mojave, and Great Basin were sampled in relation to depth and distance from the dominant plants. Comparison of results indicated differences which may be due to plant species or soil types, or to actual environmental differences between the deserts. At the warm deserts (Sonoran and Mojave), nematodes decreased with increasing depth and distance from the plants. The results of the colder Great Basin Desert sampling of *Atriplex* sp. showed little similarity to the warm deserts. Nematode densities were higher at the plant base than at the canopy edge, but did not decrease with depth. The community structure also was different from the warmer deserts, with plant feeders representing 61 percent of the nematode population density at 0–10 cm. Microbivorous nematodes were 18–22 percent of the population at the Great Basin, and 50 percent at the warmer desert sites.

Species differences between the four sites are shown in Table 3.8. The greatest numbers of species (17) were found at the Great Basin and Mojave desert sites, and the least at the Chihuahuan (11) and Sonoran (9) desert sites. The most obvious difference between these deserts is the large number of plant-feeding nematode species at the Great Basin site. This is probably attributable to the greater grass and root productivity at these sites compared to the other deserts. The large numbers of species from the Mojave Desert could be due to the greater intensity and duration of the study at this site (over 1,500 soil samples during one year), which allowed multiple replication.

To estimate the contribution (abundance, biomass, productivity) of the nematodes in the Mojave Desert, Freckman and Mankau (1986) sampled four dominant plant species weekly for one year. Using these results and data on the area of vegetation coverage and uncovered ground per hectare (Turner 1975), a model was constructed which estimated abundance of nematodes to be $1.26 \times 10^6/m^2$, 0–30 cm depth, and annual production at 26.8 kj (6.4 kcal $\cdot$ m^{-2} $\cdot$ yr^{-1}) (Freckman and Mankau 1986). Mean dry biomass was 0.07 g/m^2, with bacterivores, fungivores, and omnivores accounting for 83 percent of the biomass. This model used soil moisture and temperature data from the field samples to account for periods when the nematodes were in anhydrobiosis.

Based on these IBP studies, much of the nematode research in deserts focused on elucidating the role of the major portion of the nematode community, the fungivores, microbivores, and omnivore-predators, in nutrient cycling and decomposition processes. Nematodes are important components of organic matter decomposition processes in ecosystems (Freckman 1988). Santos, Phillips, and Whitford (1981) determined that the nematode population size was regulated by tydeid mites. When the mites were eliminated, the microbivorous nematodes overgrazed yeast and bacteria, microflora that were responsible for decomposing creosotebush (*L. tridentata*) litter. These findings initiated much of the work on soil fauna in the Chihuahuan Desert,

Table 3.8　Presence of nematode taxa at the four major North American deserts.

Trophic Group	Great Basin	Mojave	Sonoran	Chihuahuan
Fungivores				
Aphelenchus avenue	+	+	+	+
Aphelenchoides	+	+	+	+
Ditylenchus	+	+	+	+
Microbivores				
Acrobeles	+	+	+	+
Acrobeles complexus		+		+
Elaphonema		+		+
Leptonchus		+		
Plectus				+
Ominivore-Predators				
Dorylaimina	+	+	+	+
Eudorylaimus sp.		+	+	
E. monohystera		+		
Mononchus				+
Pungentus		+		+
Plant Feeders				
Apratylenchus belli	+			
Heterodera	+			
Lelpotylenchus albulbosus	+			
Megadorus	+			
Merlinius grandis		+		
Nacobbus	+			
Paratylenchus			+	
Tylenchorhynchus sp. 106		+		
T. sp. 107	+	+		
T. sp. 167		+		
T. actus	+			
T. canalis	+			
T. capitatus	+			
T. cylindricus		+	+	
T. latus	+			
Tylenchorhynchus	+	+	+	+
Tylenchoisimellus	+			

SOURCE: Data from Freckman, Sher, & Mankau (1974); Freckman, Mankau, & Sher (1975).

which was discussed earlier in the chapter and summarized in several papers (Whitford et al. 1981; Whitford et al. 1983; Parker, Freckman et al. 1984; Steinberger et al. 1984; Whitford et al. 1986; Freckman 1986; Whitford and Freckman 1988; Freckman, Whitford, and Steinberger 1987; Freckman 1988).

Using the results, Moorhead et al. (1987) developed a simulation model that suggested that rainfall frequency and soil temperature were the major determinants of nematode activity patterns, as well as of potential populations in surface litter and shallow soil layers. Moorhead's model assumed that nematodes of all life stages responded equally to drying and rainfall pulses. The nematodes, located mainly in the upper soil and organic matter layers where they are subject to highly desiccating conditions in the summer, survive by entering anhydrobiosis, thereby having a large potential population ready for activity when moisture conditions become favorable. The rainfall pulse may only reactivate the anyhdrobiotic population in the litter, rather than causing production of a large number of eggs. Because of the rapidity with which nematodes enter anhydrobiosis in surface litter, Whitford et al. (1986) suggested that activity of soil fauna in surface litter is too limited by desiccation to affect rates of decomposition.

In the Mojave Desert, long-term field studies have shown little quantitative variation in the nematode populations during the long dry and seasonal rainfall periods (Freckman and Mankau 1986). Experimental studies in the Chihuahuan Desert also indicated that both biomass of soil biota and decomposition of creosotebush litter were unaffected by irrigation (Whitford et al. 1986). Freckman, Whitford, and Steinberger (1987) estimated that in soil, about 60 percent of the nematodes were inactive and uncoupled from decomposition when soil matric potentials were -0.4 MPa. This indicates that a greater percentage of the nematode population is involved in soil decomposition processes as compared with the decomposition of surface litter.

Besides influencing decomposition processes, nematodes affect plant production as herbivores (Freckman and Caswell 1985). In results discussed above, the contribution of plant feeders to the nematode community in either biomass, abundance, or species had appeared to be of minor importance. However, more recent studies on the deep rooting (>15 m) woody legume mesquite (*Prosopis glandulosa*) at the Jornada LTER site, Chihuahuan Desert, have shown the presence of numerous plant-feeding species (Freckman and Virginia 1989). Nematode distribution and soil properties on four mesquite sites were compared to a shallow-rooted *Larrea tridentata* community. As in earlier desert studies, nematode distribution decreased with depth. However, nematodes were recovered to 11–12 m at the playa site, and 75 percent of the plant-feeding nematodes were recovered below 0.5 m (Freckman and Virginia 1989). More genera of plant-feeding nematodes and the greatest density of endoparasites were recovered at the playa site compared to the other locations (Table 3.9). Results from the deep coring study strongly revise previous concepts about the minimal contribution of nematode herbivory in desert systems, and show a similarity in the numbers of plant-feeding nematodes in the Chihuahuan and Great Basin deserts. The deep coring study suggests that in ecosystems where deep-rooted plants occur, including deserts, deep soil biota and their effects on plant growth and nutrient cycling should not be overlooked.

Summary

The structure and composition of microfaunal (microarthropod and nematode) communities within and across deserts are controlled through abiotic constraints (moisture and temperature) on activity and the interactions with the soil microflora during decomposition. As such, within and across landscape patterns of soil micro-

Table 3.9 Presence of nematode endoparasites and ectoparasites at four *Prosopis glandulosa* sites in the northern Chihuahuan Desert, New Mexico, USA.

Nematode Genus	Playa	Arroyo	Dunes	Grassland
Endoparasite				
Meloidogyne	+			+
Meloidodera chairis	+	+	+	+
Pratylenchus	+			
Ectoparasites				
Helicotylenchus	+	+		
Paratylenchus	+	+	+	+
Tylenchorhynchus	+	+	+	+
Xiphinema	+	+		+

SOURCE: Data from Freckman & Virginia (1989).

faunal species, abundance relationships and diversity are directly coupled with decomposition dynamics (i.e., quantity, quality, location, and continuity of organic matter in the system). Surface litter in deserts is subjected to extremes in temperature and moisture availability throughout the year. While decomposition rates of this material were found to be independent of rainfall, microarthropod densities, reproduction events, and activities tended to follow long-term rainfall patterns. The application of additional water to surface litter, outside these predictable periods of optimum moisture levels, was not found to increase microarthropod abundances. For buried litter or plant roots, microarthropod abundances tend to parallel soil moisture levels, as influenced by precipitation events. The occurrence of specific microarthropods in buried litter is related to the decompositional state of the material. For the Chihuahuan Desert, tydeid and paratydeid mites were the only microarthropods associated with buried litter during initial decomposition. Heterostigmatid mite populations were usually not associated with the material until 30 to 40 percent of the original material was decomposed. Subsequently, predatory taxa increased in numbers. Although this pattern of microarthropod colonization was observed for the Chihuahuan Desert, the immigration sequence and abundances of species comprising the buried litter microarthropod assemblage differed among desert systems. As with surface litter, buried litter decomposition appears to be independent of rainfall events and microarthropod densities.

The effects of soil organic matter content on the densities of microarthropods and responses of individual species were found to vary with season and desert. We determined that microarthropod species richness (number of species) was not correlated with percentage of soil organic matter, as originally reported for the Chihuahuan Desert. Similarly the relationship between species numbers and abundances of prostigmatid versus cryptostigmatid mites may result from the synergistic effects between soil moisture and microbial activity, which themselves are influenced by soil organic matter content, rather than simply by soil organic matter levels.

Irrespective of habitat (i.e., surface litter, buried organic matter, root region) and desert, the species abundance relationships of microarthropod assemblages were best described by exponential functions. Although different microarthropod taxa are found in various deserts and habitats, differences in species composition are compensatory rather than the cause of changes in functional relationships. The structure of microarthropod assemblages in deserts is a consequence of the similarity in trophic relationships (i.e., food web structure) among taxa. Cohen (1988) has indicated that for "stable" food webs, the proportion of different trophic groups remains unchanged as densities vary. In accordance with this observation, we observed no overall change in the structure of microarthropod assemblages from the North American deserts. Cohen (1988) has indicated that the law of scale invariance is one factor that determines the form and extent of terrestrial food webs. Fungivores-detritivores were the dominant microarthropod trophic group in all habitats and among all seasons. Seasonal dynamics of either nematophagous or general predators varied between habitats, with temporal declines in fungivore-detritivore densities usually associated with increased densities of predatory taxa.

Nematode abundances and trophic structure differed among the North American deserts in response to varying moisture and temperature patterns. Species richness was highest in the Great Basin and Mojave deserts and least in the Chihuahuan and Sonoran. The observed pattern in species richness contributed in part to differences in sampling efforts among these ecosystems. Nematode herbivory of root systems may play a more important role in desert systems than previously acknowledged. For deep-rooted plants (e.g., mesquite), deep soil biota may significantly affect plant growth and nutrient cycling.

Future Research Directions

One point that was evident as we gathered information to write this chapter was the paucity of data concerning the roles of microarthropods and nematodes in the functioning of desert ecosystems. Most of the published data for microarthropods and nematodes in desert ecosystems are either presence/absence data or are lists of species and abundances for specific habitats (Table 3.10). Even this information is incomplete due to problems with taxonomy, extraction efficiencies, and the amount of time and energy necessary to conduct a proper sampling. Information concerning trophic relationships and the roles of the microfauna in decomposition and nutrient cycling within the root region of desert plants is scanty and preliminary at best. Rigorous experimental work on the roles of competition, predation, and mutualistic interactions in structuring microarthropod assemblages is nonexistent, with published statements representing inferences from manipulation studies of factors affecting decomposition rates in arid and semiarid systems.

The preliminary information presented in this chapter on trophic relationships and food web structure among soil and litter microarthropods represents an important area of research from both a theoretical and an applied perspective. Recent research by Dave Walter and colleagues (e.g., Epsky, Walter, and Capinera 1988;

Table 3.10 Data available on microarthropod species richness, abundance, and trophic relationships in the North American deserts in relation to abiotic and biotic variables.

	Surface Litter		
Variable	Species Richness	Abund.[a]	Trophic Relat.[b]
ABIOTIC			
Temperature			
Spatial pattern	. . .	Smith, Smith, & Patten 1987	. . .
Experimental	MacKay et al. 1986	MacKay et al. 1986	MacKay et al. 1986
Water			
Seasonality	Wallwork 1972; Wallwork, Kamill, & Whitford 1984; Wallwork et al. 1986 Franco, Edney, & McBrayer 1979	Wallwork 1972; Wallwork, Kamill, & Whitford 1984; Wallwork et al. 1986 Franco, Edney, & McBrayer 1979 Silva, MacKay, & Whitford 1985	Zak & Freckman, analysis of Franco, Edney, & McBrayer 1979
Experimental	MacKay et al. 1986; Kamill, Steinberger, & Whitford 1985	MacKay et al. 1986; Kamill, Steinberger, & Whitford 1985	MacKay et al. 1986
ORGANIC MATTER			
Quantity	Wallwork, Kamill, & Whitford 1985	Wallwork, Kamill, & Whitford 1985	. . .
Quality	. . .	. . .	. . .
BIOTIC			
Competition	. . .	. . .	. . .
Predation	. . .	. . .	. . .
Mutualism	. . .	. . .	. . .

[a] Abund. = abundance, numbers per sampling unit.
[b] Trophic Relat. = trophic relationships.

Walter, Hunt, and Elliott 1987, 1988) has forced us to reexamine our understanding of trophic relationships among the soil microfauna and the importance of food web dynamics to the functioning of the decomposer subsystem.

From a broader perspective, the information presented in this chapter emphasizes the need to critically examine the role(s) of the soil microbiota in desertification and ecosystem response to changes in moisture and temperature patterns as a consequence of global climate changes. Wallwork (1988) has commented on the potential effects of desertification in shifting the species composition of the soil and litter

	Buried Litter			Root Region	
Species Richness	Abund.[a]	Trophic Relat.[b]	Species Richness	Abund.[a]	Trophic Relat.[b]
ntos et al. 1984	Santos et al. 1984	. . .	. . .	. . .	. . .
. . .	. . .	. . .	. . .	. . .	. . .
ntos & Whitford 1981; ntos et al. 1984	Santos & Whitford 1981; Santos et al. 1984	Santos & Whitford 1981	Zak, Freckman, & Loring n.d.	Zak, Freckman, & Loring n.d.	Zak, Freckman, & Loring n.d.
mill, Steinberger, & Whitford 1985	Kamill, Steinberger, & Whitford 1985	Kamill, Steinberger, & Whitford 1985	. . .	. . .	. . .
mill, Steinberger, & Whitford 1985 allwork, Kamill, & Whitford 1985 rker et al. 1984	Kamill, Steinberger, & Whitford 1985 Wallwork, Kamill, & Whitford 1985 Parker et al. 1984	. . . Parker et al. 1984	. . .	. . .	. . .
. . .	. . .	. . .	Zak, Freckman, & Loring n.d.	Zak, Freckman, & Loring n.d.	Zak, Freckman, & Loring n.d.
ntos & Whitford 1981	Santos & Whitford 1981	Santos & Whitford 1981	. . .	. . .	. . .
. . .	. . .	. . .	. . .	. . .	. . .

microarthropod assemblages. The restoration of these disturbed arid systems may be predicated on reestablishing the critical trophic relationships among functional groups of the soil microfauna.

As for the effects of global climate change, alterations in moisture and temperature patterns in desert systems will greatly impact the functioning of the soil microfauna in those ecosystems. Similarily, changes in soil organic matter content and nutrient input via decreased primary production could begin a process that may stress the decomposer subsystem beyond the capacity to return to a predisturbance equilibrium point. Degradation of an arid ecosystem, once it occurs, is hard to ameliorate because new patterns of soil nutrient accumulation become self-perpetuating (e.g., Wright and Honea 1986). Field studies that will examine the potential effects of global climate change on ecosystem functioning should include a desert system with emphasis given toward examining the response of the soil microfauna to these alterations.

Acknowledgments

We thank Ms. Doris Anders and Ms. Thabi Mhlongo for their efforts in the library and for calculating various data used in the tables. Mr. J. P. Miller kindly drew Figure 3.1. This chapter reflects work that was supported by NSF, BSR-8604766 and BSR-8604970.

Bibliography

Anderson, R. Y., D. C. Coleman, and C. V. Cole. 1981. Effects of saprophytic grazing on net mineralization. Pp. 201–206 in F. C. Clark and T. Rosswall (eds.), *Terrestrial Nitrogen Cycles*, Ecological Bulletins-N.F.R., vol. 33.

Clarholm, M. 1985. Interactions of bacteria, protozoa, and plants leading to mineralization of soil nitrogen. *Soil Biology and Biochemistry* 17: 181–187.

Cohen, J. E. 1988. Untangling 'an entangled bank': Recent facts and theories about community food webs. Pp. 72–91 in A. Hastings (ed.), *Lecture Notes in Biomathematics*. Berlin: Springer-Verlag.

Coleman, D. C., R. E. Ingham, J. F. McClellan, and J. A. Trofymow. 1984. Soil nutrient transformations in the rhizosphere via animal-microbial interactions. Pp. 35–58 in J. M. Anderson, A. D. M. Rayner, and D. W. H. Walton (eds.), *Invertebrate-Microbial Interactions*. Cambridge: Cambridge University Press.

Crawford, G. S., and J. S. Gosz. 1986. Dynamics of desert resources and ecosystem processes. Pp. 63–68 in N. Polunin (ed.), *Ecosystem Theory and Application*. New York: John Wiley and Sons.

Demeure, Y., D. W. Freckman, and S. D. Van Gundy. 1979. Anhydrobiotic coiling of nematodes in soil. *Journal of Nematology* 11: 189–195.

Epsky, N. D., D. E. Walter, and J. L. Capinera. 1988. Potential role of nematophagous microarthropods as biotic mortality factors of entomogenous nematodes (Rhabditida: Steinernematidae, Heterorhabditidae). *Journal of Economic Entomology* 81: 821–825.

Fisher, F. M., L. W. Parker, J. P. Anderson, and W. G. Whitford. 1987. Nitrogen mineralization in a desert soil: interacting effects of soil moisture and N fertilization. *Soil Science Society of America Journal* 51: 1033–1041.

Fisher, F. M., J. C. Zak, G. L. Cunningham, and W. G. Whitford. 1988. Water and nitrogen effects on growth and allocation patterns of creosotebush in the northern Chihuahuan Desert. *Journal of Range Management* 41: 387–391.

Franco, P. J., E. B. Edney, and J. F. McBrayer. 1979. The distribution and abundance of soil arthropods in the northern Mojave Desert. *Journal of Arid Environments* 2: 137–149.

Freckman, D. W. 1982. Parameters of the nematode contribution to ecosystems. Pp. 81–97 in D. W. Freckman (ed.), *Nematodes in Soil Ecosystems.* Austin, Tex.: University of Texas Press.

———. 1986. Ecology of dehydration in soil organisms. Pp. 157–168 in A. C. Leopold (ed.), *Membranes, Metabolism and Dry Organisms.* Ithaca, N.Y.: Cornell University Press.

———. 1988. Bacterivorous nematodes and organic matter decomposition. *Agriculture, Ecosystems and Environment* 24: 195–218.

Freckman, D. W., and J. G. Baldwin. In press. Soil nematoda. In D. L. Dindal (ed.), *Soil Biology Guide.* New York: John Wiley and Sons.

Freckman, D. W., and E. P. Caswell. 1985. Ecology of nematodes in agroecosystems. *Annual Review of Phytopathology* 23: 275–296.

Freckman , D. W., and R. Mankau. 1986. Abundance, distribution, biomass and energetics of soil nematodes in a northern Mojave Desert. *Pedobiologia* 29: 129–142.

Freckman, D. W., R. Mankau, and S. A. Sher. 1975. Biology of nematodes in desert ecosystems. In *Reports of 1974 Progress*, vol. 3 (invertebrate section). U.S. International Biological Program Desert Biome Research Memo 75-32. Logan, Utah: Utah State University.

Freckman, D. W., S. A. Sher, and R. Mankau. 1974. *Biology of Nematodes in Desert Ecosystems.* U.S. International Biological Program Desert Biome Research Memo 74-35. Logan, Utah: Utah State University.

Freckman, D. W., and R. A. Virginia. 1989. Plant feeding nematodes to depths of 12 meters in mesquite dominated desert ecosystems. *Ecology* 70: 1665–1678.

Freckman, D. W., W. G. Whitford, and Y. Steinberger. 1987. Effect of irrigation on nematode population dynamics and activity in desert soils. *Biology and Fertility of Soil* 3: 3–10.

Freckman, D. W., and C. Womersley. 1983. Physiological adaptations of nematodes in Chihuahuan Desert soil. Pp. 396–404 in Ph. Lebrun et al. (eds.), *New Trends in Soil Biology.* Louvain-La-Neuve, Belgium: Dieu-Brichard.

Gutierrez, J. R., and W. G. Whitford. 1987. Chihuahuan Desert annuals: importance of water and nitrogen. *Ecology* 68: 2032–2045.

Ingham, R. E., J. A. Trofymow, E. R. Ingham, and D. C. Coleman. 1985. Interactions of bacteria, fungi, and their nematode grazers: effects on nutrient cycling, and plant growth. *Ecological Monographs* 55: 119–140.

Kamill, B. W., Y. Steinberger, and W. G. Whitford. 1985. Soil microarthropods from the Chihuahuan Desert of New Mexico. *Journal of Zoology* (London) 205: 273–286.

Loring, S. J., D. C. Weems, and W. G. Whitford. 1988. Abundance and diversity of surface active collembola along a watershed in the northern Chihuahuan Desert. *American Midland Naturalist* 119: 21–30.

Ludwig, J. A., and J. F. Reynolds. 1988. *Statistical Ecology.* New York: John Wiley and Sons, 337 pp.

MacKay, W. P., S. Silva, D. C. Lightfoot, M. I. Pagani, and W. G. Whitford. 1986. Effect of increased soil moisture and reduced soil temperature on a desert soil arthropod community. *American Midland Naturalist* 116(1): 45–56.

MacKay, W. P., S. Silva, and W. G. Whitford. 1987. Diurnal activity patterns and vertical migration in desert soil microarthropods. *Pedobiologia* 30: 65–71.

McKell, C. M. 1985. North America. Pp. 187–231 in J. R. Goodin and D. K. Northington (eds.), *Plant Resources of Arid and Semiarid Lands: Global Perspective.* New York: Academic Press.

MacMahon, J. A. 1979. North American deserts: their floral and faunal components. Pp.

21–82 in D. W. Goodall and R. A. Perry (eds.), *Arid Land Ecosystems: Structure, Functioning and Management*, vol. 1. Cambridge: Cambridge University Press.

MacMahon, J. A., and F. H. Wagner. 1985. The Mojave, Sonoran, and Chihuahuan deserts of North America. Pp. 105–202 in M. Evenari, I. Noy-Meir, and D. Goodall (eds.), *Ecosystems of the World 12A: Hot Deserts and Arid Shrublands, A*. New York: Elsevier.

May, R. M. 1975. Patterns of species abundance and diversity. Pp. 81–120 in M. L. Cody and J. M. Diamond (eds.), *Ecology and Evolution of Communities*. Cambridge, Mass.: Belknap Press.

———. 1981. *Theoretical Ecology: Principles and Applications*. 2d ed. Sunderland, Mass.: Sinauer Associates, 489 pp.

Moorhead, D. L., D. W. Freckman, J. F. Reynolds, and W. G. Whitford. 1987. A simulation model of soil nematode population dynamics: effects of moisture and temperature. *Pedobiologia* 30: 361–372.

Noy-Meir, I. 1973. Desert ecosystems: environment and producers. *Annual Review of Ecology and Systematics* 4: 25–52.

———. 1974. Desert ecosystems: Higher trophic levels. *Annual Review of Ecology and Systematics* 5: 195–214.

Parker, L. W., D. W. Freckman, Y. Steinberger, L. Driggers, and W. F. Whitford. 1984. Effects of simulated rainfall and litter quantities on desert soil biota: soil respiration, microflora and protozoa. *Pedobiologia* 27: 185–195.

Parker, L. W., P. F. Santos, J. Phillips, and W. G. Whitford. 1984. Carbon and nitrogen dynamics during the decomposition of litter and roots of a Chihuahuan Desert annual, *Lepidium lasiocarpum*. *Ecological Monographs* 54: 339–360.

Peet, R. K. 1974. The measurement of species diversity. *Annual Review of Ecology and Systematics* 5: 285–307.

Routledge, R. D. 1984. Estimating ecological components of diversity. *Oikos* 42: 23–29.

Santos, P. F., E. Depree, and W. G. Whitford. 1978. Spatial distribution of litter and microarthropods in a Chihuahuan Desert ecosystem. *Journal of Arid Environments* 1: 41–48.

Santos, P. F., N. Z. Elkins, Y. Steinberger, and W. G. Whitford. 1984. A comparison of surface and buried *Larrea tridentata* leaf litter decomposition in North American hot deserts. *Ecology* 65: 278–284.

Santos, P. F., J. Phillips, and W. G. Whitford. 1981. The role of mites and nematodes in early stages of buried litter decomposition in a desert. *Ecology* 62: 664–669.

Santos, P. F., and W. G. Whitford. 1981. The effects of microarthropods on litter decomposition in a Chihuahuan Desert ecosystem. *Ecology* 62: 654–663.

Silva, S., W. P. MacKay, and W. G. Whitford. 1985. The relative contributions of termites and microarthropods to fluffgrass litter disappearance in the Chihuahuan Desert. *Oecologia* 67: 31–34.

Smith, S. D., W. E. Smith, and D. T. Patten. 1987. Effects of artificially imposed shade on a Sonoran Desert ecosystem: arthropod and soil chemistry responses. *Journal of Arid Environments* 13: 245–257.

Steinberger, Y., D. W. Freckman, L. W. Parker, and W. G. Whitford. 1984. Effects of simulated rainfall and litter quantities on desert soil biota: nematodes and microarthropods. *Pedobiologia* 26: 267–274.

Steinberger, Y., and W. G. Whitford. 1984. Spatial and temporal relationships of soil microarthropods on a desert watershed. *Pedobiologia* 26: 275–284.

Strojan, C. L., F. B. Turner, and R. Castetter. 1979. Litter fall from shrubs in the northern Mojave Desert. *Ecology* 60: 891–900.

Turner, F. B., ed. 1975. *Rock Valley Validation Site Report*. U.S. International Biological Program Desert Biome Research Memo 75-2. Logan, Utah: Utah State University.

Wallace, H. R. 1959. The movement of eelworms in water films. *Annals of Applied Biology* 47: 366–370.

Wallwork, J. A. 1972. Mites and other microarthropods from the Joshua Tree National Monument, California. *Journal of Zoology* (London) 168: 91–105.

———. 1982. P. 296 in *Desert Soil Fauna*. New York: Praeger Scientific.

———. 1988. The soil fauna as bioindicators. Pp. 203–215 in *Actas del Congreso de Biologio Ambientol*.

Wallwork, J. A., B. W. Kamill, and W. G. Whitford. 1984. Life styles of desert litter-dwelling microarthropods: a reappraisal based on the reproductive behaviour of cryptostigmatid mites. *South African Journal of Science* 80: 163–169.

———. 1985. Distribution and diversity patterns of soil mites and other microarthropods in a Chihuahuan Desert site. *Journal of Arid Environments* 9: 215–231.

Wallwork, J. A., M. MacQuitty, S. Silva, and W. G. Whitford. 1986. Seasonality of some Chihuahuan Desert soil oribatid mites (Acari: Cryptostigmata). *Journal of Zoology* (London) 208: 403–416.

Walter, D. E., H. W. Hunt, and E. T. Elliott. 1987. The influence of prey type on the development and reproduction of some predatory mites. *Pedobiologia* 30: 419–424

———. 1988. Guilds or functional groups? An analysis of predatory arthropods from a shortgrass steppe soil. *Pedobiologia* 31: 247–260.

Walter, D. E., and E. K. Ikonen. 1989. Species guilds and functional groups: Using taxonomy to predict trophic behavior in nematophagous arthropods. *Journal of Nematology* 21: 315–327.

West, N. E. 1979. Formation, distribution and function of plant litter in desert ecosystems. Pp. 647–660 in D. W. Goodall and R. A. Perry (eds.), *Arid Land Ecosystems: Structure, Functioning and Management*, vol. 1. Cambridge: Cambridge University Press.

Whitford, W. G. 1986. Decomposition and nutrient cycling in deserts. Pp. 93–118 in W. G. Whitford (ed.), *Patterns and Processes in Desert Ecosystems*. Albuquerque, N.Mex.: University of New Mexico Press.

Whitford, W. G., and D. W. Freckman. 1988. The role of soil biota in soil processes in the Chihuahuan Desert. Pp. 1063–1073 in E. E. Whitehead, C. F. Hutchinson, B. N. Timmerman, and R. G. Varady (eds.), *Arid Lands: Today and Tomorrow*. Tucson, Ariz.: University of Arizona Press.

Whitford, W. G., D. W. Freckman, N. Z. Elkins, L. W. Parker, R. Parmalee, J. Phillips, and S. Tucker. 1981. Diurnal migration and responses to simulated rainfall in desert soil microarthropods and nematodes. *Soil Biology and Biochemistry* 13: 417–425.

Whitford, W. G., D. W. Freckman, L. W. Parker, D. Schaefer, P. F. Santos, and Y. Steinberger. 1983. The contributions of soil fauna to nutrient cycles in desert systems. Pp. 49–50 in Ph. Lebran, H. M. Andre, A. de Midts, C. Gregoire-Wilba, and G. Wavthy (eds.), *New Trends in Soil Biology*. Louvain-La-Neuve, Belgium: Dieu-Brichard.

Whitford, W. G., D. W. Freckman, P. F. Santos, N. Z. Elkins, and L. W. Parker. 1982. The role of nematodes in decomposition in desert ecosystems. Pp. 98–116 in D. W. Freckman (ed.), *Nematodes in Soil Ecosystems*. Austin, Tex.: University of Texas Press.

Whitford, W. G., R. Repass, L. Parker, and N. Elkins. 1982. Effects of initial litter accumulation and climate on litter disappearance in a desert ecosystem. *American Midland Naturalist* 108: 105–110.

Whitford, W. F., Y. Steinberger, W. MacKay, L. W. Parker, D. W. Freckman, J. A. Wallwork, and D. Weems. 1986. Rainfall and decomposition in the Chihuahuan Desert. *Oecologia* 68: 512–515.

Whitford, W. G., K. Stinnett, and J. Anderson. 1988. Decomposition of roots in a Chihuahuan Desert ecosystem. *Oecologia* 75: 8–11.

Wright, R. A., and J. H. Honea. 1986. Aspects of desertification in southern New Mexico, USA: soil properties of a mesquite duneland and a former grassland. *Journal of Arid Environments* 11: 139–145.

Zak, J. C., D. Freckman, and S. Loring. N.d. Microfloral-microfaunal interactions within the root-region of a desert bunchgrass: temporal patterns of microarthropods. In preparation.

Zak, J. C., S. J. Loring, W. P. MacKay, and W. G. Whitford. N.d. Characteristics of woodrat middens in the Chihuahuan and Sonoran Desert. In preparation.

Zak, J. C., and W. G. Whitford. 1988. Interactions among soil biota in desert ecosystems. *Agriculture, Ecosystems and Environment* 24: 87–100.

The Community Ecology of Macroarthropod Detritivores

4

Clifford S. Crawford

To my knowledge no review has dealt explicitly with the community ecology of macroarthropod detritivores in deserts, although some authors (e.g., Crawford 1979, 1981, 1986; Crawford and Taylor 1984; Crawford and Gosz 1986; Wallwork 1982) have given the topic limited treatment. Most published works on this group of invertebrates have instead taken an autecological approach, emphasizing the behavioral and/or physiological adaptations of such arthropods to stressful physical conditions near the soil surface. Research at this level allows the community ecologist to appreciate the abiotic constraints affecting components of detritivore assemblages. More of a community-level orientation is evident in papers describing the spatial and temporal activities of small, syntopic groups of species in certain desert habitats (e.g., Faragalla and Adam 1985; Parmenter and MacMahon 1984). Studies of this kind underscore the importance of proximate factors that shape the structure of arbitrarily defined desert assemblages. Finally, a relatively small but extremely valuable set of publications describes entire faunas of surface-active desert arthropods (e.g., Fautin 1946; Pierre 1958). Several of these monographs (e.g., Holm and Scholtz 1980) consider faunal subunits that occupy a series of adjacent habitats. The latter approach provides the basis for understanding community structure in a biogeographical context.

This chapter focuses specifically on the structural and functional organization of macroarthropod detritivore assemblages in deserts. It identifies and compares organizational patterns inherent in such assemblages, and then explores processes that may explain those patterns. Finally, it considers the assemblages in the context of habitat ephemerality, resource variability, and life history strategy. The chapter does not specifically address the ecological roles of desert detritivores; that subject has been partly reviewed by Crawford (1979, 1981, 1986).

At the outset, it is essential to have a clear idea of what is meant by "macroarthropod detritivores." These widespread organisms constitute a broad array of taxa. Apart from some pulmonate gastropods, which are not considered here, and certain social insects, which are the subject of Chapter 5 (MacKay), the common groups include surface-active isopods, millipedes, thysanurans, cockroaches, camel crickets, and a great variety of beetles. The term *detritivore* refers to "animal consumers of dead matter" (Begon, Harper, and Townsend 1986). In reality, most such detritivores ought to be called omnivores, because when the opportunity arises they can consume a spectrum of both living and dead items. Nevertheless, in their natural setting they normally forage on dead, often finely particulate, matter, some of which consists of animal remains or excrement. Most, however, is of plant origin and is usually referred to as litter. Deserts appear to have less surface detritus per unit area than nearly any other type of terrestrial ecosystem (Table 4.1).

Community Patterns: Abundance and Distribution

Anyone witnessing peak activity periods of tenebrionid beetles, *Hemilepistus* isopods, or spirostreptid millipedes in arid regions can attest to the tremendous local abundance of these animals at certain times and places. Yet repeated visits to sites of reported abundance soon make clear that such abundance is limited in diel/seasonal time and habitat space. But the arthropods have not really departed; they have simply moved to a more sheltered and possibly food-rich location, usually underground.

Localized movements of this sort make it difficult to assess the absolute density—and therefore the biomass—of even large populations. Nevertheless, estimates have been made, and are usually based on pitfall trapping combined with mark-recapture calculations (e.g., Thomas and Sleeper 1977), or removal methods such as sieving (e.g., Ghabbour, Mikhaïl, & Rizk 1977) or emptying of enclosures (e.g., Mispagel and Sleeper 1983). More often, investigators have used pitfall trapping for relative estimates, which are useful for comparing activity and diversity between and within surface-active assemblages. The interpretational drawbacks of pitfall trapping have frequently been discussed in the ecological literature, and certainly apply in deserts (Greenslade and Greenslade 1983). For example, Ahearn (1971) found that "catch-densities" of Sonoran Desert tenebrionids were closely related to the densities of his pitfall traps. Inherent in any surface-collecting methodology is the additional and very real problem of accounting for individuals that are not surface-active at the time, for developmental or seasonal reasons.

Table 4.2 gives estimates of absolute densities and biomass for representative populations of desert detritivores. The table is by no means inclusive; rather, it illustrates the range that can apply to both parameters in desert environments. Of particular interest are values ascribed to tenebrionid beetles. It is unclear how Mordkovich and Afanas'ev (1980) obtained a biomass approximation of 190,000 g/ha in Kazakhstan; that figure is more than 200 times Mispagel and Sleeper's (1983) carefully derived value of 900 g/ha for about twice as many species in the Mojave Desert. This may illustrate a striking discrepancy in abundance and niche occupation

Table 4.1 World distribution of surface detritus by ecosystem type.

Ecosystem Type	Total World Surface Detritus (MT carbon × 10⁹)	World Area (ha × 10⁸)	Surface Detritus/Area
Tropical forest	3.6	24.5	0.14
Temperate forest	14.5	12.0	1.21
Boreal forest	24.0	12.0	2.0
Woodland and shrubland	2.4	8.5	0.29
Savanna	1.5	15.0	0.10
Temperate grassland	1.8	9.0	0.20
Tundra and alpine	4.0	8.0	0.50
Swamp and marsh	2.5	2.0	1.25
Cultivated land	0.7	14.0	0.05
Desert and semidesert scrub	0.2	18.0	0.01
Extreme desert, rock, & ice	0.02	24.0	0.0008

SOURCE: Values from Whittaker (1975).

between Old and New World desert tenebrionids, or it may indicate error in estimation techniques.

What do numbers of this magnitude tell us about the potential ecological significance of desert detritivores? Thomas (1979) pitfall-trapped tenebrionids for 4 years in the Mojave Desert, and obtained a 1972 dry weight estimate for a single species, *Cryptoglossa verrucosa*, of 275 g/ha. This compares with an estimate of simultaneous biomass for mammals, birds, and reptiles in his study area of 263 g/ha. Ignoring even the Kazakhstan estimate mentioned above, one can infer from other values given in Table 4.2 (e.g., those of Ghabbour and Shakir 1980) that tenebrionid assemblages (which only accounted for 12.3 percent of the total "mesofauna" [essentially, animals of the size discussed in this paper] biomass in the Egyptian coastal desert) alone undoubtedly account for much more biomass in desert habitats than do vertebrates. And, as will be made evident below, most of this arthropod biomass is composed of detritivores. Clearly, where macrodetritivores comprise the bulk of animal biomass, their contribution to a variety of processes should be comparatively great.

How constant, over seasonal time, are the densities of desert detritivores? As with most populations of terrestrial arthropods, the abundance of some species can change dramatically from year to year, while that of others remains stable for up to 4 years (Thomas 1979). Rogers and Rickard (1975) repeated an earlier study by Rickard and Haverfield (1965) of two tenebrionid species in two different shrub-steppe communities in central Washington, U.S.A. In 10 years the abundance of both species declined greatly; moreover, in both communities they exhibited a shift in relative abundance. Amount of summer rainfall was associated with the abundance of the six sympatric species in Thomas's (1979) Mojave Desert study. In contrast, Hinds and Rickard (1973) determined that the adult abundance of *Philolithus densicollis* in Washington shrub-steppe was strongly correlated with October precipitation in the year of oviposition, which occurs 2 years before adult

Table 4.2 Estimates of density and biomass of representative macroarthropod decomposers in deserts.

Taxon	Habitat, Desert	Absolute Density (no./ha)	Biomass (g/ha)	Source
Isopoda				
Hemilepistus reaumuri	Loessal plain, central Negev	480,000[a]	~19,200[e]	Shachak 1980
Venezillo arizonicus	Shrubland, northern Mojave	3,620[b]	623	Mispagel & Sleeper 1983
Diplopoda				
Orthoporus ornatus	Outwash plain, eastern Chihuahuan	1,302[b]	2,484	Crawford 1976
"millipedes"	Shrubland, northern Mojave	999[b]	223	Mispagel & Sleeper 1983
Thysanura				
Unidentified spp.	Frontal plain, Egyptian coastal	2,000[c]	8	Ghabbour & Shakir 1980
Dictyoptera: Polyphagidae				
Heterogamia spp.	Frontal plain, Egyptian coastal	116,000[c]	20,961	Ghabbour & Shakir 1980
Arenivaga sp.	Shrubland, northern Mojave	3,620	24	Mispagel & Sleeper 1983
Orthoptera				
Ammobaenetes phrixocnemoides	Small dunefield, northern Chihuahuan	18,000[b]/83,000[a]	360/1,660	Crawford unpublished
Ceuthophilus fossor	Shrubland, northern Mojave	854[b]	106	Mispagel & Sleeper 1983
Coleoptera: Tenebrionidae				
15 common spp.	Shrubland, northern Mojave	106,000[b]	900	Mispagel & Sleeper 1983
5–8 common spp.	Steppe, central Kazakhstan	...	190,000	Mordkovich & Afanas'ev 1980
All spp. associated with *Thymelaea* shrubs	Frontal plain, Egyptian coastal	103,000[c]	4,407	Ghabbour & Shakir 1980
Philolithus densicollis, Stenomorpha puncticollis	Shrub-steppe, northern Great Basin	200,000[d]	46,000	Rickard & Haverfield 1965

[a] Visual estimates.
[b] Enclosure removals.
[c] Sieving beneath shrubs.
[d] Mark-recapture from pitfall trapping.
[e] Estimated from maximum reported density as follows: (480,000 isopods/ha $\times$ 0.12 g/isopod) $\div$ 3.
Division by 3 converts live weight to approximate dry weight.

emergence. Overall, annual changes of some magnitude in the pitfall captures of many tenebrionids (e.g., Sheldon and Rogers 1984), camel crickets (Crawford unpublished), and other desert arthropods should be expected. Whether such presumably environmentally induced species-specific changes are accompanied by similar or opposite trends involving other species in the assemblage, or whether the entire detritivore assemblage undergoes major structural shifts, or both occur, is poorly understood. These remain important issues because without tracking *total* detritivore activity—relative to food availability—over time, it is difficult to fully comprehend the trophic role of these animals in deserts.

Community Patterns: Richness and Diversity

In this section I explore the structural organization of species within and between habitat assemblages in the same and different desert regions. Similarity of organizational patterns in different deserts may be explained by evolutionary history. Alternatively, similarity may be due to convergence resulting from similar ecological constraints acting selectively over evolutionary time. Regardless of cause, in assessing assemblage structure one should go beyond a comparison of only detritivores, because these normally co-occur with both carnivores and herbivores. Since carnivorous species are often abundant in soil-associated assemblages, I accord them special treatment in this chapter (see next section). Herbivorous (essentially foliage-eating) arthropods are generally less common surface dwellers in deserts (Table 4.3; Crawford 1988). Nevertheless, there may be periods when herbivorous species such as weevils (Seely and Louw 1980; Mispagel and Sleeper 1983) disperse in large numbers on the ground. Ants further complicate trophic comparisons, because they are inevitably present on desert soils, where they consume a wide variety of food items (MacKay this volume).

The importance of localized primary production to detritivore species richness and diversity has not been explicitly addressed for entire assemblages, although the influence of shrubs on segments of such assemblages has been tested (see below). A general correspondence of richness with primary production is nevertheless suggested by data in Table 4.3, which show that both detritivore and surface-active arthropod assemblage species richness can be relatively great in Mojave and Chihuahuan desert habitats. In these, shrub and other plant cover ranged from 22 to approximately 30 percent (Mispagel and Sleeper 1983; Crawford 1988). In contrast, shrub cover was less than 2 percent in the Egyptian coastal desert (Ghabbour, Mikhaïl, and Rizk 1977) and plant cover overall was probably equally low, on average, in the 15 Namib Desert sites (Crawford and Seely 1987).

Carnivore:Detritivore Ratios

For any given habitat the ratio of one trophic level to another is one measure of community structure. In three of the four desert habitats just referred to, carnivore:detritivore richness ratios were fairly even; the Mojave habitats were excep-

Table 4.3 Species richness and percentage of total individuals among detritivorous, carnivorous, and herbivorous macroarthropods in surface-active desert assemblages.

Habitat, Desert	Sampling Method	Detritivores		Carnivores		Herbivores		Source
		No. spp.	% indiv.	No. spp.	% indiv.	No. spp.	% indiv.	
Shrubland, northern Mojave	Enclosure removal (12, over 9 mo)	19	37	8	17	9	46	Mispagel & Sleeper 1983[a]
Interdune, northern Chihuahuan	Enclosure removal (1, over 11 mo)	19	80	20	19	5	1	Crawford unpublished[a]
Frontal plain, Egyptian coastal	Sieving beneath shrubs	11	93	10	4	2	3	Ghabbour, Mikhail, & Rizk 1977
Dunefield (15), Namib	Pitfall traps	4.1 ± 0.6	82	5.1 ± 1.2	18	0	0	Crawford & Seely 1987

[a]"Minimal active densities" (after Mispagel & Sleeper 1983) are given, since not all arthropods present in enclosures were necessarily in a stage or condition that allowed capture in enclosure pitfalls.

Table 4.4 Estimates of density (D = no./m^2) and biomass (B = g/m^2) among detritivorous, carnivorous, and herbivorous macroarthropods in surface-active desert assemblages.

Habitat, Desert	Detritivores		Carnivores		Herbivores		Source
	D	B	D	B	D	B	
Beneath *Thymelaea* shrubs, frontal plain, Egyptian coastal	47.0	3.3	6.5	0.1	6.5	0.1	Ghabbour & Shakir 1980
Beneath dry-farmed almond trees, frontal plain, Egyptian coastal	43.9	2.6	3.8	0.1	1.8	0.05	Ghabbour & Shakir 1982
Beneath and between shrubs, shrubland, northern Mojave[a]	10.7	0.9	4.9	0.06	13.3	0.05	Mispagel & Sleeper 1983

[a]Values recalculated from data expressed as density/ha or biomass/ha. See also footnote in Table 4.2.

tional in that about half of the species trapped there were either carnivores or herbivores. However, the picture changes when proportional and absolute density and biomass (Table 4.4) are considered. In all four desert situations the carnivore:detritivore ratios of density and biomass were substantially less than 1. Richness and relative density ratios of carnivores and detritivores also were estimated from a year-long investigation in four northern Chihuahuan Desert habitats (Crawford 1988), one of which was less than 100 m from the interdune enclosure referred to in Table 4.3. The habitats ranged from approximately 100 m to 1 km from each other. Only in the interdune did detritivore species richness exceed that of carnivores, the difference being due to tenebrionid beetles. However, as in the other studies cited above, far more *individual* detritivores than carnivores were trapped in all habitats.

A similar relationship between densities of individuals was found by Kheirallah (1986) in the arid central highlands of the Yemen Arab Republic. There, the "mesofauna" (see above) of soil and soil-surface animals consisted mainly of detritivores (87.7 percent of the individuals, 61.1 percent of the species), in contrast to carnivores (6.8 percent of the individuals, 22.2 percent of the species). The remaining three species were herbivores, which actually comprised most (59.2 percent) of the total biomass. The carnivore:detritivore biomass ratio was 0.91. Thus, Kheirallah's (1986) report adds to the evidence that among surface-active desert arthropods, detritivore densities (and often their species richness and biomass) are on balance greater than those of carnivores.

There is at least one exception to the carnivore-detritivore pattern just described. In two of the Namib habitats studied by Crawford and Seely (1987), the expected carnivore:detritivore species richness ratios—but not the ratios of individuals— were reversed. One habitat had received a rare rain of some consequence 7 months earlier while the other was in a swale next to a delta marsh. More recently, Crawford et al. (in press) detected similar richness ratios in an assemblage at the terminus of an estuary in the upper Gulf of California (lower Sonoran Desert). Spiders and/or staphylinid and carabid beetles contributed heavily to the carnivore components in both studies, and in the latter study also produced greater densities of carnivores than detritivores. Since carnivores in these investigations may have largely relied on small prey (e.g., dipteran larvae, insect eggs) not usually trapped in pitfall cups, the altered ratios may not reflect actual interactions of trapped arthropods.

Vegetation

How important is plant presence and architecture to the abundance and diversity of desert detritivores? There is no simple answer to this question; however, recent research suggests that in general the linkage may be weak. For example, Warburg, Rankevich, and Chasanmus (1978) found no good correlation between Shannon-Wiener diversity (H') in isopod populations and the plant cover/foliage height diversity of their xeric Mediterranean habitats. Likewise, Thomas (1983), also using H' as well as log series α indices in four Mojave Desert sites, concluded that diversity of local tenebrionid beetles was little influenced by vegetation heterogeneity. However, he eventually showed that three of five large species were significantly more likely to be caught in traps near shrubs (Thomas 1983). Finally, Thomas (1979) concluded that none of the tenebrionid populations he examined in the Mojave responded to radical changes in plant productivity.

Further tests of association between vegetation and certain components of surface assemblages were conducted in Wyoming shrub-steppe (Great Basin Desert). In one study, Parmenter and MacMahon (1984) compared indices of diversity (H'), evenness (J'), and similarity (Jaccard's, Schoener's), as well as dominance curves, between ground beetle assemblages. This was done in 1.25-ha plots from which shrubs had been removed, and in control plots. Both plots produced similar index patterns, and corresponded well in terms of abundance of individuals and of species richness. Later, Parmenter and MacMahon (1987) reported results of successional recolonization by these assemblages on abandoned strip mines. Again, no significant correlations were found between ground beetle richness and diversity and several vegetation parameters. Most members of the assemblages were omnivores and "scavenging feeders," which occurred in much greater numbers than carnivores and herbivores simultaneously pitfall trapped.

Despite the picture of a generally weak vegetation-detritivore association that emerges from these studies, different habitats in any given desert region are populated by qualitatively distinct assemblages. This occurs in both restricted surface taxa (e.g., Thomas 1979; Wharton and Seely 1982; Sheldon and Rogers 1984) and complete surface assemblages (e.g., Ghabbour and Mikhaïl 1978; Crawford and Seely 1987; Crawford 1988). Thus, factors *other* than vegetation—but factors that may also affect vegetation—must strongly influence patterns in question.

Season, Precipitation, and Shelter

One factor that affects vegetation is clearly "season." Not all species of tenebrionid beetles are active on the surface at the same time of year, as many studies (e.g., Rickard 1970; Kramm and Kramm 1972) have shown. Precipitation has in some instances been correlated with such seasonal variations in activity, although causative relationships are easier to infer than to document. Thomas (1979) noted, for example, that a number of tenebrionid species in the Mojave are active at times of ephemeral summer rains. Hinds and Rickard (1973), as mentioned earlier, observed a 2-year lag relationship between precipitation and adult emergence in a shrub-steppe species. In the Namib Desert, some tenebrionids increase their activity during or following light rains or precipitating fogs; however, heavy rains—unusual in the Namib—depress activity of most species (Wharton and Seely 1982).

Kheirallah (1980) observed that the surface appearance of six isopod species tended to correlate positively with seasonal rainfall in the Egyptian coastal desert. However, for entire assemblages of surface-active arthropods, seasonal rainfall is not a requirement for activity, as all studies lasting a year or more clearly show.

Another pattern-influencing factor is shelter, as was inferred from Kheirallah's (1980) isopod study. The distribution of these and many other surface-active arthropods in deserts is strongly influenced by stone coverage. Shelter of any kind is likely to shade desert arthropods from intense solar radiation (Larmuth 1978, 1979).

Smith, Smith, and Patten (1987) showed experimentally in the Sonoran Desert that prolonged shade increased the number of pitfall-trapped arthropods, including common tenebrionid beetles, for at least 6 months. However, the way shelter affects assemblage organization in arid regions remains poorly studied.

Soil as a Determinant of Pattern

Of the abiotic factors associated with richness and diversity of surface-active species, the most important may well be soil, with its attendant textural and water-related qualities. The significance of soil as a medium regulating both pattern and process among desert biotas should not be overlooked (Crawford and Gosz 1986).

Desert detritivores, as well as many other surface-foraging desert arthropods, tend to live mainly underground. This is especially true of their larvae, about which we know little. Subterranean movements of both adults and immatures reflect their various morphological adaptations (see discussion in Crawford 1981), as well as properties of their habitat soils. Reentry following surface activity involves using burrows made by other animals, digging burrows in relatively firm substrates, or "diving" directly into sand. Clearly, survival of organisms behaving in these ways must be linked to their selection of habitat substrates.

A number of studies illustrate the correspondence between substrate characteristics and population/assemblage distributions. For example, in his Mojave Desert study area, Thomas (1983) found that three tenebrionid species were mainly abundant on gravelly soil, while six others were most common on sandy soil; five more showed no clear preference for either soil type. Likewise, in Washington shrub-steppe, Sheldon and Rogers (1984) noted that certain tenebrionid species were restricted to habitats with loamy sand, sandy loam, or loam substrates. Well-documented observations of this kind have also been made for tenebrionid species by Wharton and Seely (1982) in the Namib Desert.

Although it may seem at first glance that sandy habitats are not viable habitats for many desert macroarthropods, this is far from being so (e.g., Pierre 1958; Medvedev 1965). Some species—mainly detritivores, carnivores, and ectoparasites—use burrows of vertebrates in sandy deserts (Krivokhatskiy 1985). Others use the substrate more directly. While the surfaces of sand sheets and dunes typically are dry, this has little effect on activity belowground. Thus, in the Egyptian coastal desert, Ghabbour and Mikhaïl (1978) found that sub-shrub arthropod assemblages were not influenced by the drying of surface layers in mid-summer. This is not surprising, since sand-dwelling species such as desert cockroaches tend to occupy depths at which moisture regimes permit a positive water balance (Edney 1974, 1977; Ghabbour, Mikhaïl, and Rizk 1977). Deep sand allows water to accumulate below the zone of intense evaporation (Goudie and Wilkinson 1977).

Diversity and Habitat Use

Habitat use by desert detritivores and associated arthropods has been explored with conventional diversity indices. For example, in a pitfall trap study in the Great

Table 4.5 Comparison of Hill's N_2 diversities (habitat-specific means are averaged) from high- and low-diversity assemblages of surface-active arthropods.

Habitat	Hill's N_2	
Delta swales/hummocks, Namib Desert (2 habitats)[a]	4.61 ± 1.61 (diurnal)	7.27 ± 0.73 (nocturnal)
15 other habitats, Namib Desert	1.80 ± 0.21 (diurnal)	2.80 ± 0.65 (nocturnal)
Interdune, northern Chihuahuan Desert[b]	2.38 ± 0.33	
3 other habitats < 1 km from interdune	1.23 ± 0.22	

[a]Calculated from Crawford & Seely (1987).
[b]Data from Crawford (1988); 12 monthly samples.

Basin Desert, Pietruszka (1980) obtained relatively low values for the equivalent of Hill's N_2, a diversity index with a high evenness component. Pietruszka attributed this result to an abundance of mites and ants. Similarly, Crawford and Seely (1987) found that low N_2 scores for nonsocial carnivores and detritivores in a variety of dunefield-related Namib Desert habitats were due to certain dominant species (mostly tenebrionids).

Assemblages in other desert habitats—perhaps a minority—appear to have more even species distributions. This is illustrated in Table 4.5, which shows that comparatively high average N_2 values were recorded, especially at night, from delta swales and hummocks, as opposed to 15 other dunefield-related habitats studied for 1–2 days each in the Namib Desert (Crawford and Seely 1987). High values are also shown in Table 4.5 from a Chihuahuan Desert interdune as compared with three nearby non-sandy habitats—all censused for 1 year (Crawford 1988). All of the elevated N_2 values typified sites with considerable species richness. It may be, therefore, that high species polydominance is characteristic of habitats with soils allowing quick access (via burrowing) to microsites with high moisture.

Comparative patterns of temporal and spatial heterogeneity exhibited by desert detritivores (and their resources) can be assessed with similarity indices. Assessments have sometimes involved entire surface-active assemblages; at other times only subunits of these assemblages have been compared. Employing the former approach, Pietruszka (1980) compared all possible combinations of crossed pitfall trap grid arms (each 120 m in length), using Schoener's similarity index, S. High average S-values between grid arms for all trapping periods indicated low sample heterogeneity, although declining catch abundance was felt to be evidence for increasing spatial patchiness. S-values compared over the entire trapping season by contrast were low, implying substantial faunal changes over seasonal time—even in the small area studied. Ghabbour and Shakir (1980) also used similarity indices (Sørensen's and Gleason's) to illustrate seasonal changes in the soil "mesofauna" (mainly large detritivores) under dry-farmed almond trees in the Egyptian coastal desert. Calculating from diversity values derived from nearly 2 years of examination, these authors found that while the mesofauna was relatively similar throughout one year, it was not as similar in the other, due to population shifts in spring.

Such studies formalize what is intuitively obvious, namely that arthropod assemblages on the desert surface vary qualitatively and quantitatively from season to season. Because of their relatively brief durations, however, investigations of this sort reveal little about long-term trends. Patterns of temporal and spatial diversity over decade-long periods have in fact yet to be documented. With the current emphasis on "biological diversity" in ecological and geopolitical arenas, research initiated along these lines would be timely.

The discussion in this section up to this point has centered on questions of essentially localized richness and diversity. As such it has not addressed the more far-reaching question: Are detritivore assemblages similar in terms of these properties, in similar—but widely separated—desert environments? Or, more generally still: How important is the evolutionary history of a fauna/biota to the functioning of its ecosystem (Westoby 1985)? For desert detritivores, these questions have received only limited attention, at scales ranging from meters to thousands of kilometers. Similarity indices can be applied to spatial gradients at any scale to measure species turnover (beta-diversity); however, this has only been done for desert detritivore assemblages in deserts over relatively short distances.

As an example, overlap across sharply demarcated boundaries involving vegetation and topography in the Namib dunefield system was measured in brief (1–2 day) visits by Crawford and Seely (1987). At best, only a weak depression (by day only) of overlap values was observed across dune-interdune boundaries when Sørensen's Modified Index—which adjusts for abundance of individuals—was used. However, overlap values were comparatively low at one visually striking boundary involving vegetation, suggesting differential habitat use (mainly by thysanurans). In the same study, but at a broader inter-habitat scale, species overlap (Sørensen's Index, unmodified) varied widely between adjacent habitats, averaging about 50 percent. By contrast, between nonadjacent habitats it averaged about 32 percent. Relatively species-rich habitats in a restricted portion of the desert (e.g., Skeleton Coast) were most likely to share species (Crawford and Seely 1987). Edaphic as well as topographic factors may well have been associated with these patterns, for as recent studies by Crawford (1988) (Chihuahuan Desert) and Crawford et al. (in press) (Sonoran Desert) have shown, species overlap is comparatively low between habitats with strongly differing soil properties (texture in the former study, salinity in the latter).

Species turnover along regional and intercontinental gradients has not been examined in deserts for nonsocial detritivores, to my knowledge. Louw (1984), however, trapped highly endemic, ground-dwelling beetles in similar habitats in the Namib and Kalahari deserts for 1 year. Nearly all "ecologically equivalent" beetles differed at the species level. The Kalahari site contained 37 genera and 51 species, compared with 29 genera and 45 species in the Namib, which had more tenebrionid species than the Kalahari (80 percent versus 60 percent).

Tenebrionid beetles may be the most tractable group for large-scale beta-diversity studies, simply because they are more conspicuous and better known, taxonomically, than other desert detritivores. Moreover, very rich faunas of this fifth-largest coleopteran family inhabit arid regions on all continents except Antarctica (Kryzhanovskii 1980). Most species, acording to Kryzhanovskii, are intimately associated with specific soil types. Alternatively, a functionally more interesting approach would be to examine turnover among faunal assemblages—largely detritivores and carnivores—in similar habitats of different deserts. Is pattern convergence (and therefore, presumably, process convergence) likely, or might extenuating circumstances (evolutionary histories or subtle environmental effects) be responsible for high turnover? These questions have been examined for ants and non-arthropod taxa in deserts (Mares and Rosenzweig 1978; Morton 1979, 1982), but have not been extended to detritivore faunas.

Abiotic Community Processes: Climatic and Other Habitat-Related Constraints

Behavioral and physiological adaptations to stressful surface conditions by desert detritivores are well documented (Crawford 1981; Edney 1974; Wallwork 1982). As might be expected, these animals are often better able to reduce transpiration and to tolerate or avoid high temperatures than related species in more mesic environments. Having already noted that abiotic factors—principally relating to soils and topography—affect habitat use by desert species, I now examine how climatic conditions and other abiotic constraints influence their abundance and distribution. In doing so, I recognize that small ectotherms occupying habitats defined by these limitations have evolved trade-offs regarding body size, morphology, physiology, and metabolism. Although assemblages in any desert habitat are composed of species having a wide range of such traits, to date no analyses seem to have been made of assemblage-specific trait uniqueness, irrespective of taxonomic array, that might illustrate how these trade-offs operate at the community level. Rather, studies have usually focused on how related species, in similar and/or different habitats, accommodate to constraints in question.

For example, Cohen and Cohen (1981) observed that of two species of North American burrowing desert cockroaches, one is much better adapted to high temperatures and dry air than the other. The first, *Arenivaga investigata*, inhabits sand dunes in the Mojave Desert, where it is free-ranging. The more susceptible species, *A. apacha*, is restricted to rodent mounds in the Chihuahuan-Sonoran Desert transition zone.

In another example, Parmenter, Parmenter, and Cheney (1989) concluded that species-specific temperature preferences and tolerances of field temperature regimes are associated with microhabitat selection in four sympatric species of Wyoming shrub-steppe *Eleodes* tenebrionid beetles. And, as with *Arenivaga*, differences in microhabitat use seem independent of certain inherent physiological capabilities.

The importance of physiological (and especially morphological) adaptations to spatial and temporal niche separation among Namib Desert lepismatid thysanurans may be more pronounced. Watson (1989) found that a significant amount of the variation in spatial use of habitat by seven sympatric species of these detritivores is explained by responses to detritus biomass, sand density, percent pore space, and sand grain sorting. Morphological similarity occurs between species occupying similar spatial niches. Likewise, a significant amount of variation in temporal use of habitat is explained by responses to solar radiation, temperature, humidity, and wind speed—all conditions that may be related to physiological adaptations.

Despite evidence that specific adaptations alone are sufficient to explain all aspects of habitat use by desert detritivores, results from numerous thermal studies with desert tenebrionid beetles suggest that geographically distinct assemblages of these insects may have evolved specific preferred thermal ranges for foraging. Thus, in the arid *Artemisia* steppe of Washington, U.S.A., resident species may have evolved a low range of active body temperatures in order to expand seasonal activity into earlier and later times of the year, when food resources should be most available and nutritionally valuable (Kenagy and Stevenson 1982). In contrast, tenebrionids from the Namib Desert dunes on the whole appear to have higher preferred temperatures. Seely, Roberts, and Mitchell (1988) reviewed earlier research and interpretations relating to this finding, and proposed that such preferred temperatures lie within the range of temperatures most easily attainable during the course of a year for these diurnally active insects. In the Namib dunefield, where climate is essentially aseasonal and where substrate provides a medium for rapid escape to a moderate thermal regime, this strategy should also extend the opportunity to search for food over diel and seasonal time.

Temperature alone can influence aspects of community structure, although diet and seasonal considerations are difficult to separate from thermal effects alone. Thus, both species richness and N_2 diversity values of surface-active assemblages (all species) usually correlated significantly with soil temperatures in a four-habitat, year-long study in the northern Chihuahuan Desert (Crawford 1988).

Clearly, climatic conditions and substrate-related constraints are associated with the way small-to-large species groupings of desert detritivores are functionally organized in space and time. As they move about on the desert surface, however, these consumers may also experience competition and predation—biotic processes often felt to have significant effects on the organization of biotic communities (e.g., Diamond and Case 1986).

Biotic Community Processes: Competition, Predation, and Foraging

Competition

Little evidence exists that competition contributes to the structure of detritivore assemblages in deserts, although Linsenmair (1984) alludes to intraspecific competition for burrows and mates in extremely dense populations of the social desert isopod *Hemilepistus reaumuri*. Manipulation studies by Wise (1981), in a montane tenebrionid assemblage in New Mexico, remain to date the clearest indication that interspecific competition may be uncommon among detritivores in regions with low rainfall. Likewise, results of competition experiments with larval tenebrionids in the

Namib Desert are equivocal (M. K. Seely personal communication). There should, nevertheless, be circumstances in which spatially and temporally restricted food and shelter, together with high consumer densities, render competition within certain detritivore assemblages inevitable. Conditions in which this may occur in deserts include various long-lasting accumulations of detritus in sites such as dune slopes, rock crevices, underground nests of granivorous rodents, as well as carcasses and dung of vertebrate animals.

In a Chihuahuan Desert study, Schoenly and Reid (1983) found that carrion arthropods form well-structured communities related to resource size. Rodent and rabbit carcasses were used by a total of 23 actively participating carrion species belonging to seven feeding guilds, only two of which were strictly carnivorous. More guilds, species, and individuals were supported by the heavier carcasses, and guild diversity (Simpson's index) increased in the larger carcasses in response to an increasingly even distribution of arthropod numbers in those guilds. In general, detritivore (necrovore) standing crop biomass exceeded that of carnivores, and also increased more rapidly with carcass size. When diel activities of this carrion community were recorded by Schoenly (1984), it became clear that activity patterns differed even between ecologically similar taxa. Competition was not looked for nor was it described in this study. However, in the Sonoran Desert Burger (1965) earlier observed apparent exclusion of secondarily invading blowfly species on canine and feline carcasses by the earliest blowfly colonizers. Inter- and intraspecific competition is well known for flies on carrion in other environments (Hanski 1987).

Nonsocial coprophagous arthropods in deserts are typified by scarabaeoid beetles, and a modest literature exists on their use of vertebrate feces (Crawford 1981; Schoenly 1983; Matthews and Kitching 1984), Schoenly's (1983) comparison of arthropods associated with bovine and equine dung in the Chihuahuan Desert reveals possible desert-related trends. Ants (omnivores) were numerically dominant among individuals attracted to his dung bait traps. Coprovore insects (mainly scarabaeoid and tenebrionid beetles), together with ants comprised more than five times the biomass and about four times the species richness of carnivores (mainly histerid beetles). Arthropod standing crop values declined progressively over 3 sampling days, a sequential reverse of the pattern usually recorded in temperate climates. Otherwise, dung utilization as a function of coprovore assemblages is not well known for arid regions, aside from attempts in Australia to colonize cattle dung pads with imported scarab beetles that disrupt the dung pad habitat of bushfly pests (Hughes 1981). Since bushfly larval density in a pad influences larval size, which, along with pad temperature and moisture influences larval survival (Hughes and Sands 1979), understanding the interaction of biotic and abiotic conditions in these "island" habitats offers some hope for their management.

Predation

In contrast to competition, predation by a variety of carnivores on litter-feeding and coprophagous desert detritivores is well documented (e.g., Cloudsley-Thompson 1975, 1979; McKinnerney 1978). Scorpion diets, for example, include high proportions of cockroaches, camel crickets, and tenebrionid beetles in California's Coachella Valley (Polis 1979; McCormick and Polis 1986). However, I know of only one study to date that conclusively demonstrates an effect of predation on assemblage structure. Parmenter and MacMahon (1988) trapped out rodents from experimental enclosures in a Wyoming sagebrush steppe. Five of the six large tenebrionid species subsequently increased in density by 63 percent over a 21-month period, compared to beetles in control plots. Using data from separate feeding experiments, Parmenter and MacMahon calculated that an average daily consumption of fewer than two beetles (two species in particular) per rodent per day could have produced the observed increase.

Before the results of the Parmenter and MacMahon research are used as a basis for generalizing about desert ecosystems, several qualifications should be made. One, brought up by the authors themselves, is that despite its length, the work does not address the role of rodent predation in the long-term regulation of tenebrionid populations. A second, which is certainly no criticism of this well-planned study, is that it only deals with a portion—albeit an important one—of the area's detritivores. A third qualification is that shrub-steppe represents the "wet end" of the desert climatic continuum, a place where biotic feedbacks are expected to be relatively strong (Noy-Meir 1979–80), although this hypothesis requires rigorous testing. Long-term manipulations involving entire assemblages—including lizards and rodents—need to be performed in a series of desert ecosystems before we can appreciate the role of predation in structuring detritivore assemblages.

Pursuing the question from an evolutionary point of view, Seely (1985) considered whether strong selective pressures by predators might have produced the conspicuousness and the diurnal behavior and activity patterns of the Namib Desert's tenebrionid fauna. Reviewing the evidence of occasional predation, Seely concluded that in the exceptionally arid Namib, abiotic rather than biotic factors have probably been the most significant agents of selection. Again, this agrees with the general thesis of Noy-Meir (1979–80) that biotic interactions are not so likely to affect ecosystem functioning at the dry end of the desert climatic continuum. Seely's (1985) observations are indirectly reinforced by the fact that most Namib tenebrionid species lack chemical defenses, whereas the opposite is true in the Wyoming shrub-steppe (above authors).

Finally, a case may be made for the occasional importance of cannibalism in desert detritivores. Known to occur in other scavenging and omnivorous species (Fox 1975), cannibalism seems to be common in sandy soil *Ammobaenetes phrixocnemoides* camel crickets in the northern Chihuahuan Desert (Crawford and Taylor 1984), and *A. macrobaenetes* in the Coachella Valley (G. A. Polis personal communication). While this behavior possibly evolved as an escape by these cool-season insects from warm-season predators, it also results in high localized densities at a time when the detritus pool is not being appreciably added to, and when the camel crickets themselves are therefore a prime source of food for each other. Gut content analysis (Crawford and Taylor 1984, and many subsequent observations) reveal some conspecific body parts in nearly all individuals examined. How many of these were scavenged, rather than cannibalized, is unclear. Nevertheless, cannibalism, which in some scorpions regulates population structure as well as intraguild predation on interspecific scorpions (Polis and McCormick 1987), may do the same in crowded *Ammobaenetes* assemblages. It would be interesting to follow this question with other orthopteroid detritivores, particularly polyphagid cockroaches. Cannibalism also occurs in crowded colonies of the desert isopod *Hemilepistus reaumuri*, where burrow owners kill and eat small young from other "families" that stray too close to the burrow (Linsenmair 1984).

To conclude, there is still little evidence that predation has a strong and consistent influence on the structure and function of detritivore assemblages, and therefore on the process of detritivory in arid regions. It is likely, however, that in relatively moist times and places in deserts, predation both on and within surface-active assemblages does increase. This view is supported by recorded increases in both consumer biomass and carnivore:detritivore ratios when such moisture occurs (Seely and Louw 1980; Crawford and Seely 1987), and by the results of the study by Parmenter and MacMahon (1988).

Foraging

Given the perceived lack of influence of competition and predation on detritivore assemblages in arid regions, we may ask whether this disassociation extends to their nutritional resources. Put another way, do the availability and quality of detritus in a given desert habitat affect its use by local detritivores? And if they do, how might such use in turn affect not only the detritivores, but also the surface-active communities to which they belong? Answers to these questions would help to explain much about reported patterns (see earlier discussions) of detritivore assemblages, and, more importantly, allow us to make inferences about the process of desert detritivory.

A modest beginning in this quest for answers to the community-level significance of foraging by desert detritivores has been made by M. K. Seely and S. A. Hanrahan (personal communication). In a year-long study in the dry Kuiseb River bed, these researchers used the Pianka Index of Overlap to compare habitat and food choice in three common tenebrionid species. Food types (dead leaves, flowers, animal parts) were found to overlap extensively among species throughout the year. On the other hand, there was relatively less overlap of habitat use. Resources in this vegetationally rich and productive riparian system would at first glance seem to be abundant for detritivores, yet to some degree these organisms clearly partition their resources. Whether partitioning is an indirect effect of each species' autecology, or is due to past or present competition, is not obvious.

In contrast to other biomes, there is a paucity of surface detritus in both desert scrub and extreme deserts (Table 4.1). This effect, compounded by the high spatial and temporal fluxes of particulate materials in windy and vegetationless landscapes, suggests that detrital food in deserts may at times be limiting to its consumers. One would expect this to be true—despite the view of Pimentel (1968) that many supposedly saprophagous arthropods are in reality consumers of microorganisms and fungi, rather than of plant litter—if the guts of these animals frequently contained very little ingested material. However, in my experience most field-collected desert millipedes, camel crickets, and tenebrionid beetles have relatively full guts. Moreover, while microorganisms and fungi certainly are common ingredients of detritivore gut contents (Taylor and Crawford 1982; Crawford and Taylor 1984), most items invariably are of plant origin. My impression is that during their active seasons desert detritivores typically either eat with regularity or eliminate ingested food slowly, or both. As food passage through the guts of desert millipedes and tenebrionids can take less than 24 hours at moderate temperatures (Crawford, unpublished data from dye ingestion studies), and is rapid for detritivores in general (Wotton 1988), nearly continuous ingestion in nature seems plausible. However, the hydration state of the food as well as of the consumer may also determine the rate of food passage in arthropods (Davey and Treherne 1963). Tests controlling for food availability, relative hunger, and hydration would help to answer whether food is actually limiting to detritivores in deserts.

Whether quality of detrital food relates to habitat use in desert detritivores is also poorly understood (Crawford and Taylor 1984). Other factors, such as soil type and predation (see above) may be far more important. Yet we know that a desert millipede can exert specific preferences for fungal species (Taylor 1982), that mesic-region isopods have leaf litter preferences (Hassall and Rushton 1984), and that some Namib Desert tenebrionid species also have species-specific favorite foods (Hanrahan and Seely in press) or food sizes (Crawford and Seely, unpublished). Food preference and habitat use may, however, be separate issues when detritivores are beyond the range of food attraction. The question then becomes, How do they find their food? The capacity of individual species to find food should ultimately be reflected in the organization of the assemblages they help to make up.

Judging from several studies, food location by some detritivores proceeds in steps. The first involves random wandering, as in flying silphid beetles in forest habitats (Shubeck 1968) and *Physadesmia globosa* tenebrionids in the Namib Desert (Crawford, Hanrahan, and Seely in press). While odor perception begins for the silphids at about 1 meter from the carrion source, detection and preference for *P. globosa* probably occur virtually at the level of contact with detritus. Perhaps consumption by tenebrionid beetles of debris on harvester ant trash heaps (Slobodchikoff 1979) can also be explained as a progressive switching from randomized to oriented searching.

So far, it seems reasonable to hypothesize that while assemblages of desert detritivores ordinarily are constrained in space and time by physical factors, their movements, at least on the soil surface, are sufficient to insure repeated consumption of detritus, even when specific forms of detritus are preferred over others. If so, then the qualitative distribution of desert detritus, rather than its quantitative distribution, may be an important determinant of assemblage organization.

Summary and Conclusions

This review portrays arthropod detritivores as dominant members of surface-active assemblages in most desert habitats, and as a major element of desert faunas generally. Most species forage on relatively mobile detritus during physiologically compatible "gates" in diel and seasonal time, and over landscapes with patchy and often ephemeral habitats. Vegetation usually has little influence on detritivore assemblage structure, other than as shade for surface foragers and roots for omnivorous subterranean forms. Soils, on the other hand, may have major effects on structure because of their texture, which affects burrowing and controls belowground moisture levels.

Perhaps the most inclusive and biologically significant variable in this portrayal is the ephemerality or temporal variability of many desert habitats. In the present context, ephemerality primarily refers to the rapid turnover of surface detritus and associated changes in surface relief. Resembling early seres along successional gradients, ephemeral desert habitats tend to have a scarcity of predators (Greenslade and Greenslade 1983; also citations in this review) and an abundance of detritivores. Species richness of detritivores can be overcome by the former as an ephemeral habitat ages. This happens in mesic-region carrion and dung (Hanski 1987), and apparently in desert habitats responding to moisture inputs, as discussed earlier. Presumably the shift in trophic levels in both instances is a function of a broadened food base.

The great variance in detritus (and shelter) in ephemeral desert habitats is mainly due to extensive and stochastic fluxes in microclimate. Assuming fairly random initial wandering by detritivores within and across patch boundaries, there may be occasions in some habitats when these consumers eventually overutilize their resources. Competition and even cannibalism is then to be expected and has been documented (references in this review). Yet high resource variance in such habitats, coupled with dominance by one or a few detritivore species, also suggests that these organisms specialize in exploiting that variance, and consequently undergo a relatively stable coexistence when habitat ephemerality is fairly regular and predictable (Tilman 1986). Carefully designed, long-term studies of habitat use are needed if we are to understand the role of competition in desert detritivore assemblages. Habitat manipulation and analysis of subsequent changes in food use by potential competitors should not be difficult to accomplish. Such studies would be useful background for the larger question of resource limitation as it applies to detritivores in deserts and elsewhere.

The use of detritus in any desert habitat also depends on the life history characteristics of the major detritus consumers. Earlier, I described three broad life history patterns for arthropodan desert detritivores as follows: (1) short lives and rapid responses to changing (habitat) conditions (e.g., collembolans, many mites, possibly thysanurans, cockroaches, some beetles, and many dipterans); (2) long lives and few tight linkages to environmental change (e.g., millipedes, root-boring lepidopterans, and wood-boring beetles); and (3) an intermediate set of patterns often complicated by eusociality (e.g., *Hemilepistus* isopods and social insects) (Crawford 1979). Most or all of these patterns should be represented in most desert habitats. Seasonal rates of detritus consumption in any one habitat should be a complex function of assemblage structure and pattern representation.

Detritivore life histories, within the framework of habitat use, may also be viewed as points along the familiar r–K spectrum (MacArthur and Wilson 1967). Southwood's (1977) reassessment of these selection processes emphasizes variations in habitat favorableness and predictability, and simultaneously identifies a third process, "adversity-selection," that acts in predictable but unfavorable environments (also see Greenslade 1983). Unfortunately, in order to assign a desert detritivore to a spot on Southwood's "habitat templet," one needs to know much more about its biology and habitat use than we do for all but a relatively few such species. Also, there are problems with applying templet terminology to real situations. For example, in a desert, an "unfavorable" environment may stress some species but not others, and "unpredictable" (i.e., ephemeral) habitats may be used regularly by relatively K-selected species (e.g., "sand-swimming" beetles that forage on dune surfaces).

Such difficulties raise an issue that is uncomfortable to many ecologists. Enunciated clearly by Greenslade and Greenslade (1983) for Australian soil invertebrates, it applies equally well to most surface-active desert detritivores. Basically, in order to describe (model) and analyze these groups, it is useful (essential?) to have good taxonomic keys, to appreciate how functional morphologies relate to habitat use, and to have an adequate grasp of species' natural histories. How many professionals are willing to undertake such "unfashionable" work? Perhaps the answer lies in promoting good studies by amateurs and pre-professionals, in conjunction with ecologists who enjoy working with such people.

Lack of detailed taxonomic and life history information should nevertheless not preclude pursuit of important research themes concerning desert detritivores. Here are six themes, distilled from questions and thoughts given in the text, that I think are both important and quite approachable:

1. Within-assemblage turnover (beta-diversity) of abundance and distribution parameters over seasonal, annual, and successional time
2. Convergence of assemblage structure in similar habitats occurring in different deserts or different parts of the same desert
3. Assemblage structure relative to primary production at small (patch) and large (landscape) scales
4. Trends in species dominance, and other elements of assemblage structure, along gradients of habitat "favorableness" and ephemerality (despite my reservations about these terms)
5. The roles of competition and predation in structuring macrodetritivore assemblages
6. The limiting effects of detrital foods on assemblage structure

Acknowledgments

This manuscript profited considerably from the critical reviews of Jim Brown, Gary Polis, Ken Schoenly, Mary Seely, and Charles Wisdom. Useful unpublished manuscripts were kindly supplied by Mary Seely, Robert Parmenter, and James MacMahon. I thank Irene Farmer and Claudia Crawford for help with manuscript preparation.

Bibliography

Ahearn, G. A. 1971. Ecological factors affecting population sampling of desert tenebrionid beetles. *American Midland Naturalist* 86: 385–406.

Begon, M., J. L. Harper, and C. R. Townsend. 1986. Ecology—individuals, populations, and communities. Sunderland, Mass.: Sinauer Associates.

Burger, J. F. 1965. Studies on the succession of saprophagous Diptera on mammal carcasses in southern Arizona. Manuscript, University of Arizona, Tucson.

Cloudsley-Thompson, J. L. 1975. Adaptations of Arthropoda to arid environments. *Annual Review of Entomology* 20: 261–283.

———. 1979. Adaptive functions of the colours of desert animals. *Journal of Arid Environments* 2: 95–104.

Cohen, A. C., and J. L. Cohen. 1981. Microclimate, temperature and water relations of two species of desert cockroaches. *Comparative Biochemistry and Physiology* 69(A): 165–167.

Crawford, C. S. 1976. Feeding season production in the desert millipede *Orthoporus ornatus* (Girard) (Diplopoda). *Oecologia* 24: 265–276.

———. 1979. Desert detritivores: a review of life history patterns and trophic roles. *Journal of Arid Environments* 2: 31–42.

———. 1981. *Biology of Desert Invertebrates*. Berlin: Springer-Verlag.

———. 1986. The role of invertebrates in desert ecosystems. Pp. 73–91 in W. G. Whitford (ed.), *Pattern and Process in Desert Ecosystems*. Albuquerque, N.Mex.: University of New Mexico Press.

———. 1988. Surface-active arthropods in a desert landscape: influences of microclimate, vegetation, and soil texture on assemblage structure. *Pedobiologia* 32: 373–385.

Crawford, C. S., M. L. Campbell, W. H. Schaedla, and S. Wood. In press. Assemblage organization of surface-active arthropods along horizontal moisture gradients in a coastal Sonoran Desert ecosystem. *Acta Zoologica Mexicana nueva serie*.

Crawford, C. S., and J. R. Gosz. 1986. Dynamics of desert resources and ecosystem processes. Pp. 63–87 in N. Polunin (ed.), *Ecosystem Theory and Application*. Chichester, U.K.: John Wiley & Sons.

Crawford, C. S., S. A. Hanrahan, and M. K. Seely. In press. Scale-related habitat use by *Physadesmis globosa* (Coleoptera: Tenebrionidae) in a riparian desert environment. In M. K. Seely (ed.), *Current Research on Namib Ecology—25 Years of the Desert Ecological Research Unit*. Transvaal Museum Monograph No. 8. Pretoria, South Africa: Transvaal Museum.

Crawford C. S., and M. K. Seely. 1987. Assemblages of surface-active arthropods in the Namib dunefield and associated habitats. *Revue de Zoologie Africaine* 101: 397–421.

Crawford, C. S., and E. C. Taylor. 1984. Decomposition in arid environments: role of the detritivore gut. *South African Journal of Science* 80: 170–176.

Davey, K. G., and J. E. Treherne. 1963. Studies on crop function in the cockroach. I. and II. *Journal of Experimental Biology* 40: 763–773, 775–780.

Diamond, J., and T. J. Case, eds. 1986. *Community Ecology*. New York: Harper and Row Publishers.

Edney, E. B. 1974. Desert arthropods. Pp. 311–384 in G. W. Brown, Jr. (ed.), *Desert Biology*, vol. 2. London: Academic Press.

———. 1977. *Water Balance in Land Arthropods*. Berlin: Springer-Verlag.

Faragalla, A. A., and E. E. Adam. 1985. Pitfall trapping of tenebrionid and carabid beetles

(Coleoptera) in different habitats of the Central Region of Saudi Arabia. *Zeitschrift für Angewandte Entomologie* 99: 466–471.

Fautin, R. W. 1946. Biotic communities of the northern desert shrub biome of western Utah. *Ecological Monographs* 16: 251–310.

Fox, L. R. 1975. Cannibalism in natural populations. *Annual Review of Ecology and Systematics* 6: 87–106.

Ghabbour, S. I., and W. Z. A. Mikhaïl. 1978. Ecology of soil fauna of Mediterranean desert ecosystems in Egypt. II. Soil mesofauna associated with *Thymelaea hirsuta*. *Revue Écologie et Biologie du Sol* 15: 333–339.

Ghabbour, S. I., W. Z. A. Mikhaïl, and M. A. Rizk. 1977. Ecology of soil fauna of Mediterranean desert ecosystems in Egypt. I. Summer populations of soil mesofauna associated with major shrubs in the littoral sand dunes. *Revue Écologie et Biologie du Sol* 14: 429–459.

Ghabbour, S. I., and S. H. Shakir. 1980. Ecology of soil fauna in Mediterranean desert ecosystems in Egypt. III. Analysis of *Thymelaea* mesofauna populations at the Mariut frontal plain. *Revue Écologie et Biologie du Sol* 17: 327–352.

———. 1982. Population parameters of soil mesofauna in agro-ecosystems of the Mariut Region, Egypt. I. Under dry-farmed almond. *Revue Écologie et Biologie du Sol* 19: 73–87.

Goudie, A., and J. Wilkinson. 1977. *The Warm Desert Environment*. Cambridge: Cambridge University Press.

Greenslade, P. J. M. 1983. Adversity selection and the habitat templet. *American Naturalist* 122: 352—365.

Greenslade, P. J. M., and P. Greenslade. 1983. Ecology of soil invertebrates. Pp. 645–669 in *Soils, an Australian Viewpoint*. Melbourne: CSIRO; London: Academic Press.

Hanrahan, S. A., and M. K. Seely. In press. Resource partitioning among three adesmine beetles in the dry Kuiseb river course. *South African Journal of Entomology*.

Hanski, I. 1987. Colonization of ephemeral habitats. Pp. 155–185 in A. J. Gray, M. J. Crawley, and P. J. Edwards (eds.), *Colonization, Succession and Stability*. Oxford: Blackwell Scientific Publications.

Hassall, M., and S. P. Rushton. 1984. Feeding behaviour of terrestrial isopods in relation to plant defenses and microbial activity. *Symposium of the Zoological Society, London* 53: 487–505.

Hill, M. O. 1973. Diversity and evenness: a unifying notation and its consequences. *Ecology* 54: 427–432.

Hinds, W. T., and W. H. Rickard. 1973. Correlations between climatic fluctuations and a population of *Philolithus densicollis* (Horn) (Coleoptera: Tenebrionidae). *Journal of Animal Ecology* 42: 341–351.

Holm, E., and G. H. Scholtz. 1980. Structure and patterns of the Namib Desert dune ecosystem near Gobabeb. *Madoqua* 12: 3–39.

Hughes, R. D. 1981. The Australian bushfly: a climate-dominated nuisance pest of man. Pp. 177–191 in R. L. Kitching and R. E. Jones (eds.), *The Ecology of Pests: Some Australian Case Histories*. Melbourne: CSIRO.

Hughes, R. D., and P. Sands. 1979. Modelling bushfly populations. *Journal of Applied Ecology* 16: 117–139.

Kenagy, G. J., and R. D. Stevenson. 1982. Role of body temperature in the seasonality of daily activity in tenebrionid beetles of eastern Washington. *Ecology* 63: 1491–1503.

Kheirallah, A. M. 1980. Aspects of the distribution and community structure of isopods in the Mediterranean coastal desert of Egypt. *Journal of Arid Environments* 3: 69–74.

―――. 1986. A preliminary study on soil mesofauna of Gebel Al-Nabi Shuaib, a central highland of Yemen Arab Republic. *Revue Écologie et Biologie du Sol* 23: 393–404.

Kramm, R. A., and K. R. Kramm. 1972. Activities of certain species of *Eleodes* in relation to season, temperature, and time of day at Joshua Tree National Monument (Coleoptera: Tenebrionidae). *Southwestern Naturalist* 16: 341–355.

Krivokhatskiy, V. A. 1985. On the history of formation of the nidicolous entomofauna of sandy deserts of Central Asia. *Entomologicheskoye Obozreniye* No. 4: 696–704. (Translated from Russian, New Delhi, India: Scripta Technica, Inc.)

Kryzhanovskii, O. L. 1980. Ch. 6 (Tenebrionidae) in *The Composition and Origin of the Terrestrial Fauna of Middle Asia.* Washington, D.C.: Smithsonian Institution. (Translated from Russian.)

Larmuth, J. 1978. Temperatures beneath stones used as daytime retreats by desert animals. *Journal of Arid Environments* 1: 35–40.

―――. 1979. Aspects of plant habitat as a thermal refuge for desert insects. *Journal of Arid Environments* 2: 323–327.

Linsenmair. K. E. 1984. Comparative studies on the social behaviour of the desert isopod *Hemilepistus reaumuri* and of a *Porcellio* species. *Symposium of the Zoological Society, London* 53: 423–453.

Louw, S. V. D. M. 1984. Comparison of the faunistics and ecological patterns of the ground-living beetle faunas of the Kalahari and Namib Desert. *Koedoe* (supplement): 153–165.

MacArthur, R. H., and E. O. Wilson. 1967. *The Theory of Island Biogeography.* Princeton, N.J.: Princeton University Press.

McCormick, S. J., and G. A. Polis. 1986. Comparison of the diet of *Paruroctonus mesaensis* at two sites. In *Proceedings of the IX International Congress of Arachnology.* Panama (in press).

McKinnerney, M. 1978. Carrion communities in the northern Chihuahuan Desert. *Southwestern Naturalist* 23: 563–576.

Mares, S. R., and M. L. Rosenzweig. 1978. Granivory in North and South American deserts: rodents, birds, and ants. *Ecology* 59: 235–241.

Matthews, E. G., and R. L. Kitching. 1984. *Insect Ecology.* 2d ed. St. Lucia, Australia: University of Queensland Press.

Medvedev, G. S. 1965. Adaptations of leg structure in darkling beetles. *Entomological Review* 44: 473–485.

Mispagel, M. E., and E. L. Sleeper. 1983. Density and biomass of surface-dwelling macro-arthropods in the northern Mojave Desert. *Environmental Entomology* 12: 1851–1857.

Mordkovich, V. G., and N. A. Afanas'ev. 1980. Transformation of steppe litter by darkling beetles. *Ekologiya* No. 3: 56–62.

Morton, S. R. 1979. Diversity of desert-dwelling mammals: A comparison of Australia and North America. *Journal of Mammalogy* 60: 235–264.

―――. 1982. Granivory in the Australian arid zone: diversity of harvester ants and structure of their communities. Pp. 257–262 in W. R. Barker and P. J. M. Greenslade (eds.), *Evolution of the Flora and Fauna of Arid Australia.* Frewville, Australia: Peacock Publications.

Noy-Meir, I. 1979–80. Structure and function of desert ecosystems. *Israel Journal of Botany* 28: 1–19.

Parmenter, R. R., and J. A. MacMahon. 1984. Factors affecting the distribution and abundance of ground-dwelling beetles (Coleoptera) in a shrub-steppe ecosystem: the role of shrub architecture. *Pedobiologia* 26: 21–34.

―――. 1987. Early successional patterns of arthropod recolonization on reclaimed strip mines in southwestern Wyoming: the ground-dwelling beetle fauna (Coleoptera). *Environmental Entomology* 16: 168–177.

———. 1988. Population limiting factors of arid-land darkling beetles (Coleoptera: Tenebrionidae): Predation by rodents. *Environmental Entomology* 17: 280–286.

Parmenter, R. R., C. A. Parmenter, and C. D. Cheney. 1989. Factors influencing microhabitat partitioning in arid-land darkling beetles (Tenebrionidae): temperature and water conservation. *Journal of Arid Environments* 17: 57–67.

Pierre, F. 1958. *Écologie et Peuplement Entomologique des Sables Vifs du Sahara Nord Occidental*. Paris: Centre National de la Recherche Scientifique.

Pietruszka, R. D. 1980. Observations on seasonal variation in desert arthropods in central Nevada. *Great Basin Naturalist* 40: 292–297.

Pimentel, D. 1968. Population regulation and genetic feedback. *Science* 159: 1432–1437.

Polis, G. A. 1979. Diet and prey phenology of the desert scorpion *Paruroctonus mesaensis* Stahnke. *Journal of Zoology* (London) 188: 333–346.

Polis, G. A., and S. J. McCormick. 1987. Intraguild predation and competition among desert scorpions. *Ecology* 68: 332–343.

Rickard, W. H. 1970. The distribution of ground-dwelling beetles in relation to vegetation, season, and topography in the Rattlesnake Hills, southeastern Washington. *Northwest Science* 44: 107–113.

Rickard, W. H., and L. E. Haverfield. 1965. A pitfall trapping survey of darkling beetles in desert steppe vegetation. *Ecology* 46: 873–875.

Rogers, L. E., and W. H. Rickard. 1975. A survey of darkling beetles in desert steppe vegetation after a decade. *Annals of the Entomological Society of America* 68: 1069–1070.

Schoenly, K. 1983. Arthropods associated with bovine and equine dung in an ungrazed Chihuahuan Desert ecosystem. *Annals of the Entomological Society of America* 76: 790–796.

———. 1984. Microclimate observations and diel activities of certain carrion arthropods in the Chihuahuan Desert. *Journal of the New York Entomological Society* 91: 342–347.

Schoenly, K., and W. Reid. 1983. Community structure of carrion arthropods in the Chihuahuan Desert. *Journal of Arid Environments* 6: 253–263.

Seely, M. K. 1985. Predation and environment as selective forces in the Namib Desert. Pp. 161–165 in E. S. Vrba (ed.), *Species and Speciation*. Transvaal Museum Monograph No. 4. Pretoria, South Africa: Transvaal Museum.

Seely, M. K., and G. N. Louw. 1980. First approximation of the effects of rainfall on the ecology and energetics of a Namib Desert dune ecosystem. *Journal of Arid Environments* 3: 25–54.

Seely, M. K., C. S. Roberts, and D. Mitchell. 1988. High temperature of Namib dune tenebrionids—why? *Journal of Arid Environments* 14: 131–143.

Shachak, M. 1980. Energy allocation and life history strategy of the desert isopod *H. reaumuri*. *Oecologia* 45: 404–413.

Sheldon, J. K., and L. E. Rogers. 1984. Seasonal and habitat distribution of tenebrionid beetles in shrub-steppe communities of the Hanford Site in eastern Washington. *Environmental Entomology* 13: 214–220.

Shubeck, P. 1968. Orientation of carrion beetles to carrion: Random or non-random? *Journal of the New York Entomological Society* 76: 253–265.

Slobodchikoff, C. N. 1979. Utilization of harvester ant debris by tenebrionid beetles. *Environmental Entomology* 8: 770–772.

Smith, D. S., W. E. Smith, and D. T. Patten. 1987. Effects of artificially imposed shade on a Sonoran Desert ecosystem: arthropod and soil chemistry responses. *Journal of Arid Environments* 13: 245–257.

Southwood, T. R. E. 1977. Habitat, the templet for ecological strategies. *Journal of Animal Ecology* 46: 337–365.

Taylor, E. C. 1982. Fungal preference by a desert millipede *Orthoporus ornatus* (Spirostreptidae). *Pedobiologia* 23: 331–336.

Taylor, E. C., and C. S. Crawford. 1982. Microbial gut symbionts and desert detritivores. *Scientific Reviews on Arid Zone Research* 1: 37–52.

Thomas, D. B. 1979. Patterns in the abundance of some tenebrionid beetles in the Mojave Desert. *Environmental Entomology* 8: 568–574.

———. 1983. Tenebrionid beetle diversity and habitat complexity in the eastern Mojave Desert. *Coleopterists Bulletin* 37: 135–145.

Thomas, D. B., and E. L. Sleeper. 1977. The use of pit-fall traps for estimating the abundance of arthropods, with special reference to the Tenebrionidae (Coleoptera). *Annals of the Entomological Society of America* 70: 242–248.

Tilman, D. 1986. A consumer-resource approach to community structure. *American Zoologist* 26: 5–22.

Wallwork, J. A. 1982. *Desert Soil Fauna*. New York: Praeger Publishers.

Warburg, M. R., D. Rankevich, and K. Chasanmus. 1978. Isopod species diversity and community structure in mesic and xeric habitats of the Mediterranean region. *Journal of Arid Environments* 1: 157–163.

Watson, R. T. 1989. Niche separation in Namib Desert dune Lepismatidae (Thysanura: Insecta): detritivores in an allochthonous desert ecosystem. *Journal of Arid Environments* 17: 37–48.

Westoby, M. 1985. Two main relationships between the components of species richness. *Proceedings of the Ecological Society of Australia* 14: 103–107.

Wharton, R. A., and M. K. Seely. 1982. Species composition of and biological notes on Tenebrionidae of the lower Kuiseb river and adjacent gravel plain. *Madoqua* 13: 5–25.

Whittaker, R. H. 1975. *Communities and Ecosystems*. New York: MacMillan Publishing Co.

Wise, D. H. 1981. A removal experiment with darkling beetles: Lack of evidence for interspecific competition. *Ecology* 62: 727–738.

Wotton, R. S. 1988. Dissolved organic material and trophic dynamics. *BioScience* 38: 172–178.

The Role of Ants and Termites in Desert Communities

5

William P. MacKay

Ants and termites are among the most abundant animals in most terrestrial habitats and may be the dominant insects in many ecosystems (MacKay 1981; MacKay, Zak, and Whitford 1989). Wilson (1985) estimated that one-third of the animal biomass of the Amazonian terra firma rain forest is composed of ants and termites. One ha of soil in such regions may contain more than eight million ants and one million termites. It is not only their dominance in numbers and biomass that makes these insects important in ecosystems, but also their activity. These groups can concentrate more energy at critical points than can any single competitor; they use sheer numbers to construct nests in suboptimal habitats as well as defend young and retrieve food efficiently (Wilson 1985).

Ants are one of the major components in desert ecosystems (Pisarski 1978; Crawford 1981; Briese 1982c; Reyes Lopez 1987), yet most information on their importance is purely descriptive (Baroni Urbani and Aktac 1981) and few studies have documented the relative abundance of species within a community (Marsh 1985a). Ants are the leading predators of other insects and invertebrates in many habitats (Wilson 1985). Seed harvesters are especially abundant and important in arid habitats (Briese 1982c). Schumacher and Whitford (1976) estimated ant colony densities of up to 4,000 colonies/ha in the northern Chihuahuan Desert of North America. In the Ivory Coast dry savanna there are 20 million ants/ha with a single species (*Camponotus acvapimensis*) accounting for 2 million (Levieux 1981). In the western Sahara, individual ants account for 75 percent of the whole fauna (in numbers) (Bernard 1972).

The ecology of termites has received little attention (Lee and Wood 1971b), but has recently been summarized (Krishna and Weesner 1969, 1970; Brian 1978). The limitations in knowledge about desert termites must be stressed (Harris 1970); however, they have been shown to be important in nutrient cycling and organic matter processing, and for the effects of their tunnels on soil properties and chemistry (Holt 1987; Nutting, Haverty, and La Fage 1987; MacKay, Zak, and Whitford 1989). Termites are by far the most important invertebrate decomposers in semiarid Australia (Holt 1987) and the northern Chihuahuan Desert of North America (MacKay, Zak, and Whitford 1989). *Heterotermes aureus* and *Gnathamitermes perplexus* are the dominant organisms involved in soil turnover, litter decomposition, and nutrient cycling in the Sonoran Desert of North America (Nutting et al. 1987).

Desert termites are especially difficult to study because most species are primarily subterranean. In addition to the lack of knowledge concerning the habits of termites, the taxonomy is in a state of chaos. Due to intraspecific variability and the importance of obtaining soldiers and winged reproductives for proper identification, it is difficult to identify termites and thus to study their ecology.

Both ants and termites are very important in terms of biomass (Table 5.1). In many desert locations the mass is an order of magnitude greater than the mass of the mammalian herbivore population (Lee and Wood 1971b; Wood and Sands 1978; MacKay, Zak, and Whitford 1989). The biomass of *Gnathamitermes tubiformans* exceeds that of cattle in the northern Chihuahuan Desert (MacKay, Zak, and Whitford 1989). The termite biomass in a Sahel savanna site (0.96 g/m^2) exceeds that of all other dominant aboveground arthropods (0.3 g/m^2) and birds ($0.3–0.6 \text{ g/m}^2$), but is less than the biomass of domestic stock and herbivorous mammals ($2.0–3.0 \text{ g/m}^2$) (Lepage 1974, cited in Wood and Sands 1978). Few estimates are available on the biomass of social insects in arid ecosystems, but apparently social insects are important in arid ecosystems and biomass of termites is higher than that of ants (Table 5.1).

It has been suggested that the struggle to exist in the harsh physical environments of deserts so dominates the ecology of desert organisms that interactions between species are insignificant (Brown, Reichman, and Davidson 1979). This chapter will show this is clearly not the case with social insects: interspecific interactions are very important in desert ecosystems.

Species Richness and Diversity of Social Insects in Arid Regions

Although some authors conclude that ant species richness is low in desert ecosystems (Pisarski 1978), recent studies show an amazingly diverse fauna in arid regions. Chew (1977) found 23 species in a 30.3-m^2 grid in the Chihuahuan Desert of Arizona. More than 50 species were collected along a 3-km-long transect in the Chihuahuan Desert of New Mexico, including three undescribed species (MacKay, Van Vactor, and Whitford n.d.). Wheeler and Wheeler (1973) collected 59 species of ants in a single desert canyon in California. Ant species richness is especially high in arid regions of Australia. Anderson (1983, 1986) collected 150 species in an area of less than 1 ha. Australian deserts have more than double the species richness of North American deserts, although the within-habitat diversity in Australian arid zones is about the same (Morton 1982). High species richness may be due to the low density

Table 5.1 Maximum estimates of biomasses of ants and termites in terrestrial ecosystems.

Species	Habitat	Biomass (g/m^2)	Source
ANTS			
Pogonomyrmex californicus	Chihuahuan Desert (U.S.)	0.001	Whitford 1972
Pogonomyrmex californicus	Chaparral (U.S.)	0.003	Erickson 1972
Pogonomyrmex montanus	Pine forest (U.S.)	0.01	MacKay 1981
Pogonomyrmex occidentalis	Grassland (U.S.)	0.025	Rogers, Lavigne, & Miller 1972
Pogonomyrmex rugosus	Chihuahuan Desert (U.S.)	0.028	Whitford, Johnson, & Ramirez 1976
Pogonomyrmex rugosus	Arid grassland (U.S.)	0.137	MacKay 1981
Pogonomyrmex subnitidus	Chaparral (U.S.)	0.016	MacKay 1981
Tetramorium caespitum	Heath (England)	0.193	Brian, Elmes, & Kelly 1967
TERMITES[a]			
Cubitermes exiguus	Steppe savanna (Congo)	0.19	Bouillon 1970
Cubitermes fungifaber and spp.	Rainforest (Congo)	1.10	Maldague 1964
Gnathamitermes tubiformans	Arid grassland (U.S.)	2.22	Bodine & Ueckert 1975
Gnathamitermes tubiformans	Chihuahuan Desert (U.S.)	0.30	MacKay, Zak, & Whitford 1989
Nasutitermes costalis	Rainforest (Puerto Rico)	0.01	Wiegert 1970
Nasutitermes exitiosus	Dry sclerophyll forest (Australia)	0.30	Lee & Wood 1971b

[a]Termite values were converted from wet mass to dry mass by multiplying the values by 0.1.

of rodents or other important seed predators, to the high rate of turnover between habitats, to the between-habitat diversity (Morton 1982), or even to microtopographical differences, which affect the composition of assemblages of ants in the Chihuahuan Desert of North America (Chew 1977). Australian harvester ant communities also contain a much higher diversity of polymorphic ant species (Morton 1982), the significance of which is not clear. Ants clearly play a major role in arid ecosystems due to their numerical abundance; this is especially so for the seed predators (Marsh 1986b).

Many studies have shown that precipitation indirectly determines the richness and diversity of desert ant communities (Davidson 1977b; Greenslade and Halliday

1983; Marsh 1986a). Ant species richness in the central gravel plains of the Namib Desert is strongly correlated with mean annual precipitation (Figure 5.1A). Ant species diversity is highly correlated with the amount of precipitation in North American deserts (Figure 5.1B), explaining 69 percent of the variation in species diversity among ant communities (Davidson 1977b). Mean annual precipitation is an index of primary production in arid ecosystems (Davidson 1977b; MacKay and MacKay 1984; Marsh 1986a). All desert ants depend on plants either directly or indirectly as a source of food. Harvesters and nectar feeders depend directly on plants; those that tend Homoptera, and predators, depend on plants indirectly. Rodent species diversity also is correlated with precipitation (Figure 5.1B), which explains 64 percent of the variation in species diversity among rodent communities. The slopes for rodents and ants (Figure 5.1B) are statistically indistinguishable and the intercepts differ by less than 2 percent (Davidson 1977b). Data sets for other continents suggest that species diversity in arid regions is also largely determined by the amount of rainfall. Morton (1982) found that Australian communities and North American communities with similar amounts of rainfall had similar numbers of species, although Australian communities with lower amounts of precipitation tend to be more diverse than similar communities in North America (Morton and Davidson 1988). Ant species richness of the arid Guajiran Peninsula of Colombia also is correlated with amount of precipitation (personal observation).

Unfortunately, we do not have similar data sets for desert termites. Termites are primarily a tropical group, although subterranean termites are diverse and abundant in warm arid and semiarid regions of the world (Wood and Sands 1978). There are 4 relatively common genera (4 species, one species in each genus) in the northern Chihuahuan Desert (MacKay personal observation), 6 genera in the Sonoran Desert (Haverty and Nutting 1975; MacKay personal observation), 9 species in the Coachella Valley of California (Polis this volume), 14 genera in the Kalahari Desert of southern Africa (Coaton 1963), and up to 12 genera (48 species) in central Australia (Watson, Barrett, and Lendon 1978). Termite densities actually increase with increasing aridity in tropical Australia (Holt 1987; Morton and James 1988) and Madagascar (Paulian 1970). Emerson (1955) suggested that species richness decreases with a decrease in temperature, which seems to be true based on the distribution of termites in the United States (Weesner 1965). There are fewer termite species at higher altitudes in the highlands of southwestern Kenya (Kooyman and Onck 1987).

Biogeography of Social Insects in Arid Regions

Ants arose in the tropics and remain predominantly tropical (Brown 1973). None of the genera found in deserts occurs exclusively in arid regions and all were apparently derived from tropical ancestors. Those that are most speciose in arid regions (*Pheidole, Solenopsis, Pogonomyrmex, Messor* [= *Veromessor*; see Bolton 1982], *Crematogaster, Camponotus, Aphaenogaster* [= *Novomessor*; see Brown 1974], *Conomyrma, Forelius, Ocymyrmex, Leptothorax, Cataglyphis, Tetramorium, Mono-*

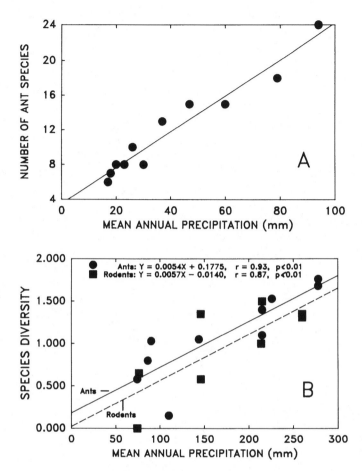

Figure 5.1. Effect of mean annual precipitation on desert ant species richness *A*, along a climatic gradient in the Namib Desert (Marsh 1986a); and *B*, on the species diversity in the deserts of southwestern United States (from Davidson 1977a), where species diversity = H = $\Sigma\, p_i \ln p_i$, where p_i are the proportions of each species.

morium, Chelaner) have representatives in mesic habitats. Most of these genera are more common in certain deserts than in others. For example, *Pogonomyrmex*, *Pheidole, Aphaenogaster, Messor*, and the subfamily Dolichoderinae (especially *Forelius* and *Conomyrma*) are most common in North American deserts, *Pheidole*, *Meranoplus, Chelaner*, and *Tetramorium* are common in Australian deserts, *Pheidole, Messor, Monomorium, Tetramorium*, and *Leptothorax* are most common in African deserts. *Pheidole* occurs in all major deserts; the common New World genus *Pogonomyrmex* does not occur in the Old World.

The composition of the African deserts is especially interesting. The fauna of the Namib Desert is similar in overall composition to that of the Sahara (Marsh 1986b).

Both deserts are dominated by the subfamilies Myrmicinae and Formicinae, but differ from North Amerian deserts as the subfamily Dolichoderinae is poorly represented. This may be because Africa separated from Gondwanaland before the subfamily Dolichoderinae had speciated to any extent (MacKay n.d.). There are some interesting differences between the two African deserts (Marsh 1986b). *Ocymyrmex* (subfamily Myrmicinae) in the Namib is replaced by *Cataglyphis* (subfamily Formicinae) in the Sahara. Although they have completely different phylogenetic origins, both are active during the heat of the day and have similarities in their foraging behavior and speed of locomotion. *Leptothorax* and *Tetramorium* have converged to occupy similar granivorous niches in two African deserts (Marsh 1986b). *Tetramorium* is rare in the Sahara where *Leptothorax* is common; in the Namib, *Tetramorium* is common and *Leptothorax* does not occur. Neither of these genera are common in North American deserts, although both occur. Perhaps they are replaced by *Pogonomyrmex*, although *Pogonomyrmex* is generally common wherever these two genera occur. Little is known of the biologies of these two genera in North American arid habitats, but *Leptothorax* is probably omnivorous. North American and Australian deserts show some similarities (Marsh 1986b). The subfamilies Formicinae and Myrmicinae dominate arid (and most other) habitats on both continents, but the subfamily Dolichoderinae is also important. Predacious ants of the subfamilies Dorylinae and Ponerinae are found in both deserts, but are uncommon in arid zones of Africa (Marsh 1986b).

Termites are predominantly tropical (Emerson 1955; Lee and Wood 1971b), although distinct faunas have evolved in the southwestern deserts of North America and in the arid regions of Madagascar (Weesner 1965, 1970; Paulian 1970). Termites are so important in arid ecosystems that the pharaohs imported termite-resistant timbers from Syria for mummy cases (Harris 1970). *Amitermes* is particularly important in desert ecosystems throughout the world (Emerson 1955; Harris 1970), especially in North Africa (Harris 1970), Australia (Gay and Calaby 1970; Lee and Wood 1971b; Holt 1987), the Arabian Peninsula (Cowie 1989), and North America (Weesner 1970). It often constructs large mounds on the soil surface in mesic habitats (Harris 1970; Holt 1987; MacKay and Whitford 1988); in arid regions it is predominantly subterranean (Gay and Calaby 1970; Harris 1970). Other common genera in arid regions of North Africa include *Anacanthotermes* (restricted to deserts; also occurs in Indian deserts), *Microcerotermes*, and the sand termites *Psammotermes* (Harris 1970). *Psammotermes* are found in areas of complete desertification where they apparently feed on subfossil relics of the humid pleistocene flora, notably trunks of large *Tamarix* (Harris 1970). The termite fauna of Madagascar is especially interesting. Most species live in the specialized xerophytic vegetation of southern arid or semiarid regions (Paulian 1970). The rain forest of Madagascar is relatively poor in species.

Most termite species in Australia are found in tropical savanna woodlands (Lee and Wood 1971b; Braithwaite, Miller, and Wood 1988). Species in arid regions are members of the genera *Anacanthotermes* and *Psammotermes* (Emerson 1955) as well as of *Drepanotermes* and *Tumulitermes* (Gay and Calaby 1970; Lee and Wood 1971b; Morton and James 1988). The family Termitidae is especially abundant in

North American deserts (Weesner 1970). Common termitid genera include *Anoplotermes*, *Gnathamitermes*, and *Tenuirostritermes*; the kalotermitids include *Paraneotermes* and *Pterotermes* (Weesner 1970; Jones, La Fage, and Wright 1981). Except for *Reticulitermes* and *Zootermopsis*, Nearctic termites have a Neotropical origin (Emerson 1955). The common *Gnathamitermes* of the North American deserts was undoubtedly derived from *Amitermes*.

Abiotic Factors in Community Patterns

Abiotic factors, especially soil temperature and moisture, are very important for desert arthropods (MacKay et al. 1986; Abushama and Al-Houty 1989). Desert ants respond to temperature and evaporative stresses in one of two major ways. They may be physiologically adapted to tolerate extremes and hence are active even at high temperatures and vapor pressure deficits, or they may avoid extremes by being active only at night, at twilight, or on cooler days (Whitford, Kay, and Schumacher 1975; Whitford 1978c; Briese and Macauley 1980; Briese 1982b; Marsh 1988; Cloudsley-Thompson 1989). Desert seed harvesters are exposed to high temperatures and desiccation stress while searching for seeds on the soil surface (Heatwole and Harrington 1989). Such ants have very low desiccation rates and high thermal maxima, especially the seed harvesters *Pogonomyrmex* spp. and the omnivore *Aphaenogaster cockerelli* (Whitford, Kay, and Schumacher 1975). Some Australian species, such as *Melophorus* spp., forage at extremely high temperatures when other ants are inactive, and are inactive during the cooler times of the day and during the cooler months (Anderson 1984). *Ocymyrmex barbiger* is a diurnal scavenger of arthropods that have succumbed to thermal and desiccation stress in the Namib Desert (Marsh 1985b; Wehner 1987). It forages in one of the hottest and most barren habitats in the world. Activity increases when the ambient temperature passes 45°C and the ant continues to forage until surface temperatures reach 70°C (Wehner 1987). The number of successful foraging excursions actually increases with higher vapor pressure deficits (Marsh 1985b). The ants are able to radiate heat by resting in the shade and by maintaining a high speed of locomotion, up to 1 m/sec (Wehner 1987). The genus *Cataglyphis* of the Saharan desert is similar in many respects to *O. barbiger* (Wehner 1987), although the colonies are much larger (Marsh 1985b). *Melophorus* has a similar "thermal niche" in Australia, although it is a seed harvester (Marsh 1985b).

Some desert ants are not physiologically adapted for extremes of temperature and low vapor pressures. *Solenopsis xyloni*, a common species in North American deserts, does not differ physiologically from the tropical *S. geminata* (Francke, Potts, and Cokendolpher 1985; Braulick, Cokendolpher, and Morrison 1988). The crepuscular exudate feeder *Formica perpilosa* has water loss rates similar to ants from more mesic environments (Whitford, Kay, and Schumacher 1975). Many diurnally active ants become nocturnal during summer, due to higher temperatures and also due to the increase in abundance of nocturnal insects used as prey (Whitford and Ettershank 1975). For example, ants in the Namib Desert tend to be diurnal in the

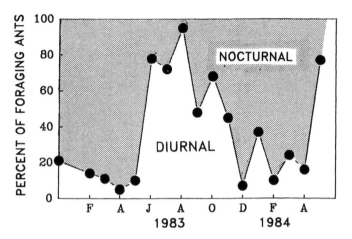

Figure 5.2. Proportion of foraging ants that were diurnally and nocturnally active over a 17-month period in the Namib Desert (from Marsh 1985b).

winter and nocturnal in the summer (December to April–May) (Marsh 1985b; Figure 5.2).

Of all the factors affecting ant foraging activity, soil temperature is the most important (Briese and Macauley 1980; Bailey and Polis 1987; Davison 1987; Porter and Tschinkel 1987; Marsh 1988; MacKay and MacKay 1989). Many diurnal desert ants demonstrate a bimodal diurnal aboveground activity pattern (Whitford 1978c; Lynch et al. 1980; Marsh 1988; MacKay and MacKay 1989), with activity reduced during the hottest part of the day (Figure 5.3). MacKay and MacKay (1989) demonstrated that ambient temperature explained most or all of the variation in foraging activity of *Pogonomyrmex rugosus* and *P. subnitidus*, but had little effect on *P. montanus*, which occurs in more mesic environments.

Seasonal foraging patterns also can be related to seasonal changes in temperature (Heatwole and Muir 1989). Lynch, Balinsky, and Vail (1980) studied seasonal foraging patterns of three common ants in Maryland hardwood forests. They found a bimodal seasonal pattern in *Aphaenogaster rudis* (Figure 5.4A), with reduced activity during the hotter summer months. *Paratrechina terricola* and *Prenolepis imparis* were active during these hotter months. Ant activity was correlated with preferred temperature. *Aphaenogaster rudis* was most active at surface temperatures below 20°C, *Pa. terricola* and *Pr. imparis* at temperatures of nearly 30°C (Figure 5.4B). Several authors have discussed reductions in foraging activity of ant species during the summer (Bernstein 1974; Davidson 1977a). One must be careful as the numbers of hours that nests are active during a particular season may not be correlated with forager effort. Reduced hours of nest activity during the summer occur during the seasonal foraging peaks (MacKay 1981), when energetic costs of the nest are higher due to the higher ambient temperature and the presence of brood (MacKay 1985). The ants are simply very active for a short period each day. Activity of desert ants

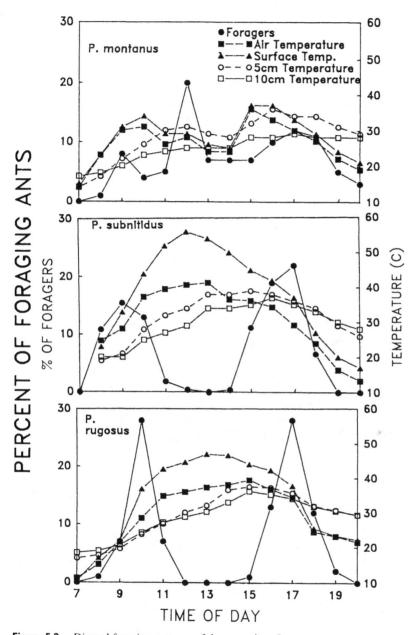

Figure 5.3. Diurnal foraging patterns of three species of
Pogonomyrmex harvester ants, the ambient temperatures in the nest
at 5-cm and 10-cm levels, the surface temperature, and the air
temperature directly above the nest surfaces (data from MacKay &
MacKay 1989).

generally increases after a rain, and nocturnal ants become diurnal when temperatures drop (Marsh 1988).

Unfortunately, we do not have many data on the diel and seasonal activity of desert termites. The desert termites *Gnathamitermes tubiformans*, *Heterotermes aureus*, *Anacanthotermes* sp., and *Psammotermes hybostoma* have unimodal seasonal foraging patterns, with most foraging occuring in the late summer and fall during the seasonal rains (Jones, Trosset, and Nutting 1987; Abushama and Al-Houty 1988; Salman et al. 1988; MacKay, Zak, and Whitford 1989). Aboveground activity is considerably reduced during the hottest periods of summer days (MacKay, Zak, and Whitford 1989). Termites are not active aboveground during the winter months in North American deserts (MacKay personal observation). Soil moisture is extremely important (together with soil temperature) for desert termites (MacKay et al. 1986; Abushama and Al-Houty 1988; Salman, Morsy, and Sayed 1988), possibly due to high rates of water loss (Collins 1969). MacKay et al. (1986) experimentally demonstrated that soil moisture at the 5-cm level was the most important factor regulating aboveground activity in *G. tubiformans* in the Chihuahuan Desert. Reducing soil temperature had no effect on termite activity. In other experiments, MacKay, Fisher et al. (1987) verified the importance of water in termite activity and showed that increased activity resulted in an increase in mass loss of fluff grass (*Erioneuron pulchellum*) litter. Soil temperatures act as an on-off switch in spring and autumn, while soil moisture regulates activity during summer (MacKay et al. 1986). Temperature may be important in other termite species. *Heterotermes aureus* attacks organic matter primarily at sites having the most vegetative cover (Jones, Trosset, and Nutting 1987). This may be due to lower temperatures as termites respond to thermal shadows (Ettershank, Ettershank, and Whitford 1980).

Activity of other desert arthropods, especially soil mites, does not depend on soil moisture (MacKay, Fisher et al. 1987; MacKay, Silva, and Whitford 1987). Soil temperature may play a more important role for such arthropods. Cooler soil temperatures had no effect on positions of ant nests, contrary to results of other studies which demonstrated that shade causes ants to move their nests (MacKay et al. 1986). However, numbers of colonies in watered plots were reduced. This may be due to soil moisture facilitating growth of brood pathogens, especially fungi.

In general, rates of ecological processes in the northern Chihuahuan Desert are not limited by water (MacKay et al. 1986; MacKay, Fisher et al. 1987; Whitford 1986). However, Strojan, Randall, and Turner (1987) suggest that, at least in the Mojave Desert, the amount of precipitation may limit decomposition rates. Primary production of natural vegetation in deserts is limited by nitrogen (Whitford 1986). Decomposition in the northern Chihuahuan Desert is not limited by nitrogen (MacKay, Zak, and Whitford 1989) and termites actually show an aversion to nitrogen-impregnated wood (Zak and MacKay n.d.).

Ants and termites are very important in soil processes. Both taxa move subterranean soil to the surface. Soil turnover by ants in arid regions ranges from 350 to 420 $kg \cdot ha^{-1} \cdot y^{-1}$ in Australia (Briese 1982c) to 842 $kg \cdot ha^{-1} \cdot y^{-1}$ in the Chihuahuan Desert (Whitford, Schaefer, and Wisdom 1986). These rates result in a deposition of from 3 mm to 2 cm of soil on the surface every century (Briese 1982a;

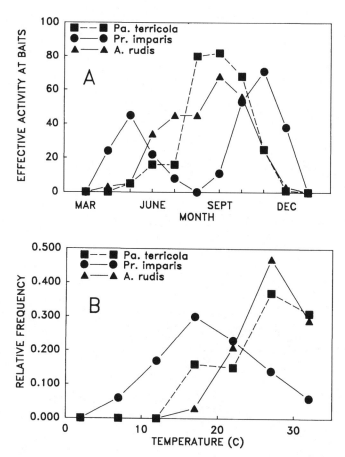

Figure 5.4. *A*, Seasonal activity patterns of three ant species
(*Paratrechina terricola*, *Prenolepis imparis*, and *Aphaenogaster rudis*) in a Maryland hardwood forest; and *B*, the air temperatures at
which the ants were most active (based on Lynch, Balinsky, & Vail
1980).

Whitford, Schaefer, and Wisdom 1986). Deposition rates for termites vary from
0.25 to 0.5 cm/century in North Queensland (Holt, Coventry, and Sinclair 1980),
4 cm/century (Williams 1968, Lee and Wood 1971a) to 7.5 to 10 cm/century in the
Ivory Coast (Lepage 1984). This mixing of plant and animal matter with soil results
in more nutrients being available to plants (Briese 1982a) and influences plant species composition (Whitford, Schaefer, and Wisdom 1986).

Termites are also important in other soil processes (Nutting, Haverty, and La
Fage 1987), possibly more important than ants. Galleries of subterranean termites
in the northern Chihuahuan Desert are very important in increasing water infiltration
rates (Elkins et al. 1986; MacKay, Zak, and Whitford 1989). Plots without termites

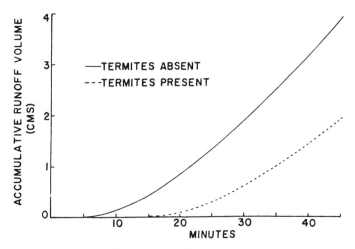

Figure 5.5. Accumulative runoff volumes from plots with and without termites in the northern Chihuahuan Desert (from MacKay, Zak, & Whitford 1989).

in the northern Chihuahuan Desert began to have sheet flow within 5 minutes of initiation of artificial irrigation, whereas plots with termites began to "run" only after 15 minutes (Figure 5.5). The total volume of runoff was much greater in plots without termites (MacKay, Zak, and Whitford 1989). Bed load concentration was three times greater from termite-free plots due to the increased bulk density of the soils after the collapse of subterranean galleries (Elkins et al. 1986).

Biotic Community Processes

Three biotic processes may be important in structuring social insect populations in deserts: competition, predation, and interactions with plants. Competition will be discussed in the first section.

Competition

Competition is well documented in social insect populations (Levings and Traniello 1981). It includes territorial conflicts and "food robbing" in which a forager of one species removes a food item from the mandibles of a forager of another species (Hölldobler 1986; Carroll 1988). For example, *Myrmecocystus mimicus* waylays returning foragers of *Pogonomyrmex* spp., *Aphaenogaster cockerelli*, and *Pheidole desertorum*, and robs food from their mandibles. *Myrmecocystus navajo* robs prey from *A. cockerelli*. This behavior may be important in the structure of desert ant communities (Hölldobler 1986).

Interspecific competition could be either less important (Brown, Reichman, and

Davidson 1979) or more important (Anderson 1986) in arid regions than in other regions. Ant species tend to be generalists with strongly overlapping resource requirements. Therefore, competition between species of desert ants would be expected to be intense, and should result in partitioning of resources (López Moreno and Díaz Betancourt 1986), although Davidson (1977b) concluded that microhabitat partitioning, interspecific aggression, and territorial defense are relatively unimportant in desert ant communities. Many ant species are unlikely to coexist locally within relatively uniform habitat types (Whitford, Johnson, and Ramírez 1976; Davidson 1977a). Thus we would expect to see patterns of resource partitioning among ant species, involving microhabitat differences, differences in food type, quality, and size, and differences in temporal foraging patterns. Ants may even compete with desert shrubs for water (Rissing 1988).

Ants of the genus *Pogonomyrmex* are especially abundant in New World arid ecosystems (MacKay 1981). Most species primarily use seeds as a food source (Figure 5.6), especially in the most arid ecosystems. *Pogonomyrmex montanus* is exceptional as it uses dead insects and vertebrate fecal material to a large extent, but occurs in somewhat more mesic habitats in clearings in pine forests. This trend also occurs in *P. occidentalis* and *P. subnitidus*, which are often found at higher elevations in somewhat more mesic environments. *Pogonomyrmex rugosus* in the Chihuahuan Desert frequently includes subterranean termites in its diet, a resource which is not as common in grasslands in Southern California, where *P. rugosus* primarily uses seeds of a single plant species, *Erodium cicutarium* (MacKay personal observation).

There is some evidence that ants are food limited. Bernstein (1974) concluded that harvester ants are seed limited, although there are large reserves of seeds in the soil even after several years during which no seeds were produced (Tevis 1958; but see Chew 1977). Davidson (1977b) found a strong correlation between ant diversity and primary production, which suggests that the number of species present in a community in an arid ecosystem is limited by the food source. This may not be true in Australia, although dominant species have a substantial impact on the remainder of the ant community, presumably due to interspecific competition (Fox and Fox 1982). Species richness in ants has also been shown to be limited by primary production (Figure 5.1), but Briese (1982b) found that areas with different seed production rates supported similar populations of ants. This suggested to Briese that some other factor limited population density of desert ants below levels set by food availability.

Ants partition resources in a number of ways. Harvester ants partition seeds on the basis of size and nutritional quality (Chew 1977; Davidson 1977a,b, 1978; Hansen 1978; Whitford 1978c). We can usually find differences in the sizes of ants in a guild of desert ants. For example, ants in a Chihuahuan Desert community were grouped within three feeding guilds (Chew 1977; Chew and De Vita 1980). The body masses within a feeding guild differed by an average ratio of 1.66; between categories the average ratio was 1.28. Davidson (1977b) demonstrated that seed size preference was highly correlated with worker body size in a number of desert harvester ants (Figure 5.7). Larger ants harvest medium and larger sized seeds, which maximizes net energy intake (Bailey and Polis 1987). Hansen (1978) found that seed size was correlated with worker mass in a group of *Pogonomyrmex* harvester ants.

The largest species (*P. rugosus*) used the largest seeds, the intermediate-sized species (*P. maricopa*) used medium-sized seeds, and the smallest species (*P. desertorum*) used the smallest seeds. Kelrick et al. (1986) were able to show the importance of nutritional quality of seeds, and pointed out that nutritional quality has been overlooked as an important characteristic in ecological studies. Omnivores may also divide resources on the basis of size (Whitford 1978c).

Seed size may be important to desert ants only in the North American deserts, and only under certain circumstances. Ants in Australian deserts tend to be smaller (Morton 1982; Morton and Davidson 1988), and differences in size among coexisting ants are not nearly as marked as they are in North America (Briese 1982b). Therefore, Australian ants presumably cannot partition seeds on the basis of size (Morton 1982). Two species in the Namib, *Tetramorium rufescens* (4.0–5.1 mm total

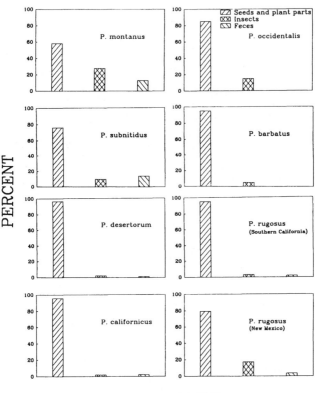

FOOD TYPE

Figure 5.6. Food types of ants of the genus *Pogonomyrmex* from arid and semiarid regions in southwestern United States. All except *P. montanus*, *P. occidentalis*, and *P. subnitidus* occur in deserts. (Data from Lavigne 1969; Whitford, Johnson, & Ramírez 1976; Whitford 1972, 1978b; MacKay 1981.)

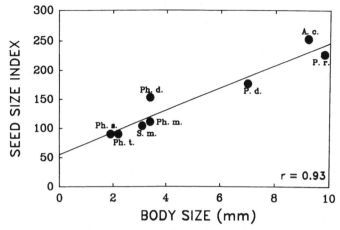

Figure 5.7. Relationship between mean worker body length and
seed size index for eight species of harvester ants in the Chihuahuan
Desert of North America (modified from Davidson 1977b). Seed size
indices were calculated as $\Sigma\ p_i d_i$, where p_i is the proportion of seeds
chosen from the ith size class and d_i is the median diameter of seeds
in that class (see Davidson 1977b for further details). The
abbreviations for the ant species are: Ph. s. = *Pheidole* sp., Ph. t. =
P. tucsonica; Ph. d. = *P. desertorum*; Ph. m. = *P. militicida*;
S. m. = *Solenopsis maniosa*; P. d. = *Pogonomyrmex desertorum*;
P. r. = *P. rugosus*; A. c. = *Aphaenogaster cockerelli*.

length) and *Messor denticornis* (5.5–11 mm total length), are very different in size,
yet both species utilize the same size of food items (Marsh 1987). In addition, the
two species are active at the same time diurnally and have similar seasonal foraging
patterns. Thus, interspecific competition for food is not important in these two
species (Marsh 1987). These data support the notion that interspecific competition
for food is not of major importance to ants in unpredictable arid environments
(Briese 1982c). Under certain circumstances (where there is an abundance or a
scarcity of seeds), North American desert harvesters collect seeds as they are encountered, without regard to mass, size, or chemical composition (Whitford 1978b).
Rissing and Pollock (1984) concluded that there was no correlation between body
size of the polymorphic species *Messor pergandei* and seed size. They state that
statistically significant correlations may have little biological impact. Although
worker size and seed size may not be correlated in all species, correlations as strong
as those shown with the variables in Figure 5.7, and which also have a reasonable
biological explanation, are difficult to refute.

Coexistence in a large community of ecologically similar desert harvester ant
species may be made possible by partitioning the times of day during which foragers
are active aboveground (Whitford, Johnson, and Ramírez 1976; Chew 1977; Hansen
1978; Briese and Macauley 1980; Baroni Urbani and Aktac 1981; Ryti and Case
1984). Baroni Urbani and Aktak (1981) demonstrated that two species of *Messor*

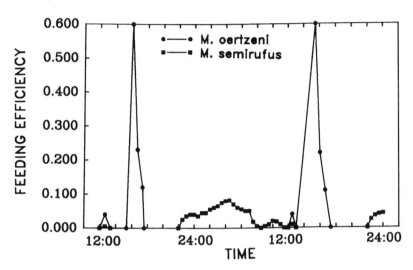

Figure 5.8. Circadian foraging patterns and feeding efficiency (E_a) in two species of *Messor* harvester ants in the desert of Turkey, where E_a = the total number of individuals of one species at one time on baits divided by the total number of ants observed on the baits (from Baroni Urbani & Aktac 1981).

harvester ants were active at different times of the day (Figure 5.8). *Messor pergandei* and the large desert *Aphaenogaster* (= *Novomessor*) spp. are usually nocturnal or crepuscular, whereas the common *Pogonomyrmex* spp. are primarily diurnal in North American deserts. *Pogonomyrmex barbatus* often forages at night, whereas *P. rugosus* never forages at night (MacKay 1981), or forages at night only occasionally (Whitford, Johnson, and Ramírez 1976; Mehlhop and Scott 1983). Temporal displacement of foraging patterns may not be important in seed harvesters, as seeds are available at all times of the day, but species active at different times would avoid aggressive encounters and could thus coexist.

Ants also exhibit differences in seasonal foraging patterns (Whitford, Johnson, and Ramírez 1976; Hansen 1978; Whitford 1978b; Briese and Macauley 1980; Anderson 1984). For example, *P. barbatus* is more active in early summer; *P. rugosus*, in midsummer (Whitford, Johnson and Ramírez 1976). *Pogonomyrmex californicus* avoids competition with its congeners by foraging in the early spring and late autumn (Whitford 1978c). It also climbs into the canopy of desert plants to harvest seeds, as do other species of desert ants (Chew 1977). *Pogonomyrmex* spp. in an upper Sonoran grassland partitioned the foraging season in a similar way (Hansen 1978). Common North American desert omnivores such as *Conomyrma insana* and *C. bicolor* co-occur and also are active at different times during the season (Whitford 1978c). Omnivores of the Australian deserts, such as ants of the *Iridomyrmex agilis* species complex, are active in the summer, and are inactive in winter when the *I. itinerans* species complex predominates (Anderson 1984). A third group, ants of the *I. dromus* species group, are active at night throughout the year, thus avoiding the other two species, which are diurnal. This may partition food resources, as omnivores may encounter different prey or foods at different times. Food type changes seasonally in harvester ants (MacKay 1981). For example, *P. subnitidus* increases the proportion of seeds in its diet in late summer. Briese (1982b) showed that niche breadth and niche overlap between species pairs continually change in response to overall availability of food resources, although it is difficult to interpret the significance of niche overlap. Ants both specialize and become generalist feeders when the opportunity occurs (Briese 1982b).

Some ants also are able to partition seed resources on the basis of the distribution and density of seeds (Chew 1977; Davidson 1977a,b, 1978; Hansen 1978; Brown, Reichman, and Davidson 1979; Kelrick et al. 1986), which are clumped in deserts (Henderson, Peterson, and Redak 1988). Species that specialize on low-density seed patches are usually individual foragers, whereas species that specialize on high density seed patches are group foragers (Davidson 1977a). (Note that Traniello [1989] objects to the use of the label "individual forager," as no ant is ever completely independent of the others, but I will continue to use the term until we have a better one.) In group foraging species, individuals locate seeds and recruit others to the source using trail pheromones or tandem running. To some degree, species tend to be characterized as either individual foragers or group foragers (Brown, Reichman, and Davidson 1979). Often the type of foraging of a nest changes depending on the situation (Curtis 1985), or some colonies of a species in an area are group foragers, whereas others are individual foragers (Whitford, Johnson, and Ramírez 1976). *Messor pergandei* may be a group forager only when high density patches of seeds are available (Davidson 1977a). *Pogonomyrmex rugosus* may group forage in areas of dense vegetation and individually forage in open areas, even along the same trunk trail (MacKay personal observation). In some assemblages of harvester ants, species of similar body sizes tend to coexist only if they differ in foraging behavior, with one being a group forager and the other an individual forager (Davidson 1977b). Group foraging species are somewhat more successful in locating food and tend to be considerably more specialized in terms of the food source (Davidson 1977a; MacKay 1981). Group foraging is metabolically more expensive due to the production of costly trail pheromones and the higher rates of predation by specialized sit-and-wait predators such as *Phrynosoma* spp. lizards, which harvest ants along trunk trails.

We can compare the costs and benefits of the two types of foraging as a function of seed density (Figure 5.9). Benefits outweigh costs at a relatively low seed density for species that forage individually, due to the lower costs involved in foraging. Seed density must be considerably higher in group foraging species before benefits outweigh costs. The benefit to group foraging species is considerably higher at high seed densities, and they may be able to outcompete individually foraging species, if food were limiting under such conditions. Species that adjust foraging behavior to seed density are probably at an advantage.

Desert harvester ants typically store seeds in their nests. Seed storage may be favored in habitats with short, infrequent pulses of primary production (Marsh 1986a), in which it serves to ensure survival during droughts (Marsh personal communication; Polis personal communication). When granaries are full, foraging ceases

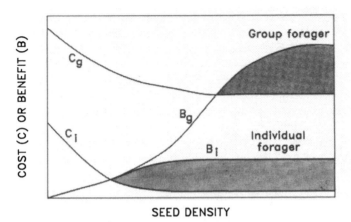

Figure 5.9. The hypothetical costs and benefits of individual foraging and group foraging vs. seed density (modified from Davidson 1977a).

in some *Pogonomyrmex* species (Whitford and Ettershank 1975; Bailey and Polis 1987). Many authors have assumed that seeds are eaten during the winter or are a form of "insurance" in an unpredictable environment. The available data do not support either of these "obvious" hypotheses (MacKay and Mackay 1984). Seed storage, in at least some *Pogonomyrmex* species, appears to be a form of protection against predation, especially by spiders. When a nest is under siege, foragers stop aboveground activity until the spiders leave.

The storage of liquids in specialized replete ants is also common in deserts. We encounter repletes in the common *Myrmecocystus* in North America, *Melophorus* in Australian deserts, *Camponotus* in the Sahara, and *Anoplolepis* in the Namib (Marsh 1986a).

Interactions between ant species can result in the formation of well-integrated communities (Davidson 1977b; Anderson 1984). Competition is especially important in some assemblages (Whitford and Ettershank 1975; Fox, Fox, and Archer 1985) and may result in habitat partitioning and variation in foraging behavior, food type preference, and worker body sizes. The outcome of competition may be difficult to predict. It undoubtedly changes as local densities of species change. Further, it can be shifted by the presence of a parasitic phorid fly, which interferes with the defensive behavior of one of the members of a species pair (Feener 1981). Even if competition can be demonstrated, it does not necessarily have a major role in community organization (Shorrocks, Rosewell, and Edwards 1984).

Other evidence for the importance of competition in desert ants is that colonies tend to be more widely spaced in more arid habitats (Davidson 1977b), although nest overdispersion may not always be a result of competition (Levings and Traniello 1981). Harvester ants are intraspecifically overdispersed and interspecifically aggregated in arid regions (Ryti and Case 1984, 1986). The spacing is due to competition between adults and expulsion (predation) on founding colonies (Ryti and Case 1986,

1988). Nests of the honeypot ants *Myrmecocystus* sp. are evenly spaced in the Chihuahuan Desert; establishment of new nests is inversely related to the existing density of nests (Chew 1987). There is some evidence for interspecific competition between *Messor pergandei* and *Pogonomyrmex californicus* in the Colorado Desert (Ryti and Case 1988). A series of nest removal and food supplementation experiments demonstrated no changes in foraging activity or diet breadth, but nests with an augmented food supply produced proportionally more reproductives.

Territoriality is common in desert ants, especially *Pogonomyrmex* (De Vita 1979), *Pheidole* (Creighton 1966), and *Myrmecocystus* (Hölldobler 1981), and can result in mortality between colonies. Apparently ant colonies can "come to an agreement" as to the boundaries of their territories. Adjacent colonies of *Pogonomyrmex barbatus* and *P. rugosus* were not observed to fight in New Mexico (Whitford, Johnson, and Ramírez 1976). Whitford et al. (1976) concluded that territoriality was relatively unimportant in relations between the two harvester ant species, although this may have been a case of "competition past." The two species rarely occur together, which would be expected of ecological equivalents (Davidson 1977b). There is a sharp boundary between the distributions of the two species in New Mexico (Whitford, Johnson, and Ramírez 1976), with *P. barbatus* occurring in slightly more mesic habitats (MacKay personal observation). *Myrmecocystus mimicus* defends territories against neighboring colonies and engages in complex display tournaments (Hölldobler 1981). Colonies also forage in different directions and avoid contact between foragers. Competitive exclusion can occur in ants (MacKay and MacKay 1982), although most examples involve displacement of native or introduced ants by other introduced ants. Some ants can be very innovative in their interactions with other species. *Tetramorium caespitum* drops soil down the nests of the alkali bees to eliminate them from the ants' territories (Schultz 1982).

Ants and rodents also compete for seeds, although to some extent they specialize on different sized seeds (Davidson, Inouye, and Brown 1984). Many species of ants, rodents, and birds may all prefer the same foods. Kelrick et al. (1986) found that all three preferred the same seeds in the Great Basin Desert, especially millet seeds, which are particularly high in percentages of soluble carbohydrates and free water. This may not be biologically meaningful, as millet does not naturally occur in the habitat. A number of investigators have studied competitive interactions between ants and rodents in arid regions. In one study in the Chihuahuan Desert, the populations of most common harvester ants initially increased in response to rodent removal (Davidson, Inouye, and Brown 1984) (Figure 5.10). Within 12 months the populations in the rodent-free plots began to decline as small-seeded plant species were competitively replaced by large-seeded annuals. After 30 months the ant populations had returned to pretreatment levels and were not different from plots with rodents present. Rodents in the ant-free plots increased slightly in density and biomass, but the long-term and short-term responses were not detectably different (Davidson, Inouye, and Brown 1984). In a similar study, Brown and Munger (1985) found the elimination of *P. rugosus* alone or of all species of granivorous ants had no effect on the granivorous rodent population in the Chihuahuan Desert of southeastern Arizona. Galindo (1986) has questioned the conclusions of Davidson, Inouye, and Brown. He states there is no firm statistical basis for the conclusion that rodents increased in density when ants were removed. See Brown and Davidson (1986) for a rebuttal. Thus, it is unclear how extensively ants and rodents compete for seeds. In fact, the net interaction of these two consumers at the same trophic level is mutualistic, as each consumer specializes on a different class of resources and those resources are in competition (Davidson, Inouye, and Brown 1984). The presence of both ants and rodents is necessary to establish moderately productive food resources for each. The selection pressure on plants to avoid predation by ants by producing larger seeds makes the seeds more attractive to rodents, and the reverse also occurs as a result of rodent predation (Mares and Rosenzweig 1978). Clearly we have not heard the last word on interphyletic competition between ants and rodents.

There is evidence of intraspecific competition between termite colonies and between different species. Some termite species are highly territorial and can eliminate competitors around their nests (e.g., Levings and Adams 1984 for tropical species); others coexist peacefully (Bouillon 1970). Territoriality can be used to define the limits of a termite colony. Termites from several parts of a study area can be mixed; if fighting occurs, the termites are from different colonies (Schaefer and Whitford 1981).

Resources may be partitioned between termite species in several ways. For example, *Gnathamitermes tubiformans* constructs galleries into the canopies of dead vegetation, whereas *Amitermes wheeleri* forages only below or at the soil surface (MacKay et al. 1985). Very subtle resource partitioning may occur in termites. Several species may attack a single piece of wood. Some may be in the hardwood, others in the wood buried by soil, others on the surface of the wood, others in partially decomposed sections of the wood, etc. Hocking (1965) found that eastern African

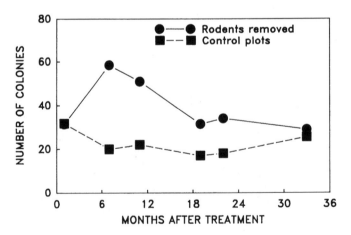

Figure 5.10. Changes in density of *Pheidole* spp. (*P. tucsonica*, *P. sitarches*, and *P. gilvescens*) on plots with and without rodents in the Chihuahuan Desert of southeastern Arizona (from Davidson, Inouye, & Brown 1984).

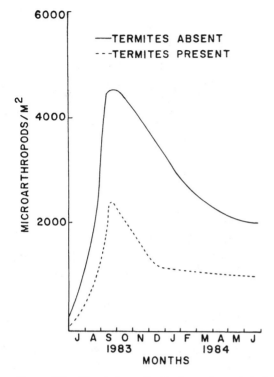

Figure 5.11.   Populations of soil microarthropods in fluffgrass litter
(*Erioneuron pulchellum*) in plots with and without subterranean
termites in the Chihuahuan Desert of North America (from MacKay,
Zak, & Whitford 1989).

termite species utilize different food sources such as dead stumps, dead branches,
earth-filled stumps, the base of the stump, etc. Some species specialize in cattle
droppings, especially in North American arid pastures. Different diurnal foraging
patterns, seasonal differences, behavioral differences, and other characteristics
allow coexistence of five species of *Trinervitermes* (Bouillon 1970).

Some evidence suggests that termites compete with soil microarthropods for organic matter (Silva, MacKay, and Whitford 1989.) Plots in the Chihuahuan Desert
in which termites had been excluded had higher densities of microarthropods (Figure
5.11). This should be expected, since mites, collembola, psocoptera, and termites
are in competition by feeding either directly or indirectly on litter (Silva, MacKay,
and Whitford 1985).

Predation

Predation is an important process in social insect populations. Such insects may
be preyed upon, especially by lizards (Munger 1984), and ants themselves are abundant and important predators (Feener 1988). Predation, especially by lizards, may

have been important in the morphological and behavioral evolution of ants (Hunt 1982).

Lizards, especially the horned lizards (*Phrynosoma* spp.) of North American deserts, are major predators of ants (Whitford and Bryant 1979; Rissing 1981; Shaffer and Whitford 1981; Munger 1984). *Phrynosoma cornutum*, a larger species of horned lizard in the Chihuahuan Desert, specializes on larger *Pogonomyrmex* harvester ants, usually foraging at the nest entrance or along the foraging trails of the ants (Whitford and Bryant 1979). They may consume an estimated 72 percent of the standing crop of the ants. *Phrynosoma modestum*, a smaller species of horned lizard in the Chihuahuan Desert, specializes on foragers of the smaller *Myrmecocystus depilis* and *M. mimicus* ants (Shaffer and Whitford 1981). *Phrynosoma modestum* may consume 50 percent of the early summer standing crop. These lizards also eat other ant species that become active after summer rains. *P. modestum* apparently partitions the ant fauna with *P. cornutum* (Shaffer and Whitford 1981). *Phrynosoma platyrhinos* in the Mojave Desert also feeds on harvester ants (Rissing 1981). It apparently prefers solitary foraging *Pogonomyrmex californicus* and avoids group foragers (*P. rugosus* and *Messor pergandei*), which "mob" lizards as a response to predation. Juveniles feed on the smaller *Pheidole gilvescens* workers in addition to the larger *Pogonomyrmex* spp. workers (Rissing 1981). Juveniles of *P. cornutum* have essentially the same diet as the adults (Whitford and Bryant 1979). Under some circumstances horned lizards may behave as "prudent predators" (preying on weaker individuals of prey populations) (Whitford and Bryant 1979; Munger 1984), although Rissing (1981) doubts this is possible. *Phrynosoma* spp. are generally not aggressive to other lizards and do not defend territories and thus could not protect themselves from "cheaters" (predators that prey on any individual of the prey population).

Other animals, such as birds (Gentry 1974) and spiders (MacKay 1982; MacKay and Vinson 1989; Porter and Eastmond 1982), are important ant predators. Spider predation on harvester ant nests may become so intense that the ants close the nest entrance with pebbles (MacKay 1982). The proportion of the nest population removed by the spiders under such circumstances may actually be low, less than 0.2 percent of the nest population per day (MacKay 1982). If spiders are removed from the vicinity of such nests, ants open the entrance and foraging resumes within 24 hours (MacKay 1982). Mammals may also be important predators of ants in desert ecosystems, especially in African deserts (Smithers 1983; Earlé and Louw 1988; Marsh personal communication). Heteromyid rodents (*Dipodomys* spp.) open harvester ant nests and remove the seed caches (Clark and Comanor 1973).

As suggested above, ants usually respond to predation. *Pogonomyrmex* harvester ants usually become immobile when nearby foragers are preyed on by horned lizards. Whitford and Bryant (1979) interpreted this behavior as avoidance of detection, but Rissing (1981) concluded that the ants are locating the predator and are preparing to attack during such times. Artificial or natural removal of part of the forager population usually causes a nest to stop aboveground activity (Gentry 1974; Whitford and Bryant 1979; MacKay 1982), although foraging activity may increase under such circumstances (Whitford and Bryant 1979). The cessation of foraging may prevent the entire nest from being destroyed by predation (Gentry 1974). Other responses of ants to predators include moving the nest entrance and the posting of guards (MacKay 1982).

Termites are also eaten by ants, lizards, birds, and humans in desert ecosystems (Smithers 1983; Redford 1984; Nkunika 1986), especially during the nuptial flights (MacKay personal observation; Marsh personal communication). During flights, termites may become the major food source for predators, even seed-harvesting ants (MacKay personal observation). The aardwolf feeds almost exclusively on termites and occurs in the eastern parts of the Namib as well as the Kalahari (Marsh personal communication). The aardvark has a similar distribution and feeds on both ants and termites. Bat-eared foxes consume considerable numbers of termites and are widespread in the Namib, occurring as far west as Gobabeb (Marsh personal communication). Termites compose up to 85 percent of the diet of *Lygodactylus* lizards in Zambia. A number of ant genera, such as *Paltothyreus* and *Megaponera* (Nkunika 1986), are termite specialists. Termites are a major food source for *Myrmecocystus mimicus* (Hölldobler 1981), *Pogonomyrmex* spp. (Hölldobler 1986), and *Pheidole titanis* (Feener 1988) in North American deserts.

Interactions Between Social Insects and Plants

Mutualism between ants and plants has been extensively investigated in tropical and temperate ecosystems (Buckley 1982; Beattie 1985), but we know little about these interactions in desert ecosystems. The few data available suggest that social insects are very important for desert flora. Myrmecochory (seed dispersal by ants) is common in drier areas of Australia and South Africa, especially in areas with poor soils (Milewski and Bond 1982). More than 1,000 species of plants of the Cape Flora of South Africa are dispersed by ants (Giliomee 1986). Ants also have been shown to be important in the structure of desert plant communities (Inouye this volume). Ants may affect desert plant communities in several ways: (1) seed harvesters remove (and move) seeds; (2) ants move tremendous amounts of subsoil to the surface during nest construction; and (3) many species clear vegetation from their mounds, or the vegetation which grows on mounds is considerably different from the surrounding vegetation. Another factor that has not been investigated is the importance of subterranean galleries on water infiltration. Presumably ant galleries would be similarly important to those of termites.

Harvester ants collect between 2 and 10 percent of the total annual seed production in the Chihuahuan Desert (Whitford 1978a), up to 20 percent in the semiarid deserts of Australia (over 20,000 seeds/m^2 produced) (Briese 1982c). The selective removal and consumption of seeds of certain species affects the relative abundance of species in a desert community (Whitford 1978b). Plants that produce seeds not consumed by ants could increase in abundance. *Filago california* (a small-seeded composite) increased to dominate habitats when ants were excluded from plots (Inouye et al. unpublished, cited in Brown, Reichman, and Davidson 1979). There was also a decrease in species diversity of the plants within such plots. In some instances, ants may exert no effect on the plant community. Briese (1982c) found little effect of ants on the plant community in semiarid New South Wales where ants are not seed limited.

Harvester ants could also change the spatial distribution of desert plants by moving seeds to other locations. Survival of the offspring of a plant could be increased by myrmecochory away from the parent plant where competition with the parent may reduce survivorship. A number of plant species have propagules made up of diaspores (seeds with protective covering) and elaiosomes (appendages that contain oils and fats and are very attractive to ants), which are collected by ants (Drake 1981). Often the elaiosome is eaten and the diaspore with seed is discarded intact. Myrmecochory is found in at least 223 genera in 61 families of plants (Nesom 1981). The majority of the herbs in a mesic forest community in New York are dispersed by ants (Handel, Fisch, and Schatz 1981). Myrmecochory was once believed to be limited to forests, but has been shown to be common in arid regions, at least in Australia (Davidson and Morton 1981b), where ants disperse the seeds of *Acacia* spp. (Mimosaceae), *Sclerolaena convexula* and *S. divaricata* (Chenopodiaceae), and *Dissocarpus biflorus* (Chenopodiaceae) (O'Dowd and Hay 1980). Although seeds are destroyed (Maddox and Carlquist 1985), myrmecochory is very important for most plant species involved. Plants compete for dispersion of seed by ants (Davidson and Morton 1981b). Myrmecochory has also been shown to assist seeds in escaping predation by rodents (Heithaus 1981). O'Dowd and Hay (1980) have shown that desert harvester ants are very important in the transportation of the seeds of *Datura discolor*. Ants collect the seeds and carry them to the nest. Ants remove the fleshy structure and discard the seed. The rate of predation by rodents is considerably reduced due to the action of the ants.

Many studies have demonstrated the importance of ant mounds in changing the

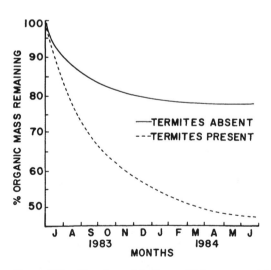

Figure 5.12. Mass loss of fluffgrass (*Erioneuron pulchellum*) in plots with and without termites in the northern Chihuahuan Desert (from Silva, MacKay, & Whitford 1985).

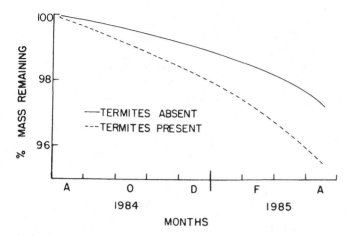

Figure 5.13.   Effect of termites on mass loss of creosotebush wood
(*Larrea tridentata*) in the northern Chihuahuan Desert of North
America (from MacKay, Zak, & Whitford 1989).

vegetation in an area (Petal 1978). Soils from mounds are higher in nutrients and
organic matter due to the concentration of resources by ants at a specific point
(Breckle 1971; Davidson and Morton 1981b; Beattie and Culver 1983). The highly
directional dispersal of diaspores by ants to nests where nutrients are concentrated
and possibly more accessible (Davidson and Morton 1981b) greatly enhances the
survival of the seedlings (Beattie and Culver 1983). As a consequence of harvesting
of seeds and the change in the chemical composition and physical properties of the
soil of the nest by the ants, floras around ant nests are quite different from floras of
surrounding areas. For example, significant differences exist between the flora on
Myrmica mounds versus that of control areas (Elmes and Wardlow 1982). There are
clear qualitative and quantitative differences in the floras of mounds of *Cataglyphis
bicolor* compared to surrounding areas in east Afghanistan (Breckle 1971). Each
nest has a narrow band of rich flora surrounding it. Some plant species grow exclusively on ant mounds (Davidson and Morton 1981b).

Social insects, especially termites, can be considered to be keystone species in
North American deserts (Whitford, Steinberger, and Ettershank 1982; Nutting, Haverty, and La Fage 1987; MacKay, Zak, and Whitford 1989). Termites are responsible
for up to 100 percent of the mass loss from organic matter in the Chihuahuan Desert
(Whitford, Steinberger, and Ettershank 1982) and can reduce the half-life of ocotillo
wood in the Colorado Desert from 29 to 17 years (Ebert and Zedler 1984). Silva,
MacKay, and Whitford (1985) were able to demonstrate that decomposition rates of
fluffgrass litter on plots with termites present were much higher than plots where
termites were excluded (Figure 5.12). Termites are also important in the mass loss
of wood of the dominant shrub in North American deserts, creosotebush (*Larrea
tridentata*), but not in mass loss of leaf litter (MacKay, Silva, Loring et al. 1987).
Wood grazed by termites lost significantly more mass than ungrazed wood (MacKay,
Zak, and Whitford 1989) (Figure 5.13).

Termite activity appears to be controlled by abiotic factors acting directly on the individual and by those factors that affect the rate of dead wood and litter production (MacKay, Zak, and Whitford 1989). Dead wood is added to the system from the effects of wind, snowfall, and insect damage (Figure 5.14). Litter falls to the soil surface as a result of drought or nutrient stress, changes in temperature, wind action, and by the browsing of mammals, especially rabbits (Steinberger and Whitford 1983). Lack of precipitation and temperature extremes also kill roots.

The effects of termites, water, and nitrogen in the composition of the Chihuahuan Desert plant community were examined by Gutierrez and Whitford (1987). Changes in the dominance of annual plant species were obtained within 3 years after termite removal (MacKay, Zak, and Whitford 1989). While some species, such as *Eriastrum diffusum*, were not affected by the treatments (Figure 5.15), others (e.g., *Baileya multiradiata*, *Descurainea pinnata*, *Eriogonum rotundifolium*, and *Lepidium lasiocarpum*) became more common in plots with termites removed. The growing period of some species (*Eriogonum tricopes* and *Tidestromia lanuginosa*) was reduced on plots without termites. Fluffgrass, the preferred food source of *Gnathamitermes tubiformans*, virtually disappeared from plots without termites by the fourth year following eradication of the termites (MacKay, Zak, and Whitford 1989). These

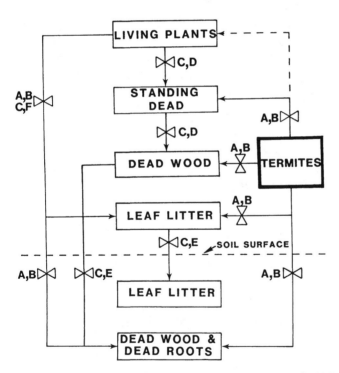

Figure 5.14. A schematic representation of the role of termites in the breakdown of organic matter in the northern Chihuahuan Desert (from MacKay, Zak, & Whitford 1989).

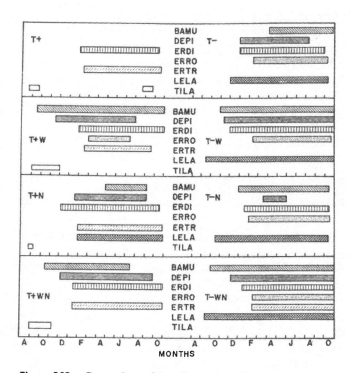

Figure 5.15. Comparison of the effects of termites, water and/or nitrogen on the phenology of a Chihuahuan Desert plant community. BAMU = *Baileya multiradiata*; DEPI = *Descurainea pinnata*; ERDI = *Eriastrum diffusum*; ERRO = *Eriogonum rotundifolium*; ERTR = *Eriogonum tricopes*; LELA = *Lepidium lasiocarpum*; and TILA = *Tidestromia lanuginosa*; T − = termites absent; T + = termites present; W = 0.6 cms of water applied to plots per week; N = nitrogen applied to plot at rate of 2.5 g/m² in a single application (from Gutierrez & Whitford 1987).

results demonstrate that termites have an important effect on the composition and dynamics of desert plant communities. The interactions between termites, water, and nitrogen on plant growth appear to be complex and dependent on the plant species examined. The type of vegetation in an area also has considerable influence on the termite fauna in semiarid woodlands of North Queensland (Holt and Coventry 1988).

Convergent Evolution among Deserts

There is considerable evidence that convergent evolution occurs in the fauna and flora of the world deserts. We must be careful in the interpretation of the data, as Australia, Africa, and South America were connected during the Cretaceous, and similarities could be due to common phylogenies (Milewski and Bond 1982).

Certain taxa compensate in diversity for the lack of other taxa. This is especially obvious with regard to granivory. Patterns of granivory in other deserts may differ from the pattern in North American deserts, although the Israeli desert seems to be similar (Abramsky 1983). Rodents are more efficient than ants at finding and harvesting seeds. Rodents find seeds below the soil surface, which may be one mechanism of coexistence, as the Israeli ants forage only on the surface. Seeds are consumed much less readily in the Monte Desert of Argentina (Mares and Rosenzweig 1978). There are apparently fewer species of granivorous ants in South American deserts than in North American deserts (Mares et al. 1977; Brown and Ojeda 1987), although the genus *Pogonomyrmex* is best represented in South America (MacKay 1981). There are also fewer seed-eating mammals in South America due to recent extinctions (Pleistocene; especially the marsupial family Argyrolagidae), and other rodents and ants have not yet responded to this reduction via species radiation (Mares and Rosenzweig 1978; Brown and Ojeda 1987). Ants are dominant seed predators in Australian deserts (Morton 1985; Morton and Davidson 1988). Birds are also important in Australian deserts, but are unimportant in the New World deserts (Mares and Rosenzweig 1978). Rodents are relatively scarce in Australian deserts, but this has not resulted in intensified seed predation by ants (Morton 1985). Birds may compensate to some extent for the lack of rodents (Morton and Davidson 1988). The high intensity of seed predation in North American deserts may be due to a more dependable precipitation regime, adequate nutrient supplies, and perhaps the rapid recycling of nutrients (Morton 1985). Thus, convergence may not occur between deserts, and North American deserts may even be peculiar when compared to others (Morton 1985). Morton (1985) concluded that the seed removal rates are highest in North America, less in Australia, and least in South America.

Conclusions

Evidence clearly shows that ants and termites are very important components of arid ecosystems. They are among the dominant organisms in terms of numbers and biomass. Yet we know little about these social insects. For example, we have only a few good estimates of populations. We might assume that certain taxa (e.g., *Pogonomyrmex*, *Pheidole*) are all seed harvesters, but this is not the case. Several species in both genera feed on other insects or vertebrate feces (MacKay personal observation). We know little of factors that influence termite species richness in arid ecosystems. These insects are important in nutrient cycling and energy flow, but again few data are available to evaluate how important they actually are. The basic biology of most species is completely unknown, including what they eat and when they are active. Systematics of ants and especially termites is in a poor state. Thus, it is always difficult and sometimes impossible to identify common species or even to put generic names on some. We need to understand the processes of evolution from studies of the convergence between different regions. What effect do social insects have on other organisms? Are mutualisms between desert plants and ants common, and if not, why not? Is the environment too dry to allow many species to nest in shrubs and trees? Are myrmecochores common in all deserts, but we simply don't know it? Morton and James (1988) have suggested that the abundance of termites in arid regions of Australia has led to a remarkable radiation of lizard species. The abundance of termites in the region may be due to the general infertility of the soils. What effect has the abundance and diversity of social insects had on the vertebrate fauna in other deserts of the world? Has the high ratio of sclerotized chitin to digestible biomass of ants (Redford and Dorea 1984) hindered predatory specialization on them? How common are chitinases in ant predators? They commonly occur among ants, even "seed predators" (MacKay personal observation). Do termites have an important effect on the vegetation in all deserts, as they do in the Chihuahuan Desert? Clearly an immense amount of work remains to be done. Unfortunately we will not understand the structure and function of arid ecosystems until we understand the roles of ants and termites in desert ecosystems.

Summary

Ants and termites are abundant in arid ecosystems of the world, both in terms of biomass as well as species numbers. They are keystone species in many ecosystems. Both ants and termites arose in the tropics and the genera present in arid regions today presumably had tropical ancestors. Primary production is the principal factor determining the number of ant species in a given habitat. Termite species richness may be determined by temperature: warmer habitats have more species. Also, areas with poorer soils have higher species richness. Ants and termites either avoid temperature extremes and vapor pressure deficits by being active during the cooler times of the day or night, or are physiologically adapted to the harsh conditions. These two taxa are especially important in altering the chemical and physical properties of the soil. Interactions between species, especially competition, are very important in structuring communities. Resources are partitioned according to differing activity patterns (seasonal or diurnal), by microhabitat differences, or based on size or quality of the food item. Both ants and termites are important in determining the composition of the plant community in a specific habitat. Termites are especially important as they greatly affect the chemical and hydrological properties of the soil, and because they also harvest and recycle plant materials.

Acknowledgments

I would like to thank Alan Marsh, Gary Polis, and an anonymous reviewer for thoughtful and careful criticisms of the manuscript. A number of colleagues, especially Walter Whitford, John Zak, and Emma MacKay helped develop many of the ideas in this paper, and participated in aspects of field work. My research discussed in this paper has been generously supported by Sigma Xi, The Scientific Research Society of North America, The American Museum of Natural History, The Chancellor's Patent Fund of the University of California, an Irwin Newell Award of the

University of California at Riverside, the Conesejo Nacional para Ciencia y Tecnología (Mexico), the Escuela Superior de Agricultura "Hermanos Escobar" (Mexico), and the National Science Foundation. Without this support the research would have been impossible.

Bibliography

Abramsky, Z. 1983. Experiments on seed predation by rodents and ants in the Israeli desert. *Oecologia* 57: 328–332.

Abushama, F., and W. Al-Houty. 1988. The foraging activity of subterranean termites in the Kuwait Desert. *Journal of Arid Environments* 14: 75–82.

———. 1989. Diurnal activity rhythms of the subterranean termite *Anacanthotermes vagans* (Hagen) under laboratory and field conditions of the Kuwait desert. *International Journal of Biometeorology* 33: 12–18.

Anderson, A. N. 1983. Species diversity and temporal distribution of ants in the semi-arid Mallee region of northwestern Victoria. *Australian Journal of Ecology* 8: 127–137.

———. 1984. Community organization of ants in the Victorian Mallee. *Victorian Naturalist* 101: 248–251.

———. 1986. Diversity, seasonality and community organization of ants at adjacent heath and woodland sites in south-eastern Australia. *Australian Journal of Zoology* 34: 53–64.

Bailey, K. H., and G. A. Polis. 1987. Optimal and central-place foraging theory applied to a desert harvester ant, *Pogonomyrmex californicus*. *Oecologia* 72: 440–448.

Baroni Urbani, C., and N. Aktac. 1981. The competition for food and circadian succession in the ant fauna of a representative Anatolian semi-steppic environment. *Bulletin de la Société de Entomologie Suisse* 54: 33–56.

Beattie, A. J. 1985. *The Evolutionary Ecology of Ant-Plant Mutualisms*. New York: Cambridge Studies in Ecology.

Beattie, A. J., and D. C. Culver. 1983. The nest chemistry of two seed-dispersing ant species. *Oecologia* 56: 99–103.

Bernard, F. 1972. Premiers résultats de dénombrement de la faune par carrés en Afrique du Nord. *Bulletin de la Société d'Historique Naturel de l'Afrique du Nord* (Algeria) 63:1–13.

Bernstein, R. A. 1974. Seasonal food abundance and foraging activity in some desert ants. *American Naturalist* 108: 490–498.

Bodine, M. C., and D. M. Ueckert. 1975. Effect of desert termites on herbage and litter in a shortgrass ecosystem in west Texas. *Journal of Range Management* 28: 353–358.

Bolton, B. 1982. Afrotropical species of the myrmicine ant genera *Cardiocondyla*, *Leptothorax*, *Melissotarsus*, *Messor*, and *Catacaulus* (Formicidae). *Bulletin of the British Museum of Natural History* (Entomology) 45: 307–370.

Bouillon, A. 1970. Termites of the Ethiopian region. Pp. 153–280 in K. Krishna and F. M. Weesner (eds.), *Biology of Termites*, vol. 2. New York: Academic Press.

Braulick, L. S., J. C. Cokendolpher, and W. P. Morrison. 1988. Effect of acute exposure to relative humidity and temperature on four species of fire ants (*Solenopsis*: Formicidae: Hymenoptera). *Texas Journal of Science* 40: 331–340.

Breckle, S. W. 1971. Die Beeinflussung der Vegetation durch hügelbauende Ameisen (*Cataglyphis bicolor* Fabricius) auf der Dasht-i-Khoshi (Ost-Afghanstan). *Bericht der Deutsch Botanikische Gesellschaft* 84: 1–18.

Brian, M. V., ed. 1978. *Production Ecology of Ants and Termites*. Cambridge: Cambridge University Press.

Brian, M. V., G. Elmes, and A. F. Kelly. 1967. Populations of the ant *Tetramorium caespitum* L. *Journal of Animal Ecology* 36: 337–342.

Briese, D. T. 1982a. The effect of ants on the soil of a semi-arid saltbush habitat. *Insectes Sociaux* 29: 375–386.

———. 1982b. Partitioning of resources amongst seed-harvesters in an ant community in semi-arid Australia. *Australian Journal of Ecology* 7: 299–307.

———. 1982c. Relationship between the seed-harvesting ants and the plant community in a semi-arid environment. Pp. 11–24 in R. C. Buckley (ed.), *Ant-Plant Interactions in Australia*. The Hague: W. Junk.

Briese, D. T., and B. J. Macauley. 1980. Temporal structure of an ant community in semi-arid Australia. *Australian Journal of Ecology* 5: 121–134.

Brown, J. H., and D. W. Davidson. 1986. Reply to Galindo. *Ecology* 67: 1423–1425.

Brown, J. H., and J. C. Munger. 1985. Experimental manipulation of a desert rodent community: food addition and species removal. *Ecology* 66: 1545–1563.

Brown, J. H., and R. A. Ojeda. 1987. Granivory: patterns, processes, and consequences of seed consumption on two continents. *Revista Chilena de Historia Natural* 60: 337–349.

Brown, J. H., O. J. Reichman, and D. W. Davidson. 1979. Granivory in desert ecosystems. *Annual Review of Ecology and Systematics* 10: 201–227.

Brown, W. L. 1973. A comparison of the Hylaen and Congo-West African rain forest ant faunas. Pp. 161–185 in B. J. Meggers, E. Ayensu, and D. Duckworth (eds.), *Tropical Forest Ecosystems in Africa and South America: A Comparative Review*. Washington, D.C.: Smithsonian Institution Press.

———. 1974. *Novomessor manni* a synonym of *Aphaenogaster ensifera* (Hymenoptera: Formicidae). *Entomological News* 85: 45–47.

Buckley, R. C. 1982, ed. *Ant-Plant Interactions in Australia*. The Hague: W. Junk.

Carroll, J. F. 1988. Worker size and piracy in foraging ants. *Proceedings of the Entomological Society of Washington* 90: 495–500.

Chew, R. M. 1977. Some ecological characteristics of the ants of a desert-shrub community in southeastern Arizona. *American Midland Naturalist* 98: 33–49.

———. 1987. Population dynamics of colonies of three species of ants in desertified grassland, southeastern Arizona, 1958–1981. *American Midland Naturalist* 118: 177–188.

Chew, R. M., and J. De Vita. 1980. Foraging characteristics of a desert ant assemblage: functional morphology and species separation. *Journal of Arid Environments* 3: 75–83.

Clark, W. H., and P. L. Comanor. 1973. The use of western harvester ant, *Pogonomyrmex occidentalis* (Cresson), seed stores by Heteromyid rodents. *Biological Society of Nevada, Occasional Papers* 34: 1–6.

Cloudsley-Thompson, J. L. 1989. Temperature and the activity of ants and other insects in central Australia. *Journal of Arid Environments* 16: 185–192.

Coaten, W. G. 1963. Survey of the termites (Isoptera) of the Kalahari thornveld and shrub bushveld of the R. S. A. *Koedoe* 6: 38–50.

Collins, M. S. 1969. Water relations in termites. Pp. 433–458 in K. Krishna and F. M. Weesner (eds.), *Biology of Termites*, vol. 1. New York: Academic Press.

Cowie, R. H. 1989. The zoogeographical composition and distribution of the Arabian termite fauna. *Biological Journal of the Linnean Society* 36: 157–168.

Crawford, C. S. 1981. *Biology of Desert Invertebrates*. New York: Springer-Verlag.

Creighton, W. S. 1966. The habits of *Pheidole ridicula* Wheeler with remarks on habit patterns in the genus *Pheidole* (Hymenoptera: Formicidae). *Psyche* 73: 1–7.

Curtis, B. A. 1985. The dietary spectrum of the Namib Desert dune ant *Camponotus detritus*. *Insectes Sociaux* 32: 78–85.

Davidson, D. W. 1977a. Foraging ecology and community organization in desert seed-eating ants. *Ecology* 58: 725–737.

———. 1977b. Species diversity and community organization in desert seed-eating ants. *Ecology* 58: 711–724.

———. 1978. Size variability in the worker caste of a social insect (*Veromessor pergandei* Mayr) as a function of the competitive environment. *American Naturalist* 112: 523–532.

Davidson, D. W., R. S. Inouye, and J. H. Brown. 1984. Granivory in a desert ecosystem: experimental evidence for indirect facilitation of ants by rodents. *Ecology* 65: 1780–1786.

Davidson, D. W., and S. R. Morton. 1981a. Competition for dispersal in ant-dispersed plants. *Science* 213: 1259–1260.

———. 1981b. Myrmecochory in some plants (F. Chenopodiaceae) of the Australian arid zone. *Oecologia* 50: 357–366.

Davison, E. A. 1987. Respiration and energy flow in two Australian species of desert harvester ants, *Chelaner rothsteini* and *Chelaner whitei*. *Journal of Arid Environments* 12: 61–82.

De Vita, J. 1979. Mechanisms of interference and foraging among colonies of the harvester ant *Pogonomyrmex californicus* in the Mojave Desert. *Ecology* 60: 729–737.

Drake, W. E. 1981. Ant-seed interaction in dry sclerophyll forest on North Stradbroke Island, Queensland. *Australian Journal of Botany* 29: 293–309.

Earlé, R. A., and S. M. Louw. 1988. Diet of the ant-eating chat *Myrmecocichla formicivora* in relation to terrestrial arthropod abundance. *S.-Afr. Tydskr. Dierk.* 23: 224–229.

Ebert, T. A., and P. H. Zedler. 1984. Decomposition of ocotillo (*Fouquieria splendens*) wood in the Colorado Desert of California. *American Midland Naturalist* 111: 143–147.

Elkins, N. Z., G. V. Sabol, T. J. Ward, and W. G. Whitford. 1986. The influence of subterranean termites on the hydrological characteristics of a Chihuahuan Desert ecosystem. *Oecologia* 68: 521–528.

Elmes, G. W., and J. C. Wardlow. 1982. A population study of the ants *Myrmica sabuleti* and *Myrmica scabrinodis* living at two sites in the south of England. II. Effect of above-ground vegetation. *Journal of Animal Ecology* 51: 665–680.

Emerson, A. E. 1955. Geographical origins and dispersions of termite genera. *Fieldiana Zoologist* 37: 465–521.

Erickson, J. M. 1972. Mark-recapture techniques for population estimates of *Pogonomyrmex* ant colonies: an evaluation of the ^{32}P technique. *Annals of the Entomological Society of America* 65: 57–61.

Ettershank, G., J. A. Ettershank, and W. G. Whitford. 1980. Location of food sources by subterranean termites. *Environmental Entomologist* 9: 645–648.

Feener, D. H. 1981. Competition between ant species: outcome controlled by parasitic flies. *Science* 214: 815–817.

———. 1988. Effects of parasites on foraging and defense behavior of a termitophagous ant, *Pheidole titanis* Wheeler (Hymenoptera: Formicidae). *Behavioral Ecology and Sociobiology* 22: 421–427.

Fox, B. J., M. D. Fox, and E. Archer. 1985. Experimental confirmation of competition between two dominant species of *Iridomyrmex* (Hymenoptera: Formicidae). *Australian Journal of Ecology* 10: 105–110.

Fox, M. D., and B. J. Fox. 1982. Evidence for interspecific competition influencing ant species diversity in a regenerating heathland. Pp. 99–110 in R. C. Buckley (ed.), *Ant-Plant Interactions in Australia*. The Hague: W. Junk.

Francke, O. F., L. R. Potts, and J. C. Cokendolpher. 1985. Heat tolerances of four species of fire ants (Hymenoptera: Formicidae: *Solenopsis*). *Southwestern Naturalist* 30: 59–68.

Galindo, C. 1986. Do desert rodent populations increase when ants are removed? *Ecology* 67: 1422–1423.

Gay, F. J., and J. H. Calaby. 1970. Termites of the Australian region. Pp. 393–448 in K. Krishna and F. M. Weesner (eds.), *Biology of Termites*, vol. 2. New York: Academic Press.

Gentry, J. B. 1974. Response to predation by colonies of the Florida harvester ant, *Pogonomyrmex badius*. *Ecology* 55: 1328–1338.

Giliomee, J. H. 1986. Seed dispersal by ants in the Cape flora threatened by *Iridomyrmex humilis* (Hymenoptera: Formicidae). *Entomologia Generalis* 11: 217–219.

Greenslade, P. J. M., and R. B. Halliday. 1983. Colony dispersion and relationships of meat ants *Iridomyrmex purpureus* and allies in an arid locality in South Australia. *Insectes Sociaux* 30: 82–99.

Gutierrez, J. R., and W. G. Whitford. 1987. Chihuahuan Desert annuals: importance of water and nitrogen. *Ecology* 68: 2032–2045.

Handel, S. N., S. B. Fisch, and G. E. Schatz. 1981. Ants disperse a majority of herbs in a mesic forest community in New York state. *Bulletin of the Torrey Botanical Club* 108: 430–437.

Hansen, S. R. 1978. Resource utilization and coexistence of three species of *Pogonomyrmex* ants in an upper Sonoran grassland community. *Oecologia* 35: 109–118.

Harris, W. V. 1970. Termites of the palearctic region. Pp. 295–313 in K. Krishna and F. M. Weesner (eds.), *Biology of Termites*, vol. 2. New York: Academic Press.

Haverty, M. I., and W. L. Nutting. 1975. Density, dispersion, and composition of desert termite foraging populations and their relationship to superficial dead wood. *Environmental Entomology* 4: 480–486.

Heatwole, H., and S. Harrington. 1989. Heat tolerances of some ants and beetles from the pre-Saharan steppe of Tunisia. *Journal of Arid Environments* 16: 69–77.

Heatwole, H., and R. Muir. 1989. Seasonal and daily activity of ants in the pre-Saharan steppe of Tunisia. *Journal of Arid Environments* 16: 49–67.

Heithaus, E. R. 1981. Seed predation by rodents on three ant-dispersed plants. *Ecology* 62: 136–145.

Henderson, C. B., K. E. Petersen, and R. A. Redak. 1988. Spatial and temporal patterns in the seed bank and vegetation of a desert grassland community. *Journal of Ecology* 76: 717–728.

Hocking, B. 1965. Notes on some African termites. *Proceedings of the Royal Entomological Society of London* (B)40: 83–87.

Hölldobler, B. 1981. Foraging and spatiotemporal territories in the honey ant *Myrmecocystus mimicus* Wheeler (Hymenoptera: Formicidae). *Behavioral Ecology and Sociobiology* 9: 310–314.

———. 1986. Food robbing in ants, a form of interference competition. *Oecologia* 69: 12–15.

Holt, J. A. 1987. Carbon mineralization in semi-arid northeastern Australia: the role of termites. *Journal of Tropical Ecology* 3: 255–263.

Holt, J. A., and R. J. Coventry. 1988. The effects of tree clearing and pasture establishment on a population of mound-building termites (Isoptera) in North Queensland. *Australian Journal of Ecology* 13: 321–325.

Holt, J. A., R. J. Coventry, and D. F. Sinclair. 1980. Some aspects of the biology and pedological significance of mound-building termites in a red and yellow earth landscape near Charters Towers, North Queensland. *Australian Journal of Soil Research* 18: 97–109.

Human, J. 1986. Seed dispersal by ants in the Cape flora threatened by *Iridomyrmex humilis* (Hymenoptera: Formicidae). *Entomologia Generalis* 11: 217–219.

Hunt, J. H. 1982. Foraging and morphology in ants: the role of vertebrate predators as agents of natural selection. Pp. 83–101 in P. Jaisson (ed.), *Social Insects in the Tropics*, vol. 2. Paris: Université Paris-Nord.

Jones, S. C., M. W. Trosset, and W. L. Nutting. 1987. Biotic and abiotic influences on foraging of *Heterotermes aureus* (Snyder) (Isoptera: Rhinotermitidae). *Environmental Entomology* 16: 791–795.

Jones, S. C., J. P. La Fage, and V. L. Wright. 1981. Studies of dispersal, colony caste and sexual composition, and incipient colony development of *Pterotermes occidentis* (Walker) (Isoptera: Kalotermitidae). *Sociobiology* 6: 221–242.

Kelrick, M. I., J. A. MacMahon, R. R. Parmenter, and D. V. Sisson. 1986. Native seed preferences of shrub-steppe rodents, birds and ants: the relationship of seed attributes and seed use. *Oecologia* 68: 327–337.

Kooyman, C., and R. F. M. Onck. 1987. Distribution of termite (Isoptera) species in southwestern Kenya in relation to land use and the morphology of their galleries. *Biology and Fertility of Soils* 3: 69–73.

Krishna, K., and F. M. Weesner, eds. 1969. *Biology of Termites*, vol. 1. New York: Academic Press.

———. 1970. *Biology of Termites*, vol. 2. New York: Academic Press.

Lavigne, R. 1969. Bionomics and nest structure of *Pogonomyrmex occidentalis* (Hymenoptera: Formicidae). *Annals of the Entomological Society of America* 62: 1166–1175.

Lee, K. E. and T. G. Wood. 1971a. Physical and chemical effects of soils of some Australian termites, and their pedobiological significance. *Pedobiologia*. 11: 376–409.

———. 1971b. *Termites and Soils*. New York: Academic Press.

Lepage, M. 1974. Les termites d'une savane sahélienne (Ferlo septentrional, Sénégal): peuplement, populations, rôle dans l'écosystème. Ph.D. dissertation, Dijon, France, 344 pp.

———. 1984. Distribution, density and evolution of *Macrotermes bellicosus* nests (Isoptera: Macrotermitinae) in the north-east of Ivory Coast. *Journal of Animal Ecology* 53: 107–117.

Levieux, J. 1982. A comparison of the ground dwelling ant population between a Guinea savanna and an evergreen rain forest of the Ivory Coast. Pp. 48–53 in M. D. Breed, C. D. Michener, and H. E. Evans (eds.), *The Biology of Social Insects*. Boulder, Colo.: Westview Press.

Levings, S. C., and E. S. Adams. 1984. Intra- and interspecific territoriality in *Nasutitermes* (Isoptera: Termitidae) in a Panamanian mangrove forest. *Journal of Animal Ecology* 53: 705–714.

Levings, S. C., and J. F. A. Traniello. 1981. Territoriality, nest dispersion and community structure in ants. *Psyche* 88: 265–319.

Lopez Moreno, I., and M. Diaz Betancourt. 1986. Foraging behavior of granivorous ants in the Pinacate Desert. Pp. 115–118 in L. C. Drickamer (ed.), *Behavioral Ecology and Population Biology*. Toulouse, France: Privat, I. E. C.

Lynch, J. F., E. C. Balinsky, and S. G. Vail. 1980. Foraging patterns in three sympatric forest ant species *Prenolepis imparis*, *Paratrechina melanderi* and *Aphaenogaster rudis* (Hymenoptera: Formicidae). *Ecological Entomology* 5: 353–371.

MacKay, W. P. 1981. A comparison of the nest phenologies of three species of *Pogonomyrmex* harvester ants (Hymenoptera: Formicidae). *Psyche* 88: 25–74.

———. 1982. The effect of predation of western widow spiders (Araneae: Theridiidae) on harvester ants (Hymenoptera: Formicidae). *Oecologia* 53: 406–411.

———. 1985. A comparison of the energy budgets of three species of *Pogonomyrmex* harvester ants (Hymenoptera: Formicidae). *Oecologia* 66: 484–494.

―――. N.d. The systematics and biogeography of the New World ants of the tribe Dolichoderini. In preparation.

MacKay, W. P., J. H. Blizzard, J. J. Miller, and W. G. Whitford. 1985. Analysis of aboveground gallery construction by the subterranean termite *Gnathamitermes tubiformans* (Isoptera: Termitidae). *Environmental Entomology* 14: 470–474.

MacKay, W. P., F. M. Fisher, S. Silva, and W. G. Whitford. 1987. The effects of nitrogen, water and sulfur amendments on surface litter decomposition in the Chihuahuan Desert. *Journal of Arid Environments* 12: 223–232.

MacKay, W. P., and E. MacKay. 1982. Coexistence and competitive displacement involving two native ant species (Hymenoptera: Formicidae). *Southwestern Naturalist* 27: 135–142.

―――. 1984. Why do harvester ants store seeds in their nests? *Sociobiology* 9: 31–47.

―――. 1989. Diurnal foraging patterns of *Pogonomyrmex* harvester ants (Hymenoptera: Formicidae). *Southwestern Naturalist* 34: 213–218.

MacKay, W. P., S. Silva, D. C. Lightfoot, M. I. Pagani, and W. G. Whitford. 1986. Effect of increased soil moisture and reduced soil temperature on a desert soil arthropod community. *American Midland Naturalist* 116: 45–56.

MacKay, W. P., S. Silva, S. J. Loring, and W. G. Whitford. 1987. The role of subterranean termites in the decomposition of above-ground creosotebush litter. *Sociobiology* 13: 235–239.

MacKay, W. P., S. Silva, and W. G. Whitford. 1987. Diurnal activity patterns and vertical migration in desert soil microarthropods. *Pedobiologia* 30: 65–71.

MacKay, W. P., S. Van Vactor, and W. G. Whitford. N.d. The ants of the long-term ecological research Jornada site, southern New Mexico. In preparation.

MacKay, W. P., and S. B. Vinson. 1989. Evaluation of the spider *Steatoda triangulosa* (Araneae: Theridiidae) as a predator of the red imported fire ant (Hymenoptera: Formicidae). *Journal of the New York Entomological Society* 97: 232–233.

MacKay, W. P., and W. G. Whitford. 1988. Spatial variability of termite gallery production in Chihuahuan Desert plant communities. *Sociobiology* 14: 281–289.

MacKay, W. P., J. Zak, and W. G. Whitford. 1989. The natural history and role of subterranean termites in the northern Chihuahuan Desert. Pp. 53–78 in J. Schmidt (ed.), *Special Biotic Relationships in the Arid Southwest*. Albuquerque: University of New Mexico Press.

Maddox, J. C., and S. Carlquist. 1985. Wind dispersal in Californian desert plants: experimental studies and conceptual considerations. *Aliso* 11: 77–96.

Maldague, M. E. 1964. Importance des populations des termites dans les sols équatoriaux. *Transactions of the 8th International Congress of Soil Science* (Bucharest) 3: 743–751.

Mares, M. A., W. F. Blair, F. Enders, D. Greegor, A. Hulse, D. Otte, R. Sage, and C. Tomoff. 1977. Strategies and community patterns of desert animals. Pp. 107–163 in O. Solbrig and G. Orians (eds.), *Evolution in Warm Deserts*. Stroudsburg, Pa.: Dowden, Hutchinson and Ross.

Mares, M. A., and M. L. Rosenzweig. 1978. Granivory in North and South American deserts: Rodents, birds and ants. *Ecology* 59: 235–241.

Marsh, A. C. 1985a. Forager abundance and dietary relationships in a Namib Desert ant community. *South African Journal of Zoology* 20: 197–203.

―――. 1985b. Microclimatic factors influencing foraging patterns and success of the thermophilic desert ant, *Ocymyrmex barbiger*. *Insectes Sociaux* 32: 286–296.

―――. 1986a. Ant species richness along a climatic gradient in the Namib Desert. *Journal of Arid Environments* 11: 235–241.

―――. 1986b. Checklist, biological notes and distribution of ants in the central Namib Desert. *Madoqua* 14: 333–344.

————. 1987. The foraging ecology of two Namib Desert harvester ant species. *South African Journal of Zoology* 22: 130–136.

————. 1988. Activity patterns of some Namib Desert ants. *Journal of Arid Environments* 14: 61–73.

Matthews, E. G. 1976. *Insect Ecology.* Brisbane: University of Queensland Press.

Mehlhop, P., and N. J. Scott. 1983. Temporal patterns of seed use and availability in a guild of desert ants. *Ecological Entomology* 8: 69–85.

Milewski, A. V., and W. J. Bond. 1982. Convergence of myrmecochory in mediterranean Australia and South Africa. Pp. 89–98 in R. C. Buckley (ed.), *Ant-Plant Interactions in Australia.* The Hague: W. Junk.

Morton, S. R. 1979. Diversity of desert-dwelling mammals: a comparison of Australia and North America. *Journal of Mammalogy* 60: 253–264.

————. 1982. Granivory in the Australian arid zone: diversity of harvester ants and structure of their communities. Pp. 257–262 in W. R. Barber and P. J. M. Greenslade (eds.), *Evolution of the Flora and Fauna of Arid Australia.* Frewville, South Australia: Peacock Publications.

————. 1985. Granivory in arid regions: comparison of Australia with North and South America. *Ecology* 66: 1859–1866.

Morton, S. R., and D. W. Davidson. 1988. Comparative structure of harvester ant communities in arid Australia and North America. *Ecological Monographs* 58: 19–38.

Morton, S. R., and C. D. James. 1988. The diversity and abundance of lizards in arid Australia: a new hypothesis. *American Naturalist* 132: 237–256.

Munger, J. C. 1984. Long-term yield from harvester ant colonies: implications for horned lizard foraging strategy. *Ecology* 65: 1077–1086.

Nesom, G. L. 1981. Ant dispersal in *Wedelia hispida* Hbk. (Heliantheae: Compositae). *Southwestern Naturalist* 26: 5–12.

Nkunika, P. O. 1986. An ecological survey of the termites (Isoptera) of Lochinvar National Park, Zambia. *Journal of the Entomological Society of South Africa* 49: 45–53.

Nutting, W. L., M. I. Haverty, and J. P. La Fage. 1987. Physical and chemical alteration of soil by two subterranean termite species in Sonoran Desert grassland. *Journal of Arid Environments* 12: 233–239.

O'Dowd, D. J., and M. E. Hay. 1980. Mutualism between harvester ants and a desert ephemeral: seed escape from rodents. *Ecology* 61: 531–540.

Paulian, R. 1970. The termites of Madagascar. Pp. 281–294 in K. Krishna and F. M. Weesner (eds.), *Biology of Termites*, vol. 2. New York: Academic Press.

Petal, J. 1978. The role of ants in ecosystems. Pp. 293–325 in M. V. Brian (ed.), *Production Ecology of Ants and Termites.* Cambridge: Cambridge University Press.

Pisarski, B. 1978. Comparison of various biomes. Pp. 326–331 in M. V. Brian (ed.), *Production Ecology of Ants and Termites.* Cambridge: Cambridge University Press.

Porter, S., and D. Eastmond. 1982. *Euryopis coki* (Theridiidae), a spider that preys on *Pogonomyrmex* ants. *Journal of Arachnology* 10: 275–277.

Porter, S. D., and W. R. Tschinkel. 1987. Foraging in *Solenopsis invicta* (Hymenoptera: Formicidae): effects of weather and season. *Environmental Entomology* 16: 802–808.

Redford, K. H. 1984. Mammalian predation on termites: tests with the burrowing mouse (*Oxymycterus roberti*) and its prey. *Oecologia* 65: 145–152.

Redford, K. H., and J. G. Dorea. 1984. The nutritional value of invertebrates with emphasis on ants and termites as food for mammals. *Journal of Zoology* (London) 203: 385–395.

Reyes Lopez, J. L. 1987. Optimal foraging in seed-harvester ants: computer-aided simulation. *Ecology* 68: 1630–1633.

Rissing, S. W. 1981. Prey preferences in the desert horned lizard: influence of prey foraging method and aggressive behavior. *Ecology* 62: 1031–1040.

———. 1988. Seed-harvesting ant association with shrubs: competition for water in the Mohave Desert? *Ecology* 69: 809–813.

Rissing, S. W., and G. B. Pollock. 1984. Worker size variability and foraging efficiency in *Veromessor pergandei* (Hymenoptera: Formicidae). *Behavioral Ecology and Sociobiology* 15: 121–126.

Rogers, L., R. Lavigne, and J. L. Miller. 1972. Bioenergetics of the western harvester ant in the shortgrass plains ecosystem. *Environmental Entomology* 1: 763–768.

Ryti, R. T., and T. J. Case. 1984. Spatial arrangement and diet overlap between colonies of desert ants. *Oecologia* 62: 401–404.

———. 1986. Overdispersion of ant colonies: a test of hypotheses. *Oecologia* 69: 446–453.

———. 1988a. Field experiments on desert ants: testing for competition between colonies. *Ecology* 69: 1993–2003.

———. 1988b. The regeneration niche of desert ants: effects of established colonies. *Oecologia* 75: 303–306.

Salman, A. G., M. A. Morsy, and A. A. Sayed. 1988. Foraging activity of the sand termite *Psammotermes hybostoma* in the New Valley, Egypt. *Journal of Arid Environments* 15: 175–177.

Schaefer, D. A., and W. G. Whitford. 1981. Nutrient cycling by the subterranean termite *Gnathamitermes tubiformans* in a Chihuahuan Desert ecosystem. *Oecologia* 58: 277–283.

Schultz, G. W. 1982. Soil dropping behavior of the pavement ant *Tetramorium caespitum* (L.) (Hymenoptera: Formicidae) against the alkali bee (Hymenoptera: Halictidae). *Journal of the Kansas Entomological Society* 55: 277–282.

Schumacher, A., and W. G. Whitford. 1976. Spatial and temporal variation in Chihuahuan Desert ant faunas. *Southwestern Naturalist* 21: 1–8.

Shaffer, D. T., and W. G. Whitford. 1981. Behavorial responses of a predator, the round-tailed horned lizard, *Phrynosoma modestum* and its prey, honey pot ants, *Myrmecocystus* spp. *American Midland Naturalist* 105: 209–216.

Shorrocks, B., J. Rosewell, and K. Edwards. 1984. Interspecific competition is not a major organizing force in many insect communities. *Nature* 310: 310–312.

Silva, S. I., W. P. MacKay, and W. G. Whitford. 1985. The relative contributions of termites and microarthropods to fluffgrass litter disappearance in the Chihuahuan Desert. *Oecologia* 67: 31–34.

———. 1989. Temporal patterns of microarthropod population densities in fluffgrass (*Erioneuron pulchellum*) litter; relationship to subterranean termites. *Pedobiologia* 33: 333–338.

Steinberger, Y., and W. G. Whitford. 1983. The contribution of shrub pruning by jackrabbits to litter input in a Chihuahuan Desert ecosystem. *Journal of Arid Environments* 6: 183–187.

Strojan, C. L., D. C. Randall, and F. B. Turner. 1987. Relationship of leaf litter decomposition rates to rainfall in the Mojave Desert. *Ecology* 68: 741–744.

Tevis, L. 1958. Interrelations between the harvester ant *Veromessor pergandei* (Mayr) and some desert ephemerals. *Ecology* 39: 695–704.

Traniello, J. F. A. 1989. Foraging strategies of ants. *Annual Review of Entomology* 34: 191–210.

Watson, J. A. L., R. A. Barrett, and C. Lendon. 1978. Termites. Pp. 101–108 in W. A. Low (ed.), *The Physical and Biological Features of Kunoth Paddock in Central Australia.* Melbourne: CSIRO Division of Land Resources Management.

Weesner, F. M. 1965. *The Termites of the United States, a Handbook.* Elizabeth, N.J.: National Pest Control Association.

———. 1970. Termites of the Nearctic region. Pp. 477–525 in K. Krishna and F. M. Weesner (eds.), *Biology of Termites*, vol. 2. New York: Academic Press.

Wehner, R. 1987. Spatial organization of foraging behavior in individually searching desert ants, *Cataglyphis* (Sahara Desert) and *Ocymyrmex* (Namib Desert). *Experientia Supplementum* 54: 15–42.

Westoby, M., B. Rice, J. M. Shelley, D. Haig, and J. L. Kohen. 1982. Plants' use of ants for dispersal at West Head, New South Wales. Pp. 75–87 in R. C. Buckley (ed.), *Ant-Plant Interactions in Australia.* The Hague: W. Junk.

Wheeler, G. C., and J. Wheeler. 1973. *Ants of Deep Canyon, Colorado Desert, California.* Berkeley, Calif.: University of California Press.

Whitford, W. G. 1972. Demography and bioenergetics of herbivorous ants in a desert ecosystem as functions of vegetation, soil type and weather variables. Manuscript.

———. 1976. Foraging behavior of Chihuahuan Desert harvester ants. *American Midland Naturalist* 95: 455–458.

———. 1978a. Foraging by seed-harvesting ants. Pp. 107–110 in M. V. Brian (ed.), *Population Ecology of Ants and Termites.* Cambridge: Cambridge University Press.

———. 1978b. Foraging in seed-harvester ants, *Pogonomyrmex* spp. *Ecology* 59: 185–189.

———. 1978c. Structure and seasonal activity of Chihuahuan Desert ant communities. *Insectes Sociaux* 25: 79–88.

———. 1986. Pattern in desert ecosystems: water availability and nutrient interactions. Pp. 109–117 in Z. Dubinsky and Y. Steinberger (eds.), *Environmental Quality and Ecosystem Stability*, vol. 3 A/B. Ramat Gay, Israel: Bar-Ilan University Press.

Whitford, W. G., and M. Bryant. 1979. Behavior of a predator and its prey: the horned lizard (*Phrynosoma cornutum*) and harvester ants (*Pogonomyrmex* spp.). *Ecology* 60: 686–694.

Whitford, W G., and G. Ettershank. 1975. Factors affecting foraging activity in Chihuahuan Desert harvester ants. *Environmental Entomology* 4: 689–696.

Whitford, W. G., P. Johnson, and J. Ramírez. 1976. Comparative ecology of the harvester ants *Pogonomyrmex barbatus* (F. Smith) and *Pogonomyrmex rugosus* (Emery). *Insectes Sociaux* 23: 117–132.

Whitford, W. G., C. A. Kay, and A. M. Schumacher. 1975. Water loss in Chihuahuan Desert ants. *Physiological Zoology* 48: 390–397.

Whitford, W. G., D. Schaefer, and W. Wisdom. 1986. Soil movement by desert ants. *Southwestern Naturalist* 31: 273–274.

Whitford, W. G., Y. Steinberger, and G. Ettershank. 1982. Contributions of subterranean termites to the "economy" of Chihuahuan Desert ecosystems. *Oecologia* 55: 298–302.

Wiegert, R. G. 1970. Energetics of the nest-building termite *Nasutitermes costalis* (Holm.) in a Puerto Rican forest. Pp. 57–64 in H. T. Odum (ed.), *A Tropical Rain Forest. A Study of Irradiation and Ecology at El Verde, Puerto Rico*, vol. 1. Division of Technical Information, United States Atomic Energy Commission.

Williams, M. A. J. 1968. Termites and soil development near Brock's Creek, Northern Territory. *Australian Journal of Science* 31: 153–154.

Wilson, E. O. 1985. The sociogenesis of insect colonies. *Science* 228: 1489–1495.

Wood, T. G., and W. A. Sands. 1978. The role of termites in ecosystems. Pp. 245–292 in M. V. Brian (ed.), *Population Ecology of Ants and Termites.* Cambridge: Cambridge University Press.

Zak, J., and W. P. MacKay. N.d. Factors influencing wood selection by the subterranean termite *Gnathamitermes tubiformans*. In preparation.

Patterns of Heterogeneity in Desert Herbivorous Insect Communities

6

Charles S. Wisdom

Herbivorous insects in desert regions face considerable environmental obstacles to successful growth and reproduction. These organisms must deal directly with the excessively high temperatures, vapor pressure deficits, and drought found in the arid regions of the world. Many studies have demonstrated the physiological and behavioral traits that desert-adapted arthropod herbivores use in dealing with these environmental variables, such as temperature tolerance (Edney 1967; Cloudsley-Thompson 1975; Mares et al. 1977), behavioral thermoregulation (Stower and Griffiths 1966; Cloudsley-Thompson 1975; Casey 1976, 1977; Mares et al. 1977; Anderson, Tracy, and Abramsky 1979; May 1979; Parker 1982), morphological adaptations (Cloudsley-Thompson 1975; Willmer and Unwin 1981), regulation of evaporative water loss (Edney 1967; Cloudsley-Thompson 1975; Edney 1977; May 1979; Toolson 1987), and mechanisms to prevent water loss (Edney 1967; Cloudsley-Thompson 1975; Edney 1977; Massion 1983).

Community organization of herbivorous insects is also affected indirectly by the environmental extremes of desert habitats through the responses of their host plants. Plants in desert climates face similar conditions of high temperature and irradiance and the unpredictable availability of water (Neales, Patterson, and Hartney 1968; Balding and Cunningham 1974; Beatley 1974; Caldwell 1985; Ehleringer 1985; Nobel 1985). Additionally, plants must deal with the occasional scarcity of some elements (e.g., nitrogen and phosphorus) (Turner et al. 1966; West and Skujins 1978) or overabundance of others (e.g., salts) (Moore, Breckle, and Caldwell 1972; Goodman 1973; Caldwell 1985).

Consequently, the community composition of herbivorous insects results directly from responses to the abiotic environment and indirectly from the reactions of host plants to these same parameters. Changes in the community organization of host plants, such as year-to-year variation in the presence or absence of annual populations and the deciduous nature of perennial plant vegetation, can amplify the influence of the abiotic environment.

This dual character of environmental influences (both direct and indirect) plus the biotic pressures of predation and competition on the community structure of desert herbivorous insects makes them particularly interesting for studies of community organization. Also, a common characteristic of desert systems is the extreme values expressed for these different factors. One consequence of these different pressures on the structure of desert communities of herbivorous insects is a substantial level of heterogeneity in community composition. Species abundances of herbivorous insects in deserts can vary both spatially and temporally. The degree of this heterogeneity and the factors that generate it will be the focus of this chapter.

Desert Herbivorous Insects

Herbivorous insects found in the desert regions of the world reflect those orders commonly found in other habitats (Brues 1946; Borror, DeLong, and Triplehorn 1981; Crawford 1981). In this review, I examine only insects that consume leaf, stem, and root tissue, excluding granivores and nectar and pollen feeders. Species of the four most abundant groups of herbivorous insects (Hemiptera, Homoptera, Lepidoptera, and Orthoptera) are all found in considerable numbers in desert regions. Hemipteran and homopteran families are found frequently in desert faunas (Beck and Allred 1968; Edney 1974; Crawford 1981). Butterfly families are irregularly present (Schmoller 1970; Crawford 1981) but families of moths are often well-represented (Cloudsley-Thompson and Chadwick 1964; Matthews 1976; Crawford 1981). Orthopterans are some of the most abundant herbivorous insects found in deserts (Uvarov 1957; Matthews 1976; Otte 1976; Crawford 1981). Migratory locusts frequently reach plague levels and are capable of migrating long distances (White 1976; Cheke 1978; Riley and Reynolds 1983) and have been subjected to many control efforts (Gunn 1960; Roffey and Popov 1968; Barron 1972).

Other orders with herbivorous families (Coleoptera, Diptera, and Thysanoptera) are also represented in all desert systems. Coleopterans are frequently found in desert environments and may best illustrate the adaptations evolved by insects to desert life (Cloudsley-Thompson and Chadwick 1964; Crawford 1981). Important herbivorous beetle families present in desert ecosystems are the Bruchidae, Cerambycidae, Chrysomelidae, Curculionidae, and Meloidae. Herbivorous species of flies (in particular, many families of gall-forming flies) are present on desert plants (Crawford 1981). Finally, the species richness of thrips (Thysanoptera) appears to reflect the overall plant species diversity of the different deserts (Lewis 1973; Crawford 1981). Ants and termites can be significant herbivores in deserts and are discussed in detail in Chapter 5.

Factors Influencing Distribution and Abundance

The presence or absence of a particular species of herbivorous insect in a particular geographic location has been attributed by investigators to a host of causative factors. In this section, I review studies that investigated relationships between species distributions and both abiotic and biotic parameters.

Abiotic Factors

Andrewartha and Birch (1954, 1960) helped initiate the study of the influence of abiotic factors on the distribution and abundance of insects. For example, they documented the influence of temperature and relative humidity on population fluctuations of thrips (Andrewartha and Birch 1954) and the geographic distribution of locusts (Andrewartha 1940, 1944; Birch and Andrewartha 1942). Additional studies have explored the relationships of both the total environment and specific microclimates on insect behavior and population levels (Edney 1967; Cloudsley-Thompson 1975; May 1979; Willmer 1982; Andrewartha and Birch 1984).

Herbivorous insects in xeric environments must deal with extreme temperatures and high vapor pressure deficits, resulting in the potential for high water loss rates. Insects in these environments usually show high resistance to water loss (Edney 1977; Anderson, Tracy, and Abramsky 1979; Massion 1983), but some do have high water loss rates (Toolson 1987). Additionally, many desert insects exhibit behavioral and morphological traits that allow them to maintain optimal (or close to optimal) body temperatures under ambient temperatures usually higher than found in mesic environments (Stower and Griffiths 1966; Abushama 1968; Casey 1976, 1977; Willmer and Unwin 1981; Parker 1982). Precipitation most likely influences communities of herbivorous insects through the productivity of host plants. These effects will be discussed below.

Competition, Predation, and Disease

Competition, both past and present, frequently has been proposed as a major structuring force in community organization (Cody and Diamond 1975; Connell 1980; Diamond and Case 1986) and just as frequently dismissed (Lawton and Strong 1981; Jeffries and Lawton 1984; Lawton 1984b). Studies on leaf miners (Bultman and Faeth 1986; Faeth 1986) and leafhoppers (Stiling 1980) demonstrate that competition can structure the communities of herbivorous insects (particularly when insect populations are abundant). Mechanisms of competition could be interference (such as territorial behavior) or exploitation of the common resource of the host plant (either actual plant tissue or the limited carbon budget of the plant) (Janzen 1973).

Some dispersion patterns of desert insect species appear related to territorial behavior (Otte and Joern 1975; Baker 1983; Alcock and O'Neill 1986; Greenfield, Shelley, and Downum 1987). Distribution of the grasshopper *Ligurotettix coquilletti* within and between communities in the Sonoran Desert appears to be determined by a combination of resource-based territorial defense by males of specific creosotebushes (*Larrea tridentata*) and differential palatability of bushes within a host plant population (Otte and Joern 1975; Greenfield, Shelley, and Downum 1987).

Biotic pressures (predation, parasitism, and diseases) can affect population dynamics, patterns of extinction and colonization, and spatial and temporal heterogeneity of herbivorous insects (Goeden and Louda 1976). Predators can limit the number of species of herbivorous insects (Jeffries and Lawton 1984; Joern 1986) and their distributions and behavior within a particular community (Schultz 1981). Buildup of parasites, predators, and disease is proposed to play a role in the population fluctuations in locust plagues (Clark 1953; Uvarov 1957; Dempster 1963). Disease outbreaks are often correlated with periods of competitive stress due to overcrowding or poor resources such as dead and decaying plants (Dodd 1940, 1959).

Parasitic organisms infesting herbivorous insects are a regular feature in desert communities; common parasitoids are tachinid flies (Polk and Ueckert 1973; Arnaud 1978; DeLoach 1982, 1983; Wisdom 1985). Seasonal use of the host plant *Encelia farinosa* by larvae of the beetle *Trirhabda geminata* (Chrysomelidae) is constrained by tachinid parasites. These parasites increase significantly in the larval population across a season (Wisdom 1982, 1985). The buildup of parasites and diseases could be one mechanism promoting temporal and geographical patterns of extinction and recolonization in herbivorous insects in deserts.

Finally, biotic factors of competition, predation, and disease can also interact with abiotic environmental parameters. For example, the development of fungal diseases on insects can be related to relative humidity (Balfour-Browne 1960; Schaerffenberg 1966), but can also be independent of relative humidity (Marcandier and Khachatourians 1987).

Host Plant Factors

Appropriate vegetation must first be available for non-diapause, reproductively active species to be present in a particular community (Labeyrie 1978). Consequently, the phenological patterns of leaf production in desert plants constrain the potential activity patterns of herbivorous insects. The overriding factor in seasonal patterns of plant production is water availability (Beatley 1974; Ehleringer 1985). The growth of perennial vegetation and the germination of annuals both respond to critical levels of water (Went 1948, 1949; Juhren, Went, and Phillips 1956; Beatley 1974; Cunningham et al. 1979; Ehleringer 1985; Gutierrez and Whitford 1987a,b). The difficulty with considering water as solely a plant cue is that herbivorous insects likely respond directly to this same cue for initiating growth (Farrow and Longstaff 1986).

Host plants also consist of nutritional factors, for example, plant water and nitrogen, and anti-nutritional factors, for example, secondary metabolites such as phenols and terpenes. As a limiting resource in desert environments, nitrogen plays an important role in the growth of both plants and their herbivorous insects (Mattson 1980; Gutierrez and Whitford 1987a; Lightfoot and Whitford 1987). The general absence of water can result in substantial metabolic changes that can enhance the palatability of plant species to herbivorous insects (Lewis 1982a,b, 1984; White 1969, 1976, 1984; Bernays and Lewis 1986). White (1976, 1984) has proposed that changes in the form of nitrogen present in water-stressed plants, particularly those plants exposed to multi-year droughts, can result in increases in plant palatability with resultant increases in the number of associated herbivorous insects. However, Bernays and Lewis (1986) suggest that changes in plant palatability to the locust *Schistocerca gregaria* more likely result from changes in deterrent compounds (anti-nutritional agents produced by the plant) or feeding stimulants than simple plant osmotic responses to drought. Additionally, the rearing temperature of the grasshopper *Aulocara elliotti* and the growth temperatures of its host plant, western wheat grass (*Agropyron smithii*), reportedly affected the reproduction and longevity of the insect (Visscher, Lund, and Whitmore 1979).

Herbivorous insects gain important resources from their host plant, such as all of their nitrogen intake and most, if not all, of their water intake. Consequently, the distribution of these elements in host plant tissue can influence insect spatial and temporal distributions (Southwood 1973; McNeil and Southwood 1978; Scriber and Slansky 1981). Additionally, the desert locust (*S. gregaria*) apparently uses terpenes from host vegetation as a signal to initiate sexual maturation (Ellis, Carlisle, and Osborne 1965).

Anti-nutritional factors in plant tissue that serve as defense against insect herbivory are ubiquitous (Rosenthal and Janzen 1979) and will not be reviewed here. However, I will discuss a few points related to this and community patterns of herbivorous insects. The predictability of host plants in time and space to herbivorous insects (related to environmental variability) and plant chemistry can affect the preferences and population levels of herbivorous insects (Cates 1980, 1981). Individual numbers of insect species eating mesquite (*Prosopis glandulosa*) correlate positively with nitrogen content and negatively with tannin and phenolic acid content (Wisdom 1988). Finally, patterns of heterogeneity in insect species distributions and abundances could be further influenced by the secondary metabolite content of plants, which can vary qualitatively (the type of compound present) and quantitatively (the amounts of a specific compound present) (Denno and McClure 1983; Whitham 1983; Greenfield, Shelley, and Downum 1987; Shelley, Greenfield, and Downum 1987).

Community Diversity and Species Abundance

Many different definitions of communities have been offered by ecologists (see the various chapters in Strong, Lawton, and Southwood 1984 and Diamond and Case 1986). In general, "a community is a group of species that share the same habitat" (Strong, Lawton, and Southwood 1984). Operationally, different ecologists have defined communities of herbivorous insects as all species of insects that occur on all plant species in a particular area or, alternatively, all insect species on a specific host plant species. Some investigators have considered species diversity, which can include evenness, or relative abundance. Studies that have examined the causative factors of species diversity of herbivorous insects typically have focused on those insects using a specific host plant species. In this section, I will discuss several investigations that have examined species diversity and abundances in desert communities of herbivorous insects, based on single plant species as well as multiple plant use.

Insect Communities on Mesquite

Prosopis glandulosa (mesquite) is a widespread and abundant plant in both the Sonoran and Chihuahuan deserts of North American (Figure 6.1) (Simpson 1977; Leakey and Last 1980). *P. glandulosa* (honey mesquite) is a perennial legume which exhibits both shrub and tree growth forms (Shearer et al. 1983), with a much higher level of primary production than predicted for desert regions, most likely due to a deep root system and its nitrogen fixing capabilities (Whittaker 1970; Hadley and Szarek 1981; Nilsen, Rundel, and Sharifi 1981; Sharifi, Nilsen, and Rundel 1982). Over the geographical range of mesquite, these trees support a fauna of approximately 200 species of herbivorous insects which consume the stems, leaves, and fruits of this plant (Table 6.1) (Cates and Rhoades 1977; Kingsolver et al. 1977; Ward et al. 1977).

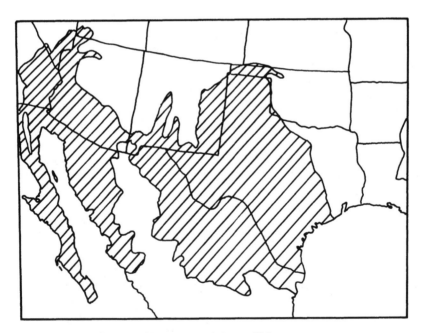

Figure 6.1. Distribution of honey mesquite trees (Fabaceae: *Prosopis glandulosa*) in North America (redrawn from Simpson 1977).

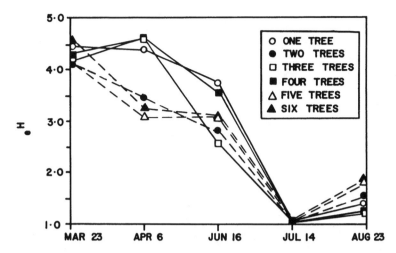

Figure 6.2. Changes in the species diversity index (using the Hill's number one species index; Ricklefs 1979; Ludwig and Reynolds 1988) for herbivorous insects on branches, leaves, and fruits of *Prosopis glandulosa* over the active growing period in 1983 when the diversity index is calculated with one, two, three, four, five, or six total trees. See text for further details.

In a study of the nutritional factors present in mesquite leaves and the herbivorous insects that feed on mesquite, I sampled five branches each of six mesquite trees over the 1983 growing season at Harper's Well, California (approximately 15 km west of the Salton Sea), in the Colorado Desert (Wisdom 1988). Forty families in six orders were collected over the season (Table 6.1). Number one of the Hill index (defined as e^H, a transformation of the Shannon-Weiner diversity index) was used to examine differences in species diversity (Ludwig and Reynolds 1988). Hill's number one represents the number of abundant species in the sample, excluding very rare species, and has units of numbers of species (Ludwig and Reynolds 1988).

The general pattern of species diversity, calculated from the Hill index, varied greatly through the season (Figure 6.2). Species number peaked ($e^H = 4.7$) in early April, and dropped off quickly in mid-July ($e^H = 1.2$). Species number increased again in August, reflecting the second leaf flush produced by mesquite trees at this time period. Additionally, within sampling dates, the calculated value of species number varied with sample size (Figure 6.2). The number of species was calculated first from the branches of only one tree, then recalculated using tree 1 and tree 2, adding trees progressively until all six were used to calculate the final value. The influence of sample size varied between sampling dates, being greatest when species diversity was highest and diminishing with decreasing species numbers (Figure 6.2). This is, in part, due to substantial changes in evenness (the relative proportions of species present) for sampling dates with high species diversity (Table 6.2). Otte (1976) demonstrated similar effects of area sampled on the number of species represented in a community survey.

Table 6.1   Genera and species of herbivorous insects that occur on *Prosopis glandulosa* throughout its range and those found at Harper's Well, Calif.

Order and Family	Total		Harper's Well	
	Genera	Species	Genera	Species
Coleoptera				
Bostrichidae	1	2	0	0
Bruchidae	2	3	1	1
Buprestidae	6	19	1	1
Cerambycidae	11	14	1	1
Chrysomelidae	13	17	0	0
Curculionidae	6	9	1	1
Elateridae	1	1	0	0
Meloidae	4	7	1	1
Total	44	72	5	5
% of total			11	7
Hemiptera				
Berytidae	1	1	0	0
Coreidae	4	6	0	0
Cydnidae	3	3	0	0
Largidae	1	2	0	0
Lygaeidae	3	3	0	0
Miridae	8	8	2	2
Pentatomidae	9	15	0	0
Rhopalidae	2	2	0	0
Total	31	40	2	2
% of total			6	5
Homoptera				
Acanaloniidae	2	2	0	0
Cicadellidae	6	6	0	0
Cicadidae	6	7	0	0
Cixiidae	2	4	0	0
Coccidae	5	5	0	0
Diaspididae	7	7	0	0
Flatidae	4	4	0	0
Fulgoridae	1	1	0	0

SOURCE: Total range data: Ward et al. (1977); Harper's Well data: Wisdom (1988).

Examining species diversity in this fashion reveals the dependence of what we perceive as a community upon the sampling strategy. Differences in plant nutritional quality or the isolation of the plant, reflected in differences in herbivore preference, could strongly affect interpretations of community. The herbivores of *P. glandulosa* in this area show both a significant negative correlation with concentrations of natural products (tannins and simple phenolic acids), and a positive correlation with nitrogen amount (Wisdom 1988).

Depending on the number of trees included in the index, the inferred seasonal pattern for insects eating *Prosopis* vegetation would change from one peaking in April to a pattern that declines continuously over the course of the season of first

Order and Family	Total		Harper's Well	
	Genera	Species	Genera	Species
Issidae	1	1	0	0
Membracidae	6	7	1	1
Psyllidae	2	2	0	0
Total	42	46	1	1
% of total			2	2
Lepidoptera				
Aegeriidae	1	1	0	0
Geometridae	1	1	1	1
Lycaenidae	1	1	0	0
Noctuidae	6	7	1	1
Psychidae	3	3	0	0
Saturniidae	2	3	0	0
Tineidae	1	1	0	0
Tortricidae	1	1	0	0
Washiidae	1	1	0	0
Total	17	19	2	2
% of total			12	11
Orthoptera				
Acrididae	10	13	0	0
Phasmatidae	1	1	0	0
Tetigoniidae	5	7	1	1
Total	16	21	1	1
% of total			6	5
Thysanoptera				
Heterothripidae	1	1	1	1
Thripidae	1	1	0	0
Total	2	2	1	1
% of total			50	50
Grand totals	152	200	12	12
% Grand total			8	6

flush of leaves starting in March (Figure 6.2). Additionally, the magnitude of increase in the diversity of herbivorous insects corresponding with the second flush of vegetation in August depends strongly upon sample size.

The genera represented in the Harper's Well sample are only a small subset of the total herbivorous insect fauna that are known to use *P. glandulosa* over its geographic range (Table 6.1). The percentage of species present ranges from 50 (Thysanoptera) to 2 (Homoptera), with an average of 6 percent of all possible species actually represented in the sample survey from Harper's Well. The initial and simplest explanation for this low figure is that some insect species are missing due to geographic restrictions. *P. glandulosa* is widely distributed in two different desert systems,

Sonoran and Chihuahuan, with associated variation in abiotic and biotic factors, including climatic restrictions such as temperature and precipitation (Andrewartha and Birch 1954). Additionally, the presence or absence of various competitors or predators or differences in *Prosopis* secondary metabolites over its geographic range may prevent the utilization of the Harper's Well population of host plant by particular members of the potential species pool of herbivorous insects.

Herbivorous Insect Communities on Ragweed

Investigations by Goeden and Ricker (1974, 1975, 1976a,b) have focused on the insect communities of different species of the genus *Ambrosia*. Ragweeds (*Ambrosia* spp.), annual or perennial herbs or subshrubs in the Asteraceae, are extremely common in most arid parts of the Colorado Desert (Goeden and Ricker 1974; Munz 1974).

The seven species of *Ambrosia* in the Sonoran Desert of Southern California differ in numbers of herbivorous insect orders, families, genera, and species that feed on them (Table 6.3). The most abundant ragweed species in the survey area (Munz 1974; Goeden and Ricker 1974, 1975, 1976a,b) have the highest number of herbivore species and genera (Table 6.3).

These data appear to support predictions that the number of species of herbivorous insects associated with a particular host plant reflects species-area relationships, rather than the historical time of association between insects and hosts (Strong, Lawton, and Southwood 1984). *Ambrosia* species with wider distributions hosted more herbivorous species than those with more restricted distributions. Because *Ambrosia* apparently originated in the desert southwest of North America, the time of association is most likely not significantly different between the different plant species and the associated herbivorous arthropod communities (Payne 1964; Raven and Axelrod 1977).

Grasshopper Communities

Orthopterans (grasshoppers and locusts) are prominent members of desert insect communities and have been well studied because their populations can reach economically destructive levels (Uvarov 1957; Gunn 1960; Roffey and Popov 1968; Barron 1972; Matthews 1976; Otte 1976; White 1976; Cheke 1978; Riley and Reynolds 1983). In this section, I will discuss local scale distributions of grasshoppers and locusts for different communities that are using more than one species of host plant.

A northern mixed-grass prairie had very high grasshopper densities during outbreak years ($36/m^2$ in 1982 and $60/m^2$ in 1981, as compared to a high of $32/m^2$ in the 5 years of the New Mexican survey reviewed in the next section) (Pfadt 1984). The Pfadt survey found a very high grasshopper diversity ($e^H = 7.39$, compared to the highest e^H for the *Prosopis* community of 4.7). The grasshopper diversity did not change significantly with sampling date within a year, but was greater in 1981 than in 1982, apparently due to changes in abundance of the most common species (Pfadt

Table 6.2 Individual abundances and species number of herbivorous insects for each of six individuals of *Prosopis glandulosa* surveyed on April 6, 1983, at Harper's Well, Calif.

Tree No.	No. of Species	No. of Individuals
1	7	33
2	6	23
3	4	24
4	5	76
5	3	54
6	4	13

SOURCE: Wisdom (1988).

Table 6.3 Distributions of phytophagous insects for seven desert species of *Ambrosia* found in the Sonoran Desert, Southern California.

Ambrosia species	No. of Orders	No. of Families	No. of Genera	No. of Species	Distri- bution[a]	G & R Source[b]
acanthicarpa	7	31	73	87	com.	1974
chenopodiifolia	6	17	24	31	uncom.	1976b
confertifolia	6	33	74	88	com.	1975
dumosa	5	32	58	89 +	com.	1976a
eriocentra	6	21	31	33	uncom.	1976b
ilicifolia	5	14	19	19	uncom.	1976b
psilostachya	7	36	91	113 +	com.	1976c

SOURCE: Data abstracted from Goeden & Ricker (1974, 1975, 1976a,b,c).
[a]Species distributions estimated from Munz (1974) and Goeden & Ricker (1974, 1975, 1976a,b,c). com. = Common in survey area; uncom. = Uncommon in survey area.
[b]Goeden & Ricker.

1984). These higher values quite likely relate to the greater plant productivity in semiarid grasslands over those of true desert.

Grasshoppers in an Arizona desert grassland reached highs during outbreaks of approximately 48 grasshoppers/m^2, with the number of species varying from 10 to 17 in a perennial grass/low disturbance site (species diversity, e^H = 3.35) to 5–6 in a forb and annual/high disturbance grass site (species diversity, e^H = 1.84) (Pfadt 1982). Grasshoppers in these communities apparently only reach high densities in perennial grasses, because the longer availability of foliage allows late-hatching species to be supported.

Grasshopper Regional (Intercommunity) Distributions

To examine intercommunity herbivore distributions, I used data on the regional abundances of orthopterans from the United States Department of Agriculture.

Unfortunately, no direct measure of community species diversity can be made for these insects because individual insects were not identified to species. However, much useful information on the regional distributions of herbivorous insects can still be extracted from these data.

Collections for identification were made for each county of the New Mexico survey (Table 6.4). No one species of grasshopper was found in all counties of the survey (77 species reported from 10 counties in the southeast arid region of New Mexico; see Figure 6.3F). The area represented in Figures 6.3A–6.3F consists of plant communities that are dominated by Great Plains flora (semiarid grasslands), Chihuahuan Desert flora (desert scrub), or by a mixture of both floristic types (see last entries in Table 6.4) (Martin and Hutchins 1980).

Of the three different floristic regimes, the presence of grasshopper species was lowest in the four desert counties (18–47 percent of total species number). In comparison, counties with mixed floristic regimes ranged from relatively low (36 percent) to the highest species representation (71 percent) (Table 6.4). The three semiarid grassland counties varied from 40 to 53 percent of the total number of grasshopper species found in the survey. In contrast, those counties with the most species did not consistently record the greatest number of individuals within or between years (1981–1985; Figures 6.3A–6.3F). For example, in the first year of the survey (1981), abundances were highest in the southern, Chihuahuan-dominated portion of Lea County and in the purely Great Plains areas of Roosevelt and De Baca counties.

During the 5 years of the survey, distributions of grasshopper density changed considerably from year to year and from habitat to habitat (Figures 6.3A–6.3F). Areas that may have high levels of grasshoppers one year may have very low populations the following year (e.g., 1982–1983, Figures 6.3B–6.3C, in the lower left of each figure). One noticeable pattern is that the eastern portion of this distribution appears to be functioning as a reservoir, always with some grasshoppers present (these four eastern counties contain approximately 85 percent of all grasshopper species in the survey). Thus, we can see at larger spatial scales a heterogeneous pattern of increases and decreases of grasshoppers in specific locales, with the potential for certain areas to allow regional species continuance. Additionally, local populations can decrease to extremely low abundances, passing through bottlenecks, and could be reestablished both by local population growth and immigration.

Figure 6.3. The broad-scale distribution of different species of grasshoppers in the arid southeast corner of New Mexico. (See Table 6.4 for species distribution by county.) Key to distributions: Pluses = 0–3 grasshoppers/yd^2; Diagonal lines = 4–7 grasshoppers/yd^2; Crosshatched lines = 8 or more grasshoppers/yd^2. A: 1981 survey distribution. B: 1982 survey distribution. C: 1983 survey distribution. D: 1984 survey distribution. E: 1985 survey distribution. F: County names for the area of the survey (southeastern New Mexico). (Data from the fall adult grasshopper survey conducted by the USDA/PPQ.)

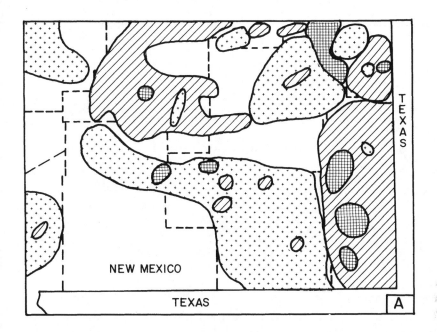

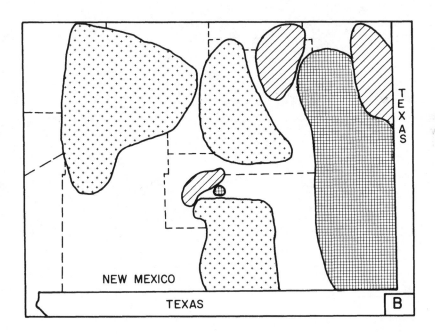

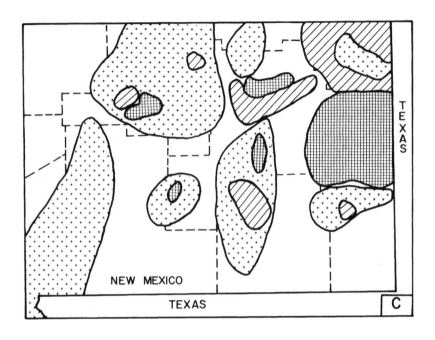

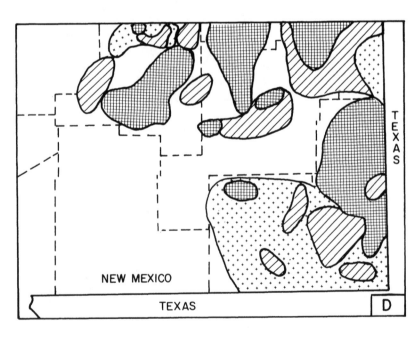

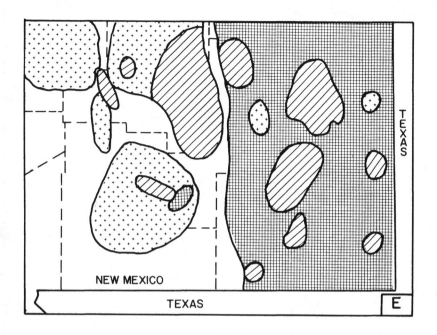

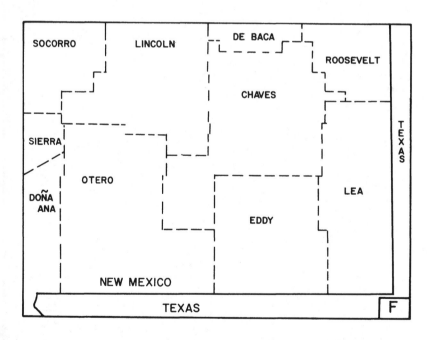

Table 6.4 Distributions of grasshoppers found in counties of southeastern New Mexico during the annual survey conducted by the Plant Protection and Quarantine Service.

Species	County									
	Chaves	De Baca	Doña Ana	Eddy	Lea	Lincoln	Otero	Roosevelt	Sierra	Socorro
Acantherus piperatus	X									
Acrolophitus hirtipes	X				X			X		
Acrolophitus nevadensis		X		X			X			
Aeoloplides rotundipennis									X	X
Aeoloplides trunbulli	X	X	X	X	X		X	X		X
Ageneotettix deorum	X	X		X	X	X		X		X
Amphitornus coloradus	X	X		X	X	X	X	X	X	X
Arphia pseudonietana	X					X			X	X
Aulocara elliotti	X	X		X	X	X	X	X	X	X
Barytettix humpreysii	X								X	
Boopedon numbilum	X	X		X	X	X		X	X	X
Brachystola magna	X			X	X	X		X	X	X
Cibolacris parviceps			X					X		X
Cordillacris crenulata	X	X		X	X	X	X	X	X	X
Cordillacris occipitalis	X	X	X	X	X	X	X	X	X	X
Dactylotum variegatum	X	X	X	X	X	X	X	X		
Derotmema haydenii	X	X		X		X	X	X	X	
Dissosteira carolina	X						X	X		
Dissosteira longipennis	X	X				X	X	X	X	
Encoptolophus sordidus						X	X			
Eremiacris virgata	X									
Eritettix simplex	X	X			X	X				X
Goniatron planum				X						
Hadrotettix trifasciatus	X	X		X	X	X		X		X
Heliastis benjamini							X		X	
Heliaula rufa	X				X					
Hesperotettix speciosus	X			X	X		X	X		X
Hesperotettix viridis	X	X	X	X	X		X	X		X
Hippopedon capito	X								X	

Lactista gibbosus
Leprus wheeleri
Melanoplus aridus
Melanoplus arizonae
Melanoplus augustipennis
Melanoplus bivittatus
Melanoplus bowditchi
Melanoplus differentialis
Melanoplus femur-rubrum
Melanoplus flavidus
Melanoplus foedus
Melanoplus franciscanus
Melanoplus gladstoni
Melanoplus glaucipes
Melanoplus lakinus
Melanoplus occidentalis
Melanoplus packardii
Melanoplus pictes
Melanoplus regalis
Melanoplus sanguinipes
Melanoplus thomasi
Melanoplus tristus
Melanoplus viridipedes
Mermiria bivitatta
Mermiria picta
Mestrobregma plattei
Metator pardalinus
Opeia obscura
Parapomala pallida
Parapomala wyomingensis
Philbostroma quadrimaculatum
Phoetaliotes nebrascensis
Phrynotettix tschivavensis
Psoloessa delicatula
Psoloessa texana

Table 6.4 Continued

Species	Chaves	De Baca	Doña Ana	Eddy	Lea	Lincoln	Otero	Roosevelt	Sierra	Socorro
						County				
Schistocerca lineata	X			X	X			X	X	
Schistocerca shoshone	X	X		X	X			X	X	
Spharagemon cristatum		X						X		X
Spharagemon equale						X		X		
Syrbula admirabilis	X			X	X					
Syrbula montezuma	X						X			
Trachyrhachys kiowa	X	X			X					
Trimerotropis californica	X	X	X	X		X	X	X	X	X
Trimerotropis pallidipennis	X	X	X	X		X	X	X	X	X
Trimerotropis strenua	X	X						X	X	X
Trimerotropis texana	X									
Tripidolophus formosus	X		X			X	X	X		
Xanthippus corallipes	X	X	X			X	X	X		X
Total no. of species/county	55	40	14	36	37	34	23	38	26	28
% of all possible species	71	52	18	47	48	44	30	49	34	36
Floral composition[a]	MIX	G.P.	CHI	CHI	MIX	G.P.	CHI	G.P.	CHI	MIX

SOURCE: Ward et al. (1977).

[a]Floral composition abbreviations: CHI = Chihuahuan Desert flora; G.P. = Great Plains flora; MIX = mixed composition of Great Plains and Chihuahuan Desert floras.

These findings are in accord with Otte's (1976) conclusions on comparisons of the Monte Desert in South America with the Sonoran Desert in North America. He concludes that "recruitment of [grasshopper] species through immigration and colonization may be greater in North America where the deserts have richer source areas" (Otte 1976). Similar patterns were demonstrated in a survey of desert grasslands; once again grasshopper densities and relative abundances varied from year to year (6- to 10-fold differences in abundance) (Joern and Pruess 1986). Interestingly, the composition of grasshopper assemblages relative to host plant usage over several years suggested constancy of community structure (i.e., the rank order of species abundances remained similar) (Joern and Pruess 1986).

In a survey of grasshopper species from two desert grassland communities, the feeding patterns and community niche breadth distributions of the grasshopper species present varied little from site to site (8 km apart) (Joern 1979). Host plant use by grasshoppers was not strongly correlated with plant abundance in this study: one site showed a significant relationship while the other did not, with 6.5 percent of the plant species accounting for over 50 percent of the plant biomass consumed (Joern 1979). Several factors could account for these observed patterns: phylogenetic constraints on the herbivores, host plant predictability, chemical defenses, or competition (Joern 1979; Joern and Lawlor 1980). The structures of feeding guilds are very similar across arid grassland communities, even when there is little species overlap (Joern and Lawlor 1981).

Species Persistence and Local Extinctions

A pattern emerging from these studies is that herbivorous insects persist in arid environments at the regional level, but can vary extensively in numbers and community composition at the local level. For herbivorous insects in deserts, the environment consists of patches of appropriate habitat connected by migration (Wiens 1976; Caswell 1978). In such systems, environmental factors will have distinct influences on community structure. Predation, for example, can allow for coexistence of species of herbivorous insects which otherwise could be driven to local extinction (Caswell 1978). The patchy nature of the environment can thus maintain species that in more uniform circumstances would be absent (e.g., *Opuntia* and *Cactoblastis cactorum* in Australia; Dodd 1959). Additionally, the density and diversity of host plant species in such patches can influence movement and diversity of insect species (Kareiva 1983).

Another explanation for the heterogeneity of patterns of colonization and extinction of desert herbivorous insects on their host plants is that of island biogeography (MacArthur and Wilson 1963, 1967). Developed initially for fauna and flora on real islands (land surrounded by bodies of water), it has also been applied to resources that are patchily distributed spatially (i.e., habitat islands), particularly host plants (Janzen 1968, 1973; Brown and Kodric-Brown 1977; Kuris, Blaustein, and Alio 1980). This theory predicts that the equilibrium number of species on an "island," meaning the balance of species between increases from immigration and losses from extinction, is related to both island area and distance from a source of potential colonists.

Island biogeography has been used to predict herbivore species diversity (Lawton 1984a; Strong, Lawton, and Southwood 1984), but many potential additional criteria for host plant "islands" must be considered in order to more accurately predict these relationships. Host plants can act as islands at several levels (as individuals, populations, and species) (Kuris, Blaustein, and Alio 1980), so it is very important to identify what level an investigator is studying. "Island" size in terrestrial habitats may not be the sole factor influencing turnover rates of individual species, as seen in the arguments over reserve sizes in conservation biology, but can affect community composition through loss of "keystone" species (Wilcox and Murphy 1985).

Additionally, host plants differ from islands in several important criteria (Kuris, Blaustein, and Alio 1980). Plants produce toxin defenses that potentially can exclude immigrants (Kuris, Blaustein, and Alio 1980; Strong, Lawton, and Southwood 1984) and are capable of evolving in response to immigrant presence (Futuyma and Slatkin 1983), neither of which occurs in islands. Host plants also present a complex architecture that can influence the number of species present on a particular plant (Lawton 1983; Strong, Lawton, and Southwood 1984). Finally, in predicting the expected species diversity of herbivorous insects, it is important to recognize that cross-taxa comparison of turnover rates of organisms appears to vary directly with generation time (Schoener 1983). Consequently, insects with differing voltinism patterns may differ in turnover rates.

Conclusions

In discussing the patterns of distribution and abundance of herbivorous insects in desert environments, it is necessary to consider characteristics of the individual species and the heterogeneity of its environment. For a species of herbivorous insect to be maintained in a desert community, it must first exhibit an appropriate degree of autecological adaptation to the abiotic environment of the host plant range (i.e., temperature and precipitation regimes). Additionally, such a species must demonstrate the ability to successfully overcome the biochemical and mechanical defenses expressed by its host plant. These adaptations will determine the broad-scale (region-wide) distributions of different insect species.

At smaller scales (specific community locations), the response of herbivorous insects to the extreme patchiness of the abiotic and biotic environment will determine which herbivore species are present in a particular community. These patterns of heterogeneity can be the result of the dynamic interplay of predation, competition, diseases, immigration, and extinction rates, resulting in a patchwork distribution of species and communities across the landscape.

This patchwork distribution can have significant consequences for community structure. For example, patches could provide local refugia, spreading the risk of species extinction and allowing species to be maintained at the larger regional scale

(Wiens 1976). Measurements of the species diversity of a locality will differ significantly between determinations of what species are present at a specific time or if they include the total number of species of herbivorous insects that colonize and go extinct in that locality over time. Thus, high heterogeneity in desert communities can actually support much higher species diversity and more complex communities of herbivorous insects.

Directions for Future Research

If patterns of species distributions in different communities are strongly heterogeneous, differing geographically on a seasonal and annual basis, then populations may undergo considerable changes in genetic composition in these time frames. As herbivorous insects recolonize host plant patches, the newly reconstituted populations can differ in gene frequencies. Lande (1985) has suggested that such changes in subdivided populations, together with local extinction and colonization, can accelerate the spread of chromosomal mutations within a species.

Such changes can have important consequences for the prey (host plants), the competitors (other herbivorous insects), and the predators of these herbivorous insects. Different herbivorous insect genotypes may respond quite differently to plant nutritional and anti-nutritional characteristics (Gallun 1972), changing the selection environment for the host plant. This subsequently could affect potential competitor herbivore species (Faeth 1986). Additionally, as new herbivore genotypes colonize an area, they could select host plant species not used before, making them harder to locate by predators that use host plant cues to search for prey. Thus, changes in the genetic makeup of herbivorous insects could influence several trophic levels in community organization.

The communities of herbivorous insects in desert habitats could be composed of different species with differing persistence characteristics, resulting in varying species and genetic compositions of the community through time. Studies to determine the persistence versus extinction/recolonization of different herbivorous insect species and temporal changes in genetic makeup in these same insects in desert communities will assist in determining the relevance of these potential influences. Along with this, the relationships of regional species diversity and local communities should be investigated by long-term measurements of species composition and abundance in specific localities.

Finally, the biogeographic distributions of insect species in desert communities should be compared to the underlying distributions of their host plant species. Further examination of the overlap of the distributions of herbivorous insects with respect to plant distributions will contribute to resolving the controversy "Do vacant niches exist in different plant species?" (Lawton 1984a,b). Investigations of the changes in communities of herbivorous insects across the geographic range of a particular species of host plant are important and have not been done systematically, particularly for temporal changes within and between years. Continuing research in these areas should be encouraged.

Acknowledgments

I wish to acknowledge the Plant Protection and Quarantine Service, United States Department of Agriculture, Albuquerque, New Mexico, for providing me the data on grasshopper distributions in New Mexico. A. Williams assisted greatly in the library searches for information on additional insect communities. I thank J. Brown, N. Cobb, C. Crawford, K. Ernest, G. Polis, K. Schoenly, S. Sommer, and an anonymous reviewer for their comments and criticisms of this manuscript. The research on the mesquite insect communities was supported by the United States Department of Energy, Contract No. DE-AC03-SF00012, administered by the University of California at Los Angeles.

Bibliography

Abushama, F. T. 1968. Rhythmic activity of the grasshopper *Poecilocerus hieroglyphicus* (Acrididae: Pyrogomorphinae). *Entomologia Experimentalis et Applicata* 11: 341–347.

Alcock, J., and K. M. O'Neill. 1986. Density-dependent mating tactics in the Grey hairstreak, *Strymon melinus* (Lepidoptera: Lycaenidae). *Journal of Zoology* (London) 209: 105–113.

Anderson, R. V., C. R. Tracy, and Z. Abramsky. 1979. Habitat selection in two species of short-horned grasshoppers. *Oecologia* 38: 359–374.

Andrewartha, H. G. 1940. The environment of the Australian plague locust (*Chortoicetes terminifera* Walk.) in South Australia. *Transactions of the Royal Society of South Australia* 64: 76–94.

———. 1944. The distribution of plagues of *Austroicetes cruciata* Sauss (Acrididae) in Australia in relation to climate, vegetation and soil. *Transactions of the Royal Society of South Australia* 68: 315–326.

Andrewartha, H. G., and L. C. Birch. 1954. *The Distribution and Abundance of Animals.* Chicago: University of Chicago Press.

———. 1960. Some recent contributions to the study of the distribution and abundance of insects. *Annual Review of Entomology* 5: 219–236.

———. 1984. *The Ecological Web. More on the Distribution and Abundance of Animals.* Chicago: University of Chicago Press.

Arnaud, P. H. 1978. *A Host-Parasite Catalogue of North American Tachinidae (Diptera).* U. S. Department of Agriculture Science and Education Administration Miscellaneous Publication no. 1319.

Baker, R. R. 1983. Insect territoriality. *Annual Review of Entomology* 28: 65–69.

Balding, F. R., and G. L. Cunningham. 1974. The influence of soil water potential on the perennial vegetation of a desert arroyo. *Southwestern Naturalist* 19: 241–248.

Balfour-Browne, F. L. 1960. The green muscardine disease of insects, with special reference to an epidemic in a swarm of locusts in Eritrea. *Proceedings of the Royal Entomological Society, London* (A) 35: 65–74.

Barron, S. 1972. *The Desert Locust.* New York: Charles Scribner's Sons.

Beatley, F. C. 1974. Phenological events and their environmental triggers in Mohave Desert ecosystems. *Ecology* 55: 856–863.

Beck, D. E., and D. M. Allred. 1968. Faunistic inventory—BYU ecological studies at the Nevada Test Site. *Great Basin Naturalist* 28: 132–141.

Bernays, E. A., and A. C. Lewis. 1986. The effect of wilting on palatability of plants to *Schistocerca gregaria*, the desert locust. *Oecologia* 70: 132–135.

Birch, L. C., and H. G. Andrewartha. 1942. The influence of moisture on the eggs of *Austroicetes cruciata* Sauss (Orthoptera) with reference to their ability to survive desiccation. *Australian Journal of Experimental Biological and Medical Sciences* 20: 1–8.

Borror, D. J., D. M. DeLong, and C. A. Triplehorn. 1981. *An Introduction to the Study of Insects*. New York: Saunders College Publishing.

Brown, J. H., and A. Kodric-Brown. 1977. Turnover rates in insular biogeography—effect of immigration on extinction. *Ecology* 58: 445–449.

Brues, C. T. 1946. *Insects, Food and Ecology*. New York: Dover Publications.

Bultman, T. L., and S. H. Faeth. 1986. Experimental evidence for intraspecific competition in a Lepidopteran leaf miner. *Ecology* 67: 442–448.

Caldwell, M. 1985. Cold desert. Pp. 162–180 in B. F. Chabot and H. A. Mooney (eds.), *Physiological Ecology of North American Plant Communities*. New York: Chapman and Hall.

Casey, T. M. 1976. Activity patterns, body temperature and thermal ecology in two desert caterpillars (Lepidoptera: Sphingidae). *Ecology* 57: 485–497.

———. 1977. Physiological responses to temperature of caterpillars of a desert population of *Manduca sexta* (Lepidoptera: Sphingidae). *Comparative Biochemistry and Physiology* 57A: 53–58.

Caswell, H. 1978. Predator-mediated coexistence: A nonequilibrium model. *American Naturalist* 112: 127–154.

Cates, R. G. 1980. Feeding patterns of monophagous, oligophagous and polyphagous insect herbivores: The effect of resource abundance and plant chemistry. *Oecologia* 46: 22–31.

———. 1981. Host plant predictability and the feeding patterns of monophagous, oligophagous and polyphagous insect herbivores. *Oecologia* 48: 319–326.

Cates, R. G., and D. F. Rhoades. 1977. *Prosopis* leaves as a resource for insects. Pp. 61–83 in B. B. Simpson (ed.), *Mesquite: Its Biology in Two Desert Scrub Ecosystems*. Stroudsburg, Pa.: Dowden, Hutchinson and Ross.

Cheke, R. A. 1978. Theoretical rates of increase of gregarious and solitarious populations of the desert locust. *Oecologia* 35: 161–171.

Clark, L. R. 1953. An analysis of the outbreak of the Australian plague locust *Chortoicetes terminifera* (Walker) during the seasons 1940–41 and 1944–45. *Australian Journal of Zoology* 1: 70–107.

Cloudsley-Thompson, J. L. 1975. Adaptations of Arthropoda to arid environments. *Annual Review of Entomology* 20: 261–283.

Cloudsley-Thompson, J. L., and M. J. Chadwick. 1964. *Life in Deserts*. Philadelphia, Pa.: Dufour Editions, 218 pp.

Cody, M. L., and J. M. Diamond. 1975. *Ecology and Evolution of Communities*. Cambridge, Mass.: Harvard University Press, 543 pp.

Connell, J. H. 1980. Diversity and the coevolution of competitors, or the ghost of competition past. *Oikos* 35: 131–138.

Crawford, C. S. 1981. *Biology of Desert Invertebrates*. New York: Springer-Verlag.

Cunningham, G. L., J. P. Syvertsen, J. F. Reynolds, and J. M. Willson. 1979. Some effects of soil-moisture availability on above-ground production and reproductive allocation in *Larrea tridentata* (DC) Cov. *Oecologia* 40: 113–123.

DeLoach, C. J. 1982. Biological studies on a mesquite leaf tier, *Tetralopha euphemella*, in central Texas. *Environmental Entomology* 11: 261–267.

———. 1983. Field biology and host range of a mesquite looper, *Semiothisa cyda* (Lepidoptera: Geometridae), in central Texas. *Annals of the Entomological Society of America* 76: 87–93.

Dempster, J. P. 1963. The population dynamics of grasshoppers and locusts. *Biological Review* 38: 490–529.

Denno, R. F., and M. S. McClure, eds. 1983. *Variable Plants and Herbivores in Natural and Managed Systems*. New York: Academic Press.

Diamond, J., and T. J. Case, eds. 1986. Pp. 333–343 in *Community Ecology*. New York: Harper and Row Publishers.

Dodd, A. P. 1940. *The Biological Campaign Against Prickly Pear*. Brisbane, Australia: Commonwealth Prickly Pear Board, 177 pp.

———. 1959. The biological control of prickly pear in Australia. In A. Keast, R. Crocker, and C. Christian (eds.), *Biogeography and Ecology in Australia*. Monographs in Biology, vol. 8. The Hague: W. Junk.

Edney, E. B. 1967. Water balance in desert arthropods. *Science* 156: 1059–1066.

———. 1974. Desert arthropods. Pp. 311–384 in G. W. Brown, Jr. (ed.), *Desert Biology*, vol. 2. London and New York: Academic Press.

———. 1977. *Water Balance in Land Arthropods*. New York: Springer-Verlag.

Ehleringer, J. 1985. Annuals and perennials of warm deserts. Pp. 162–180 in B. F. Chabot and H. A. Mooney (eds.), *Physiological Ecology of North American Plant Communities*. New York: Chapman and Hall.

Ellis, P. E., D. B. Carlisle, and D. J. Osborne. 1965. Desert locusts: Sexual maturation delayed by feeding on senescent vegetation. *Science* 149: 546–547.

Faeth, S. H. 1986. Indirect interactions between temporally separated herbivores mediated by the host plant. *Ecology* 67: 479–494.

Farrow, R. A., and B. C. Longstaff. 1986. Comparison of the annual rates of increase of locusts in relation to the incidence of plagues. *Oikos* 46: 207–222.

Futuyma, D. J., and M. Slatkin, eds. 1983. *Coevolution*. New York: Sinauer and Associates.

Gallun, R. L. 1972. Genetic interrelationships between host plants and insects. *Journal of Environmental Quality* 1: 259–265.

Goeden, R. D., and S. M. Louda. 1976. Biotic interference with insects imported for weed control. *Annual Review of Entomology* 21: 325–342.

Goeden, R. D., and D. W. Ricker. 1974. The phytophagous insect fauna of the ragweed, *Ambrosia acanthicarpa*, in southern California. *Environmental Entomology* 3: 827–834.

———. 1975. The phytophagous insect fauna of the ragweed, *Ambrosia confertifolia*, in southern California. *Environmental Entomology* 4: 301–306.

———. 1976a. The phytophagous insect fauna of the ragweed, *Ambrosia dumosa*, in southern California. *Environmental Entomology* 5: 45–50.

———. 1976b. The phytophagous insect fauna of the ragweeds, *Ambrosia chenopodifolia, A. ericentra*, and *A. ilicifolia* in southern California. *Environmental Entomology* 5: 923–930.

———. 1976c. The phytophagous insect fauna of the ragweed, *Ambrosia psilostachya*, in southern California. *Environmental Entomology* 5: 1169–1177.

Goodman, P. J. 1973. Physiological and ecotypic adaptations of plants to salt desert conditions in Utah. *Journal of Ecology* 61: 473–494.

Greenfield, M. D., T. E. Shelley, and K. R. Downum. 1987. Variation in host plant quality: Implications for territoriality. *Ecology* 68: 828–838.

Gunn, D. L. 1960. The biological background of locust control. *Annual Review of Entomology* 5: 279–300.

Gutierrez, J. R., and W. G. Whitford. 1987a. Chihuahuan Desert annuals: Importance of water and nitrogen. *Ecology* 68: 2032–2045.

———. 1987b. Responses of Chihuahuan Desert herbaceous annuals to rainfall augmentation. *Journal of Arid Environments* 12: 127–129.

Hadley, N. F., and S. R. Szarek. 1981. Productivity of desert ecosystems. *BioScience* 31: 747–753.

Janzen, D. H. 1968. Host plants as islands in evolutionary and contemporary time. *American Naturalist* 102: 592–595.

———. 1973. Host plants as islands. II. Competition in evolutionary and contemporary time. *American Naturalist* 107: 786–790.

Jeffries, M. J., and J. H. Lawton. 1984. Enemy free space and the structure of ecological communities. *Biological Journal of the Linnean Society* 23: 269–286.

Joern, A. 1979. Feeding patterns in grasshoppers (Orthoptera: Acrididae): Factors influencing diet specialization. *Oecologia* 38: 325–347.

———. 1986. Temporal constancy in grasshopper assemblies (Orthoptera: Acrididae). *Ecological Entomology* 11: 379–385.

Joern, A., and L. R. Lawlor. 1980. Food and microhabitat utilization by grasshoppers from arid grasslands: Comparisons with neutral models. *Ecology* 61: 591–599.

———. 1981. Guild structure in grasshopper assemblages based on food and microhabitat resources. *Oikos* 37: 93–104.

Joern, A., and K. P. Pruess. 1986. Temporal constancy in grasshopper assemblies (Orthoptera: Acrididae). *Ecological Entomology* 11: 379–385.

Juhren, M., F. W. Went, and E. Phillips. 1956. Ecology of desert plants. IV. Combined field and laboratory work on germination of annuals in the Joshua Tree National Monument, California. *Ecology* 37: 318–330.

Kareiva, P. 1983. Influence of vegetation texture on herbivore populations: Resource concentration and herbivore movement. Pp. 259–289 in R. F. Denno and M. S. McClure (eds.), *Variable Plants and Herbivores in Natural and Managed Systems*. New York: Academic Press.

Kingsolver, J. M., C. D. Johnson, S. R. Swier, and A. Teran. 1977. *Prosopis* fruits as a resource for invertebrates. Pp. 108–125 in B. B. Simpson (ed.), *Mesquite: Its Biology in Two Desert Scrub Ecosystems*. Stroudsburg, Pa.: Dowden, Hutchinson and Ross.

Kuris, A. M., A. R. Blaustein, and J. J. Alio. 1980. Hosts as islands. *American Naturalist* 116: 570–586.

Labeyrie, V. 1978. The significance of the environment in the control of insect fecundity. *Annual Review of Entomology* 23: 69–89.

Lande, R. 1985. The fixation of chromosomal rearrangements in a subdivided population with local extinction and colonization. *Heredity* 54: 323–332.

Lawton, J. H. 1983. Plant architecture and the diversity of phytophagous insects. *Annual Review of Entomology* 28: 23–39.

———. 1984a. Herbivore community organization: General models and specific tests with phytophagous insects. Pp. 329–352 in P. W. Price, C. N. Slobodchikoff, and W. S. Gaud (eds.), *A New Ecology: Novel Approaches to Interactive Systems*. New York: John Wiley & Sons.

———. 1984b. Non-competition populations, non-convergent communities, and vacant niches: The herbivores of Bracken. In D. R. Strong, D. Simberloff, L. G. Abele, and

A. B. Thistle (eds.), *Ecological Communities: Conceptual Issues and the Evidence*. Princeton, N.J.: Princeton University Press.

Lawton, J. H., and D. R. Strong. 1981. Community patterns and competition in folivorous insects. *American Naturalist* 118: 317–338.

Leakey, R. R. B., and F. T. Last. 1980. Biology and potential of *Prosopis* species in arid environments with particular reference to *P. cineraria*. *Journal of Arid Environments* 3: 9–24.

Lewis, A. 1982a. Conditions of feeding preference for wilted sunflower in the grasshopper *Melanoplus differentialis*. Pp. 49–56 in J. Visser and A. Minks (eds.), *Proceedings of the 5th International Symposium on Insect-Plant Relationships*. Wageningen: Pudoc.

———. 1982b. Leaf wilting alters a plant species ranking by the grasshopper *Melanoplus differentialis*. *Ecological Entomology* 7: 391–395.

———. 1984. Plant quality and grasshopper feeding: Effects of sunflower condition on preference and performance in *Melanoplus differentialis*. *Ecology* 65: 836–843.

Lewis, T. 1973. P. 349 in *Thrips—Their Biology, Ecology and Economic Importance*. New York: Academic Press.

Lightfoot, D. C., and W. G. Whitford. 1987. Variation in insect densities on desert creosote bush: Is nitrogen a factor? *Ecology* 68: 547–557.

Ludwig, J. A., and J. F. Reynolds. 1988. Pp. 76–103 in *Statistical Ecology. A Primer on Methods and Computing*. New York: John Wiley & Sons.

MacArthur, R. H., and E. O. Wilson. 1963. An equilibrium theory of insular zoogeography. *Evolution* 17: 373–387.

———. 1967. *The Theory of Island Biogeography*. Princeton, N.J.: Princeton University Press.

McNeil, S., and T. R. E. Southwood. 1978. The role of nitrogen in the development of insect/plant relationships. Pp. 77–98 in J. B. Harborne (ed.), *Biochemical Aspects of Plant and Animal Coevolution*. New York: Academic Press.

Marcandier, S., and G. G. Khachatourians. 1987. Susceptibility of the migratory grasshopper, *Melanoplus sanguinipes* (Fab.) (Orthoptera: Acrididae), to *Beauveria bassiana* (Bals.) Vuillemin (Hyphomycete): Influence of relative humidity. *Canadian Entomologist* 119: 901–907.

Mares, M., W. Blair, F. Enders, D. Greegor, A. Hulse, J. Hunt, D. Otte, R. Sage, and C. Tomoff. 1977. The strategies and community pattern of desert animals. Pp. 107–163 in G. Orians and O. Solbrig (eds.), *Convergent Evolution in Warm Deserts*. Stroudsburg, Pa.: Dowden, Hutchinson and Ross.

Martin, W. C., and C. R. Hutchins. 1980. *A Flora of New Mexico*, vol. 1. Hirschberg, Germany: Strauss and Cramer.

Massion, D. D. 1983. An altitudinal comparison of water and metabolic relations in two acridid grasshoppers (Orthoptera). *Comparative Biochemistry and Physiology* 74(A): 101–105.

Matthews, E. G. 1976. P. 226 in *Insect Ecology*. St. Lucia, Australia: University of Queensland Press.

Mattson, W. J. 1980. Herbivory in relation to plant nitrogen content. *Annual Review of Ecology and Systematics* 11: 119–161.

May, M. L. 1979. Insect thermoregulation. *Annual Review of Entomology* 24: 313–349.

Moore, R. T., S. W. Breckle, and M. M. Caldwell. 1972. Mineral ion composition and osmotic relations of *Atriplex confertifolia* and *Eurotia lanata*. *Oecologia* 11: 67–78.

Munz, P. A. 1974. *A Flora of Southern California*. Los Angeles: University of California Press.

Neales, T. F., A. A. Patterson, and V. J. Hartney. 1968. Physiological adaptation to drought in the carbon assimilation and water loss of xerophytes. *Nature* 219: 469–472.

Nilsen, E. T., P. W. Rundel, and M. R. Sharifi. 1981. Summer water relations of the desert phreatophyte *Prosopis glandulosa* in the Sonoran Desert of southern California. *Oecologia* 50: 271–276.

Nobel, P. 1985. Desert succulents. Pp. 162–180 in B. F. Chabot and H. A. Mooney (eds.), *Physiological Ecology of North American Plant Communities*. New York: Chapman and Hall.

Otte, D. 1976. Species richness patterns of New World desert grasshoppers in relation to plant diversity. *Journal of Biogeography* 3: 197–209.

Otte, D., and A. Joern. 1975. Insect territoriality and its evolution: Population studies of desert grasshoppers on creosote bushes. *Journal of Animal Ecology* 44: 29–54.

Parker, M. A. 1982. Thermoregulation by diurnal movement in the barberpole grasshopper (*Dactylotum bicolor*). *American Midland Naturalist* 107: 228–237.

Payne, W. W. 1964. A re-evaluation of the genus *Ambrosia* (Compositae). *Journal of the Arnold Arboretum* 45: 401–438.

Pfadt, R. E. 1982. Density and diversity of grasshoppers (Orthoptera: Acrididae) in an outbreak on Arizona rangeland. *Environmental Entomology* 11: 690–694.

———. 1984. Species richness, density, and diversity of grasshoppers (Orthoptera: Acrididae) in a habitat of the mixed grass prairie. *Canadian Entomologist* 116: 703–709.

Polk, K. L., and N. Ueckert. 1973. Biology and ecology of a mesquite twig girdler, *Oncideres rhodosticta*, in west Texas. *Annals of the Entomological Society of America* 66: 411–417.

Raven, P. H., and D. I. Axelrod. 1977. P. 34 in *Origin and Relationships of the California Flora*. University of California Publications in Botany, vol. 72. Berkeley, Calif.: University of California Press.

Ricklefs, R. E. 1979. *Ecology*. New York: Chiron Press.

Riley, J. R., and D. R. Reynolds. 1983. A long-range migration of grasshoppers observed in the Sahelian zone of Mali by two radars. *Journal of Animal Ecology* 52: 167–183.

Roffey, J., and G. B. Popov. 1968. Environmental and behavioural processes in a desert locust outbreak. *Nature* 219: 446–450.

Rosenthal, G. A., and D. H. Janzen, eds. 1979. *Herbivores: Their Interactions with Secondary Plant Metabolites*. New York: Academic Press.

Schaerffenberg, B. 1966. Konnen Pilzepidemien bei Insekten kunstlich hervorgerufen werden? *Zeitschrift fur Angewandte Entomologie* 58: 362–372.

Schmoller, R. R. 1970. Terrestrial desert arthropods: Fauna and ecology. *Biologist* 52: 77–98.

Schoener, T. W. 1983. Rate of species turnover decreases from lower to higher organisms: a review of the data. *Oikos* 41: 372–377.

Schultz, J. C. 1981. Adaptive changes in antipredator behavior of a grasshopper during development. *Evolution* 35: 175–179.

Scriber, J. M., and F. Slansky, Jr. 1981. The nutritional ecology of immature insects. *Annual Review of Entomology* 26: 183–211.

Sharifi, M. R., E. T. Nilsen, and P. W. Rundel. 1982. Biomass and net primary production of *Prosopis glandulosa* (Fabaceae) in the Sonoran Desert of California. *American Journal of Botany* 69: 760–768.

Shearer, G., D. H. Kohl, R. A. Virginia, B. A. Bryan, J. L. Skeeters, E. T. Nilsen, M. R. Sharifi, and P. W. Rundel. 1983. Estimates of N_2 fixation from variation in the natural abundance of ^{13}N in Sonoran Desert ecosystems. *Oecologia* 56: 365–373.

Shelley, T. E., M. D. Greenfield, and K. R. Downum. 1987. Variation in host plant quality: Influences on the mating system of a desert grasshopper. *Animal Behavior* 35: 1200–1209.

Simpson, B. B. 1977. *Mesquite: Its Biology in Two Desert Scrub Ecosystems.* Stroudsburg, Pa.: Dowden, Hutchinson and Ross.

Southwood, T. R. E. 1973. The insect/plant relationship—an evolutionary perspective. Pp. 3–30 in *Symposium of the Royal Entomological Society of London*, vol. 6.

Stiling, P. D. 1980. Competition and coexistence among *Eupteryx* leafhoppers (Hemiptera: Cicadellidae) occurring on stinging nettles (*Urtica dioica*). *Journal of Animal Ecology* 49: 793–805.

Stower, W. J., and J. F. Griffiths. 1966. The body temperature of the desert locust (*Schistocerca gregaria*). *Entomologia Experimentalis et Applicata* 9: 127–178.

Strong, D. R., J. H. Lawton, and T. R. E. Southwood. 1984. *Insects on plants—Community patterns and mechanisms.* Cambridge, Mass.: Harvard University Press.

Strong, D. R., D. Simberloff, L. G. Abele, and A. B. Thistle, eds. 1984. *Ecological Communities: Conceptual Issues and the Evidence.* Princeton, N.J.: Princeton University Press.

Toolson, E. C. 1987. Water profligacy as an adaptation to hot deserts: Water loss rates and evaporative cooling in the Sonoran Desert cicada, *Diceroprocta apache* (Homoptera: Cicadidae). *Physiological Zoology* 60: 379–385.

Turner, R. M., A. M. Alcorn, G. Olin, and J. A. Booth. 1966. The influence of shade, soil, and water on saguaro seedling establishment. *Botanical Gazette* 127: 95–102.

Uvarov, B. P. 1957. The aridity factor in the ecology of locusts and grasshoppers of the Old World. Pp. 164–187 in *Arid Zone Research VIII, Human and Animal Ecology.* Paris: UNESCO.

Visscher, S. N., R. Lund, and W. Whitmore. 1979. Host plant growth temperatures and insect rearing temperatures influence reproduction and longevity in the grasshopper, *Aulocara elliotti* (Orthoptera: Acrididae). *Environmental Entomology* 8: 253–258.

Ward, C. R., C. W. O'Brien, L. B. O'Brien, D. E. Foster, and E. W. Huddleston. 1977. *Annotated Checklist of New World Insects Associated with* Prosopis (*Mesquite*). U.S. Department of Agriculture Technical Bulletin No. 1557, 115 pp.

Went, F. W. 1948. Ecology of desert plants. I. Observations on germination in the Joshua Tree National Monument, California. *Ecology* 29: 242–253.

———. 1949. Ecology of desert plants. II. The effect of rain and temperature on germination and growth. *Ecology* 30: 1–13.

West, N. E., and J. J. Skujins, eds. 1978. *Nitrogen in Desert Ecosystems.* Stroudsburg, Pa.: Dowden, Hutchinson and Ross.

White, T. C. R. 1969. An index to measure weather-induced stress of trees associated with outbreaks of psyllids in Australia. *Ecology* 50: 905–909.

———. 1976. Weather, food and plagues of locusts. *Oecologia* 22: 119–134.

———. 1984. The abundance of invertebrate herbivores in relation to the availability of nitrogen in stressed food plants. *Oecologia* 63: 90–105.

Whitham, T. G. 1983. Host manipulation of parasites: Within-plant variation as a defense against rapidly evolving pests. Pp. 15–41 in R. F. Denno and M. S. McClure (eds.), *Variable Plants and Herbivores in Natural and Managed Systems.* New York: Academic Press.

Whittaker, R. H. 1970. *Communities and Ecosystems.* New York: Macmillan.

Wiens, J. A. 1976. Population responses to patchy environments. *Annual Review of Ecology and Systematics* 7: 81–120.

Wilcox, B. A., and D. D. Murphy. 1985. Conservation strategy: The effects of fragmentation on extinction. *American Naturalist* 125: 879–887.

Willmer, P. G. 1982. Microclimate and the environmental physiology of insects. *Advances in Insect Physiology* 16: 1–57.

Willmer, P. G., and D. M. Unwin. 1981. Field analyses of insect heat budgets: Reflectance, size and heating rates. *Oecologia* 50: 250–255.

Wisdom, C. S. 1982. Effects of natural product variation on the interaction between a herbivore and its host plant. Ph.D. dissertation, University of California, Irvine.

———. 1985. Use of chemical variation and predation as plant defenses by *Encelia farinosa* against a specialist herbivore. *Journal of Chemical Ecology* 11: 1553–1565.

———. 1988. Comparisons of insect use and chemical defense patterns of two Sonoran Desert shrubs. Pp. 36–49 in R. G. Zahary (ed.), *Desert Ecology 1986: A Research Symposium.* Southern California Academy of Sciences and the Southern Desert Studies Consortium.

The Ecology and Importance of Predaceous Arthropods in Desert Communities

7

Gary A. Polis and
Tsunemi Yamashita

Predaceous arthropods (PA) form a taxonomically and trophically diverse assemblage in deserts. They range from species that exclusively kill and eat other animals to those that do so opportunistically, from parasitoid insects whose larvae develop by eating whole living arthropods to holometabolous insects that are predaceous only during part of their life cycles. The great diversity and high densities of these predators suggest that they may be a key component in desert ecosystems. Unfortunately, we know little of their ecology. We possess much more quantitative data on arachnids than on predaceous insects. Consequently, in spite of all our efforts, this paper will be biased with spider and scorpion data. Despite this limitation, a number of general patterns emerge.

We will present evidence that argues that desert PA are a relatively "successful" group. They are diverse, some taxa are extraordinarily abundant, and, as a group, they form a large proportion of the biomass of all desert arthropods and easily exceed the biomass of all desert vertebrates. Their success arises because they exhibit several physiological and ecological traits that (pre-)adapt them to the low and unpredictable food levels of deserts. These traits make PA relatively more important in deserts than PA in more mesic habitats.

We have several objectives in this chapter. First, we characterize the diversity, feeding ecology, and life history of PA. Second, we discuss factors influencing the distribution and abundance of PA. Third, we evaluate the importance of PA in desert communities by assessing their role in energy flow and how they influence the diversity, distribution, and abundance of their prey and competitors.

Characteristics

Diversity

Deserts contain an array of arthropod taxa that are generally similar to those found in temperate or tropical communities. PA include several orders of insects, arachnids, centipedes, and mites (see Appendix 13.2 in Chapter 13 for some examples). Some taxa are quite diverse. Successful groups include prostigmatic mites; scorpions, spiders, and solpugids; mutillid, tiphiid, pompillid, sphecid, and vespid wasps; asilid, tachinid, therevid, and bombyliid flies; meloid and carabid beetles; antlions, reduviid bugs, and certain predaceous or omnivorous ants (see also Schmoller 1970; Crawford 1981). Although the relative importance of these taxa varies from desert to desert, these groups appear to be important in most deserts, with a few notable exceptions; for example, solpugids are absent in Australian deserts.

Some data allow estimates to be made of the diversity of insect predators. In the University of California Deep Canyon Desert Research Station and Reserve (Riverside County, California), both insect predators (30 families) and parasitoids (>15 families) are common (Frommer 1986). Some families are speciose: Carabidae (31 species), Cleridae (15), Meloidae (15), Staphylinidae (20), Asilidae (56), Bombyliidae (147), Ichneumonidae (41), Mutillidae (34), Pompilidae (34), Sphecidae (123), Vespidae (35), and Myrmeleontidae (28). Polis (this volume) tabulated 32 families of insect predators in the sandy desert on the floor of the Coachella Valley (Riverside County, California). At least 20 families of parasitoids also occur. Pitfall traps captured 19 families of insect predators and parasitoids in Nevada's Mojave Desert (Pietruszka 1980). A survey by Andrews, Hardy, and Giuliani (1979) of the beetle fauna in five sand dunes in California deserts showed seven families of predators and two of parasitoids (Table 7.1). Each dune averaged 22.8 species of beetle predator and 5.4 species of beetle parasitoid. Diversity at the Nevada Test Site, recorded only for some insects, is quite high for certain taxa, with, for example, 53 species of ants and 33 species of fleas (Allred, Beck, and Jorgensen 1963). The sand dune habitat of the Namib Desert has the lowest recorded number of predaceous insect taxa: between 16–20 species (Holm and Scholtz 1980) (with a few more species known to be present; M. Seely and E. Holm personal communication).

Some arachnids are particularly successful in deserts, including scorpions, spiders, solpugids, and acarines. Pseudoscorpions occur in most deserts, whereas uropygids and amblypygids are found only in some deserts (e.g., the southern section of the Sonoran Desert). We quantified the diversity of predaceous arachnids from many deserts (Table 7.2). (Diversities are much better known for arachnids than insects or acarines.) Species diversities of scorpions (average = 7.1 species) and spiders (43.3 species) are high. The number of species in North American deserts ranges from 2 to 13 scorpions, 30 to more than 100 spiders, and 6 to 28 solpugids (Polis and McCormick 1986a). Scorpions and solpugids are more speciose in deserts than other habitats (Polis 1990; M. Muma personal communication). Although rigorous comparative data are unavailable, spider diversity is lower than in tropical areas and likely the same or less than that in most temperate habitats.

Guilds

Desert PA can be divided into a number of guilds based on their foraging. For example, Hatley and MacMahon (1980) recognize five guilds of desert spiders: nocturnal hunters, runners, ambushers, agile hunters, and web-builders. Alternatively, Bultman and Uetz (1982) proposed five guilds divided into two "macroguilds." They partitioned web-building spiders (a macroguild) into three guilds: scattered line web builders, sheet web builders, and vagrant web builders. Hunting spiders (a macroguild) subsume a guild of sit-and-wait (ambush) predators that often change foraging location and a second guild of mobile predators that actively search for and pursue prey. Humphreys (1988) adds a sixth guild consisting of sit-and-wait predators that rarely change foraging site. This guild primarily consists of burrowing spiders.

Non–web-building spiders constitute the majority of desert spiders: 70 percent in the Great Basin Desert (Fautin 1946), 75 percent at the Nevada Test Site (Mojave Desert; Allred, Beck, and Jorgensen 1963), 80 percent in the Chihuahuan Desert (Chew 1961), 59 percent in the Coachella Valley (Colorado Desert; Polis and McCormick 1986a), 66 percent (in shrubs) to 87 percent (on ground) in sagebrush steppe of the Great Basin Desert (Abraham 1983), and 63 percent of those in the Namib Desert sand dunes (J. Henschel personal communication). By comparison, non-web builders form about 50 percent of spiders in temperate forests (Schmoller 1970).

Table 7.1 Species diversity of predaceous and parasitoid Coleoptera from California sand dunes.

| Taxon | Location | | | | |
	Owens	Cadiz	Rice	Palen	Algodones
Carabidae	3	3	4	11	18
Cicindelidae	2				1
Cleridae			3	2	7
Coccinellidae	3	5	3	5	4
Histeridae	2		3	5	7
Staphylinidae	1	1	3	7	9
Trogositidae				1	1
Meloidae[a]	1	4	5	12	6
Rhipiphoridae[a]					1
Total predators	11	9	16	31	47
Total parasitoids	1	4	5	12	7

SOURCE: Andrews, Hardy, & Giuliani (1979).
[a]Parasitoids.

Table 7.2 Species diversity of arachnids in deserts.

Location	Number of Species			Source
	Spiders	Scorpions	Solpugids	
Etosha Park, S.W. Africa (12 sites)	12–50[a] (35.8)	2–7[a] (4.8)	2–8[a] (5.6)	E. Griffin pers. comm.
Namib Desert (plains) (6 sites)	30–60[a] (46)	5–6[a] (5.4)	5–12[a] (8.6)	E. Griffin pers. comm.
Namib Desert (dunes)	26	2	8	J. Henschel pers. comm. Holm & Scholtz 1980
Repetek Reserve (sand), Central Asia (Kara Kum Desert)	90	6	10	V. Fet pers. comm.
Coachella Valley, CA (sand), (Colorado Desert)	>54	4	11	Polis & McCormick 1986a
Deep Canyon Reserve, CA[b] (Colorado Desert)	71	6	9	Frommer 1986
Nevada Test Site[b] (Mojave Desert)				Allred, Beck, & Jorgensen 1963
All habitats	87	8	31	
Larrea-Franseria habitat	25	6	9	
Grayia-Lycium habitat	34	7	18	
White Sands, Alamogordo, NM (Chihuahuan Desert)	23–31	1	4–5	Muma 1975
Carrizozo Malpais, NM (Chihuahuan Desert)	90	. . .	5–7	Gertsch & Reichert 1976; M. Muma pers. comm.
Chihuahuan Desert, NM (6 sites)	. . .	. . .	5–7[a] (6.1)	M. Muma pers. comm.
Portal, S.E. AZ (Chihuahuan Desert)	50	8	10	Chew 1960; Polis unpub. data; M. Muma pers. comm.
Great Basin Desert, UT	23–42 80–100	1–4[a] (2.4)	7	Fautin 1946; Allred & Gertsch 1975; Hatley & MacMahon 1980; Allred & Kaston 1983; M. Muma pers. comm.
Baja California, Mexico (Sonoran Desert) (10 sites)	. . .	5–8[a] (6.5)	. . .	Due & Polis 1986; M. L. Jimenez pers. comm.
(Vizcaino Desert) (12 sites)	. . .	4–13[a] (8.5)	. . .	
(Cape region) (5 sites)	86	7–9[a] (8.0)	. . .	
Average (± s.d.)	54.4 ± 26.7	7.1 ± 2.5[c]	9.9 ± 6.7	
Number of sites	31	58	37	

SOURCE: Polis 1990.
[a]Range and average.
[b]Total number of species for entire area; totals summed over several habitats, including some non-desert areas.
[c]Includes diversity from several sites not listed; see Polis 1990 for sources.

Diurnal jumping spiders (Salticidae) and nocturnal Gnaphosidae are well-represented taxa that hunt without a web. Crab spiders (Thomisidae) are common sit-and-wait predators that change foraging sites. Important spiders that hunt near or from their burrows include giant crab spiders (Heteropodidae, especially in the Namib Desert), burrowing wolf spiders (*Geolycosa*), and various mygalomorphs (e.g., Ctenizidae, Theraphosidae). Burrowing spiders from several families are particularly important in Australia (Main 1981; Humphreys 1988).

Some web builders are either fairly diverse or have species that are abundant in deserts: comb-footed spiders (Theridiidae; particularly black widows, *Latrodectus* spp.), cobweb spiders (Pholcidae), funnel web spiders (Agelenidae), and (at least in North America) Dictynidae and Diguetidae. The first two families belong to the guild of scattered line web builders, the last three families to the guild of sheet web builders. Pholcids and other web spiders are abundant in holes and abandoned burrows (with theridiids in North America; Fowler and Whitford 1985; agelenids and pisaurids in southern Africa; Heidger 1988; gnaphosids and zodariids in the Kara Kum; Krivokhatskii and Fet 1982). Such spiders occur in more than 90 percent of burrows. Vagrant web builders (various orb weavers, e.g. Araneidae) are not speciose or abundant in deserts, possibly because of the paucity of high web sites (trees, large shrubs).

Some families appear numerically dominant (but this may be a function of trap protocol). Lycosids, gnaphosids, and pholcids form 68 percent and 74 percent, respectively, of all spiders in White Sands and Alamogordo, New Mexico (Muma 1975). Salticids form 78 percent of the spiders recorded in the Great Basin Desert by Fautin (1946).

Different families of scorpions are dominant on different continents. In North American deserts, vaejovid scorpions are speciose and scorpionids are absent. For example, vaejovids form 84 percent of the 61 species in Baja California (Williams 1980). In Australia and Africa, scorpions in the family Scorpionidae are most common and vaejovids are absent. They form 75 percent of the Australian fauna (28 species; Koch 1977) and 43 percent of the scorpion fauna (56 species; Lamoral 1979) of Namibia. Scorpions of the family Buthidae (the oldest extant family of scorpions) are present in all deserts. Some buthid genera form dense populations (2,000 to >5,000/ha) in deserts (e.g., *Centruroides* in North America, *Uroplectes* in southern Africa, *Leiurus* in northern Africa). Other species that are quite dense include some species of Australian *Urodacus* (Scorpionidae), southern African *Opisthophthalmus* (Scorpionidae), and North American *Vaejovis, Vejovoidus*, and *Paruroctonus* (Vaejovidae) (Polis 1990).

Scorpion taxa can be classified into three guilds based on foraging behavior. Most vaejovids leave their burrows and sit motionless on the surface, ready to ambush prey. Vaejovids are included with similar scorpions (e.g., some chactids) in a guild of sit-and-wait species that forage on the ground within several meters of their burrows. Scorpionids ambush prey primarily from the entrance of their burrows and form a guild of sit-and-wait predators that rarely change foraging sites. Scorpions from several other families (e.g., Diplocentridae) are also in this guild. The third guild consists of species that have no burrow and forage on vegetation or bark, for example, many genera of buthids. These species either remain motionless or alternate bouts of sit-and-wait foraging with active searching. Some nonburrowing buthids do occupy "home sites" for long periods (>1 year), for example, *Centruroides exilicauda* (C. Myers personal communication).

Insect predators can be similarly classified into guilds according to foraging tactics. Sit-and-wait (ambush) predators include antlions, reduviids, and mantids. The guild of actively foraging predators includes asilids, many predaceous ants and wasps, coccinellids, carabids, and chyrosopids. These guilds could be subdivided according to activity periods (diurnal, nocturnal) or microhabitat (ground, shrub, or aerial).

Omnivory and predator-predator feeding links (intraguild predation) (see below) make unclear the importance of guilds in providing ecological separation among desert PA. Root (1967) defined guild as "a group of species that exploit the same class of environmental resources in a similar way." The term can be used more broadly to include all taxa in a community that use similar resources (food or space) and may compete, regardless of differences in tactics of resource acquisition (Polis, Myers, and Holt 1989). Guilds based on hunting tactics, microhabitat, or foraging phenology provide trophic separation only if predators capture different subsets of prey (e.g., sit-and-wait versus widely foraging predators; Huey and Pianka 1981). Predators may belong to apparently distinct compartments (diurnal versus nocturnal species, surface versus subsurface species) but not form discrete trophically based guilds. This occurs if prey overlap among predators from different compartments is high (Polis this volume).

Many factors contribute to decompartmentalization and increase in prey overlap. Diel activity patterns change from day to night as a function of temperature; nocturnal (diurnal) prey are eaten by predators in their resting places during the day (night); consequently, nocturnal species often eat diurnal prey and vice versa. For example, nocturnal scorpions eat diurnal prey that have landed for the night (e.g., honeybees, robberflies, and mantids; McCormick and Polis 1990). Diurnal predators feed on resting nocturnal prey, for example, wasps on spiders, and lizards (*Cnemidophorus tigris*, *Nucras tessellata*) on scorpions (Polis, Sissom, and McCormick 1981). Surface and sub-surface species exchange energy and food; many (most?) desert arthropods spend time both below and above ground. They feed and are eaten by predators in both microhabitats (e.g., tenebrionid larvae are eaten by soil predators, and adults by surface predators). Such trophic crossover complicates clear recognition of guild structure (based on use of different prey) among PA.

Abundance and Biomass

PA appear to be a particularly important component of the arthropod fauna in deserts. They constitute a high proportion of the total number of all species and all individuals when compared with other trophic groups of arthropods (Chew 1961; Schmoller 1970; Seely and Louw 1980; Crawford 1981, 1986, this volume). PA form 17.4 percent of the species and 17 percent of the total individual arthropods in the northern Mojave; 45.5 percent and 19 percent in the northern Chihuahuan; 43.5

percent and 4 percent in Egyptian coastal deserts; and 55.4 percent and 18 percent in Namib Desert sand dunes (Crawford 1981, 1986, this volume; N.B., these statistics are based on limited samples of the entire arthropod fauna). Individual spiders represented 28 percent of the total number of all arthropods collected by Chew (1961) in the Chihuahuan Desert near Portal, Arizona. Predaceous desert soil mites exhibit similarly high ratios (Crawford 1981). However, PA represent only 6.1 percent of all individuals in the invertebrate mesofauna in Egyptian deserts (Ghabbour and Shakir 1982).

Predators also can represent a large fraction of the total arthropod biomass. For example, the average biomass of spiders near Portal was estimated to be about 50 percent of the biomass of the aboveground insects (N.B., termites and ants were not included; Chew 1961; Schmoller 1970).

Analysis of the diet of arthropodivorous consumers also provides evidence that PA constitute a large proportion of all arthropods. PA formed an average of 41 percent (range: 28–71 percent) of all arthropods eaten by 15 vertebrates and 51.5 percent (range: 45–60 percent) of those eaten by six PA (Polis this volume). (These numbers may be biased toward inclusion of PA if these predators exhibit preference for arthropods from the larger end of the body size scale; PA are generally larger than the arthropods on which they prey.)

These statistics suggest that at least some PA may be relatively more important in deserts than in other habitats (Chew 1961; Crawford 1981, 1986). Wagner and Graetz (1981) found that the proportion of predators in deserts is generally greater than that in less arid habitats. For example, spiders represent only 5–17 percent of the arthropods in more mesic habitats (Chew 1961). Scorpions and solpugids are more diverse and dense in deserts than other habitats (Polis 1990; M. Muma personal communication).

It appears that aridity somehow promotes this relative success of PA. For example, Péfaur (1981) found a gradient in Peruvian coastal deserts: at low elevations and more xeric sites, most arthropod species are predaceous; the proportion of PA progressively decreased with more mesic and productive sites at higher elevations. The numbers of PA species and their relative abundance were significantly greater at the relatively more arid of two sites in the Coachella Valley of Southern California (McCormick and Polis 1985). The biomass ratio (PA:other arthropods) decreased from 1:1.2 to 1:7.4 from a dry to a wet period (Seely and Louw 1980; N.B., some vertebrates were included). Although these data suggest that the relative importance of PA increases with aridity, no comprehensive study has rigorously compared the relative importance (diversity, density, biomass) of PA in desert versus more mesic habitats.

How do the density and biomass of desert arthropods measure against those of other groups of consumers? To approach this question, we compiled studies from the literature reporting the density and biomass of various groups of desert macroarthropods and vertebrates (Tables 7.3 and 7.4; see Yamashita and Polis n.d. for details). These data are not without bias and other problems. For example, studies are not usually conducted in areas where the focal taxon is absent or rare; nor are rare species studied as often as common ones. Consequently, the statistics in Tables 7.3

Table 7.3 Estimated average density and diversity of major taxa of desert macrofauna.

Group	Density (# of indiv./ha)[a]	Diversity[b]	(# of species)
Mammals	43 ± 130	28.5 ± 21.3	14
Birds	0.6 ± 1.1	46.7 ± 27.3	14
Lizards	36.8 ± 43.0[c]	12.0 ± 4.6	12
"Insects"[d]	35,100 ± 54,600	592 ± 796	10
Termites	4,025,000 ± 2,793,000	9.1 ± 15	7
Isopods	241,810 ± 336,850	1.3 ± 0.8	7
Millipedes	1,150 ± 214	2.4 ± 2.1	7
Spiders	3,220 ± 8,800	54.4 ± 26.7	31
Solpugids	. . .	9.9 ± 6.7	37
Scorpions	3,210 ± 3,500	7.1 ± 2.5	58

SOURCE: Yamashita & Polis (n.d.).
NOTE: These statistics should be viewed as very rough approximations rather than absolute values.
[a]Mean density estimates with s.d. Density estimates collected from the literature, standardized to a per hectare basis.
[b]Diversity = mean number of species per taxon from local sites in different deserts (e.g., as in Table 7.2 this chapter).
[c]Calculated from 40 studies that reported only density of adults.
[d]Data for "insects" come from a number of sources, each reporting different (and often incomplete) components of the insect fauna. Ants and termites were usually excluded in these reports and/or reported separately.

and 7.4 overestimate density and biomass and should be taken as a first approximation of actual parameters. However, the data are instructive.

The best data are for scorpions (see Polis 1990 for references). Scorpion populations are often quite dense, averaging more than 3,200 individuals/ha with several species maintaining populations of 5,000 to more than 10,000 individuals/ha (e.g., Shorthouse 1971; Lamoral 1978; Polis and Farley 1980; Polis and McCormick 1986a; Bradley 1986; Polis unpublished data). On the average, reported numbers of scorpions show them to be more dense than all other macroscopic animal taxa in these North American, Australian, and African deserts except "insects" (lumped, excluding ants and termites), termites, and isopods (ants alone are undoubtedly also more dense). Since scorpions are among the largest of all terrestrial arthropods (adults of desert species weight from 0.5 to 10 g; Polis and Farley 1980), these high densities produce rather large estimates of standing biomass (density of individual species × mass of individual animal). Populations of each species of desert scorpion average 7.15 kg/ha. Only termites (and possibly ants) support a greater population biomass per species or per taxon (from the ordinal level down) than scorpions.

From the data we compiled we found that the biomass of all scorpions living in desert areas (50.8 kg/ha; calculated as mass of individual species × average number of sympatric species) is greater than that of any one other taxon except termites (113.4 kg/ha) or the sum of all other insects without termites (521.2 kg/ha; Table 7.4). Note also that the population biomass of scorpions is higher than that of any one group of vertebrates (e.g., mammals and lizards average 39.9 and 6.8 kg/ha,

Table 7.4 Estimated population biomass for taxa
reported in the literature.

Taxon	Population Biomass (kg/ha)		N^c
	per species[a]	per taxon[b]	
Mammals	1.40	39.9	29
Birds	0.02	0.9	25
Lizards	0.57	6.8	46
"Insects"	0.88	521.2	31
Termites	12.45	113.4	7
Isopods	9.91	12.8	2
Millipedes	1.15	2.8	2
Spiders	0.13	7.1	4
Scorpions	7.15	50.8	17
All vertebrates		47.7 kg/ha	
All arthropods		708.1 kg/ha	
All macrofauna		755.8 kg/ha	

SOURCE: Primary references and additional information reported in
Yamashita & Polis (n.d.).
NOTE: These statistics are only gross approximations of reality.
[a] Average wet weight of all members of one species per hectare. These
statistics were sometimes reported; however, in many cases these
statistics were not reported and we calculated them by multiplying the
average mass of an individual × density.
[b] Population biomass per species × average diversity of taxon.
[c] Number of species per taxon for which biomass data exist.
[d] Data for "insects" come from a number of sources, each reporting different
(and often incomplete) components of the insect fauna. Ants and termites
were usually excluded in these reports and/or reported separately.

respectively) or all vertebrates combined (47.7 kg/ha). Overall, scorpions form 6.7 percent of the biomass of all macrofauna species combined, 7.1 percent of the biomass of all macroarthropods, and 106 percent of the biomass of all vertebrates.

Limited information (biomass data on only four species) is available for spiders in deserts. Spiders are much more diverse (12 to >100 species; mean = 43.3) than scorpions (4–13 species; mean = 7.1). Undoubtedly spider populations are more dense: surveys of published studies from non-desert areas report means of 130.8 spiders/m^2 (Turnbull 1973) and 82 spiders/m^2 (Humphreys 1988). In deserts, reports have shown average spider densities of only 2.5/m^2 (desert creosote communities; Hadley 1980) and 0.81/m^2 (sand community in the Coachella Valley; Polis and McCormick 1986b). By contrast, the density of desert scorpions averages 0.32/m^2. However, many desert species are relatively dense, at least in local patches, for example, on shrubs or in burrowing spider colonies. Populations of heteropodids in the Namib vary from 9 to 302/ha (*Leucorchestris arenicola*; Henschel in press) and 1.1 to 556/ha (*Carparachne* spp.; J. Henschel personal communication). The number of web spiders on *Acacia erioloba* trees in the desert riparian habitat of the Namib varies from about 10 to more than 300 spiders/trunk (Polis and Seely unpublished data).

However, desert spiders do not form extraordinarily high biomasses. This may be due to their relatively small body mass, ranging from 0.05 to 10 g, with most averaging less than 0.3 g. Thus, spiders (all species combined) only average 7.1 kg/ha in deserts, an order of magnitude less than scorpions. However, spider biomass in deserts is still greater than that of birds (0.9 kg/ha). Spiders represent 0.9 percent of the biomass of all desert animal species combined, 1 percent of the biomass of all arthropods, and their totals measure 15 percent of the total biomass of all vertebrates.

Few data exist on the densities and biomass of the many species of solpugids, predaceous insects, and insect parasitoids. Solpugids may be important in some deserts. Their individual body mass is intermediate between scorpions and spiders; they are more diverse than scorpions, but less so than spiders. Scattered reports and trapping data indicate that they may be quite abundant (Muma 1975; Muma and Muma 1988; Polis and McCormick 1986b; Wharton 1987). Some parasitoids (e.g., tiphiid, mutillid, and chalcidoid wasps, bombyliid flies) and predators (e.g., asilids, carabids, myrmeleontids) are diverse and sometimes quite abundant (Polis this volume). Yet most are small and their biomass may not be great. Facultatively predaceous ants are often quite dense and their population biomass is expected to be high.

Life History

Desert arthropods can be placed into two categories of divergent life history strategies: opportunistic, fugitive, or *r*-selected species, versus equilibrium, stable, or *K*-selected species (Polis and Farley 1980; Louw and Seely 1982). In spite of well-known limitations, this dichotomy is useful to indicate general differences in the life history of many species (Polis and Farley 1980; Louw and Seely 1982; Polis 1990).

Equilibrium species develop slowly to a relatively large size and produce (for a series of years) relatively few offspring, each of which has proportionally more parental care allocated to it than offspring of opportunistic species; survivorship is Type I or II; mortality is predominantly density-dependent and the magnitude of population fluctuation is relatively small. Equilibrium species inhabit a relatively predictable environment in which species interactions are a paramount force, and they usually occupy a narrow, specialized niche. Opportunistic species exhibit the inverse of these characteristics.

Most desert arthropods (predator and prey species) exhibit opportunistic life histories. They are characterized by populations that fluctuate widely within and between years. Populations respond to benign conditions (i.e., precipitation and subsequent productivity) by increasing, often dramatically (Seely and Louw 1980). Conversely, populations decrease precipitously during unfavorable conditions. Opportunistic arthropod predators from deserts include species that live less than 1 year, such as most mites, most spiders, hemimetabolic insects, and a majority of holometabolic insects.

A few desert PA conform to our conceptions of equilibrium species. Most scorpions, some spiders (mygalomorph and burrowing), some centipedes, and some mites live for many years (as do many important prey taxa, e.g., many tenebrionids and other large beetles). Their populations are relatively stable, even in harsh and unpredictable desert environments (Polis and Farley 1980; Louw and Seely 1982; Polis 1990). Further, the age structure of long-lived species is often more defined as compared to species that live less than 1 year. Overall, the age structure of desert PA (and its many ecological manifestations) may be more important than that of PA in more mesic areas.

These life history differences are important in understanding the community ecology of desert PA. To a high degree, life history characteristics govern susceptibility and response to favorable and unfavorable environmental perturbations. These characteristics partially determine the distribution and abundance of individual species and thus contribute to community structure. Further, opportunistic and equilibrium species exhibit different predator-prey dynamics. These differences are discussed below.

Diet Desert PA display a variety of feeding modes. Some (e.g., scorpions, web spiders, and antlions) are sit-and-wait predators that forage by remaining immobile and waiting for prey to move within detection range. Others (e.g., hunting spiders, centipedes, ants, and wasps) actively search for relatively immobile prey. Most eat the prey they capture; however, the adults of some (e.g., pompilid, sphecid, and vespid) wasps actively hunt prey but eat only nectar and pollen; captured prey are immobilized and fed to developing larvae or placed into egg chambers. In spite of their name, parasitoids function similarly to predators. They hunt arthropod prey. However, the prey are not consumed by the hunter; rather, an egg is placed on or in the prey and the parasitoid's larvae develop by eating the still-living host. Some parasitoids (e.g., pompilid and sphecid wasps) actually capture their prey (like predators) to provision nests containing developing larvae. Unlike with true parasites, the host is always killed as the parasitoid completes its life cycle. Some "parasitoids" (e.g., mutillids, clerid beetles) are simply egg predators: the adults oviposit on eggs and developing larvae feed externally on the eggs and young.

Consumers are classified according to resource specialization (i.e., number of species eaten within a similar group of prey) or trophic specialization (i.e., the number of different groups eaten, e.g., plants, detritus, arthropods, vertebrates). Omnivores (trophic generalists) take prey from different trophic levels. All gradations of resource and trophic specialists and generalists occur in deserts.

It appears that most desert PA are resource generalists that eat many species of animal prey. Centipedes and most arachnids opportunistically capture whatever prey they are able to subdue. Consequently, they eat many species of prey. For example, the scorpion *Paruroctonus mesaensis* is recorded to eat more than 125 prey species; the spider *Diguetia mohavea*, more than 70 species; and the black widow, *Latrodectus hesperus*, 35 species (Polis 1979; McCormick and Polis 1985; Polis and Sculteure unpublished). Both *P. mesaensis* and *Latrodectus* were recorded to eat solpugids, scorpions, spiders, insects, isopods, and vertebrates (small snakes; see McCormick and Polis 1982). In general, spiders (Riechert and Harp 1987) and scorpions (McCormick and Polis 1990) show little or no discrimination in choice of prey.

Acarine predators, at least in soils and detritus, are mainly generalists (Edney et al. 1974; Franco, Edney, and McBrayer 1979; see Zak and Freckman this volume). Predatory mites eat nematodes, Collembola, and other mites from all feeding categories. Hemimetabolic insect predators (e.g., mantids, phymatids, reduviids, and other Hemiptera) are resource generalists. A McKinnerney (1978) study of the fauna associated with rabbit carrion revealed many generalist predators. Reduviids, asilids, and (hister, carabid, and staphylinid) beetles ate all types of necrophagous insects. Ants, silphid beetles, and Opiliones were omnivorous, eating both carrion and larvae. These and other predators (spiders and solpugids) in carrion included other predators in their diets.

Many parasitoids and nest predators are catholic in diet (Powell and Hogue 1979; Austin 1985; Wasbauer and Kimsey 1985; Coville 1987). The desert sphecid wasp *Solierella* places about 200 prey items in eight to nine larval cells; prey include Hemiptera, Lepidoptera, Orthoptera (grasshoppers, cockroaches, katydids, mantids), and Coleoptera (M. Cazier personal communication). Species of desert pompilid wasps hunt certain categories of spiders (burrowers, web builders) regardless of species or family (Wasbauer and Kimsey 1985). Desert sand wasps of the genus *Photopsis* (Mutillidae) are generalist parasitoids; they eat all species of ground-nesting wasps and bees in addition to functioning as facultative hyperparasitoids on the large variety of parasitoids of these Hymenoptera (Ferguson 1962).

A few desert arachnids tend to specialize. *Mimetus* spiders (Mimetidae) primarily prey on other spiders. Some spiders, for example, some theridiids, salticids, oecobiids, and zodariids, specialize on ants (e.g., Main 1957; Hölldobler 1970; Krivokhatskii and Fet 1982; MacKay 1982; Ryti and Case 1986). *Minosiella intermedia* (Gnaphosidae), the most abundant spider in rodent burrows in the Kara Kum Desert, specialized and accepted only fleas and small scarab beetles (Krivokhatskii and Fet 1982). Two species of scorpions are specialists: the Australian *Isometroides vescus* on trapdoor spiders (Main 1956), and the northern African *Scorpio maurus*, with 20–77 percent of its diet being the isopod *Hemilepistus reaumuri* (Shachak 1980; Krapf 1986 personal communication; *Scorpio* actively selects its homesite to be within 10 cm of isopod burrows). Some solpugids appear to specialize on termites (Muma and Muma 1988). Adult velvet mites, *Dinothrombium*, feed almost exclusively on termites (however, larvae parasitize Orthoptera; Tevis and Newell 1962).

Insects that specialize on particular prey include coccinelids (on aphids and/or scales, such as *Dactylopius*, the homopteran cochineal scale on *Opuntia* cactus) and chrysopid lacewings (on aphids) (Powell and Hogue 1979). Many parasitoids prefer certain species of insect or spider host (Powell and Hogue 1979; Hawkins and Goeden 1984; Austin 1985; Wasbauer and Kimsey 1985; Coville 1987; M. Crazier personal communication). For example, some sphecid wasps hunt specific prey (*Larropsis* preying on camel crickets, *Sphecius* on cicadas, *Aphilanthops* on *Pogonomyrmex*). Many specialists occur in rabbit carrion; in one study tachinid flies and braconid wasps specifically attacked dipteran larvae, and the larvae of clerid beetles attacked those of dermestid beetles (McKinnerney 1978).

Most arachnids, centipedes, and predaceous hemimetabolic insects and some holometabolic insects feed only on other arthropods. However, opportunism and age structure produce trophic generalism in many taxa. Opportunism takes many forms. For example, some scorpions and spiders, quintessential obligate predators, gain energy from sources other than live prey. Species of both scavenge dead arthropods (Knost and Rovner 1975; McCormick and Polis 1990). More remarkable, some spiderlings are aerial plankton feeders, eating pollen and fungal spores trapped by their webs (Smith and Mommsen 1984).

Fully omnivorous taxa are occasionally but regularly predaceous. This is particularly true for ants. For example, California harvester ants take seeds, plant parts, spiders, and insects from at least six orders, including four ant species (Ryti and Case 1988). On occasion, live arthropods contribute greatly to their diet; for example, *Messor pergandei* opportunistically ate termites (89.7 percent of their diet) during a short period of alate emergence after rains (Gordon 1978). Normally arthropods form 2–10 percent of the diet of these granivores. Most desert ants normally eat arthropods. Nine of 10 Chihuahuan Desert ant species included both seeds and arthropods in their diets (Chew and DeVita 1980). Three species (the "arthropod foraging guild") primarily ate arthropods (87–98 percent of their diet); arthropods formed 2–75 percent of the diet of the seven other species. Termites (14–57 percent of the arthropods) and ants (30–40 percent) were the main prey.

Camel crickets, grasshoppers, and other desert Orthoptera regularly supplement their herbivorous diets with conspecifics (Polis this volume). (Many herbivorous insects, especially Orthoptera, are normally opportunistic cannibals; Polis 1981, 1984b, C. Crawford personal communication). Jerusalem crickets (Gryllacrididae) from California sand dunes eat scorpions in addition to plants and detritus (scorpions also eat the crickets; Polis, Sissom, and McCormick 1981).

Finally, the marked age structure characteristic of most arthropods often results in marked trophic generalization. Life stages and age classes often use different foods, thereby greatly expanding the diet of a species. Pimm and Rice (1987) refer to such changes as "life history omnivory." "Ontogenetic diet shifts" are characteristic of those species that undergo radical metamorphosis during growth (Polis 1984a; Werner and Gilliam 1984). This is illustrated during the complex life cycle of holometabolous insects. Many families of Coleoptera (e.g., Cleridae, Meloidae), Diptera (e.g., Tachinidae), and Hymenoptera (e.g., Tiphiidae) are predaceous as larvae and herbivorous as adults. For example, *Pherocera* (Therevidae) fly larvae are predators on beetle, fly, and moth larvae (including con- and heterospecific therevids); adults are nectar feeders (Irwin 1971). (See appendix 13.2 in Chapter 13 for more examples.) Hymenoptera that hunt live prey to feed to their larvae are also trophic generalists. Adult honeypot ants (*Myrmecocystus*) feed on nectar, and larvae are fed insects scavenged and preyed upon by adults (Wheeler and Wheeler 1973). Some parasitic Hymenoptera (e.g., sphecids, pompilids) function as predators: adults eat pollen and nectar but capture prey to feed larvae.

Age structure generally contributes to resource generalization in taxa other than holometabolic insects. For equilibrium species that grow slowly through a "wide size range" (Polis 1984a), diet changes gradually as prey size increases with predator size (e.g., arachnids, mites, hemimetabolic arthropods; Polis 1984a, 1988a). For example, growing *Paruroctonus mesaensis* scorpions increase 60- to 80-fold in mass;

instar 2 scorpions eat prey that average 5 mm in length, whereas adults use prey averaging three times larger, two-thirds of which are different species. In fact, the differences in body sizes and resource use for the three age classes of *P. mesaensis* are equivalent to or greater than the differences between most real "biological species" (Polis 1984a). This magnitude of change is typical of hemimetabolic insects, larvae of holometabolic insects, and other arachnids. (In addition, foraging behavior, home range, microhabitat use, seasonal and diel patterns, and intra- and interspecific interactions change greatly during growth, at least in scorpions; Polis 1979, 1984a; Polis and McCormick 1987).

Overall, age structure and trophic opportunism allow PA to eat a wide variety of prey. Such trophic flexibility provides access to a wider variety of foods, an especially important option in deserts where food stress is both chronic and unpredictable (see below). Such diet generalization is probably an important mechanism that stabilizes predator populations. Further, age structure and trophic opportunism facilitate cannibalism and intraguild predation on other PA (Polis 1981; Polis and McCormick 1987; Polis, Myers, and Holt 1989; see below). Omnivory, predator-predator feeding links, and age structure also complicate both predator-prey dynamics (see below) and community level patterns, for example, guild (see above) and food web structure (Polis this volume).

Intermediate-Level Predators   PA are intermediate-level predators, themselves eaten by a variety of other predators. Smaller PA (including young age classes of larger species) are eaten by larger PA. Larger PA are eaten by vertebrate predators. All sizes are subject to parasitoids. These sources of mortality are a central feature in the evolution of these species. Many aspects of their behavior, ecology, morphology, and life history evolved to counter predation. To illustrate, coloration (cryptic and aposematic), noxious (repugnatorial) chemicals, and potent venoms are considered to be anti-predator adaptations (see McCormick and Polis 1990 for discussion of anti-predator traits of scorpions).

Several behaviors reduce the range of times and places that desert PA hunt prey. For example, nocturnality has been characterized as a tactic that avoids diurnally active predators such as birds and lizards (Cloudsley-Thompson 1975). Polis and McCormick (1987) present evidence that the spatial patterns and foraging phenology of small species and age classes of desert scorpions evolved to avoid predation by larger scorpions (see below; but see Bradley 1988).

Modification of activity may exert profound consequences on the evolution of life history traits. Some desert PA (scorpions, mygalomorph spiders, large scolopendrous centipedes) are "time minimizers" (Schoener 1971; Polis 1980a; Bradley 1986). Time minimization is a foraging strategy defined by short feeding periods, the duration of which lasts only until a set amount of energy is assimilated. The contrasting strategy is that of an "energy maximizer" in which energy assimilated is maximized during feeding periods, which are bounded only by the physical constraints of the environment. Schoener (1971) argues that time minimizers are forced into this strategy because increased foraging decreases fitness, for example, by increasing the probability of death from predation or harsh abiotic conditions. Essentially, time minimization is a low-risk, low-reproductive-gain strategy. It can exert a strong influence on the evolution of life history. Time minimizers exhibit many traits that characterize equilibrial species, for example, long times to maturity, high longevity, determinate growth, relatively constant brood size from year to year, and low reproductive capacity.

In later sections, we will show that the fact that PA are intermediate-level predators exerts profound influence on their distribution, abundance, and effectiveness as predators. In particular, intermediate predators (and prey in general) often restrict their foraging to certain times and places in which the risk of predation is relatively low. Foraging limited to enemy-free refuges can, in theory, alter the dynamics of the prey community (e.g., Gilliam and Fraser 1988). For example, depletion of food can occur in refuges; depletion can cause predators to switch to less-preferred prey or to leave their refuges to forage elsewhere under a higher risk of predation (Polis 1988b).

Factors Influencing Diversity, Distribution, and Abundance

Two major questions have dominated much of the ecological research over the last several decades. Hutchinson (1959) posed one: Why are there so many different species? Andrewartha and Birch (1954) framed the other: What factors limit the distribution and abundance of species? The answers to these questions lie at the interface between population and community ecology. Population ecology focuses on how abiotic and biotic factors impinge on a species to mold the number of individuals that exist at any one time and place. Community ecology considers the joint distribution and abundance of a set of species and what factors determine the number and identity of species that coexist. A major goal is to describe "community structure" and to determine the identity and strength of various factors that produce the observed structure. By community structure we mean such attributes as species diversity, the abundance of member species, their use of important resources (water, food), and their spatial and temporal distribution.

Questions posed by population and community ecologists have spawned similar debate over the extent that species interactions (especially competition) influence community diversity and the distribution and abundance of species. The distribution and abundance of a particular species can be strongly dependent or largely independent of other species. When distribution and abundance are independent, "community structure" reduces to the product of the individual autecologies of the summed set of species. When communities are dependent on biotic factors, we tend to think of them as co-evolved sets of strongly interacting species. This debate over the relative importance of autecological versus biotically interactive factors has been applied to desert communities (see Noy-Meir 1979–80; Crawford 1986; Polis, Wiens this volume).

There are seven general hypotheses about potential contributory factors to the community structure of desert PA (see Polis this volume). They are not mutually exclusive. (Hypotheses four, five, and six posit a community that is structured by biotic processes.)

1. Productivity: levels of diversity and abundance are controlled by the limited water supply and consequent low productivity in deserts.
2. Autecology: assemblages of species are the sum of the individual biologies of the member species. Those species whose autecologies (e.g., foraging biology, response to physical factors) allow them to be present in a particular time and place form the assemblage.
3. History and chance: species assemblages are a random draw of available species with a specific combination of species in any one area being a product of chance (e.g., climatic and biogeographic history, dispersal, random colonization, or extinction).
4. Resource limitation via exploitation competition: competition for limited resources (usually food, occasionally space or water) selects for species to diverge from one another in the use of resources. Such divergence decreases overlap and allows coexistence.
5. Interference competition: interference selects for subordinate species to avoid more dominant species in space and time. Such avoidance also decreases overlap and may allow coexistence.
6. Enemies (predators, parasites, and pathogens): mortality caused by enemies limits the distribution and abundance of species.
7. Heterogeneity in space and time.

Each of these hypotheses as applied to deserts is discussed by Polis (this volume). Unfortunately, we know little about how some of these factors influence desert PA. For example, the impact of historical factors is largely unstudied. Below we focus on how autecology, food limitation, and interactive biotic factors contribute to producing observed community patterns.

Autecology

In this section we discuss how community structure is affected by the manner that arthropods respond to the "harsh" abiotic conditions of the desert. First, life history shapes how populations respond. Some species (opportunistic) track perturbations (e.g., precipitation); other species (equilibrial) are relatively insensitive to short-term changes. Such differences in population-level responses to changes in productivity can exert a strong influence on the ability of these predators to limit prey populations. Second, the autecology of a taxon governs its success (diversity, density). Some taxa are extremely well adapted to the abiotic extremes of the desert; others, less so. These two factors in part determine the abundance and composition of community assemblages. Along a cline in aridity, we expect to see relatively mesic species become less abundant and finally drop out, whereas "desert" species will become increasingly more abundant. Present in the most arid areas are only those populations able to cope with the vagaries of the desert environment and only those species that possess the requisite behavioral and ecophysiological adaptations to withstand the extremes of heat and water stress.

In general, we should note that arthropods, as a group, are considered to be more susceptible to abiotic factors than vertebrates (Price 1975; Crawford 1981). We expect (and find) great fluctuations in populations of many desert arthropods as they respond to such events as precipitation (Andrewartha and Birch 1954; Noy-Meir 1974; Seely and Louw 1980; Crawford 1981, 1986; Louw and Seely 1982). (However, the populations of many desert PA do not fluctuate dramatically; see below).

Such fluctuations by desert biota prompted Noy-Meir (1973, 1979–80) to propose his "pulse and reserve" paradigm/hypothesis. This hypothesis states that pulses of water input promote productivity; this productivity is then stored as "reserves" by plants and animals in the form of propagules or tissue reserves. In its most stringent form, a correlate of this hypothesis argues that the diversity, distribution, and abundance of desert species are totally functions of nutrient and water input and not biotic factors (Noy-Meir's [1979–80] autecological hypothesis). A similar view that the distribution and abundance of arthropod populations are primarily (almost exclusively) determined by abiotic factors was espoused by Andrewartha and Birch (1954). This polemic view is clearly not correct, except possibly in the most extreme deserts (e.g., parts of the Atacama, Iranian, and Namib deserts). But it does serve to remind us that populations of desert arthropods (both prey and predators) are products, to a greater or lesser degree, of water availability and subsequent primary productivity. However, in the next section we see that desert PA in general are much less responsive (compared to nonpredaceous desert arthropods) to annual changes in precipitation and productivity.

Life History and Population Dynamics   The magnitude of population fluctuations in response to changes in the abiotic environment is a function of life history characteristics. Desert PA exhibit either opportunistic or equilibrial life histories (see earlier). These life histories influence population-level responses of both PA and their prey (see below).

Many desert (and non-desert) taxa of short-lived (<1 yr), opportunistic arthropods respond quickly (in the same season) to favorable conditions by rapid population growth (see references above). Populations of these species then decrease during unfavorable times. Consequently, densities are quite variable. We would expect that populations of some univoltine insect and acarine predators should increase markedly as a response to rains either in the same season or the following year. The increased productivity after rains should provide more prey to these predators, thus creating less food (and water) stress, increasing survivorship and population levels.

Surprisingly, we could find no evidence that any short-lived (or long-lived) desert PA increased dramatically immediately after rains. For example, PA from Namib sand dunes increased their densities only slightly or not at all following heavy rains (Holm and Scholtz 1980; Seely and Louw 1980). The total biomass of "carnivorous" arthropods only increased 34 percent above levels before rains; in contrast, herbivores increased 110 percent and omnivores (some of which are partially predaceous), 1,534 percent (Seely and Louw 1980). Riechert (1974) showed that populations of the desert *Agelenopsis aperta* (funnel-web) spiders did not increase after rains, nor did populations of the desert spider *Diguetia mohavea* increase (Sculteure and Polis unpublished data). In fact, rains destroyed many smaller *Agelenopsis* and *Diguetia*. Very little other data are available to test the speculation that the populations of short-lived PA opportunistically increase immediately after rains.

A similar paucity of data exists for annual changes in PA. Muma (1979) trapped several species of solpugids in arid grasslands; he found that some species fluctuated greatly from year to year (e.g., the number of *Eremobates* captured at the Hurley site varied from 9 to 74). However, other species varied little (e.g., *Eremobates* at the Lordsburg site varied from 32 to 59, with numbers in five of the six years ranging from 32 to 40). Muma notes that heavy rains actually depressed solpugid populations at Hurley because of extensive flooding.

Long-lived equilibrium species do not show population-level responses to either pulses of productivity or prolonged periods of poor conditions (Polis and Farley 1980). Scorpions exemplify the dynamics of such species. Examples are *Paruroctonus mesaensis* (Polis and Farley 1979, 1980; Polis and McCormick 1987) and *P. utahensis* (Bradley 1984, 1988). These species are intermediate in their size and longevity. Adult *P. mesaensis* weigh 2–5 g and individuals can live 3–5 years. Gestation lasts 10–14 months, and maturity, 18–24 months. Densities of *P. mesaensis* fluctuate little in spite of order-of-magnitude differences in seasonal and annual primary productivity and prey abundance. Over 4 years, the maximum population at 22 months of age was only 1.15 times greater than the minimum population; the maximum adult population in June was twice that of the minimum June population. Reproduction is also independent of yearly changes in productivity. Further, food supplementation failed to increase the number of young produced (Bradley 1984; Polis and McCormick 1987). Food deprivation (starvation for months) reduces embryo size but not number. Thus, *Paruroctonus* scorpions "wait out" low productivity and do not respond to high productivity.

It is likely that other long-lived desert PA (mygalomorph and other burrowing spiders, uropygids, large centipedes) exhibit similar life histories (Polis and Farley 1980). For example, large (2–5 g) heteropodid crab spiders (e.g., *Leucorchestris arenicola*) that burrow in the sands of the Namib Desert dunes are similar to scorpions in characteristics of their life history and population dynamics (Henschel in press). This spider develops slowly (24 months to maturity), produces relatively small broods (mean of 78 eggs) iteroparously (three times/year for several years), and provides extensive maternal care of offspring (young stay in maternal burrow for an average of 75 days). Burrowing mygalomorphs in Australian deserts are also large (3 to >5 g), take 5 to 8 years to mature, and several species can live more than 20 years (Main 1978, 1981). Species of *Geolycosa* (burrowing wolf) spiders in the Chihuahuan (Conley 1985) and Australian (Humphreys 1988) deserts are other examples of relatively large PA that live several years. We predict that all these taxa consist of populations that are relatively stable, being influenced relatively little by seasonal and annual changes in productivity.

Ecophysiology and Individual Adaptations   Responses on the individual level to abiotic factors also influence community structure by promoting differential success among taxa. Certain taxa are well represented in deserts because they possess a suite of characteristics that is particularly well adapted to desert conditions. Again, we use scorpions to illustrate.

Scorpions are one of the most important and successful groups of desert predators in terms of their diversity, densities, and biomass (Tables 7.3 and 7.4). Such success can be attributed to several autecological features. Scorpions can withstand higher temperatures than any other arthropod (Hadley 1990). Their rate of water loss is the lowest published for any arthropod (Hadley 1990; however, Namib pseudoscorpions are recorded to exhibit even lower water loss; L. McClain personal communication). Finally, their metabolic rates are among the lowest in the animal kingdom, an order of magnitude less than insects and slightly lower than spiders (McCormick and Polis 1990; see Table 7.5). Such low energy needs in concert with production efficiencies the highest ever recorded (68 percent; see Table 7.6) allow scorpions to function at extremely low rates of feeding (1–5 percent of individuals/night; McCormick and Polis 1990). Scorpions function normally under chronic and intermittent food stress, resisting starvation for 2–12 months (Polis 1988b). These physiological traits do much to explain the great success of scorpions in deserts.

Autecology is a key factor in understanding the success of other taxa of desert PA. For example, the diversity of spiders (Table 7.2) in arid areas has been attributed to such traits as low metabolism, high production efficiency, and thermoregulatory abilities (Cloudsley-Thompson 1983; Riechert and Lockley 1984; Riechert and Harp 1987; Humphreys 1988). Spiders are particularly adapted to feast-and-famine cycles of deserts. They possess a number of traits that allow them to withstand long periods of starvation and to increase their body weight 33–100 percent by gorging on available prey. The distribution and abundance of desert species are explained, in part, by their physiological and life history adaptations (Main 1978; Humphreys 1988).

However, the autecology of certain taxa may exclude them from deserts. For example, species that originate in more mesic areas may not possess the adaptations necessary for efficient water conservation (e.g., many amphibians, some birds and mammals; see Wiens, Woodward and Mitchell this volume). Orb-weaving spiders,

Table 7.5   Metabolic rates of various animal taxa.

Taxon	Metabolic Rate $(\text{ml } O_2 \cdot g^{-1} \cdot hr^{-1})$
Homeotherms (basal rates)	
Mammals	0.07–7.4
Shrew	7.4
Rodents	1.80
Elephant	0.07
Birds	2.3–4.7
Poikilotherms (at 25°C)	
Insects[a]	1.665 ± 1.25
Spiders[b]	0.92 ± 0.92
Scorpions[c]	0.057 ± 0.048

SOURCE: Data taken from many sources (see Yamashita & Polis n.d.).
[a]Sample size: 82 species.
[b]Sample size: 8 species.
[c]Sample size: 7 species.

Table 7.6 Ecological (production) efficiency of various animal taxa.

Group	Production Efficiency[a]
Insectivorous mammals	0.86
Birds	1.29
Small Mammals	1.51
"Other" mammals	3.14
"Homeotherms"	3.1
Fish	9.77
Social insects	10.3
Terrestrial invertebrates[b]	25.0
Solitary herbivorous insects	40.7
Solitary detritivorous insects	47.0
Solitary carnivorous insects	55.6
Spiders	45–60
Scorpions	68.2

SOURCE: Humphreys (1979); McCormick & Polis (1990).
[a]Production efficiency = proportion of assimilated energy that is incorporated into new biomass (production/assimilation ratio).
[b]Does not include insects or arachnids.

not well represented in deserts, are not particularly well adapted for water conservation and require a more complex vegetational architecture for web placement than is often available in deserts.

Autecological adaptations to abiotic factors also influence local distribution and abundance of desert species. For example, the biogeographic and microhabitat distribution of the desert iguana (*Dipsosaurus dorsalis*) is determined by moisture levels in the soil in which their eggs incubate (Muth 1980). Although no examples exist for PA, similar factors limit other desert arthropods, for example, desert grasshoppers (Andrewartha and Birch 1954).

Edaphic factors also limit distribution. Many species of desert scorpions and spiders are limited to certain substrate types (Polis, Myers, and Quinlan 1986). For example, psammophilic species of scorpions and spiders occur only in sand; they cannot burrow in harder soils (Polis 1990; Henschel in press). Lithophilic species are never found away from large rocks. Congeneric species of *Opisthophthalmus* segregate spatially on a scale of meters according to differences in soil hardness (Lamoral 1978). Other features influence local densities. For example, lower densities of scorpions occur in alluvial areas, particularly after floods (Bradley 1983). The abundance of some spiders (e.g., pholcids and theridiids) is limited by the number of abandoned burrows available (Fowler and Whitford 1985; Polis personal observation). All artificial cavities added to this habitat are rapidly colonized by these spiders. Burrows presumably supply a refuge from heat and predators.

In general, we speculate that the number of hiding places (refuges) or specialized foraging sites (e.g., shrubs for some web spiders) may place limits on the abundance of some desert arthropods, including PA (see also Riechert and Harp 1987). For example, manipulation of the architecture and foliage density of sagebrush (*Artemisia tridentata*) shrubs significantly altered the abundance and local species composition of desert spiders (Hatley and MacMahon 1980; see also Chew 1961). The number of suitable homesites is a major factor limiting the abundance of *Agelenopsis aperta* spiders (Riechert 1981). Muma (1980, personal communication) found that some species of solpugids use surface debris (e.g., boards) placed in the desert as refuges. Populations of the scorpion *C. exilicauda* increase with the addition of more home-sites (C. Myers personal communication). More experiments are needed to deter-mine if suitable homesites and physical refuges are resources in short supply and thereby limit the populations of desert PA.

Thus, abiotic factors obviously are important in shaping communities. It is against this environmentally established background that biotic factors have the opportunity to contribute to community patterns. Species-specific traits (life history and physio-logical adaptations) help govern the distribution and abundance of desert species. Such autecological factors screen species membership in assemblages that exist in particular places and times.

Heterogeneity in Space and Time

Heterogeneity in space and time as a category contains many processes and events that combine with other factors to affect the structure of desert communities (see Polis this volume). For example, processes promoting temporal heterogeneity (e.g., droughts, storms, freezes; see next section) may restrict the distribution and/or abundance of certain species to a greater degree than of other, more adapted species. Such disturbances may promote coexistence by not allowing sufficient time for biotic processes (competition and predation) to proceed to the point where species are lost or excluded from a community (the intermediate disturbance hypothesis). Similarly, enemies may differentially exploit competitively superior species and therefore allow coexistence by preventing exclusion of inferior species. Spatial heterogeneity provides different microhabitats, thus allowing coexistence if species are differen-tially adapted to these niches. These effects are probably quite marked in deserts, one of the most heterogeneous habitats on the planet (Polis this volume).

Historical Factors and Chance

History, both recent and distant, influences community structure. Geological, climatic, and biogeographic events combine with the vagaries of dispersal, coloniza-tion, and extinction to supply sets of different species (each with its distinct biology) to PA assemblages that occur in one place at one time. For example, paleohistorical events have shaped Australian deserts so they exhibit a high diversity of burrowing spiders, a low diversity of scorpions, and the complete absence of solpugids. Such events produced scorpion faunas in Australia and southern Africa characterized by a high proportion of fossorial species that forage from their burrows; in North America, a taxon absent on the other two continents (the Vaejovidae) has radiated such that most American species forage primarily on the surface. Such differences in diversity and macrogeographic distributions present the possibility of "vacant niches" and niche expansion by existing species. No research comparing the structure of assemblages of PA in different deserts has been conducted. Such research, conducted on ants, rodents, birds, and lizards, has been fruitful (Polis this volume).

More recent historical events are also important, particularly as they affect local assemblages. For many PA, the desert is best viewed as a series of patches, largely isolated from one another yet connected in various degrees by dispersal. These patches are often refuges from physical conditions (e.g., abandoned burrows) or "hotbeds" of primary and secondary production (e.g., runoff channels). Plants can be viewed as islands of productivity with high diversities of prey and PA; for example, 60 herbivores and 26 spider species occur on creosote (*Larrea tridentata*) (Schultz, Otte, and Enders 1977); more than 100 arthropod species and more than 25 spider species occur on saltbush (*Atriplex canescens*) (Polis and Sculteure unpublished data); 67 species (30 predators/parasitoids) occur in galls of *Atriplex* (Hawkins and Goeden 1984); and more than 200 arthropod species occur on mesquite (*Prosopis glandulosa*) (Wisdom this volume).

The composition and abundance of member species in these habitat islands are functions of colonization, extinction, and stochastic physical events. Thus, similar patches may support different assemblages. Such physical factors as extreme temperatures, wet soils, fires, flash floods, and severe wind (sand) storms can produce considerable mortality in local populations of scorpions (references in Polis 1990). For example, entire populations of *Uroplectes* scorpions in southern Africa were exterminated by local fires (Eastwood 1978). An abnormal prolonged freeze in the deserts of central Baja California decimated local populations of the scorpion *Centruroides exilicauda* (Polis unpublished data). Heavy rains and flooding produced a marked decline of localized populations of *Paruroctonus utahensis* scorpions (Bradley 1983) and solpugids (Muma 1979). Protected populations and species are unaffected, for example, populations of *Uroplectes* in unburned patches and *Centruroides* on thermally buffered beaches bordering the Sea of Cortez. Neither *Uroplectes* nor *Centruroides* dwell in burrows; burrowing species were totally unaffected by the fire and freeze. Similarly, floods differentially influence populations along channels with and without heavy runoff.

Thus, not all suitable patches contain the full array of local species capable of existing within the patch. A good example of how such irregular local distributions affect the structure of PA assemblages occurs with trees as patches in the Namib Desert (Polis and Seely unpublished data). Two arachnids, the scorpion *Uroplectes otjimbinguensis* and the spider *Gandanimeno eresus* (Eresidae), live under loose bark on larger *Acacia erioloba* trees and are the major arboreal predators on such trees. Trees grow along the banks of dry rivers and become less dense and more sporadic with increasing distance from the riverbed. The local abundance of arachnids on each tree is a function of differential dispersal, extinction, and a predator-prey relationship between the two species. Although both species are found on river trees, *U. otjimbinguensis* is, with no exceptions, the numerically dominant species (20 to >50 scorpions/tree) and *G. eresus* is relatively uncommon (5–20/tree). This occurs because scorpions are effective predators on *G. eresus* spiders (see below). However, this outcome is more variable on isolated trees farther from the river and along smaller washes entering the river. On some trees close to the river, the abundance of scorpions and spiders is similar to that found in the river. However, some trees have no scorpions and great numbers of spiders (50–400/tree), some trees have neither scorpions nor spiders, and some no scorpions and few spiders (<20). These distributions exist because neither *U. otjimbinguensis* nor *G. eresus* disperse far from the river, yet spiders disperse farther than scorpions. Dense spider populations occur only in more isolated trees where scorpions are absent. In trees quite distant from the river, neither species occurs. Thus, differential dispersal and semi-deterministic biotic interactions combined with differing isolation of patches are major determinants of the distribution and abundance of these species. These distributions then influence the distribution and abundance of their prey (see below).

Caswell (1978) modelled a similar situation and argues that historical and stochastic dispersal events in patchy environments are a paramount factor explaining the distribution and abundance of predators and their prey and species of competitors. Such conditions can produce local extinctions via deterministic biotic interactions, but promote global coexistence. We suspect that such situations are normal among PA living in the notoriously heterogeneous desert. Many such assemblages occur in patches; differential dispersal, local extinctions, and "hide-and-seek" dynamics are undoubtedly extremely important in determining the exact structure of local assemblages. Unfortunately, little research has focused on these processes.

Food Limitation, Exploitation Competition, and Relative Food Shortages

Individuals of different species interact in a number of ways that may influence their distribution and abundance: as food, consumers, competitors for similar resources, or via commensalism and mutualism. Consumers may be predators, parasitoids, or parasites. We limit this discussion to the effects of predation and competition on the community structure of desert PA. We first discuss food limitation, a concept central to understanding the effects of both these processes.

At a population level, food limitation produces a rate of population increase (r) less than the maximum capacity to increase (r^{max}) due to decreased reproduction or survivorship. However, since many other factors affect r^{max}, ecologists generally focus on individual-level effects of food limitation. On this level, a population is thought to be food limited if growth or reproduction proceeds at less than maximum because individuals do not ingest sufficient nutrients and energy. Hairston, Smith, and Slobodkin (1960) proposed, as a general rule, that terrestrial predators are limited by food. Plants are abundant in the "green world"; herbivores are limited not by food but by predators; predators are limited by the amount of food that they capture.

It is likely that the vast majority of all predators are limited by food availability, at least seasonally and at sites that are less productive or that exhibit relatively high average temperatures. The high ambient temperatures of deserts increase the metabolic rates of poikilothermic PA, thus increasing energy requirements, further exacerbating the effects of low food availability, and making food limitation more probable (see Benton and Uetz 1986; Humphreys 1988). It is well known that the abundance of arthropod prey fluctuates dramatically during the year (Noy-Meir

1974; Pietruszka 1980; Seely and Louw 1980; Polis 1988b, this volume). Typically, prey abundance peaks in spring with a smaller peak in fall. Mid-summer and mid-winter are periods when prey are particularly scarce. Presumably food is limiting to PA, at least during these unproductive periods (food may or may not be limiting during productive periods). Similar arguments apply to spatial differences in food availability along the geographic and microhabitat distribution of a species's range.

Food limitation for PA in time and space is illustrated with scorpions and spiders. Much evidence indicates that food levels affect the performance of desert scorpions (specifically, *P. mesaensis*; Polis 1979, 1988b; Polis and McCormick 1987a); percent feeding, growth rate, and body size vary between years, months, and sites with changing levels of prey abundance. In the studies cited, scorpions grew faster, larger, and fed significantly more often at more productive sites. Further, the percent feeding was significantly correlated with prey availability, and differed more than sixfold between the lowest and highest levels of prey availability. Finally, feeding experiments show that the rates of embryological development (but not clutch size) are functions of food availability (Bradley [1984] shows similar results for the Chihuahuan Desert *P. utahensis*).

Humphreys (1988) and Wise (1984) review the extensive evidence that food frequently limits growth and reproductive output of spiders in various habitats. It is reasonable to speculate that food limitation should be particularly acute in desert species, especially during periods of low productivity. Riechert (1981) and Riechert and Harp (1987) present many lines of evidence to show that food limits those desert funnel web spiders (*Agelenopsis aperta*) that inhabit more arid areas. The prey available to three desert populations of *A. aperta* was sufficient to fulfill optimal energy requirements on only 9, 13, and 17 percent of all days; in contrast, prey available to more mesic populations of *A. aperta* was sufficient on 31 and 96 percent of the days. Desert orb-weaving spiders, *Metepeira spinipes*, were smaller, less fecund, and produced less total "reproductive biomass" than those from more mesic areas (Benton and Uetz 1986). These authors argue that higher temperatures in the desert are suboptimal and cause more energy to be channeled toward respiration and less toward growth and reproduction. Further, prey availability was lowest in desert populations of *M. spinipes*. Most desert *Geolycosa* spiders are apparently food limited, for example, the Australian *G. godeffroyi* (Humphreys 1973) and the Sonoran *G. carolinensis* (Shook 1978; fewer than 0.5 percent of all *G. carolinensis* were observed feeding). However, Conley (1985) experimentally demonstrated that food did not limit survival of adult *G. rafaelana*, a burrowing spider in the Chihuahuan Desert.

Two major factors account for food limitation (Polis and McCormick 1986a, 1987). First, inherent limitations in the foraging biology of the consumer may not allow sufficient capture or ingestion of food. Consumers may not have enough time to forage, may not be efficient under all concentrations of prey, or may not be able to handle enough prey to provide the energy necessary for maximum growth or reproduction. Foraging time limitations may be exerted by constraints that are either abiotic (e.g., high temperatures) or biotic (e.g., predator avoidance; see below). Such inherent limitations in foraging biology were described by Andrewartha and

Birch (1954) as "relative food shortage" (see also Andrewartha and Browning 1961; White 1978).

Alternatively, food limitation may result when exploitative competitors deplete the supply of food. Exploitation competition is the negative effect exerted by one individual or population on a second due to a differential ability to harvest food. Two factors are necessary and sufficient to demonstrate exploitation competition in nature. Food levels must affect growth, survival, and/or reproduction ("food limitation condition"), and lowered performance or abundance of species A must be mediated through a change in food availability clearly due to the presence or absence of species B. This "resource depletion condition" would occur if predation or harvesting of food by species B sufficiently decreased food levels, thus suppressing the reproduction, growth, and/or percent feeding of species A.

Although food is apparently limiting to PA in the desert (at least to scorpions and spiders), evidence suggests that these taxa normally do not compete for food. Field experiments with desert scorpions (Polis and McCormick 1986a, 1987) and with spiders from mesic habitats (Wise 1975; Horton and Wise 1983; Riechert and Cady 1983; Turner 1983; see Wise 1984 for a review) failed to demonstrate exploitation competition. In fact, the only evidence that we could find in the literature that different species of terrestrial PA compete for food comes from Spiller's (1984, 1986) experiments on coastal orb-weaving spiders, and research on facultatively predaceous desert ants (however, ants compete for seeds; see MacKay this volume).

Thus, although many conditions (food limitation, dense populations, niche partitioning) could be interpreted to suggest that PA compete for food, apparently they do not, at least not via differential harvesting of food (i.e., exploitation competition). Note that competitive effects due to food exploitation and depletion may occur on a very local scale within patches (see above) when one individual captures food, making it unavailable to its neighbors. Although it is reasonable to believe that this may occur, especially within dense populations of PA (e.g., burrowing arachnids), research has not focused on exploitation competition among PA at this scale. However, evidence indicates that PA often interact/compete more directly, via the extensive use of various forms of interference.

Interference Competition

Competition via interference is much easier to document than that via exploitation because discrete acts usually are performed. Interference among PA commonly manifests itself both among and between species as aggression, territoriality, and predation. The importance of these interactions to population dynamics and community structure is largely unstudied. However, the intensity of some types of interference suggests that they may exert significant influence.

Interference, both intra- and interspecific, tends to limit populations of subordinates (species and age classes; see below). In some systems, it is likely that interference keeps populations below the point where strong effects of exploitation competition occur. This has been suggested to operate among desert spiders (Riechert and Harp 1987) and scorpions (Polis and McCormick 1987).

Aggresive interference between the black widow *Latrodectus hesperus* and the pholcid *Artema atlanta* is a major factor shaping the distribution and abundance of desert spiders living in natural cavities (e.g., rodent burrows; Fowler and Whitford 1985). Groups of *A. atlanta* behaviorally dominate solitary *L. hesperus* by cutting their webs and otherwise restricting web construction. Fowler and Whitford suggest that cavities are the major factor limiting populations of these spiders. Similarly, another pholcid (*Holocnemus pluchei*, introduced from Mediterranean areas) interferes with and displaces local populations of *L. hesperus* in arid areas of Southern California (W. Icenogle personal communication).

Such aggression for space intergrades into territoriality. In over a decade of work, Riechert (1981; Riechart and Harp 1987) has shown the importance of territoriality as a major mechanism of population limitation among funnel web *Agelenopsis aperta* desert spiders. She maintains that the frequent occurrence of regular distribution among individuals of many species suggests that territoriality is relatively widespread among spiders.

Territoriality is maintained either by contests and fights (e.g., by *A. aperta*) or via predation on conspecifics (cannibalism) and heterospecifics (intraguild predation). For example, territorial boundaries between ant colonies of the same and different species are maintained by "wars" that include fighting, killing, and predation (for desert species see MacKay, this volume; Ryti and Case 1986, 1988; in general, see Polis, Myers, and Holt 1989). Desert ants also kill new queens that attempt to establish colonies within the boundaries of their territories.

Predation among potentially competing desert PA commonly occurs among spiders, scorpions, and parasitoids, between unrelated arachnids, and between arachnids and predaceous insects. This predation (like most forms of interference) is usually a function of size: larger individuals (species and age classes) are aggressive and territorial dominants or eat smaller individuals. For example, in more than 150 observations of scorpion-scorpion predation and more than 500 cases of scorpion predation on other PA, the predator was always larger (Polis 1988a, unpublished). Predation among PA may be relatively more frequent and important in deserts than in other habitats; many desert PA are long-lived and their populations are characterized by several discrete age classes (see above). Age/size structure facilitates cannibalism and intraguild predation and makes all species of PA vulnerable to predation by other PA, at least as juveniles (Polis 1981; Polis, Myers, and Holt 1989). In the next section, we will show that such predation can act as a major source of mortality capable of shaping much of the behavior and ecology of species and age classes of PA.

The frequency of within-group predation in spiders and scorpions is high (see Polis, Myers, and Holt 1989). Humphreys (1988) provides an estimate of spider-spider predation in 11 non-desert species: other spiders formed 2.5–38.2 percent (average = 19 percent) of the diet (cannibalism = 0–15 percent; intraguild predation = 1.3–18.2 percent). In deserts, spider-spider predation is common among and within species of heteropodid crab spiders in the Namib Desert (Henschel in press) and among many species in the Coachella Valley desert (Polis this volume).

Similarly, scorpions frequently eat each other; in a Polis and McCormick study

(1987), scorpions formed 12–46.7 percent of the diet of four desert species (cannibalism = 4–33.3 percent; intraguild predation = 8–21.9 percent). Scorpion-scorpion predation is frequent in a number of other assemblages of desert scorpions (e.g., Polis, Sissom, and McCormick 1981; Bradley 1983; Bradley and Brody 1984; Polis and McCormick 1987; Polis 1990).

Predation also occurs among unrelated PA in many deserts. For example, in the Namib, *Leucorchestris arenicola* (Heteropodidae) spiders eat scorpions, solpugids, and arthropodivorous geckos (Henschel in press); arboreal *Uroplectes* scorpions eat several potential competitors (smaller spiders and solpugids; Polis and Seely unpublished). In the deserts of Southern California's Coachella Valley, *Diguetia mohavea* spiders eat 14 families of predatory/parasitoid insects (clerids, mantids, reduviids, nabids, phymatids, asilids, mutillid and tiphiid wasps, ants, adult antlions, chrysopids, carabids, melyrids, staphylinids) and eight spider species (Sculteure and Polis unpublished data). Predators constitute 54 percent of the diet of Coachella *L. hesperus* (three scorpions, three solpugids, three spiders, eight insects). The diet of the Coachella scorpion *Paruroctonus mesaensis* consisted of 8.1 percent spiders and 14.4 percent solpugids (Polis and McCormick 1986b). In total, 47 percent of this scorpion's diet was PA and parasitoids (59 species; see Table 13.3 in Chapter 13). Overall, the diet of three scorpions and three spiders in the Coachella averaged 51.5 percent predators and parasitoids.

Parasitoids likewise attack other parasitoids via facultative hyperparasitoidism. For example, *Photopsis* (Mutillidae) and other hymenopteran parasitoids commonly feed on one another (Ferguson 1972; see Figure 13.4 in Chapter 13). Many of the 26 species of parasitoids and the four insect predators within galls on saltbush (*Atriplex*) feed on each other (Hawkins and Goeden 1984; see Figure 13.3 in Chapter 13).

Looping mutual predation is common among PA in the Coachella: scorpions eat spiders and solpugids; spiders eat scorpions and solpugids; solpugids eat scorpions and spiders; parasitoids prey on one another; and spiders eat the parasitoids and egg predators that parasitize/eat these same spiders (Polis and McCormick 1986b; Polis, Myers, and Holt 1989; see Figures 13.4–13.6 in Chapter 13). For example, *Latrodectus* falls prey to three species which it eats (mutual predators are *Steatoda grossa*, *S. fulva*, *P. mesaensis*, and other *L. hesperus*). *Diguetia* is involved in mutual predation with several species of salticids, a clerid, and *Mimetus*, an araneophagous spider.

Changes in relative size among species promote mutual predation among desert PA. Henschel (in press) provides several examples of age-dependent mutual predation between Namib spiders, scorpions, and lizards. In the Coachella, age-dependent loops were observed among 10 pairs of species (see also Bradley 1983 for looping among Chihuahuan Desert PA). Thus, young *P. mesaensis* scorpions are eaten by relatively larger adult *P. luteolus* (6.7 percent of its diet) and *Vaejovis confusus* (8 percent of its diet); adult *P. mesaensis* prey on the (now) relatively smaller adults of these species (*P. luteolus*, 0.5 percent of its diet, *V. confusus*, 8.3 percent of its diet; Polis and McCormick 1987). Loops also can occur independent of age structure: black widow spiders (*L. hesperus*) catch these Coachella scorpions by using web silk to pull them off the ground; black widows traveling on the ground are captured by the same scorpions.

Impact of Predation on Community Structure

It is well known that predation can greatly influence the distribution and abundance of prey. Desert PA are all intermediate-level predators that must both capture food and avoid predation from a diverse suite of vertebrate and invertebrate arthropodovores. The intensity of predation on PA prompted Schoener and Toft (1983) to suggest that intermediate predators may be an exception to Hairston, Smith, and Slobodkin's (1960) generalization that all predators compete for food. Schoener and Toft suggested that (vertebrate) predation was the major factor limiting spider populations.

Predation influences the community structure of PA in several ways. First, it decreases their populations; in theory this may make more food available and lessen or eliminate competition (both exploitation and interference). Predation also modifies foraging by causing intermediate-level predators to seek refuges or modify foraging activity or phenology to avoid predation. Such behavior often puts intermediate predators in suboptimal times and places; in theory, this decreases their food intake and, in some cases, may increase competition by locally depleting food in refuges. Finally, the time minimization foraging/life history strategy (see above) has been attributed to predation (Schoener 1971; Polis 1980a; Bradley 1988). Thus, predation, by promoting the evolution of time minimization, indirectly functions to decrease interactions (exploitation and interference) among PA.

Evidence, although limited, strongly suggests that predation is a potent force affecting the distribution and abundance of desert PA. In general, predation by vertebrates (Rypstra 1984), lizards (Spiller and Schoener 1988), birds (Askenmo et al. 1977; Gunnarsson 1983), and other spiders (Riechert and Cady 1983; Turner 1983) is a major mortality factor known to greatly reduce spider populations. Several studies show that predation likewise causes major reductions among desert spiders. Bird exclusion reduces mortality by 30 percent in natural populations of *A. aperta* (Riechert and Harp 1987). Predation by gerbils caused 64 percent of known deaths of Namib Desert *Leucorchestris arenicola* crab spiders (Henschel in press). Sporadic foraging on sand dunes by riverine chacma baboons also causes substantial mortality to local populations of *L. arenicola*. A great variety of vertebrate predators feed on desert scorpions (Polis et al. 1981) and on desert insects (Polis this volume). PA formed an average of 41 percent of the diet of 14 desert vertebrates (birds, lizards, and mammals; see Table 13.3 in Chapter 13).

Arthropod predation on other PA is perhaps an even greater source of mortality than that by vertebrates. It is a very frequent interaction because of the size disparity between age classes and species and because most PA are generalists that choose any prey that they can subdue, regardless of the prey's trophic position (see above and Figures 13.4–13.6 in Chapter 13 for many examples from the Coachella Valley).

Intraspecific predation is likely ubiquitous and may be particularly important among populations of desert PA that live in dense local patches. It is the major mortality factor among many scorpion populations and may regulate the population of *P. mesaensis* (Polis 1980b, 1990; Polis and Farley 1980). Cannibalism produced 19 percent of all known deaths in *L. arenicola* (Henschel in press). *Photopsis* parasitoids cannibalize and possibly self-regulate each other (Ferguson 1972).

Parasitoids also can kill many spider and insect predators. For example, a pompilid wasp killed 50–65 percent (winter) to 4–5 percent (summer) of all adult female *Geolycosa rafaelana* (Conley 1985). Populations of the Namib Desert heteropodid crab spider *Carparachne aureoflava* are subject to intense predation by pompilid wasps (Henschel n.d.).

In some desert communities, predation by scorpions on spiders is a key determinant of spider densities. For example, populations of the spider *Gandanimeno eresus* are severely reduced when they co-occur locally on *Acacia* trees with the scorpion *Uroplectes otjimbinguensis* (see earlier; Polis and Seely unpublished). Additions of *Uroplectes* over a 1-year period to scorpion-free trees reduced *G. eresus* populations to 42 percent of that on control trees. Removal of scorpions produced a 2.9-fold increase in *G. eresus* as compared to controls. In the Coachella Valley, removal of more than 6,000 individual *P. mesaensis* scorpions from 300 (100 m²) plots over a 29-month period produced a more than twofold increase in the number of spiders (Polis and McCormick 1986b). The authors determined that changes in the level of (intraguild) predation—versus exploitation competition—were the key process in both the Namib and Coachella spider populations.

Scorpion populations are also limited by intraguild predation from heterospecific scorpions (Polis and McCormick 1986a, 1987). Intraguild predation by *P. mesaensis* killed 8 percent and 6 percent of two smaller species (*P. luteolus* and *Vaejovis confusus*, respectively). Populations of these uncommon scorpions responded to the removal of *P. mesaensis* in the above experiments by increasing significantly compared to controls (*P. luteolus*, more than 600 percent; *V. confusus*, 135 percent). Again, increases were due to the cessation of predation, not exploitation competition.

Since predation is apparently a key factor in the population dynamics of these species of PA, natural selection is expected to favor adaptations that reduce the probability that an individual will encounter its predator. Indeed, prey often avoid places and times that their predators frequent or where the probability of predation is high (see references in Polis and McCormick 1987). For example, the surface activity of several species of scorpions is relatively low on nights with a full moon, probably because the risk of predation by vertebrates is higher at such times (Warburg and Polis 1990).

Prey also avoid predation from potential competitors (i.e., cannibalism and intraguild predation). The temporal and spatial distribution of smaller age classes and species often reflect avoidance (in ecological or evolutionary time, or both) of larger age classes and species (Polis 1981; Polis, Myers, and Holt 1989), for example, Namib spiders (Henschel in press, personal communication) and North American desert scorpions (Polis and McCormick 1986a, 1987; Polis 1990; but see Bradley 1988). Typically, the large predatory entity (species or age class, or both) occurs in productive periods and microhabitats, whereas smaller entities coexist by spatial segregation in a heterogeneous habitat and by temporal displacement.

Coachella Valley scorpions illustrate such shifts. The overall distribution of *P. luteolus* and *V. confusus* tends to place these species on the surface during times (in winter, late fall) and in places (off sand) characterized by relatively low surface populations of adult *P. mesaensis*. These times and microhabitats support significantly less prey than those used by adult *P. mesaensis*; consequently, *P. mesaensis* has a feeding rate (2.95 percent) significantly greater than that of all other species combined (1.70 percent). Further, the minority of *P. luteolus* and *V. confusus* that forage when and where *P. mesaensis* is active suffer a disproportionately greater chance of being eaten by *P. mesaensis*. Intraspecific predation has produced similar patterns of temporal distribution, feeding, and mortality patterns among age classes of *P. mesaensis* (Polis 1980b, 1984). Many other assemblages of desert scorpions exhibit patterns that suggest that scorpion-scorpion predation is a major process shaping distribution and abundance (Polis and McCormick 1987; Polis 1990; but see Bradley 1988).

We suspect that similar processes and patterns occur among many assemblages: PA are generally intolerant of each other and most are nonspecialists that will capture any smaller prey including other PA. Cannibalism and intraguild predation affect population dynamics and community structure primarily when a relatively large species is particularly abundant.

Resource Partitioning and Guild Structure

A major thrust in community ecology from the 1960s to the 1980s was to describe how species differed from one another. Not surprisingly, we learned that species differed in their use of critical resources, were separated ecologically in their temporal and spatial patterns, and/or exhibited important differences in trophic organs or body size. Indeed, species of desert PA show such differences (e.g., scorpions: Polis and McCormick 1986a, 1987; Polis 1990; spiders: Chew 1961; Hatley and MacMahon 1980; Abraham 1983; predaceous ants: Chew and DeVita 1980). One of the major assertions of the exploitation competition paradigm in vogue at that time was that such competition produced the observed resource partitioning, niche segregation, and character displacement. However, pattern does not imply process, and data on niche differences are not sufficient evidence for exploitation competition.

Several other processes can produce the observed differences. Most simply, autecological tracking of resources by consumers could produce differences because each species adapts to a different set of resources (for a good example, see Bloom 1981). Predation also can produce similar patterns, as shown by Holt's (1984) work on "apparent competition." Finally, as shown for cannibalism and intraguild predation, interference can produce niche differences (see above). In fact, because such extreme forms of interference actually cause death, selective pressure to avoid being eaten is much greater than if the interaction were either exploitation competition or a milder form of interference (Polis 1988).

We suggest that many of the niche differences among desert PA are products of natural selection to avoid interference. Crawford (1986) also suggested that desert PA tend to "space optimize," thereby reducing interference competition.

Finally, we note that studies of niche differences are of limited value unless age classes are incorporated (see Polis and McCormick 1986a). Although spatial and temporal differences may allow individuals to use different prey, body size differences may not divide prey except among different sized adults. However, because adults of big species must grow from a small size at birth, all species overlap in size (and associated size-related prey use) during some parts of their lives. Such "developmental overlap" is generally unrecognized and is difficult to study. Nevertheless, it greatly limits the degree that size differences divide resources among PA (Polis 1984a, 1988a; Polis and McCormick 1986a). (Note that size is paramount to interference.)

The Role of Predaceous Arthropods in Desert Communities

From the previous discussion on the diversity, abundance, and biomass of PA, it is obvious that PA play a potentially important role in desert communities. "Ecological importance" can be expressed in terms of either energetics and nutrient cycling (the quantity of matter and energy flowing through a population or functional group) or regulation of community structure (the impact on diversity, distribution, and abundance of other populations). PA can influence desert communities in many ways: as predators affecting characteristics of their prey, as prey of other predators, and as competitors to other arthropodivores.

Energetics and Nutrient Cycling

The importance of desert PA to energy and nutrient cycling is a function of the quantity of prey biomass captured. The amount of biomass captured is a function of the density, population biomass, metabolism, and efficiency of energy transfer. Several characteristics of PA suggest that they could be a major link in the flow of energy through desert communities. Earlier, we showed that PA are speciose, that individuals and species form a relatively high proportion of all arthropods, that many PA are quite dense, and that PA often maintain population biomasses high relative to other taxa (see Tables 7.3 and 7.4).

However, two characteristics lessen their importance. First, they exhibit the highest ecological (production) efficiencies (percent assimilated energy incorporated into new biomass) of all taxa that live in the desert (Table 7.5). Second, they exhibit metabolisms that are low (spiders) to extremely low (scorpions) relative to other poikilotherms and endotherms (Table 7.6). Although these features are powerful adaptations that allow efficient use of food and partially explain the success of PA in deserts, they also function to decrease the amount of energy transferred. Thus, a gram of PA does not process as much food as a gram of arthropodivorous vertebrate. Overall then, desert PA are less important in energy and nutrient cycling than their diversity, abundance, and biomass suggest.

How much energy do they process? We have two estimates, both for desert scorpions. Polis (1988a) calculated that average populations of *P. mesaensis* used more than 9,000 grams of prey $\cdot$ ha^{-1} $\cdot$ yr^{-1}. The Australian *Urodacus yaschenkoi* requires 7,900 g $\cdot$ ha^{-1} $\cdot$ yr^{-1}; this translates to 98,400 ants or 31,570 medium-sized spiders eaten per hectare per year (Shorthouse 1971; Marples and Shorthouse 1982). It is uncertain exactly how such figures for individual PA or for the sum of all PA compare to those for vertebrates. We do know that spiders in mesic communities are one of the most important energy processors of animal tissue (Van Hook 1971; Moulder and Reichle 1972). This is an interesting question to pursue.

Finally, it appears paradoxical that PA are so abundant relative to their prey (a similar situation—an inverted biomass pyramid—occurs in some aquatic systems). Such predator:prey ratios occur if there is a high turnover in prey, if predators show high energetic efficiencies processing prey, or if predators feed on other predators in their own "trophic level." Although we are ignorant of prey turnover, we know that cannibalism and intraguild predation are frequent and important interactions and that many PA are very efficient metabolically.

Regulation and Stabilization of Prey

It is unclear to what extent desert PA influence their prey. We do know that predators and parasitoids from many communities regulate the abundance and determine the diversity of their prey. Nevertheless, predators are often considered to be relatively unimportant in desert communities (Noy-Meir 1974; Louw and Seely 1982). In particular, Crawford (1986) states that desert arthropods probably seldom severely affect prey populations. Low metabolic rates, high production efficiencies, and the general inactivity of the many species of time minimizers reduce the amount of prey eaten. The ubiquity of interference decreases the effectiveness of PA. For example, self-limiting effects due to interference limit the numerical response to prey of most PA (see Riechert and Lockley 1984, Provencher and Vickery 1988 for spiders). Further, in other systems, intraguild predation often acts to decrease the numbers of more effective, smaller predators and actually indirectly produces an increase in prey numbers (Polis 1988a; Polis, Myers, and Holt 1989). The frequency of interference among spiders prompted Spiller (1986) to note that under some conditions, a subset of the predator assemblage could be more effective at reducing prey populations than an entire guild. Finally, their generalist diets tend to reduce the impact that any one species of PA exerts on a particular prey population. Single generalist species normally are not capable of regulating their prey at all densities (Riechert and Lockley 1984).

However, theory and empirical research suggest that desert PA significantly influence the structure of their prey communities. A priori, their diversity, abundance, and their high ratios compared to other arthropods argue that they could process a large mass of prey, as exemplified in the scorpion examples above. Although their opportunistic feeding habits do not allow for tight regulation of particular species, such habits do facilitate a quick functional response to prey fluctuations.

The high diversity of desert PA may, in itself, limit prey populations. Riechert and Lockley (1984) note that it may be more profitable to envision prey regulation as a function of the sum of many species of PA (with different foraging and life history strategies) rather than assume that a single species will exert a major impact on prey dynamics. Thus, opportunistic and equilibrium species exhibit different but complementary predator-prey dynamics. Opportunistic species provide a rapid numerical response to prey, whereas long-lived equilibrium species do not show immediate responses to either prey eruptions or to prolonged periods of low food availability (Polis and Farley 1980). Equilibrium species possess a suite of adaptations to the feast-and-famine nature of deserts that allow them to maintain relatively stable and large populations throughout long periods of unfavorable conditions (e.g., some arachnids; see earlier). Thus traits of both opportunistic and equilibrium species each tend to stabilize prey populations, especially during periods of increase.

Our knowledge of how PA influence desert prey communities comes mainly from spiders and scorpions. Many desert spiders are quite dense and can remove large quantities of prey (Cloudsley-Thompson 1983; Humprheys 1988; Henschel in press). For example, in the Namib, the average adult *Gandanimeno eresus* webs contain more than 20 adult tenebrionids (Polis and Seely unpublished data); in the Coachella Valley, adult *Diguetia mohavea* webs average more than 50 individual prey items (Polis and Sculteure unpublished data). Many desert spiders specialize on ants (see earlier). In the deserts of Southern California, spiders greatly influence the dynamics of harvester ants, both directly as a mortality factor and indirectly by limiting foraging (Hölldobler 1970; MacKay 1982; Ryti and Case 1986). Spider predation caused a 36 percent decrease in workers per day in *Pogonomyrmex californicus* colonies (Ryti and Case 1986). Although no other study indicates how desert spiders affect prey populations, we expect that spiders are quite important in many desert systems. Long-term and experimental field studies are needed to evaluate this speculation. Spiders in other habitats and in agrosystems can cause high mortality among some prey (Riechert and Lockley 1984; Riechert and Harp 1987; Humphreys 1988; Provencher and Vickery 1988).

We know more about scorpions. The effects of predation by *Scorpio maurus* on their isopod prey (*Hemilepistus*) were estimated by Shachak (1980) in the Negev Desert and Krapf (1986 personal communication) in Tunisia. They calculated that scorpions annually ate an average of from 10.9 percent (in the Negev; range: 1.7–19.3 percent over 5 years) to 27 percent (in Tunisia) of the entire isopod population. Neither speculated if such predation regulated or even limited isopod numbers.

Other research with scorpions has shown the effects of intraguild predation on potential competitors (see earlier); for example, *Uroplectes* scorpions greatly reduce the density of *Gandanimeno* spiders on *Acacia* trees in the Namib (Polis and Seely unpublished data). Polis and McCormick (1986b, 1987) quantified the impact of predation by the abundant scorpion *P. mesaensis*. The diet of this scorpion consisted of 8 percent spiders, 14 percent solpugids, 61 percent insects, 8 percent heterospecific scorpions, and 7 percent conspecifics. Removal of more than 6,000 *P. mesaensis* (3.2 kg) allowed spiders to more than double and two species of smaller scorpion to increase 130 percent and 600 percent as compared to controls (see earlier). Surprisingly, neither solpugids nor all insects combined increased in removal plots. These taxa did not increase either because individuals dispersed from removal areas, because the increase of spiders and smaller scorpions compensated for the removal of *P. mesaensis* by eating surplus arthropods, or simply because scorpions exerted little impact on insect populations. The first explanation is likely true for widely foraging solpugids and is possible but unlikely for the less mobile insects. The second explanation is unlikely: the biomass increase of spiders and scorpions represented less than 10 percent of the removed *P. mesaensis* biomass. The third explanation, difficult to accept, is nonetheless a real possibility: *P. mesaensis* may take such a small proportion of all insects that its removal does not affect insect density.

Our knowledge of the impact of predaceous insects is quite limited. Ryti and Case (1986, 1988) argue that predation by ants on foundress queens and during colony "wars" is an important interaction limiting the number and size of sympatric ant colonies. Such predation may be a common feature structuring many ant assemblages (Polis, Myers, and Holt 1989).

The rich literature on biological control demonstrates that many taxa of parasitoids regularly limit and regulate their host insect species. In the Coachella Valley, populations of the eggs of the grasshopper *Bootettix punctatus* occasionally suffer high mortality from ovipositing *Mythicomia* bombyliid flies (Mispagel 1978). Up to 37 percent of the population of *Photopsis* may be destroyed when these parasitoids are parasitized by some of the same parasitoids (e.g., sphecids) that fall host to *Photopsis* (Ferguson 1972). Finally, spider parasitoids (e.g., pompilid wasps) may be the major mortality factor limiting some species of desert spiders (Conley 1985; Henschel n.d.; see Polis this volume).

Desert PA affect other characteristics of their prey. Anti-predator traits among desert arthropods are near-universal, for example, crypsis, sclerotization, escape-behaviors, burrows, noxious chemicals, venom, and conglobation by isopods. Although such traits confer protection against predation by both vertebrates and PA, some of these characteristics are attributed specifically to arthropod predation (e.g., burrowing in spiders; Conley 1986; Henschel n.d.; modified movement or foraging activity by prey of scorpions; see McCormick and Polis 1990). Predation also affects other characteristics. For example, sex-biased predation by three species of Namib spiders on male tenebrionid beetles (they are more active than females) apparently skews the sex ratio toward females (Polis and Seely unpublished data). Similar sex-biased predation by scorpions on other scorpions also alters the sex ratio (McCormick and Polis 1990).

Interactions with Vertebrates

PA regularly interact with vertebrates as prey, occasionally as their predators, and possibly as competitors for common prey. Since PA form such a large proportion of all arthropods, they are expected to be a major diet item of many arthropod-eating predators. In the Coachella Valley, the diets of *Ammospermophilus leucurus* (antelope ground squirrel) and *Vulpes macrotus* (kit fox) consist of 71 percent and 67 percent PA, respectively, and 15 Coachella vertebrates averaged 41 percent PA in their diets (Polis this volume). In fact, some species specialize on PA; for example, several lizards, snakes, and mammals primarily eat scorpions (Polis, Sissom, and McCormick 1981). For these and many other arthropodivores, PA constitute an important energy source.

Some PA regularly eat vertebrates (McCormick and Polis 1982). Such predation occurs because many PA possess effective venoms directed at vertebrates and several species are relatively larger than their vertebrate prey. Small nocturnal reptiles are particularly at risk from being eaten by large scorpions, mygalomorph spiders (e.g., tarantulas), and scolopendrid centipedes. Williams (1971) suggests that large scorpions (*Hadrurus*) may be the top predator on small islands in the Gulf of California. In the Coachella, regular predation by *P. mesaensis* and *H. arizonensis* scorpions on the worm snake *Leptotyphlops humilis* is likely a major mortality factor (McCormick and Polis 1990).

Perhaps the greatest effect that PA exert on vertebrates is via potential exploitation competition. Arthropodivorous PA and vertebrates eat many of the same prey taxa, and the biomass of desert spiders and scorpions alone is more than 160 percent of that of all vertebrates (herbivores and predators combined). Does consumption by PA, which are greater in diversity, abundance, and biomass, negatively influence lizards, birds, and other arthropod-eating vertebrates? We know that granivorous arthropods and vertebrates compete with one another in deserts (see MacKay, Reichman this volume) and that spiders, lizards, and birds may compete for prey in tropical environments (see Spiller and Schoener 1988; Polis, Myers, and Holt 1989 for references). It is likely that, at least during times of low prey availability, arthropodivorous PA and vertebrates negatively influence each other via resource depletion. However, no data address this issue. This is a very important topic for future research. Reciprocal removals of lizards and arachnids (scorpions and many spiders) in replicated quadrats would be one fruitful approach.

Conclusions

Predaceous desert arthropods are a taxonomically and trophically diverse assemblage of insects, parasitoids, mites, and arachnids. Some groups are quite "successful." Success is exhibited in their great diversity, abundance, and in that they form a relatively large share of animal biomass in desert communities, in particular relative to vertebrates and other arthropods. These PA are relatively successful because they possess autecological traits that are particularly suited to desert conditions. Their low metabolisms, high production efficiencies, and generalist feeding habits combine to allow them to prosper under conditions of unpredictable and low food supplies. It is likely that, as a group, PA have replaced groups less successful in deserts (i.e., predaceous birds and mammals). Morton and James (1988) use similar autecological arguments to explain the relative success of Australian lizards compared to homeotherms; Wiens (this volume) maintains that autecology also allows certain taxa of birds to succeed in deserts.

As a group, PA are probably very important conduits of energy flow in deserts. Further, they may exert considerable impact as predators, prey, and competitors of other functional groups. However, their high metabolic and production efficiencies, key factors promoting the success of PA, also lessen their impact on energetics, prey, and competitors.

Theory and empirical evidence suggest that PA, as a group, regularly limit and tend to stabilize the numbers of their prey in many desert systems. However, the occasional occurrence of prey eruptions, self-limiting behaviors, and generalist feeding habits argue that desert PA do not tightly regulate the populations of their prey.

It is likely that their exact importance varies in time and space as a function of productivity, history, and chance (dispersal) events. Thus PA are unimportant in some systems and at some times. Further, patchy distributions combined with hide-and-seek dynamics may allow (non-equilibrium) global stability of prey populations.

These same conditions partially explain the coexistence of these intermediate-level predators with the array of species that eat them. Further, heterogeneity contributes to coexistence among PA by allowing separation of antagonistic species and age classes. Many desert PA interact strongly via various forms of interference (territoriality, cannibalism, and intraguild predation). Patchy distributions in space and time allow subordinate age classes and species to coexist with their dominants.

In conclusion, we present several speculations on how desert PA may differ from PA in more mesic habitats.

1. They are apparently relatively more important in deserts in terms of their biomass as compared to PA in other habitats and to vertebrates. This may be a product of an autecology that preadapts them particularly well to the harsh and variable climate and productivity of deserts (see also 5 below).
2. They may be more generalized and opportunistic in their diet than other PA.
3. Age structure may be more marked because of the relatively large proportion of long-lived equilibrium species.
4. A generalized diet and the existence of clearly separated age classes may predispose desert PA to interfere with and eat one another more frequently than PA in other habitats.
5. Generalized diets, "life history omnivory" due to age structure, and frequent cannibalism and intraguild predation may explain why desert PA are abundant relative to their non-predatory arthropod prey.
6. Although desert PA are regularly limited by food, it appears that they seldom interact via exploitation competition for food.

Acknowledgments

We thank the many people who have contributed to this work. The following supplied their papers or unpublished data: Joh Henschel, Dieter Krapf (now Dieter Mahsberg), Martin Muma, Chris Myers, Randy Ryti, Wendell Icenogle, Richard Bradley, Erin Griffin, Victor Fet, Liz McClain, and Maria Luisa Jiminez. The following read the manuscript and offered valuable critiques: David McCauley, Todd Jackson, Victor Fet, Chris Myers, and Amy Rosemond.

Bibliography

Abraham, B. 1983. Spatial and temporal patterns in a sagebrush steppe spider community. *Journal of Arachnology* 11: 31–50.

Allred, D., D. Beck, and C. Jorgensen. 1963. Biotic communities of the Nevada Test Site. *Brigham Young University Science Bulletin* 2(2): 1–52.

Allred, D., and W. Gertsch. 1975. Spiders and scorpions from Northern Arizona and Southern Utah. *Journal of Arachnology* 3: 87–100.

Allred, D., and B. J. Kaston. 1983. A list of Utah spiders, with their localities. *Great Basin Naturalist* 43: 493–522.

Andrewartha, H. G., and L. C. Birch. 1954. *The Distribution and Abundance of Animals.* Chicago: University of Chicago Press.

Andrewartha, H. G., and T. Browning. 1961. An analysis of the idea of resources in animal ecology. *Journal of Theoretical Biology* 8: 83–97.

Andrews, F. G., A. R. Hardy, and D. Giuliani. 1979. *The Coleopterous Fauna of Selected California Sand Dunes.* California Department of Food and Agriculture Report, 142 pp.

Askenmo, C., A. Von Bromssen, J. Ekman, and C. Jansson. 1977. Impact of some wintering birds on spider abundance in spruce. *Oikos* 28: 90–94.

Austin, A. D. 1985. The function of spider egg sacs in relation to parasitoids and predators, with special reference to the Australian fauna. *Journal of Natural History* 19: 359–376.

Benton, M. J., and G. W. Uetz. 1986. Variation in life history characteristics over a clinal gradient in three populations of a communal orb weaving spider. *Oecologia* 68: 395–399.

Bloom, S. 1981. Specialization and noncompetitive resource partitioning among sponge-eating dorid nudibranchs. *Oecologia* 49: 305–315.

Bradley, R. 1983. Complex food webs and manipulative experiments in ecology. *Oikos* 41: 150–152.

——. 1984. The influence of the quantity of food on fecundity in the desert grassland scorpion (*Paruroctonus utahensis*) (Scorpionida, Vaejovidae): an experimental test. *Oecologia* 62: 53–56.

——. 1988. The influence of weather and biotic factors on the behaviour of the scorpion (*Paruroctonus utahensis*). *Journal of Animal Ecology* 57: 533–551.

Bradley, R., and A. Brody. 1984. Relative abundance of the three vaejovid scorpions across a habitat gradient. *Journal of Arachnology* 11: 437–440.

Bradley, R. A. 1986. The relationship between population density of *Paruroctonus utahensis* (Scorpionida: Vaejovidae) and characteristics of its habitat. *Journal of Arid Environments* 11: 165–171.

Bultman, T., and G. Uetz. 1982. Abundance and community structure of forest floor spiders following litter manipulation. *Oecologia* 55: 34–41.

Caswell, H. 1978. Predator-mediated coexistence: A nonequilibrium model. *American Naturalist* 112: 127–154.

Chew, R. M. 1961. Ecology of spiders of a desert community. *Journal of the New York Entomological Society* 69: 5–41.

Chew, R. M., and J. DeVita. 1980. Foraging characteristics of a desert ant assemblage: functional morphology and spider separation. *Journal of Arid Environments* 3: 75–83.

Cloudsley-Thompson, J. L. 1975. Adaptations of Arthropoda to arid environments. *Annual Review of Entomology* 20: 261–283.

——. 1983. Desert adaptations in spiders. *Journal of Arid Environments* 6: 307–317.

Conley, M. Reeves. 1985. Predation versus resource limitation in survival of adult burrowing wolf spiders (Araneae: Lycosidae). *Oecologia* 67: 71–75.

Coville, R. 1987. Spider-hunting sphecid wasps. In W. Nentwig (ed.), *Ecophysiology of Spiders.* New York: Springer-Verlag.

Crawford, C. S. 1981. *Biology of Desert Invertebrates.* New York: Springer-Verlag.

——. 1986. The role of invertebrates in desert ecosystems. Pp. 73–92 in W. Whitford (ed.), *Pattern and Process in Desert Ecosystems.* Albuquerque, N.Mex.: University of New Mexico Press, 139 pp.

Due, A. D., and G. A. Polis. 1986. Trends in scorpion diversity along the Baja California peninsula. *American Naturalist* 128: 460–468.

Eastwood, E. B. 1978. Notes on the scorpion fauna of the Cape. IV. The burrowing activity of some scorpionids and buthids (Arachnida, Scorpionida). *Annals of the South African Museum* 74: 249–255.

Edney, E. B., J. F. McBrayer, P. J. Franco, and A. W. Phillips. 1974. *Distribution and Abundance of Soil Arthropods in Rock Valley, Nevada*. U.S. International Biological Program Desert Biome Research Memo 74-32.

Fautin, R. W. 1946. Biotic communities of the northern desert shrub biome in western Utah. *Ecological Monographs* 16: 251–310.

Ferguson, W. 1962. Biological characteristics of the mutillid subgenus *Photopsis* Blake and their systematic value. *University of California at Berkeley Publications in Entomology* 27: 1–82.

Fowler, H., and W. Whitford. 1985. Structure and organization of a winter community of cavity-inhabiting, web-building spiders (Pholcidae and Theridiidae) in a Chihuahuan Desert habitat. *Journal of Arid Environments* 8: 57–65.

Franco, P. J., E. B. Edney, and J. F. McBrayer. 1979. The distribution and abundance of soil arthropods in the northern Mojave Desert. *Journal of Arid Environments* 2: 137–149.

Frommer, S. I. 1986. *A Hierarchic Listing of the Arthropods Known to Occur Within the Deep Canyon Desert Transect*. Riverside, Calif.: Deep Canyon Publications.

Gertsch, W., and S. Riechert. 1976. The spatial and temporal partitioning of a desert spider community; with descriptions on new species. *American Museum Novitiates* 2604: 1–25.

Ghabbour, S. I., and S. H. Shakir. 1982. Population parameters of soil mesofauna in agro-systems of the Mariut Region, Egypt. I. Under dry farmed almond. *Revue Ecologie et Biologie du Sol* 19: 73–87.

Gilliam, J., and D. Fraser. 1987. Habitat selection under predation hazard: test of a model for foraging minnows. *Ecology* 68: 1856–1862.

———. 1988. Resource depletion and habitat segregation by competitors under predation hazard. Pp. 173–184 in B. Ebenman and L. Persson (eds.), *Size-Structured Populations, Ecology and Evolution*. New York: Springer-Verlag.

Gordon, S. 1978. Food and foraging ecology of a desert harvester ant, *Veromessor pergandei* (Mayr). Ph.D. dissertation, University of California at Berkeley, 158 pp.

Gunnarsson, B. 1983. Winter mortality of spruce-living spiders: effect of spider interactions and bird predation. *Oikos* 40: 226–233.

Hadley, N. 1980. Productivity of desert ecosystems. Section B in *Handbook of Nutrition*. West Palm Beach, Fla.: Chemical Rubber Company Press.

———. 1990. Physiological ecology of desert scorpions. In G. A. Polis (ed.), *Biology of Scorpions*. Stanford, Calif.: Stanford University Press.

Hairston, N. G., F. E. Smith, and L. B. Slobodkin. 1960. Community structure, population control, and competition. *American Naturalist* 94: 421–425.

Hatley, C., and J. MacMahon. 1980. Spider community organization: Seasonal variation and the role of vegetation architecture. *Environmental Entomology* 9: 632–639.

Hawkins, B., and R. Goeden. 1984. Organization of a parasitoid community associated with a complex of galls on *Atriplex* spp. in southern Calfornia. *Ecological Entomology* 9: 271–292.

Heidger, C. 1988. Ecology of spiders inhabiting abandoned mammal burrows in South African savanna. *Oecologia* 76: 303–306.

Henschel, J. In press. The biology of *Leucorchestris arenicola* (Araneae: Heteropodidae), a burrowing spider of the Namib Dunes. In M. K. Seely (ed.), *Current Research on Namib*

Ecology—25 years of the Desert Ecological Research Unit. Transvaal Museum Monograph 8. Pretoria: Transvaal Museum.

Henschel, J. N.d. Spiders that cartwheel to escape predation. In preparation.

Hölldobler, B. 1970. *Steatoda fulva* (Theridiidae), a spider that feeds on harvester ants. *Psyche* 77: 202–208.

Holm, E., and C. Scholtz. 1980. Structure and pattern of the Namib Desert dune ecosystem at Gobabeb. *Madoqua* 12: 3–39.

Holt, R. 1984. Spatial heterogeneity, indirect interactions, and the coexistence of prey species. *American Naturalist* 124: 377–406.

Horton, H. S., and D. H. Wise. 1983. The experimental analysis of competition between two syntopic species of orb-web spiders (Araneae: Araneidae). *Ecology* 64: 929–944.

Huey, R., and E. Pianka. 1981. Ecological consequences of foraging mode. *Ecology* 62: 991–999.

Humphreys, W. F. 1973. The environment, biology, and energetics of the wolf spider *Lycosa godeffroyi* (L. Koch 1865). Ph.D. dissertation, Australian National University, Canberra.

———. 1979. Production and respiration in animal populations. *Journal of Animal Ecology* 48: 427–454.

———. 1988. The ecology of spiders with special reference to Australia. Pp. 1–22 in A. Austin and N. Heather (eds.), *Australian Arachnology.* Miscellaneous Publication No. 5 of the Australian Entomological Society, Brisbane.

Hutchinson, G. E. 1959. Homage to Santa Rosalia, or why are there so many kinds of animals? *American Naturalist* 93: 145–149.

Irwin, M. 1971. Ecology and biosystematics of the pherocine Thereviidae (Diptera). Ph.D. dissertation, University of California at Riverside, 263 pp.

Knost, S., and J. Rovner. 1975. Scavenging by wolf spiders (Araneae: Lycosidae). *American Midland Naturalist* 93: 239–244.

Koch, L. 1977. The taxonomy, geographic distribution and evolutionary radiation of Australo-Papuan scorpions. *Records of the Western Australian Museum* 5: 83–367.

Kraft, D. 1986. Predator-prey relations in diurnal *Scorpio maurus* L. *Actas X Congreso Internacional de Arachnologia* 1: 133.

Krivokhatskii, V. A., and V. Ya. Fet. 1982. The spiders (Aranei) from rodent burrows in East Kara Kum. *Problems of Desert Development* (U.S.S.R.) 4: 68–75.

Lamoral, B. 1978. Soil hardness, an important and limiting factor in burrowing scorpions of the genus *Opisthophthalmus* C. L. Koch, 1837 (Scorpionidae, Scorpionida). *Symposium of the Zoological Society of London* 42: 171–181.

———. 1979. The scorpions of Namibia (Arachnida: Scorpionida). *Annals of the Natal Museum* 23: 497–784.

Louw, G. W., and M. K. Seely. 1982. *Ecology of Desert Organisms.* London: Longman.

Ludwig, J. 1987. Primary productivity in arid lands: myths and realities. *Journal of Arid Environments* 13: 1–7.

McCormick, S. J., and G. A. Polis. 1982. Arthropods that prey on vertebrates. *Biological Reviews* 57: 29–58.

———. 1986. Comparison of the diet of *Paruroctonus mesaensis* at two sites. *Proceedings of the IX International Arachnological Congress*: 167–171.

———. 1990. Prey, predators and parasites. Chapter 7 in G. A. Polis (ed.), *Biology of Scorpions.* Stanford, Calif.: Stanford University Press.

MacKay, W. P. 1982. The effect of predation of western widow spiders (Araneae: Theridiidae) on harvester ants (Hymenoptera: Formicidae). *Oecologia* 53: 406–411.

McKinnerney, M. 1978. Carrion communities in the northern Chihuahuan Desert. *Southwestern Naturalist* 23: 563–576.

Main, B. Y. 1956. Taxonomy and biology of the genus *Isometroides* Keyserling (Scorpionida). *Australian Journal of Zoology* 4: 158–164.

———. 1957. Biology of aganippine trapdoor spiders (Mygalomorpha: Ctenizidae). *Australian Journal of Zoology* 5: 402–473.

———. 1978. Biology of the arid-adapted Australian trapdoor spider *Anidops villosus* (Rainbow). *Bulletin of the British Arachnological Society* 4: 161–175.

———. 1981. Australian spiders: diversity, distribution and ecology. Pp. 23–41 in A. Keast (ed.), *Ecological Biogeography of Australia*, vol. 2. The Hague: W. Junk.

Marples, T. G., and D. J. Shorthouse. 1982. An energy and water budget for a population of arid zone scorpion *Urodacus yaschenkoi* (Birula 1903). *Australian Journal of Ecology* 7: 119–127.

Mispagel, M. 1978. The ecology and bioenergetics of the acridid grasshopper, *Bootettix punctatus* on creosotebush, *Larrea tridentata*, in the northern Mojave Desert. *Ecology* 59: 779–788.

Mittlebach, G. 1988. Competition among refuging sunfishes and effects of fish density on littoral zone invertebrates. *Ecology* 69: 614–623.

Morton, S., and C. James. 1988. The diversity and abundance of lizards in arid Australia: a new hypothesis. *American Naturalist* 132: 237–256.

Moulder, B. C., and D. E. Reichle. 1972. Significance of spider predation in the energy dynamics of forest-floor arthropod communities. *Ecological Monographs* 42: 473–498.

Muma, M. 1975. Long-term can trapping for population analyses of ground-surface, arid-land arachnids. *Florida Entomologist* 58: 257–270.

———. 1979. Arid-grassland solpugid population variations in southwestern New Mexico. *Florida Entomologist* 62: 320–328.

———. 1980. Comparison of three methods for estimating solpugid (Arachnida) populations. *Journal of Arachnology* 8: 267–270.

Muma, M., and K. Muma. 1988. *The Arachnid Order Solpugida in the United States*. Supplement 2, *A Biological Review*. Silver City, N. Mex.: Southwest Offset.

Muth, A. 1980. Physiological ecology of desert iguana (*Dipsosaurus dorsalis*) eggs: temperature and water relations. *Ecology* 61: 1335–1343.

Noy-Meir, I. 1974. Desert ecosystems: higher trophic levels. *Annual Review of Ecology and Systematics* 5: 195–214.

———. 1979–80. Structure and function of desert ecosystems. *Israel Journal of Botany* 28: 1–19.

Péfaur, J. 1981. Composition and phenology of epigeic animal communities of the Lomas of southern Peru. *Journal of Arid Environments* 4: 1–42.

Pietruszka, R. D. 1980. Observations on seasonal variation in desert arthropods in central Nevada. *Great Basin Naturalist* 40: 292–297.

Pimm, S. L., and J. Rice. 1987. The dynamics of multispecies, multi–life-stage models of aquatic food webs. *Theoretical Population Biology* 32: 303–325.

Polis, G. A. 1979. Diet and prey phenology of the desert scorpion, *Paruroctonus mesaensis* Stahnke. *Journal of Zoology* (London) 188: 333–346.

———. 1980a. Seasonal and age specific variation in the surface activity of a population of desert scorpions in relation to environmental factors. *Journal of Animal Ecology* 49: 1–18.

———. 1980b. The significance of cannibalism on the population dynamics and surface activity of a natural population of desert scorpions. *Behavioral Ecology and Sociobiology* 7: 25–35.

———. 1981. The evolution and dynamics of intraspecific predation. *Annual Review of Ecology and Systematics* 12: 225–251.

———. 1984a. Age structure component of niche width and intraspecific resource partitioning: can age groups function as ecological species? *American Naturalist* 123: 541–564.

———. 1984b. Intraspecific predation and "infant killing" among invertebrates. Pp. 87–104 in G. Hausfater and S. Hrdy (eds.), *Infanticide: Comparative and Evolutionary Perspectives.* New York: Aldine Publishing.

———. 1988a. Exploitation competition and the evolution of interference, cannibalism and intraguild predation in age/size structured populations. In L. Persson and B. Ebenmann (eds.) *Size Structured Populations: Ecology and Evolution.* New York: Springer-Verlag.

———. 1988b. Foraging and evolutionary responses of desert scorpions to harsh environmental periods of food stress. *Journal of Arid Environments* 14: 123–134.

———. 1990. Ecology. Chapter 6 in G. A. Polis (ed.), *Biology of Scorpions.* Stanford, Calif.: Stanford University Press.

———. In press. Complex trophic interactions in deserts: an empirical assessment of food web theory. *American Naturalist.*

Polis, G. A., and R. D. Farley. 1979. Characteristics and environmental determinants of natality, growth and maturity in a natural population of the desert scorpion, *Paruroctonus mesaensis* Stahnke. *Journal of Zoology* (London) 187: 517–542.

———. 1980. Population biology of a desert scorpion: survivorship, microhabitat, and the evolution of life history strategy. *Ecology* 61: 620–629.

Polis, G. A., and S. J. McCormick. 1986a. Patterns of resource use and age structure among species of desert scorpions. *Journal of Animal Ecology* 55: 59–73.

———. 1986b. Scorpions, spiders and solpugids: predation and competition among distantly related taxa. *Oecologia* 71: 111–116.

———. 1987. Intraguild predation and competition among desert scorpions. *Ecology* 68: 332–343.

Polis, G. A., C. A. Myers, and R. Holt. 1989. The ecology and evolution of intraguild predation: Potential competitors that eat each other. *Annual Review of Ecology and Systematics* 20: 297–330.

Polis, G. A., C. A. Myers, and M. A. Quinlan. 1986. Burrowing biology and spatial distribution of desert scorpions. *Journal of Arid Environments* 10: 137–146.

Polis, G. A., W. D. Sissom, and S. J. McCormick. 1981. Predators of scorpions: Field data and a review. *Journal of Arid Environments* 4: 309–327.

Powell, J., and C. Hogue. 1979. *California Insects.* Berkeley, Calif.: University of California Press, 339 pp.

Price, P. 1975. *Insect Ecology.* New York: Wiley Interscience.

Provencher, L., and W. Vickery. 1988. Territoriality, vegetation complexity, and biological control: the case for spiders. *American Naturalist* 132: 257–266.

Reichman, O. J. 1984. Spatial and temporal variation of seed distributions in Sonoran Desert soils. *Journal of Biogeography* 11: 1–11.

Riechert, S. 1974. The pattern of local web distribution in a desert spider: mechanisms and seasonal variation. *Journal of Animal Ecology* 43: 733–746.

———. 1981. The consequences of being territorial: spiders, a case study. *American Naturalist* 117: 871–892.

Riechert, S., and A. Cady. 1983. Patterns of resource use and tests for competitive release in a spider community. *Ecology* 64: 889–913.

Riechert, S., and J. Harp. 1987. Nutritional ecology of spiders. Pp. 645–672 in F. Slansky and J. G. Rodriguez (eds.), *Nutritional Ecology of Insects, Mites, and Spiders*. New York: John Wiley and Sons.

Riechert, S., and T. Lockley. 1984. Spiders as biological control agents. *Annual Review of Entomology* 29: 299–320.

Root, R. 1967. The niche exploitation pattern of the blue-grey gnat catcher. *Ecological Monographs* 37: 317–50.

Rypstra, A. 1984. A relative measure of predation on web spiders in temperate and tropical forests. *Oikos* 43: 129–132.

Ryti, R., and T. Case. 1986. Overdispersion of ant colonies: a test of hypotheses. *Oecologia* 69: 446–453.

———. 1988. Field experiments on desert ants: testing for competition between colonies. *Ecology* 69: 1993–2003.

Schmoller, R. 1970. Terrestrial desert arthropods: Fauna and ecology. *Biologist* 52: 77–98.

Schoener, T. 1971. Theory of feeding strategies. *Annual Review of Ecology and Systematics* 2: 369–404.

Schoener, T., and C. Toft. 1983. Spider populations: extraordinary high densities on islands without top predators. *Science* 219: 1353–1355.

Schultz, J., D. Otte, and F. Enders 1977. *Larrea* as a habitat component for desert arthropods. In *Creosote Bush*. U.S. International Biological Program Synthesis Series 6. Stroudsburg, Pa.: Dowden, Hutchinson, and Ross.

Seely, M., and G. Louw. 1980. First approximation of the effect of rainfall on the ecology and energetics of a Namib Desert dune ecosystem. *Journal of Arid Environments* 3: 25–54.

Shachak, M. 1980. Energy allocation and life history strategy of the desert isopod *H. reaumuri*. *Oecologia* 45: 404–413.

Shook, R. 1978. Ecology of the wolf spider *Lycosa carolinensis* Walckenaer (Araneae: Lycosidae) in a desert community. *Journal of Arachnology* 6: 53–64.

Shorthouse, D. 1971. Studies on the biology and energetics of the scorpion, *Urodacus yaschenkoi* (Birula, 1904). Ph.D. dissertation, Australian National University, Canberra, 163 pp.

Smith, R., and T. Mommsen. 1984. Pollen feeding in an orb-weaving spider. *Science* 226: 1330–1332.

Spiller, D. 1984. Competition between two spider species: experimental field study. *Ecology* 65: 909–919.

———. 1986. Interspecific competition between spiders and its relevance to biological control by general predators. *Environmental Entomology* 15: 177–181.

Spiller, D., and T. Schoener. 1988. An experimental study of lizards on webspider communities. *Ecological Monographs* 58: 57–77.

Tevis, L., and I. Newell. 1962. Studies on the biology and seasonal cycle of the giant red velvet mite, *Dinothrombium pandorae* (Acari, Trombidiidae). *Ecology* 43: 497–505.

Turnbull, A. L. 1973. Ecology of the true spiders (Araneomorphae). *Annual Review of Entomology* 58: 305–348.

Turner, M. 1983. Mechanisms structuring a guild of raptorial spiders. Ph.D. dissertation, University of Tennessee, Knoxville, 97 pp.

Van Hook, R. 1971. Energy and nutrient dynamics of spider and orthopteran populations in a grassland ecosystem. *Ecological Monographs* 41: 1–26.

Wagner, F., and R. Graetz. 1981. Animal-animal interaction. Pp. 51–83 in D. W. Goodall and R. A. Perry (eds.), *Arid-Land Ecosystems: Structure, Functioning and Management*, vol 2. IBP No. 17. Cambridge, Mass.: Cambridge University Press.

Warburg, M., and G. A. Polis. 1990. Behavioral responses, rhythms and activity patterns. Chapter 5 in G. A. Polis (ed.), *Biology of Scorpions*. Stanford, Calif.: Stanford University Press.

Wasbauer, M., and L. Kimsey. 1985. California spider wasps of the sub-family Pompilinae. *Bulletin of the California Insect Survey* 26: 1–130.

Werner, E., and J. Gilliam. 1984. The ontogenetic niche and species interactions in size-structured populations. *Annual Review of Ecology and Systematics* 15: 393–425.

Wharton, R. 1987. Biology of the diurnal *Metasolpuga picta* (Kraepelin) (Solifugae, Solpugidae) compared with that of nocturnal species. *Journal of Arachnology* 14: 263–284.

Wheeler, G., and J. Wheeler. 1973. *Ants of Deep Canyon*. Berkeley, Calif.: University of California Press, 192 pp.

White, T. 1978. The importance of relative shortage of food in animal ecology. *Oecologia* 33: 76–86.

William, S. 1980. Pp. 1–127 in *Scorpions of Baja California, Mexico, and Adjacent Islands*. Occasional Papers of the California Academy of Science no. 135.

Williams, S. 1971. In search of scorpions. *Pacific Discovery* 24: 1–10.

Wise, D. 1984. The role of competition in spider communities: insights from field experiments with a model organism. Pp. 42–53 in D. R. Strong et al. (eds.), *Ecological Communities: Conceptual Issues and the Evidence*. Princeton, N.J.: Princeton University Press.

Yamashita, T., and G. A. Polis. N.d. The diversity, density and population biomass of desert fauna. In preparation.

The Community Ecology of Desert Anurans

8

Bruce D. Woodward and
Sandra L. Mitchell

The nature of community ecology has changed greatly over its long and tumultuous history (Pound and Clements 1898; Odum 1957; Hutchinson 1959; Hardin 1960; Whittaker 1967; Paine 1969; MacArthur 1972; Connell 1975; Wiens 1977; Strong 1984; Price, Slobodchikoff, and Gaud 1984; Wilbur 1984; Wiens et al. 1986). Although one goal of community ecologists has remained to determine patterns of distribution and abundance of organisms, the way community ecologists have studied these patterns, and the scale at which they have searched for patterns, have changed tremendously in recent years. We believe that community ecology is currently in the process of making yet another change, a change that will lead to the incorporation of many aspects of what is now considered population biology into the field of community ecology.

Community ecologists have long suspected that the distribution and abundance of a species depended upon interspecific interactions and interactions with the abiotic environment (traditional community ecology). Historically, community ecologists have regarded species characteristics as fixed species-wide attributes. The change we foresee for community ecology stems from the dawning realization that a species's characteristics, and thus how that species interacts with other species or the abiotic environment, are greatly influenced by local-scale phenomena that have traditionally been regarded as part of population biology. We agree with Colwell (1984) that combining population biology with community ecology will greatly improve our understanding of community dynamics. However, the enhanced realism of these studies will come at the expense of decreased generality. Thus, as we begin to incorporate local scale phenomena into our studies of communities, we may lose some generality.

In this paper we review the community ecology of desert amphibians. Because anurans are, with rare exceptions, the only amphibians in the desert, we will emphasize anurans in our discussion. We will review the history of anuran community ecology, snythesize current knowledge about desert anuran communities, and then suggest areas that deserve attention in the future. In the final section we suggest how incorporating aspects of population biology may lead to a better understanding of the structure and dynamics of desert anuran communities.

Historical Perspective

We suggest community ecology has passed through a "natural history" stage, an "emergent properties" stage, and an "experimental" stage over its history. We suggest that each stage yielded useful information, and that approaches from each stage should be part of any current community ecology study. We argue that the methodologies employed in each stage have had a large effect on the nature of the questions that could be examined, and suggest that too few ecologists realize how their methodologies constrain their viewpoint of what defines their system.

Until the mid- to late-1970s community ecology focused on natural history and biogeography; investigators looked at patterns across large taxa and large geographic areas (e.g., Savage 1973; Blair 1976). In these studies, associations between species were used to infer the underlying processes that generated the patterns of species distributions and abundances. Physiological or anatomical traits, such as traits that prevent desiccation or heat overload in desert regions, were often used to infer selection pressures responsible for determining species distributions (e.g., Warburg 1965; Bentley 1966; Mayhew 1968). These traits were common to many members of a taxon and easy to measure in the laboratory. "Adaptation" and "adaptiveness" were predominant themes in this literature. This approach favors examination of easily measured traits and discourages study of more complex traits, both because of the time required for study of more complex traits and because simple traits are present in more species and may thus yield a more general pattern than complex traits.

This early natural history approach revealed that some species tended to co-occur in certain broad habitat types such as deserts (for desert amphibians in North America see Bragg 1965 and Stebbins 1975; for desert amphibians worldwide see Low 1976 and Pomeroy 1981). Relative to other areas, few amphibian species are found in deserts (Mayhew 1968). Worldwide distributional patterns led early natural historians to conclude that water retention problems play a major role in preventing many amphibians from existing in deserts. Only the Salientia (frogs and toads) are represented to any degree in deserts, with urodeles (salamanders) and gymnophians (caecilians) restricted to moister areas.

Following the natural history period, community ecology went through a brief period when community attributes, as opposed to species characteristics, were the major focus of study. Community characteristics such as the number of species, number of individuals per species, species evenness, and number of species per taxon were used to compare communities (Heatwole 1982; Scott and Campbell

1982). In this context "communities" followed MacArthur's definition of closely related co-occurring species of interest to the investigator. At this time, island biogeography theory, emergent properties, and holism versus reductionism received much attention. Ecologists debated whether communities were mere sums of their component parts, or if there were characteristics of communities that derive from a group of species being a community. Although island biogeography theory did appear to predict species number (e.g., MacArthur and Wilson 1967; Simberloff and Wilson 1970), a major conclusion from this period was that differences among communities could not be understood without considering the component species themselves. In part as a reaction to this conclusion, community-level studies again returned to an examination of the characteristics of the individuals or species that make up the community. Unlike the natural history phase of community studies, however, the recent emphasis is on the role that biotic interactions have on structuring communities. As a consequence, the scale of investigation has been greatly reduced, because although abiotic conditions are often similar across large geographic areas (e.g., high temperatures across a warm desert), biotic factors, such as competition or predation, tend to vary over a smaller spatial scale.

The "emergent properties approach" revealed that desert communities contained relatively few species, irrespective of taxa. In desert amphibians, two to five species are common on a local scale (Bragg 1965; Main 1968; Mayhew 1968; Low 1976; Blair 1976; Woodward 1981; Pomeroy 1981). This approach and the natural history approach (e.g., Bragg 1965) also revealed that the number of species and number of individuals per species varied a great deal from year to year in desert amphibian communities. This same pattern occurs in most desert taxa.

The hallmark of the most recent approach to community ecology is the use of experiments to examine species interactions. Because the experimental approach is context specific, and the emphasis is on interactions among particular species, the scope of these studies is very limited relative to the earlier approaches. The experimental nature of these studies and the current focus on biological interactions have shifted the emphasis away from explaining why some species of a taxon commonly co-occur in one regional area, to the question of why, among the species occurring regionally, a particular species grouping occurs on a local scale.

Many early experimental studies asked whether two or more species interacted, by determining whether the presence of one species affected the other (Connell 1983; Schoener 1983; Endler 1986). Competition was the interaction most commonly examined (Glassner 1979; Wilbur 1980; Schoener 1983; Strong 1984), but predation also received a considerable amount of attention in anurans (Wilbur 1972; Calef 1973: Heyer, McDiarmid, and Weigmann 1975; Heyer 1976). More recent studies have evaluated the relative importance of different types of interspecific interactions on local species distributions or abundances (e.g., Wilbur 1972; Price, Slobodchikoff, and Gaud 1984; Morin 1987; Wilbur 1987). Within the last 5 years we have started to examine how interspecific interactions and interactions with abiotic factors can jointly affect community structure. The work on anuran community structure has been especially innovative and in many respects is far ahead of the work on other taxa (see below).

The three approaches to the study of community ecology (natural history, emergent properties, and experimental) have asked superficially similar, but actually quite different questions about the distributions of organisms. The early natural history approach established that there were patterns of association between some species groups and various gross habitat types (e.g., deserts). Because certain traits were implicated as factors important in determining species distribution and abundances *across* gross habitat types, it has commonly been assumed that these same factors are important in determining species distributions or abundances *within* habitat types as well. Although abiotic factors appear to play important roles in determining species distributions on a continental scale (i.e., allowing access to deserts), we feel that abiotic factors have far too often been oversold as the sole causal factor explaining species distributions on a local scale.

Although the early natural history approach was applied on a large geographic scale, more recent approaches have operated on a smaller scale, on the order of magnitude of home ranges of the individuals in the species making up the community. The experimental approach in particular has concentrated on the nature and strengths of the interactions among the members of the group, reflecting the modern MacArthurian concept of a community as a group of strongly interacting species. At the same time, this approach downplays the importance of factors that may affect distributions on a large geographic scale, such as desiccation tolerances or heat resistance. The point we wish to make here is that while "communities" have been studied for a long time, the approach and the focus of these community studies has changed a great deal over time. Factors that may have very important effects on species distributions at one scale may have little to do with what structures a "community" at a different scale. The most thorough studies of community ecology will incorporate factors from all scales, and recognize the significance of scale when interpreting the results. The local distributions and abundances of organisms will depend on both the large-scale processes of speciation, colonization, and extinction and the local-scale interactions that are currently the major locus of investigation.

Amphibian Life Histories and the Nature of Amphibian Communities

Amphibians have complex life histories (Wilbur 1980) with the young stages often morphologically and ecologically quite distinct from adults. This is especially true in anurans, which form the vast majority of amphibians in desert systems. These differences between young and adults have important implications for the study of community dynamics. One striking difference is that often young are restricted to water, whereas adults can move between water sources. This dichotomy means the source of strong interactions, and thus the nature of the community, may differ greatly for larval and adult amphibians. The important species a larva interacts with are individuals present in its pond, but the adult may interact with organisms in several ponds, or even organisms that live between ponds. To an ecologist studying amphibian communities this means "a community" could be recognized at three scales: species present in a single pond, species present in all ponds in which a given species breeds locally, or species present in all local ponds. These three types of communities can also be viewed as occurring at one point in time, or as having a longer temporal component (e.g., all species that breed in this pond this year or over several years). All of these types of communities are communities in the MacArthurian sense that they include only species with strong interactions among themselves. All these approaches to studying anuran communities have been used in the recent past. How a community is defined has important implications for how we interpret the significance of interactions among organisms. In general, the more restricted (temporally or spatially) a study is, the more caution we should use when interpreting the effects of interactions for long-range community dynamics, because the unstudied portions of the system may have important but unknown effects on the dynamics of the system that we are calling "the community." Of particular concern is the possibility that the unstudied parts of the system may reverse the conclusions drawn from the studied portion of the system (Morin 1981; Bender, Case, and Gilpin 1984; see below).

Most experimental work with anurans has been performed with the larval (tadpole) stage rather than the adults. This is largely a practical consideration as the adults in many species are widely dispersed and hard to sample for much of the year, and the larvae are well suited to experimental manipulation. This is especially true for many desert species that breed in temporary ponds. In these species adults are concentrated at breeding ponds for a night or two (Bragg 1965; Creusere and Whitford 1976; Low 1976; Woodward 1982a), feed intermittently for a month or two above ground (Bragg 1965; Creusere and Whitford 1976), and then remain inactive underground for several months each year (Bragg 1965; Low 1976; Blair 1976). There are also some biological reasons why the adult stage may have lesser effects on community structure than the larvae, and thus why the emphasis has been on the larvae. Clutch sizes are extremely high in most anurans, ranging from hundreds to thousands of eggs in most species (Crump 1974; Low 1976; Woodward 1987a,b). Larval density can also be high, because of the large clutch sizes, high hatching success, and small size of most larval ponds. The fact that adults breed in localized ponds but disperse widely to live as adults suggests that resource demands are more likely to be exceeded by the larvae than by the adults. Mortality during the larval stage is also high (Bragg 1965; Herreid and Kinney 1966; Newman 1987), suggesting that slight changes in the ability of larvae to deal with their environment could have large effects on population sizes. Finally, in many desert species larvae may interact with more selective contexts (see Table 8.1) for a longer period of time than the adult stage since the adults estivate for much of the year. However, we should bear in mind that most current studies of anuran communities do not deal with the adult stage at all, and thus factors operating on adults could have strong, unknown effects on community dynamics (Woodward, Travis, and Mitchell 1988).

Amphibian Community Structure

Most modern community studies attempt to discover what "structures" or determines the distribution and relative abundance of species in that community. Much of

Table 8.1 Factors that may influence population size or growth rate of individuals from the various life history stages of desert anurans.

Stage	Source
EGG STAGE	
Size, hatching probability, and no./female probably influenced by size of female and/or time since last breeding bout; directly contributes to density	Woodward 1987a,b; Salthe & Duellman 1973
No. oviposited in pond strongly influenced by no. females breeding, which is a function of factors that influence adult survival	Woodward 1987b
Hatching success probably depends on ability of females to pair with males of correct species and perhaps with males of a specific genotype	Creusere & Whitford 1976
Hatching strongly influenced by desiccation due to rapid water drop	Bragg 1965
Hatching rarely influenced by predators in temporary ponds (but tadpoles may be important predators)	Woodward 1987d
Hatching occasionally influenced by O_2 depletion due to cattle feces in water	Woodward manuscript e
TADPOLE STAGE	
Growth and survival influenced by:	
1. Tadpole density, competition, desiccation, cannibalism, predation	Woodward 1982a,b,c, 1983a,b, 1987a; Morin 1987; Wilbur 1987; Newman 1987
2. O_2 levels	
3. Paternity through direct effects on survival and indirect size-mediated effects on competitive and predator-avoiding abilities	Woodward 1986, 1987a,b; Woodward, Travis, & Mitchell 1988; Travis, Emerson, & Blauin 1987; Mitchell 1990

what is known about anuran community structure comes from anuran communities from the eastern (non-desert) United States. We will review the extensive work performed on eastern anuran communities and then speculate on how these results would apply to the largely unstudied desert anuran communities.

Most anuran communities in the eastern United States occur in forested areas that receive moderate to heavy rainfall with fair regularity (relative to desert systems) during the warmer months when anurans are active. Ponds are relatively common and most of the ones studied to date usually last for a long time relative to the duration of the larval period; however, local factors such as sandy soils can cause the ponds to be highly ephemeral. Ephemeral ponds that form after heavy summer thundershowers are also common and are often used by anurans as breeding spots.

The forest habitat, high rainfall, and short distances between ponds in the East mean adults probably can move between ponds with far greater ease than is possible between isolated desert ponds. Evidence adults move between ponds in the East is largely anecdotal, but changes in local species compositions have been reported several times (e.g., Walters 1975; Morin 1987). Ten or more species commonly breed

Stage	Source
4. Tadpole size (probably related to egg size and to chance encounters with food)	Woodward 1982c
5. Temperature	
6. Food availability, which is influenced by density, competition, and storm frequency (which adds nutrients to pond), and prior presence of other species	Alford & Wilbur 1985; Wilbur & Alford 1985; Woodward 1982b,c; Newman 1987; Wilbur 1984, 1987
7. Growth inhibitors released by other tadpoles	Steinwascher 1978b
JUVENILE STAGE	
Survival influenced by food available to fuel overwinter growth, and size of juvenile at metamorphosis	Creusere & Whitford 1976; Martof 1956; Smith 1987
Survival influenced by duration of winter and time of metamorphosis	Smith 1987
Time and size at metamorphosis determined by paternity, which influences overwinter survival	Mitchell 1990
Survival influenced by safe sites for overwintering	Creusere & Whitford 1976
ADULT STAGE	
Female fitness influenced by female size-fecundity relationships	Salthe & Duellman 1973; Woodward 1987a
Male fitness influenced by male size-mating success relationship	Woodward 1982a
Male and female body sizes influenced by many of the factors listed under the other life history stages	
Adult survival influenced by predation	Woodward & Mitchell in press

or co-occur as tadpoles in any local area—considerably more than occupy any local area in desert systems (Conant 1975; Stebbins 1975). In the East, different species occupy different types of ponds (Conant 1975), but several species often coexist in any pond at one time. Most eastern species have a definite breeding period, during which they breed if a pond is available. Ponds usually persist for several months, allowing a series of species to sequentially occupy any one pond (Conant 1975; Wiest 1982). Because larval periods of these eastern species are relatively long, tadpoles of early breeders often coexist with tadpoles of later breeders. Thus, when a temporary pond fills with water and how long it persists may influence the types of tadpoles or tadpole predators with which a later breeding species will coexist (Alford 1989).

Desert species typically occupy widely scattered ponds isolated by habitats much less conducive to travel by frogs or many of their predators. Although some ponds along streams or springs are permanent, most desert ponds are highly ephemeral. These ponds receive very little rain, which occurs unpredictably and locally. Breeding in temporary ponds is usually concentrated over a short period immediately after pond formation. Species using permanent ponds have longer breeding seasons, and may breed sequentially in one pond, as is more typical in eastern ponds. Few anuran species occur in any local area in the desert (around five per local area; Bragg 1965), and one to three species typically co-occur in any one pond (Stebbins 1975; Creusere and Whitford 1976; Low 1976; Pomeroy 1981; Woodward 1982a, 1987a). These species breed whenever water is available, but usually breed in each pond at most once each year (Woodward 1984). Unlike in the East, where breeding occurs in most years, breeding is not an annual event in any particular area in the desert (Creusere and Whitford 1976).

Taken together, this information suggests eastern anuran communities are potentially much more complex than desert anuran communities. Spatial and temporal shifts in the number of anurans and anuran predators that commonly occur in the East make possible a greater number of potential species interactions. On the other hand, the limited number of breeding sites and the extremely short duration of most ponds may mean that the larval interactions are more intense in western desert ponds.

Most eastern breeding ponds are structurally complex with a dense leaf or litter area on the bottom, and emergent and sometimes floating vegetation contributing to the number of places that tadpoles can hide. Predators are often abundant. Tadpoles are usually not very visible, and appear to be at relatively low densities (Morin 1987) in all but the most ephemeral temporary ponds. In contrast, tadpole predators tend to be relatively scarce (Woodward 1983a) but important in desert temporary ponds (Woodward 1983a,b; Woodward and Johnson 1985; Newman 1987). Tadpole densities typically are high (Woodward 1982b; Newman 1987), and tadpoles are conspicuous in most desert ponds that contain very little vegetation where tadpoles might hide. Permanent ponds in the desert contain more predators and more places for tadpoles to hide than do temporary ponds (Woodward 1981). These differences between eastern and desert areas suggest predation may play a greater role in structuring eastern anuran communities than it does in desert communities.

Relative to eastern ponds, a higher proportion of desert ponds dry up before any tadpoles transform into juveniles (Bragg 1965; Newman 1987). Temperature fluctuations are extreme in exposed and shallow desert ponds. The young of desert species transform at a smaller fraction of adult size relative to many eastern species. Finally, in contrast to many eastern species, which breed at an early age and have short lifespans, many desert species take 2 or more years to reach reproductive size, and live a long time (Woodward manuscript a, based on bone growth ring estimated ages). These life history patterns suggest abiotic factors, especially pond drying rate, may be of great importance in desert systems.

Most work on eastern anuran communities has concentrated on interactions among tadpoles within a single pond. Early studies of the structure of eastern anuran communities tested for the existence of competitive interactions over food. Most tadpoles have rasping or filtering mouthparts or both and ingest many types of food. Many "food items" ingested pass through the tadpoles either undigested or with only some nutrients leached from the cells. Thus, studies of diet overlap, or diet shifts resulting from introduction of a second species, are difficult to perform and interpret. Consequently, these types of studies, which were very common in other vertebrates (Pianka 1974; Pyke, Pulliam, and Charnov 1977; Reynolds and Scott 1982), have rarely been performed in larval anurans. Instead, the short larval period favors direct measurement of life history traits that affect survival or reproductive success.

Much early work tested for the effect of intraspecific competition by raising tadpoles at several densities. Although competition could be either by exploitation or interference (Woodward 1982c), most investigators interpreted their results as exploitative competition over food. Increases in density almost always had a detrimental effect. At higher densities, tadpoles suffered reduced survival, increased larval period duration, and reduced growth rates, larval size, or metamorphic size (Brockelman 1969; Wilbur 1976, 1977a,b; Steinwascher 1978a). Experimental densities commonly fell within natural ranges, suggesting competition was an important factor in nature that could influence tadpole distribution or abundance (Wilbur 1980).

Interspecific competition also is an important force in some eastern anuran communities. Much of the early work on interspecific competition asked whether species that co-occurred on a local scale but did not occur in the same pond were strong competitors (e.g., Wilbur 1972; DeBenedictis 1974; Wiltshire and Bull 1977). Evidence of strong competitive effects was taken as one explanation for why some species did not occupy the same ponds simultaneously. Other interspecific experiments tested for competitive interactions among species that commonly co-occurred in the same pond. Such experiments (Wilbur 1972; Smith-Gill and Gill 1978; Morin 1981, 1983, 1986, 1987) also often documented strong negative effects, raising the question of how these competing species could coexist in the face of such strong negative interactions. One potential solution to this problem was that predation on tadpoles in nature was so strong that densities rarely reached levels at which competition occurred (Heyer 1976; Roughgarden and Diamond 1986). Indeed, one school of thought argued that predation was the major factor responsible for structuring anuran communities (Neish 1970; Calef 1973; Walters 1975; Heyer, McDiarmid, and Weigmann 1975; Heyer 1976; Caldwell, Thorp, and Jervey 1980). Predators are abundant in some anuran ponds and appear to be capable of drastically reducing tadpole densities (Neish 1970; Calef 1973; Wilbur, Morin, and Harris 1983; Morin 1987; Wilbur 1987).

Another possible solution was that some unstudied factor counterbalanced the negative effects of competition. Morin (1981, 1987) showed experimentally that competitively superior species tended to be more vulnerable to predation, a result similar to Paine's (1966, 1969) classic work on intertidal communities. Morin (1987) has suggested that yearly changes in predation intensities, in conjunction with occasionally strong competitive interactions, may allow coexistence in a single pond of species that could not coexist in the presence of either consistently strong predation or competition.

Recently, ecologists have examined the joint effects of both abiotic and biotic factors (e.g., Smith 1983; Wilbur 1987; Newman 1988a). Wilbur (1987) showed that pond duration can determine which species complete metamorphosis in a given pond. His results also indicated that interactions between biotic and abiotic factors could influence important parameters such as survival and growth. These important results demonstrate that understanding community structure at a local scale may require understanding not only how species respond to each other or to abiotic factors, but also how they respond to both types of factors simultaneously. Roth and

Jackson's (1987) study likewise suggests that abiotic factors may have strong effects on biotic interactions. Morin, Lawler, and Johnson's (1988) demonstration that interspecific competition between tadpoles and insects is as strong as competition between two anuran species means that the range of interspecific interactions we examine must be broadened considerably. The incorporation of several factors into experimental tests of anuran communities is advanced relative to studies of other vertebrate communities in which single-factor studies are still the norm (but for invertebrates see Price 1983 and Polis and McCormick 1987).

Eastern anuran communities are complex communities in which tadpole species composition may change a fair amount from year to year in a single pond. Changes in the presence or absence of tadpole predators, or types of tadpoles present in the pond, may shift the relative magnitudes of competition and predation and have large effects on the number of individuals in each anuran species that survive through the larval period (Morin 1987; Wilbur 1987). Three aspects of work on eastern anuran communities have especially important implications for studies of other communities, including desert anuran communities. First, species that perform well in one selective context may perform quite poorly in another (Wilbur 1980, 1987; Morin 1981, 1987). This means that changes in the suite of species that co-occur can drastically change the relative survival probabilities of each species, both because their own abilities change and because of the indirect effects of the changing fate of other species (Morin 1987). For instance, the presence of a predator may indirectly benefit one anuran species by greatly reducing the densities of that species's strongest competitor (Morin 1981, 1987).

Secondly, recent history may affect current community dynamics (Morin 1987; Wilbur 1987). Wilbur and Alford (1985) and Alford and Wilbur (1985) have demonstrated that the presence or absence of a species in a pond early in the year can affect the life history traits of other species that occur in the pond later in the season long after the early breeders have transformed. Thus, early arriving species can affect the nature of the community, perhaps by changing the nature of resources available or predators present.

Finally, Morin, Lawler, and Johnson (1988) demonstrated that tadpoles and aquatic insects compete to the same degree as do anurans. This suggests that we should not limit our study of communities to narrow taxonomic entities. All these ideas imply that understanding the distribution and abundance of species in nature requires considerably more than examining two closely related species in one interactive context at one point in time. They also imply that community dynamics are largely a consequence of local, recent conditions, and thus that generalizations based on studies of a single community should be well qualified.

Desert anuran communities have received considerably less attention than eastern communities. Early workers in this area (Bragg 1965; Low 1976) documented that several species did occupy the desert, and that certain species tended to co-occur in a single pond (Conant 1975; Stebbins 1975; Pomeroy 1981; Woodward 1987a). Spadefoot toads (*Scaphiopus*) and true toads (*Bufo*) primarily occupy temporary ponds in the deserts of the western United States. Permanent water sources are locally rare in desert areas, and in the southwestern United States these areas contain a different suite of frogs (mainly *Rana* and *Bufo*) than temporary ponds (Creusere and Whitford 1976; Woodward 1987a). In some desert areas a third suite of frogs (*Rana*, *Bufo*, and *Hyla*) occupy intermittent streams that flow from the rocky foothills onto desert lowlands (Stebbins 1975; Creusere and Whitford 1976).

Bragg's (1965) early natural history work revealed that desert anurans breed whenever heavy rains create a breeding pond during the warm parts of the year. These species usually have a short breeding season of a few nights (Bragg 1965; Wells 1977; Pomeroy 1981; Woodward 1982a). Unlike eastern anuran communities, sequential occupation of a single temporary pond by several species rarely occurs in temporary ponds in the desert. There are three possible reasons why this does not occur. First, the pond may be of such short duration that breeding at any time other than immediately after the pond forms will not allow sufficient time for larval transformation. Second, anuran larvae in temporary ponds are commonly omnivorous and voracious foragers, and thus might consume any eggs deposited in their pond (see Pomeroy 1981; Woodward 1982c). Finally, Woodward's (1987d) results suggest that hatchling tadpoles grow considerably faster and have higher survival if they grow with similarly sized tadpoles than if they grow with larger tadpoles, as would be the case for offspring of females that deposited eggs later than others.

Tadpole densities and biomass are quite high in temporary ponds in the desert (Bragg 1965; Woodward 1982b; Newman 1987), and total die-off of tadpoles in drying ponds is common (Bragg 1965; Pomeroy 1981; Newman 1987; Woodward manuscript b). Newman (1987) found that supplemental feeding increased survival in only the lowest density *Scaphiopus couchi* ponds that he manipulated in Texas. These results may mean that food is in such short supply that only a tremendous increase in resource level could increase survival. If this is the case, the combination of high tadpole density and low resource level may often lead to strong selection for tadpoles that are superior competitors for a limited resource. Woodward's (1981, 1982c) observations of scramble competition over food, and experimental demonstration of negative effects of high larval densities, also imply that tadpole densities in temporary desert ponds are high enough for competition over food to influence important life history parameters such as survival and growth rate.

Direct tests for interspecific interactions in desert anuran communities are almost nonexistent. Woodward's (1982b,c) competition experiments and Woodward's (1983a,b) and Woodward and Johnson's (1985) predation experiments are the only experimental studies that deal specifically with desert species. Woodward considered the anuran community to be all anurans breeding in ponds in a local area. In the spring, *Bufo woodhousei* breeds primarily in permanent ponds, but occasionally in temporary ponds. Summer breeders include *Rana catesbeiana* and *Rana pipiens*, which breed exclusively in permanent ponds, and four species (*Scaphiopus couchi*, *S. multiplicatus*, *S. bombifrons*, and *Bufo cognatus*) that breed exclusively in temporary ponds. *Bufo woodhousei* tadpoles can overlap temporarily with other species if Woodhouse's toad breeds late and the pond persists until later species breed in it.

Woodward's (1987a) desert anuran study was an attempt to determine why all anurans in a local area did not co-occur in all pond types. In several desert areas in the western United States, temporary and permanent ponds occur within meters of each other; thus, anurans have access to both pond types, but each species typically breeds only in one pond type. Woodward's study was similar to many early studies of eastern anuran communities in that it was designed to ascertain whether interspecific interactions could cause the locally disjunct distribution of species.

Woodward (1987a) found that temporary and permanent ponds differ in two important respects: Temporary ponds contain significantly fewer types of predators and smaller numbers of individual predators than permanent ponds (Woodward 1981, 1983a), but they also have higher tadpole densities than permanent ponds (Woodward 1981, 1982b). High tadpole densities characteristic of temporary ponds, the broad omnivorous and sometimes cannibalistic diet of temporary pond species, and observations of tadpoles physically pushing each other away from food, all suggest that competitive interactions over food might be an important selection pressure in temporary desert ponds. Similarly, abundant predators in permanent ponds suggested that predation might be a more significant selection pressure in permanent than in temporary ponds.

The existence of two types of ponds with two different selective regimes, and two groups of species, each restricted to a single pond type, suggests that divergent selection pressures in the two pond types might explain local distributions of species and thus explain community structure at a local scale. To determine if competition might be responsible for keeping permanent pond species from using temporary ponds, Woodward (1982c) raised temporary and permanent pond species together in artificial temporary ponds at natural densities. He found that temporary pond species had strong negative effects on survival and/or developmental rate of *R. pipiens*, the native permanent pond species, and on *B. woodhousei*, which breeds in both temporary and permanent ponds. Somewhat surprisingly, the temporary pond species had a positive effect on the survival and developmental rate of *Rana catesbeiana*, an introduced permanent pond species. These results suggest that competitive interactions could prevent *R. pipiens* and *B. woodhousei* from occupying temporary ponds. Although the mechanism responsible for lowering survival or reducing developmental rate could not be ascertained, exploitation competition over food and direct predation on other tadpoles by carnivorous *S. multiplicatus* tadpoles appear to be the most probable mechanisms (Woodward 1982c). *Rana catesbeiana* apparently cannot colonize temporary ponds because insufficient time exists for its larvae to complete development (>11 months needed) before the ponds dry.

Woodward (1983a,b) reasoned that predation could prevent temporary pond species from occupying permanent ponds if temporary pond species were more vulnerable to predation than permanent pond species. He found that most natural predators (insects, fish, snakes) consistently bias predation toward temporary pond tadpoles (Woodward 1983a), suggesting that predation might prevent temporary pond tadpoles from occupying permanent ponds. Although many factors, including defensive chemistry (Formanowicz and Brodie 1982), relative sizes of predator and prey (Calef 1973; Cooke 1974; Heyer, McDiarmid, and Weigmann 1975; Low 1976; Woodward 1983a,b), or behavior (Heyer, McDiarmid, and Weigmann 1975) can influence a tadpole's susceptibility to predation, the relatively frequent movement of temporary pond tadpoles appears to be a major mechanism responsible for the observed bias

(Woodward 1983a). Increased movement probably makes the tadpoles more obvious to visually searching predators and increases the contact rate between sit-and-wait predators and their prey.

Woodward's (1982c, 1983b, 1987a) study suggested that the community derived its structure from the fact that local areas are subdivided into two types of ponds with very different selective regimes, one with strong competitive interactions and another with strong predator-prey interactions. Wiens (1985) suggested that this type of community structuring may be common for vertebrates, as their environment consists of distinct patches over which the relative success of any group of species varies greatly. Although many have interpreted the success or failure of various species in different patch types as an outcome of interspecific competitive interactions (e.g., Schroeder and Rosenzweig 1975; Glass and Slade 1980), Glassner (1979) and Wiens (1977, 1985) have pointed out that other factors such as predation are equally tenable explanations, given the information currently available in many studies.

Patchiness is especially likely to influence community structure of aquatic organisms because nonaquatic interpatch areas form effective dispersal barriers to most aquatic organisms. Because of this, aquatic habitats in a local area are experienced in a coarse-grained manner by the aquatic organisms. The coarse-grained nature of the system means that individual organisms are seldom, if ever, exposed to selective pressures in aquatic environments other than the one in which they were born. Consequently, individuals with traits that enhance performance in other habitats only rarely experience the fitness benefits associated with possession of the traits in the other habitat. Rather, the bearer is more likely to experience any negative fitness consequences associated with having the trait in its natural habitat. Woodward (1981) suggested that communities composed of organisms that experience their environment in a coarse-grained manner may be structured in a slightly different manner than communities of organisms that experience their environment in a more fine-grained manner. Over their evolutionary history, species that experience their environment in a coarse-grained manner have rarely interacted with species from different patches. To the degree that their extant traits reflect a history of selection, these traits will be shaped by factors inside their patch. Species that experience their environment in a fine-grained manner probably possess traits resulting from a history of interactions with individuals from many patches. Thus, for a fine-grained species inequalities in abilities of two species to use a given resource probably have evolved in the context of divergent selection pressures over one resource (MacArthur 1972). For these species the same selection pressures that have been responsible for keeping the species apart in the past currently prevent the two species from using the same patch or resource. Coarse-grained species may differ considerably in this respect. That is, factors that have caused the species in different patches to have different characteristics may not be the same factors that currently prevent them from occupying the same patch or using the same resource. Thus, although inequalities in abilities of two species to interact in the contested habitat currently prevent coexistence, the inequalities have resulted from histories of divergent selection pressures in the two habitat types and not from a history of interactions over the contested habitat type or resource (Woodward 1981). Intraspecific interactions may commonly be a major factor shaping a coarse-grained species's characteristics in many desert temporary ponds, as often only a single anuran species occupies a pond. Thus intraspecific interactions may play a major role in influencing the outcome of interspecific interactions when they do occur.

Although Woodward's desert anuran study shed some light on how interspecific interactions could prevent coexistence of some species, it largely ignored abiotic factors and interactions among species within a pond type. These are important factors that may have strong effects on community structure, as outlined below. Woodward's study, like most early experimental studies of community structure, asked whether species interacted in a pattern consistent with the observed distribution of species. Because the interactions existed and were consistent with species distributions, Woodward (1982c, 1983a, 1987a) concluded that these interactions could be responsible for the community's structure. This in no way implies that other factors are not important or that other factors may not prevent the interactions he examined from being important. Woodward argued only that the strength of the species interactions was strong enough to prevent the species from different pond types from coexisting. The temporal scale of most experimental studies is less than a year and thus most of these studies produce results in which the experimenter must speculate concerning the impact of the interactions on community dynamics. This is quite different from studies performed during the natural history phase of community ecology when most traits considered (such as desiccation tolerance) were regarded as absolute requirements for admission to the community.

We think that the differences in selection pressures that Woodward (1982c, 1983a, 1987a) reported between temporary and permanent ponds probably exist in a qualitatively similar manner in most desert areas, although local factors may mitigate these patterns. One important local mitigating factor is the proximity of temporary ponds to permanent water, and thus to a source of predators. Close proximity to a permanent pond tends to increase predator densities in temporary ponds (Woodward manuscript c), and thus potentially reduce tadpole densities to levels where competitive interactions may not occur (Calef 1973). Man's irrigation practices in the desert may well be increasing the influence of predation in the desert anuran system.

Predation is probably relatively unimportant in temporary ponds because both the location and timing of occurrence of prey vary greatly from year to year, making the prey difficult to locate for predators that move between ponds. Most tadpole predators move between ponds rather than existing in dry temporary ponds in some sort of resting stage. Most of these predators and their eggs are small and may be incapable of storing enough energy to survive through long, sometimes multi-year, dry spells. Predation can become severe late in the season in some temporary ponds because many predators have arrived by this time (Pomeroy 1981). Predation is probably more important in permanent ponds because tadpoles are present every year and thus constitute a predictable resource (Low 1976). In general, the predator population is extant when tadpoles hatch, and thus individual predators are probably capable of consuming the prey for a long period of time before the tadpoles become too large to eat.

Competition is probably an important selection pressure in most temporary ponds because tadpole densities are often high and time available for growth is very short. Tadpole densities are high because of the large clutch sizes of anurans (Salthe and Duellman 1973; Kuramoto 1978; Crump and Kaplan 1979; Woodward 1987a,b), the small (and commonly shrinking) size of ponds, and because frogs from a large area congregate to breed in any one pond. Resource levels are probably low relative to demand in temporary ponds in the desert, as implied by Newman's (1987) food addition experiments. Temporary ponds usually form immediately before they are occuped by tadpoles and there is little time for algal communities to build up. Further, there is little living vegetation when the pond first forms because pond duration is preceded by a long dry season. The short duration of temporary ponds means that there is a premium on rapid growth, and slow growth rates cannot be compensated for by a longer larval period in temporary ponds. We suspect that competitive interactions are usually of lesser importance in permanent than temporary ponds for two reasons. The resource base is probably larger because of the pond's persistence, and this also allows a predator base to build up. High resource levels and predators that decrease tadpole densities may combine to make competition less important in permanent ponds (Calef 1973). Theorists studying predator-prey interactions in other taxa have come to similar conclusions about how predictability will influence exposure to predators. Janzen (1968), Levin (1971), Cates and Orians (1975), and Orians et al. (1977) suggest that plants whose presence is unpredictable in time and space are able to escape from specialized predators. In the absence of predators, densities may reach levels at which competitive interactions become important.

Unstudied Factors of Potential Importance

Many factors that could affect presence, absence, or relative abundance of anurans in desert communities have not been investigated and at this point we suggest some important unstudied factors that could affect desert anuran communities. We draw from previous work on eastern anuran communities (Wilbur 1980, 1987), and from our knowledge of similarities and differences of eastern and desert anurans.

Some of the more striking differences between eastern and desert anuran communities are the longer breeding periods and the greater number of species that occupy eastern anuran communities. The presence of many species and frequent occurrence of sequential breeding in the eastern communities mean that the potential for interspecific interactions is greater in eastern communities. Priority effects (Paine 1977; Morin 1987; Wilbur 1987) and indirect effects (Morin 1987) appear to be very important in structuring eastern anuran communities, but are probably less important in desert anuran communities simply because of the lesser number of species present, and lack of temporal subdivision of pond use, especially in desert temporary ponds. Priority effects may influence desert anurans in permanent ponds, in a manner similar to that in the east, although in most areas the species pool consists of only two or three species (Stebbins 1975). Changes in the temporal sequence of species occupation of a desert temporary pond within any year are unlikely, as breeding is essentially synchronous. However, year-to-year changes in which species breed in a desert temporary pond are not unknown (Creusere and Whitford 1976), and shifts in which species share a pond are common.

Year-to-year shifts in species co-occurrence may be common in desert temporary ponds if the species require different amounts of rainfall or different pond-water depths to breed (Bragg 1965). Shifts in which species occupy a pond may influence the relative success of these species in different years. For instance, *S. couchi* breeds in more ephemeral ponds than *S. multiplicatus*. Coexistence of these species in a local area may depend on the frequency of years in which ponds are ephemeral enough to preclude the presence of *S. multiplicatus* and thus remove *S. couchi* tadpoles from the presence of the predaceous *S. multiplicatus* tadpoles. In a similar manner, co-occurrence of tadpole shrimp or fairy shrimp and desert anurans also varies considerably from pond to pond (Woodward manuscript d). The numbers of these crustaceans may be high enough to have important effects on anuran community structure (Morin, Lawler, and Johnson 1988). Long-term coexistence of temporary pond species in the desert may depend on these year-to-year shifts in community composition in a manner similar to eastern communities (Morin 1987). Similarly, yearly shifts in pond duration may also influence the number of individuals of each species that transform before the pond dries (Wilbur 1987). This potentially exerts an important effect on the dynamics of desert anuran community structure. Because of the long lifetimes of adult desert anurans, current conditions, through an effect on which individuals survive, probably have long-lasting effects on community composition. Presently, these factors have not been studied in desert communities. Effects due to short-term shifts in community composition may also be important in other desert organisms that have relatively poor dispersal abilities and that can be strongly affected by local changes in community composition. Desert plants and insects are two groups likely to be affected in this manner.

Other changes in local ponds may also change the nature of interspecific interactions and thus change relative species abundance. The effect predators have on anurans may be affected by tadpole density. When tadpoles occur at high densities (a function of many adults breeding, large clutch size, or small pond size), harvesting costs for tadpole predators may be low and predators may grow rapidly. The most probable outcome of larger predator sizes is that more tadpoles are consumed (Woodward 1983a) and that the size, and consequently the species, of prey consumed may shift (Woodward 1983a). The relative abundance of alternative prey types may also influence predator growth rates. Although increased abundance of more palatable or catchable prey may increase predator growth rate, increased abundance of species such as *S. multiplicatus*, that probably have diets similar to their predators, may reduce the predators' growth rate, and thus reduce the effect of predation on some species. Changes in water temperature may also influence the effect of predation if changes in temperature differentially affect the growth or survival of the predator or any of its prey. Similarly, changes in tadpole densities, relative abundances of different anuran tadpoles, or temperatures can influence the outcome of competitive interactions if the species respond differentially to the changes in conditions. Changes could act directly on an anuran species (e.g., slower growth of its predator enhances its own survival) or indirectly (e.g., slower predator growth increases the survival of the anurans' competitors, reducing the amount of food available and lowering the anurans' survival because of competitive losses), as discussed by Morin (1987). Tadpole densities, tadpole composition, and temperature vary tremendously between desert ponds. How these factors interact to influence survival of different species is a wide-open question for the desert system.

Intraspecific Interactions and Community Structure

At the beginning of this chapter we suggested that studies of how local phenomena affect community structure was a rapidly developing area of community ecology. We have already briefly discussed how some local factors such as priority effects or historical effects could influence community structure. We close by suggesting some ways that intraspecific interactions may influence anuran community structure, and developing one of these ideas in some detail. A large and growing body of data suggests that tadpole density (Wilbur 1980, 1987) and body size (Steinwascher 1978a,b; Wilbur 1980; Woodward 1983a,b, 1987d) affect interspecific interactions by influencing the ability of tadpoles to compete or avoid predators. Many types of intraspecific interactions can affect interspecific interactions by influencing tadpole survival and thus tadpole densities, or by influencing tadpole growth rate and thus body size (see Table 8.1 for numerous examples of how intraspecific factors can influence densities or body sizes of anurans). Understanding how intraspecific interactions shape characteristics of larvae or other life history stages should help us to better understand interspecific interactions.

The mating system is one intraspecific factor that can affect interspecific interactions. Study of mating systems traditionally has been the province of population biologists; until recently (Thornhill 1987), community ecologists have devoted little attention to mating systems and their effects on community structure. Part of this neglect probably stems from the fact that most community ecologists regard the traits of organisms as uniform within a species (Colwell 1984) and attempt to deduce community structure by looking at where these traits place species along a continuum of resource use. Because mating systems determine who mates and thus which subset of intraspecific trait variation is expressed in the following generation, it determines the traits individuals use in interspecific interactions in the next generation.

The first indication that mating systems might affect anuran community structure came from the field of ecological genetics. A series of studies of anurans that breed in temporary ponds reported considerable differences among fullsibling and halfsibling families in larval performance (Travis 1980a,b, 1981, 1983a,b; Travis, Keen, and Julianna 1985; Travis and Trexler 1986; Travis, Emerson, and Blouin 1987; Berven 1982a,b, 1987; Berven and Gill 1983; Woodward 1986, 1987c; Newman 1988a,b; Woodward, Travis, and Mitchell 1988; Mitchell 1990). Extensive interfamily variation in larval survival, larval and juvenile growth rates, metamorphic size, and larval period duration suggests that who mates may influence the nature of interspecific interactions in the next generation. Larval period duration, growth rate, and metamorphic size affect the nature of predator-tadpole interactions by influencing how long a tadpole is exposed to predation (Travis, Emerson, and Blouin 1987) and by influencing a tadpole's size and thus its susceptibility to predation (Caldwell, Thorp, and Jervey 1980; Woodward 1983a,b; Woodward and Johnson 1985; Cronin and Travis 1986). Tadpole size also probably influences interspecific competitive ability (Steinwascher 1978a,b; Woodward 1987d). Finally, all of the above traits affect tadpole survival and thus tadpole densities, which can determine the relative importance of competition and predation in an anuran community (Wilbur 1987).

Although part of the variation among fullsibling families may be due to maternal nutritive effects or cytoplasmic genes, much is due to chromosomal genes (Travis 1981; Berven 1982a,b, 1987; Berven and Gill 1983; Travis, Emerson, and Blouin 1987; Woodward 1986, 1987c; Woodward, Travis, and Mitchell 1988). Further, although part of the genetic variation in offspring larval period duration, metamorphic weight, developmental rates, and survival is due to dominance variation, a considerable portion is due to additive genetic effects and is thus heritable (Travis, Emerson, and Blouin 1987; Berven 1987; Newman 1988a,b).

Several recent studies have established that there is a positive relationship between male mating success, or correlates of male mating success, and components of offspring fitness. Woodward (1986, 1987c) found that correlates of male mating success in *Scaphiopus* predict juvenile growth rate (*S. multiplicatus*) and tadpole survival (*S. couchi*). Mitchell (1990) has found that a correlate of male mating success in Woodhouse's toad (*B. woodhousei*) predicts offspring metamorphic weight. In the only study of its sort, Woodward, Travis, and Mitchell (1988) found that male *Hyla crucifer* who obtained mates in nature produced tadpoles with higher larval growth rates than males that did not obtain mates in the field. These data suggest that anuran mating systems can influence offspring traits that could eventually affect community structure. These mating systems appear to favor as mates those males that enhance larval abilities in interspecific interactions (Woodward 1986, 1987c; Woodward, Travis, and Mitchell 1988). Whether females are choosing their mates to obtain genes superior in the context of interspecific interactions (Hamilton and Zuk 1982; Kodric-Brown and Brown 1984) remains to be seen. It is clear, however, that the mating system may be sorting among variation within a species and affect that species's success in interspecific interactions in a way that could shape community structure.

Another aspect of the mating system can also affect community structure. Female anurans do not lay their eggs indiscriminately around the environment, but restrict oviposition to a subset of available sites. It has long been known in desert species (Bragg 1965) that males arrive at a shallow temporary pond and call but that females do not breed until subsequent rains fill it to a greater depth. Thus, females appear to control where and when they breed. If as Bragg (1965) and our own observations suggest, different temporary pond species breed in ponds of differing depths, then female "choice" of pond site may determine which species currently co-occur as tadpoles in any pond. Thus the mating system may not only influence characteristics of the tadpoles that interact with other species (Woodward 1986, 1987c; Woodward, Travis, and Mitchell 1988) but also determine whether certain species co-occur in a pond in a given year. Resetarits and Wilbur (1989) report that some female anurans choose oviposition sites to avoid competitors and predation on their young.

Clearly, we have just touched the surface of the numerous factors that influence the nature of interspecific interactions and thus structure desert anuran communities. The list of factors that influence anuran community structure is already long (Wilbur 1980, 1984, 1987; Morin 1981, 1987) but will undoubtedly grow longer. Studies of community structure in most other taxa are not as well advanced, most likely because tadpoles are so amenable to manipulation. Many factors demonstrated to be important in anuran communities will probably also prove to be important in other taxa with limited powers of dispersal, such as plants and many insects.

Conclusions

1. Desert anuran communities are quite depauperate, with a single pond containing 1–5 species, and any local area usually containing 10 or fewer species (Bragg 1965; Pomeroy 1981; Woodward 1981).
2. Only a small fraction of the world's amphibians, mainly anurans, occupy deserts.
3. Each desert species usually breeds only in temporary or only in permanent ponds, but not in both.
4. Very little is known about why any of these species occur where they do on a local scale. Woodward's (1982c, 1983a, 1987a) and Newman's (1987) studies suggest that predation and competition may play major roles in organizing desert anuran communities.
5. Whether competition or predation are important local selection pressures appears to depend on how predictable a prey resource the tadpoles will be. When tadpoles are a predictable resource, predation appears to be a major selection pressure. If tadpoles are not a predictable resource, predation appears to be of lesser importance, and densities reach levels at which competitive interactions may have a major effect on community structure. Morin, Lawler, and Johnson's (1988) results suggest that we should broaden our tests for competitive interactions to include interphyletic comparisons.
6. Deserts are inherently variable systems and conditions in a pond can change greatly from one year to the next. Changes in anuran species composition, tadpole densities, presence or absence of predators, pond duration, or pond temperatures can alter relative survival probabilities of anurans. These changes may temporarily shift the success of different species, allowing coexistence of several species in one area over time (Morin 1987). Much work remains to be performed on these topics in desert systems.
7. Species that perform well in one selective context may perform poorly in another (Woodward 1982c, 1983a; Morin 1987). Although understanding how a species performs in one selective context is useful information, under-

standing community dynamics requires an examination of a species's performance across the range of natural selective contexts.

8. The ideas discussed in numbers 5 and 6 above suggest that understanding community dynamics requires considerable knowledge about local conditions both in the present and recent past. This emphasis on local conditions and history means that to a large degree, each community represents a unique case. This does not mean that general patterns may not emerge from community studies, but it implies that the patterns will probably most typically deal with relative strengths of different factors and with absence of some factors in specific systems.

9. Recent studies by Wilbur (1987) suggest not only that both abiotic and biotic factors affect anuran community structure, but they also imply that the interaction of these factors may also have an important effect. Studies examining the relative magnitudes of effects of abiotic and biotic factors are long overdue for desert systems.

10. Several types of intraspecific interactions, especially those involved with mating systems, have the potential to affect the characteristics of individuals and thus influence interspecific interactions that structure the community. Thus, understanding the nature of these intraspecific interactions should assume a more important role in community ecology.

11. More attention needs to be paid to the relative magnitudes of effects (e.g., intra- or interspecific competition, predation, interaction with abiotic factors) and to the temporal and spatial scales over which these effects apply.

Acknowledgments

We would like to thank J. Brown, L. Marshall, G. Polis, and N. Scott for comments on the manuscript, and our colleagues at the University of New Mexico for contributing to our understanding of community ecology. We also gratefully acknowledge financial support from the U.S. Fish and Wildlife Service and the National Science Foundation (grants BNS-8204329, and BNS-8300138 to B. Woodward; and grant BSR84-13223 to J. Travis and B. Woodward) that have made some of the work for this paper possible.

Bibliography

Alford, R. A. 1989. Variation in predator phenology affects predator performance and prey community composition. *Ecology* 70: 206–219.

Alford, R. A., and H. M. Wilbur. 1985. Priority effects in experimental pond communities: competition between *Bufo* and *Rana. Ecology* 66: 1097–1105.

Bender, A. E., T. J. Case, and M. E. Gilpin. 1984. Perturbation experiments in community ecology: theory and practice. *Ecology* 65: 1–13.

Bentley, P. J. 1966. Adaptations of amphibia to arid environments. *Science* 152: 619–623.

Berven, K. A. 1982a. The genetic basis of altitudinal variation in the wood frog, *Rana sylvatica*. I. An experimental analysis of life history traits. *Evolution* 36: 962–983.

————. 1982b. The genetic basis of altitudinal variation in the wood frog, *Rana sylvatica*. II. An experimental analysis of larval traits. *Oecologia* 52: 360–369.

————. 1987. The heritable basis of variation in larval developmental patterns within populations of the wood frog (*Rana sylvatica*). *Evolution* 41: 1088–1097.

Berven, K. A., and D. E. Gill. 1983. Interpreting geographic variation in life history traits. *American Zoologist* 23: 85–97.

Blair, W. F. 1976. Adaptations of anurans to equivalent desert scrub of North and South America. Pp. 197–222 in D. W. Goodall (ed.), *Evolution of Desert Biota*. Austin, Tex.: University of Texas Press.

Bragg, N. 1965. *Gnomes of the Night*. Philadelphia, Pa.: University of Pennsylvania Press.

Brockelman, W. Y. 1969. An analysis of density effects and predation in *Bufo americanus* tadpoles. *Ecology* 50: 632–644.

Caldwell, J. P., J. H. Thorp, and T. O. Jervey. 1980. Predator-prey relationships among larval dragonflies, salamanders, and frogs. *Oecologia* 46: 285–289.

Calef, G. W. 1973. Natural mortality of tadpoles in a population of *Rana aurora*. *Ecology* 54: 741–758.

Cates, R. G., and G. H. Orians. 1975. Successional status and the palatability of plants to generalized herbivores. *Ecology* 56: 410–418.

Colwell, R. K. 1984. What's new? Community ecology discovers biology. Pp. 387–396 in P. W. Price, C. N. Slobodchikoff, and W. S. Gaud (eds.), *A New Ecology: Novel Approaches to Interactive Systems*. New York: John Wiley and Sons.

Conant, R. 1975. A field guide to reptiles and amphibians of eastern and central North America. Boston: Houghton Mifflin.

Connell, J. H. 1975. Some mechanisms producing structure in natural communities: a model and evidence from field experiments. Pp. 460–490 in M. L. Cody and J. M. Diamond (eds.), *Ecology and Evolution of Communities*. Cambridge, Mass.: Belknap Press, Harvard University.

————. 1983. On the prevalence and relative importance of interspecific competition: evidence from field experiments. *American Naturalist* 122: 661–696.

Cooke, A. S. 1974. Differential predation by newts on anuran tadpoles. *British Journal of Herpetology* 5: 386–390.

Creusere, F., and W. Whitford. 1976. Ecological relationships in a desert anuran community. *Herpetologica* 32: 7–18.

Cronin, J. T., and J. Travis. 1986. Size-limited predation on larval *Rana areolata* (Anura: Ranidae) by two species of backswimmer (Hemiptera: Notonectidae). *Herpetologica* 42: 171–174.

Crump, M. 1974. Variation in propagule size as a function of environmental uncertainty for tree frogs. *American Naturalist* 117: 724–737.

Crump, M., and R. Kaplan. 1979. Clutch energy partitioning of tropical tree frogs (Hylidae). *Copeia* 1979: 826–834.

DeBenedictis, P. A. 1974. Interspecific competition between tadpoles of *Rana pipiens* and *Rana sylvatica*: an experimental field study. *Ecological Monographs* 44: 129–151.

Endler, J. A. 1986. *Natural Selection in the Wild*. Princeton University Press Monographs on Population Biology 21: 1–337.

Formanowicz, D. R., and E. D. Brodie, Jr. 1982. Relative palatabilities of members of a larval amphibian community. *Copeia* 1982: 91–97.

Glass, G. E., and N. A. Slade. 1980. The effect of *Sigmodon hispidus* on spatial and temporal activity of *Microtus ochrogaster*: evidence for competition. *Ecology* 61: 358–370.

Glassner, J. W. 1979. The role of predation in shaping and maintaining the structure of communities. *American Naturalist* 113: 631–641.

Hamilton, W. D., and M. Zuk. 1982. Heritable true fitness and bright birds: a role for parasites? *Science* 218: 384–387.

Hardin, G. 1960. The competitive exclusion principle. *Science* 131: 1292–1297.

Heatwole, H. 1982. A review of structuring in herpetofaunal assemblages. Pp. 1–19 in N. J. Scott, Jr. (ed.), *Herpetological Communities*. U.S. Fish and Wildlife Service Wildlife Research Report 13, Washington, D.C.

Herreid, C. F., and S. Kinney. 1966. Survival of Alaskan woodfrog (*Rana sylvatica*) larvae. *Ecology* 47: 1039–1041.

Heyer, W. R. 1976. Studies in larval amphibian habitat partitioning. *Smithsonian Contributions to Zoology* 242: 1–27.

Heyer, W. R., R. W. McDiarmid, and D. L. Weigmann. 1975. Tadpoles, predation, and pond habitats in the tropics. *Biotropica* 7: 100–111.

Hutchinson, G. E. 1959. Homage to Santa Rosalia, or Why are there so many kinds of animals? *American Naturalist* 93: 145–159.

Janzen, D. H. 1968. Host plants as islands in evolutionary and contemporary time. *American Naturalist* 102: 592–595.

Kodric-Brown, A., and J. H. Brown. 1984. Truth in advertising: the kinds of traits favored by sexual selection. *American Naturalist* 124: 309–323.

Kuramoto, M. 1978. Correlations of quantitative parameters of fecundity in amphibians. *Evolution* 32: 287–296.

Levin, D. A. 1971. Plant phenolics: an ecological perspective. *American Naturalist* 105: 157–181.

Licht, L. E. 1967. Growth inhibition in crowded tadpoles: intraspecific and interspecific effects. *Ecology* 48: 736–745.

Low, B. S. 1976. The evolution of amphibian life histories in the desert. Pp. 149–195 in D. W. Goodall (ed.), *Evolution of Desert Biota*. Austin, Tex.: University of Texas Press.

MacArthur, R. A. 1972. *Geographical Ecology*. New York: Harper and Row.

MacArthur, R. A., and E. O. Wilson, 1967. *The Theory of Island Biogeography*. Princeton University Press Mongraphs on Population Biology 1: 1–203.

Main, A. R. 1968. Ecology, systematics, and evolution of Australian frogs. *Advances in Ecological Research* 5: 37–87.

Martof, B. 1956. Factors influencing size and composition of populations of *Rana clamitans*. *American Midland Naturalist* 56: 244–245.

Mayhew, W. W. 1968. Biology of desert amphibians and reptiles. Pp. 195–356 in G. W. Brown (ed.), *Desert Biology*, vol. 1. New York: Academic Press.

Mitchell, S. 1990. Genetic effects of the mating system on offspring traits in Woodhouse's toad (*Bufo woodhousei*): a comparison of lab and field results. *Evolution* 44: 502–519.

Morin, P. J. 1981. Predatory salamanders reverse the outcome of competition among three species of anuran tadpoles. *Science* 212: 1284–1286.

———. 1983. Predation, competition, and the composition of larval anuran guilds. *Ecological Monographs* 53: 119–138.

———. 1986. Interactions between intraspecific competition and predation in an amphibian predator-prey system. *Ecology* 67: 713–720.

———. 1987. Predation, breeding asynchrony, and the outcome of competition among treefrog tadpoles. *Ecology* 68: 675–683.

Morin, P. J., S. P. Lawler, and E. A. Johnson. 1988. Competition between aquatic insects and vertebrates: interaction strength and higher order interactions. *Ecology* 69: 1401–1409.

Neish, I. C. 1970. A comparative analysis of the feeding behaviour of two salamander populations in Marion Lake, B.C. Ph.D. dissertation, University of British Columbia, Vancouver.

Newman, R. 1987. Effects of density and predation on *Scaphiopus couchi* tadpoles in desert ponds. *Oecologia* 71: 301–307.

———. 1988a. Adaptive plasticity in development of *Scaphiopus couchi* tadpoles in desert ponds. *Evolution* 42: 774–783.

———. 1988b. Genetic variation for larval anuran (*Scaphiopus couchi*) development time in an uncertain environment. *Evolution* 42: 763–774.

Odum, H. T. 1957. Trophic structure and productivity of Silver Springs, Florida. *Ecological Monographs* 27: 55–112.

Orians, G. H., R. G. Cates, M. A. Mares, A. Moldenke, J. Neff, D. F. Rhoades, M. L. Rosenzweig, B. B. Simpson, J. C. Schultz, and C. S. Tomoff. 1977. Resource utilization systems. In G. Orians and O. Solbrig (eds.), *Convergent Evolution in Warm Desert Systems*. Stroudsburg, Pa.: Dowden, Hutchinson and Ross.

Paine, R. T. 1966. Food web complexity and species diversity. *American Naturalist* 100: 65–85.

———. 1969. The *Pisaster-Tegula* interaction: prey patches, predator food preference, and intertidal community structure. *Ecology* 50: 950–961.

———. 1977. Controlled manipulations in the marine intertidal zone, and their contributions to ecological theory. Pp. 245–270 in *The Changing Scenes in Natural Sciences, 1776–1976*. Academy of Natural Sciences Special Publication no. 12, Philadelphia.

Pianka, E. R. 1974. *Evolutionary Ecology*. New York: Harper and Row.

Polis, G. A., and S. J. McCormick. 1987. Intraguild predation and competition among desert scorpions. *Ecology* 63: 332–343.

Pomeroy, L. 1981. Developmental polymorphisms in the tadpoles of the spadefoot toad *Scaphiopus multiplicatus*. Ph.D. dissertation, University of California, Riverside.

Pound, R., and F. E. Clements. 1898. *The Phytogeography of Nebraska*. Lincoln: University of Nebraska Botanical Survey.

Price, P. W. 1983. Hypotheses on organization and evolution in herbivorous insect communities. Pp. 559–596 in R. F. Denno and M. S. McClure (eds.), *Variable Plants and Herbivores in Natural and Managed Systems*. New York: Academic Press.

Price, P. W., C. N. Slobodchikoff, and W. S. Gaud, eds. 1984. *A New Ecology: Novel Approaches to Interactive Systems*. New York: John Wiley and Sons.

Pyke, G. H., H. R. Pulliam, and E. L. Charnov. 1977. Optimal foraging: a selective review of theory and tests. *Quarterly Review of Biology* 52: 137–154.

Resetarits, W. J., and H. M. Wilbur. 1989. Choice of oviposition site by *Hyla chrysoscelis*: role of predators and competitors. *Ecology* 70: 220–228.

Reynolds, R. P., and N. J. Scott, Jr. 1982. Use of a mammalian resource by a Chihuahuan snake community. Pp. 99–118 in N. J. Scott, Jr. (ed.), *Herpetological Communities*. U.S. Fish and Wildlife Service Wildlife Research Report 13, Washington, D.C.

Roth, A. H., and J. F. Jackson. 1987. The effect of pool size on recruitment of predatory insects and on mortality in a larval anuran. *Herpetologica* 43: 224–232.

Roughgarden J., and J. Diamond. 1986. Overview: The role of species interactions in community ecology. Pp. 333–343 in J. Diamond and T. J. Case (eds.), *Community Ecology*. New York: Harper and Row.

Salthe, S., and W. Duellman. 1973. Quantitative constraints associated with reproductive mode in anurans. Pp. 229–249 in J. Vial (ed.), *Evolutionary Biology of the Anurans*. Columbia, Mo.: University of Missouri Press.

Savage, J. M. 1973. The geographic distribution of frogs: patterns and predictions. Pp. 351–445 in J. L. Vial (ed.), *Evolutionary Biology of the Anurans*. Columbia, Mo.: University of Missouri Press.

Schoener, T. W. 1983. Field experiments on interspecific competition. *American Naturalist* 122: 240–285.

Schroeder, P. M., and M. L. Rosenzweig. 1975. Perturbation analysis of competition and overlap in habitat utilization between *Dipodomys ordii* and *Dipodomys merriami*. *Oecologia* 19: 9–28.

Scott, N. J., Jr., and H. W. Campbell. 1982. A chronological bibliography, the history and status of studies of herpetological communities, and suggestions for future research. Pp. 221–239 in N. J. Scott, Jr. (ed.), *Herpetological Communities*. U.S. Fish and Wildlife Service Wildlife Research Report 13, Washington, D.C.

Simberloff, D. S., and E. O. Wilson. 1970. Experimental zoogeography of islands: a two year record of colonization. *Ecology* 51: 934–937.

Smith, D. C. 1983. Factors controlling tadpole populations of the chorus frog (*Pseudacris triseriata*) on Isle Royale, Michigan. *Ecology* 64: 501–510.

———. 1987. Adult recruitment in chorus frogs: effects of size and date at metamorphosis. *Ecology* 68: 344–350.

Smith-Gill, S. J., and D. E. Gill. 1978. Curvilinearities in the competition equations: an experiment with ranid tadpoles. *American Naturalist* 112: 557–570.

Stebbins, R. C. 1975. *A Field Guide to Western Reptiles and Amphibians*. Boston: Houghton Mifflin.

Steinwascher, K. 1978a. Competitive interactions among tadpoles: responses to resource level. *Ecology* 60: 1172–1183.

———. 1978b. The effect of coprophagy on the growth of *Rana catesbeiana* tadpoles. *Copeia* 1978: 130–134.

Strong, D. R., Jr. 1984. Exorcising the ghost of competition past: phytophagous insects. Pp. 28–41 in D. R. Strong, Jr., D. Simberloff, L. G. Abele, and A. B. Thistle (eds.), *Ecological Communities: Conceptual Issues and the Evidence*. Princeton, N.J.: Princeton University Press.

Thornhill, R. 1987. The relative importance of intra- and interspecific competition in scorpionfly mating systems. *American Naturalist* 130: 711–729.

Travis, J. 1980a. Genetic variation for larval specific growth rate in the frog *Hyla gratiosa*. *Growth* 44: 167–181.

———. 1980b. Phenotypic variation and the outcome of interspecific competition in Hylid tadpoles. *Evolution* 34: 40–50.

———. 1981. Control of larval growth variation in a population of *Pseudacris triseriata* (Anura: Hylidae). *Evolution* 37: 496–512.

———. 1983a. Variation in development patterns of larval anurans in temporary ponds. I. Persistent variation within a *Hyla gratiosa* population. *Evolution* 37: 496–512.

———. 1983b. Variation in growth and survival of *Hyla gratiosa* larvae in experimental enclosures. *Copeia* 1983: 232–237.

Travis, J., S. B. Emerson, and M. Blouin. 1987. A quantitative genetic analysis of fitness-related traits in *Hyla crucifer*. I. Larval development patterns. *Evolution* 41: 145–156.

Travis, J., W. H. Keen, and J. Julianna. 1985. The effects of multiple factors on viability selection in *Hyla gratiosa* tadpoles. *Evolution* 39: 1087–1099.

Travis, J., and J. C. Trexler. 1986. Interactions among factors affecting growth, development, and survival in experimental populations of *Bufo terrestris* (Anura: Bufonidae). *Oecologia* 69: 110–116.

Walters, B. 1975. Studies of interspecific predation within an amphibian community. *Journal of Herpetology* 9: 267–279.

Warburg, M. R. 1965. Studies on the water economy of some Australian frogs. *Australian Journal of Zoology* 13: 317–330.

Wells, K. D. 1977. The social behavior of anuran amphibians. *Animal Behaviour* 25: 666–693.

Whittaker, R. H. 1967. *Communities and Ecosystems*. New York: Macmillan.

Wiens, J. A. 1977. On competition and variable environments. *American Scientist* 65: 590–597.

———. 1985. Vertebrate responses to environmental patchiness in arid and semiarid ecosystems. Pp. 169–193 in S. T. A. Pickett and P. S. White (eds.), *The Ecology of Natural Disturbance and Patch Dynamics*. New York: Academic Press.

Wiens, J. A., J. F. Addicott, T. J. Case, and J. Diamond. 1986. Overview: The importance of spatial and temporal scale in ecological investigations. Pp. 145–153 in J. Diamond and T. J. Case (eds.), *Community Ecology*. New York: Harper and Row.

Wiest, J. A., Jr. 1982. Anuran succession at temporary ponds in a post oak-savanna region of Texas. Pp. 39–47 in N. J. Scott, Jr. (ed.), *Herpetological Communities*. U.S. Fish and Wildlife Service Wildlife Research Report 13, Washington, D.C.

Wilbur, H. M. 1972. Competition, predation and the structure of the *Ambystoma-Rana sylvatica* community. *Ecology* 53: 3–21.

———. 1976. Density-dependent aspects of growth and metamorphosis in *Ambystoma* and *Rana sylvatica*. *Ecology* 57: 1289–1296.

———. 1977a. Density-dependent aspects of growth and metamorphosis in *Bufo americanus*. *Ecology* 58: 196–200.

———. 1977b. Interactions of food level and population density in *Rana sylvatica*. *Ecology* 58: 206–209.

———. 1980. Complex life cycles. *Annual Review of Ecology and Systematics* 11: 67–94.

———. 1984. Complex life cycles and community organization in amphibians. Pp. 196–224 in P. W. Price, C. N. Slobodchikoff, and W. S. Gaud (eds.), *A New Ecology: Novel Approaches to Interactive Systems*. New York: John Wiley and Sons.

———. 1987. Regulation of structure in complex systems: experimental temporary pond communities. *Ecology* 68: 1437–1452.

Wilbur, H. M., and R. A. Alford. 1985. Priority effects in experimental pond communities: responses of *Hyla* to *Bufo* and *Rana*. *Ecology* 66: 1106–1114.

Wilbur, H. M., P. J. Morin, and R. N. Harris. 1983. Salamander predation and the structure of experimental communities: anuran responses. *Ecology* 64: 1423–1429.

Wiltshire, F. J., and C. M. Bull. 1977. Potential competitive interactions between larvae of *Pseudophryne bibroni* and *P. semimarmorata* (Anura: Leptodactylidae). *Australian Journal of Zoology* 25: 229–234.

Woodward, B. D. 1981. The roles of competition and predation in organizing a desert anuran community. Ph.D. dissertation, University of New Mexico, Albuquerque, 222 pp.

———. 1982a. Sexual selection and nonrandom mating patterns in desert anurans (*Bufo woodhousei, Scaphiopus couchi, S. multiplicatus,* and *S. bombifrons*). *Copeia* 1982: 351–355.

———. 1982b. Tadpole interactions in the Chihuahuan Desert at two experimental densities. *Southwestern Naturalist* 27: 119–121.

———. 1982c. Tadpole competition in a desert anuran community. *Oecologia* 54: 96–100.

————. 1983a. Predator-prey interactions and breeding pond use of temporary pond species in a desert anuran community. *Ecology* 64: 1549–1555.

————. 1983b. Tadpole size and predation in the Chihuahuan Desert. *Southwestern Naturalist* 28: 470–471.

————. 1984. Operational sex ratios and sex biased mortality in *Scaphiopus* (Pelobatidae). *Southwestern Naturalist* 29: 232–233.

————. 1986. Paternal effects on juvenile growth in *Scaphiopus multiplicatus* (the New Mexican spadefoot toad). *American Naturalist* 128: 58–65.

————. 1987a. Clutch parameters and pond use in some Chihuahuan Desert anurans. *Southwestern Naturalist* 32: 13–19.

————. 1987b. Intra- and interspecific variation in spadefoot toad (*Scaphiopus*) clutch parameters. *Southwestern Naturalist* 32: 127–156.

————. 1987c. Paternal effects on offspring traits in *Scaphiopus couchi* (Anura: Pelobatidae). *Oecologia* 73: 626–629.

————. 1987d. Tadpole interactions and breeding season duration of Woodhouse's toad (*Bufo woodhousei*). *Copeia* 1987: 380–385.

————. Manuscript a. Interpopulation variation in ages of adult spadefoot toads. Albuquerque: University of New Mexico.

————. Manuscript b. Total population failure in *Scaphiopus* breeding ponds. Albuquerque: University of New Mexico.

————. Manuscript c. Pond spatial distributions and predator densities in spadefoot toad breeding ponds. Albuquerque: University of New Mexico.

————. Manuscript d. The local distribution of anuran larvae and crustaceans in desert and prairie ponds. Albuquerque: University of New Mexico.

————. Manuscript e. Bovine feces and egg death in *Scaphiopus*. Albuquerque: University of New Mexico.

Woodward, B. D., and P. Johnson. 1985. *Ambystoma tigrinum* (Ambystomatidae) predation on *Scaphiopus couchi* (Pelobatidae) tadpoles of different sizes. *Southwestern Naturalist* 30: 460–461.

Woodward, B. D., and S. L. Mitchell. In press. Predation on frogs in breeding choruses. *Southwestern Naturalist*.

Woodward, B. D., J. Travis, and S. L. Mitchell. 1988. Nonrandom genetic consequences of the spring peeper (*Hyla crucifer*) mating system. *Evolution* 42: 784–794.

9 Desert Reptile Communities

Laurie J. Vitt

Since the pioneering studies on thermoregulation by Cowles and Bogert (1944), desert reptiles have caught the attention of physiologists, behaviorists, and ecologists. Among reptiles, desert species are particularly amenable to ecological studies because (1) compared to tropical forest, chaparral, savanna (cerrado), or temperate forest reptiles, most desert species can be easily observed, marked, monitored, collected, and manipulated; (2) for most deserts, the reptile fauna is well known and good reference collections exist; (3) many desert species often occur at high local densities and have relatively high reproductive rates, and mass sampling affects populations minimally; (4) an individual reptile (particularly lizards) can provide information on morphology, activity temperature, diet, reproduction, and microhabitat utilization; and (5) because desert populations remain relatively stationary (desert reptiles do not migrate), there is a continuous local history as a background for development of ecological relationships of resident species. Indeed, many ecologists consider lizards "model organisms" for ecological studies (Huey, Pianka, and Schoener 1983) and the vast majority of ecological work by Eric Pianka and his co-workers has been done on desert lizards (see Pianka 1986).

Birds have assumed a central role in theoretical ecology (MacArthur 1965, 1972; Cody 1974, 1975; and others), and it is not entirely clear whether avian-based models apply to relatively small, less vagile terrestrial poikilotherms such as lizards and snakes. Reptiles may respond in a different manner to abiotic and biotic variables than endothermic and highly mobile birds. Small reptiles, particularly snakes, for example, appear buffered from extended periods of low resource availability because of their relatively low metabolic rates and their ability to become inactive or fast for long periods. Most reptiles are primary or secondary carnivores, and it has been suggested that they should experience keen interspecific competition due to their trophic level (Pianka 1975). Consequently, studies on desert reptiles have the potential to contribute significantly to development and testing of ecological theory.

In this review, I will attempt to summarize some of the literature on the ecology of desert reptiles, focusing on selected studies that describe communities of reptiles, addressing diversity at the local, continental, or global level, studies that present ecological data pertinent to understanding how communities are organized, and studies that examine species interactions through manipulation. As a prelude, a few general comments about higher taxa of reptiles are in order. Most studies have been comparative; these will be addressed next. The few experimental studies, one in particular, will then be considered, as the results bear directly on interpretations of comparative studies. I will attempt to summarize the many factors that might influence desert reptile communities and to examine evidence in support of each factor. Finally, I will comment on potentially profitable avenues of future research on desert reptile communities.

General Comments on Reptiles in Deserts

There are no crocodilians that can be considered desert species, although some enter deserts in river systems. They will not be further considered.

The order Testudinata is poorly represented in deserts, generally with not more than one species occurring at any one locality. Consequently, there is no such thing as a desert tortoise community, and likewise no interesting interactions between species. It is interesting that desert tortoise species are strictly herbivorous even though there are many carnivorous aquatic turtles (Pritchard 1967). This may simply reflect a difference between aquatic and terrestrial turtles, however, because most terrestrial species are herbivorous and in a single family (Pritchard 1967).

The few amphisbaenians (a primarily tropical and subtropical group) that occur in deserts are generally restricted to mesic areas and cannot be considered xeric-adapted. They will not be considered further.

Snakes are well represented in deserts, although the actual number of species at any given locality is often unknown. Twenty-two species occur at one locality near the Verde River in the Sonoran Desert of Arizona (from unpublished appendix in Vitt 1986), whereas only seven species occur together in the Great Basin Desert of Utah (Woodbury 1951; Brown and Parker 1982). The number of species of desert snakes appears negatively correlated with latitude, but the cause of this correlation may be the positive relationship between numbers of snake species and numbers of potential prey species (Arnold 1972; Vitt 1986). The number of small vertebrates (potential prey species) also correlates negatively with latitude (Schall and Pianka 1978). It appears that prey diversity in general determines snake species diversity (Arnold 1972; Schoener 1974a; Vitt 1986), partially as a consequence of the high trophic position of snakes. There are no known herbivorous snakes, and a large proportion of snake species feed on other vertebrates. All snakes swallow their prey whole and, in general, a substantial proportion of species in a snake assemblage are specialists.

Lizards are the best known desert reptiles. Most desert species are insectivorous or feed on small vertebrates, and most species tend to take a wide variety of prey items (see Pianka 1986). Species may be diurnal or nocturnal, terrestrial, arboreal, or fossorial. Species are typically either "sit-and-wait" or "wide" foragers and there are many ecological, behavioral, physiological, and life history correlates of foraging mode (Vitt and Congdon 1978; Huey and Pianka 1981).

The number of species of desert lizards in North America correlates negatively with latitude, similar to the correlation of small vertebrate species with latitude (Schall and Pianka 1978), and the habitat characteristic that best explains this relationship is spatial heterogeneity (Pianka 1966, 1967). However, the number of species of desert lizards in Australia is much greater than predicted by latitude or spatial heterogeneity alone and much greater than other deserts of the world (Pianka 1986). The reasons for this will be considered in more detail near the end of this treatment.

Comparative Studies of Desert Reptiles

Most ecological studies of desert reptiles have been comparative. Approaches vary from examinations of taxonomic subunits of reptile communities (e.g., Mitchell 1979; Pianka and Pianka 1976) to continental comparisons of entire lizard communities (Pianka 1973, 1975, 1986). Most desert reptile studies were designed to keep habitat characteristics relatively constant, either by working in one place (e.g., Mitchell 1979; Vitt, Van Loben Sels, and Ohmart 1981) or by selecting similar habitats for repetitions (e.g., Pianka 1967, 1971; and many more). In most instances, samples from throughout the activity season are pooled for analysis of resource utilization patterns among species. Underlying all of the comparative approaches is the notion that the communities are in some sort of equilibrium and that organization is maintained by competition (a controversial notion). Given these assumptions, the expectation is that sympatric species should separate on the basis of at least one niche dimension.

Most comparative studies of desert reptile communities compare three niche axes: time, space, and food (Pianka 1973). A major assumption is that resources are limited either regularly, or on a frequent enough basis to influence species interactions, and that interactions between those species are more important than interactions with distant taxa (birds, for example). Because there may be alternative explanations for the same results, there has been considerable controversy over the importance of competition in structuring communities (see, e.g., Wiens 1977; Quinn and Dunham 1983; Connor and Simberloff 1979; Strong 1983; and others). The problem would be more serious in lower trophic level organisms. I will sidestep these issues here, as they have been heavily debated (Simberloff 1983; Connell 1983).

The most widely used quantitative approach to desert reptile community analysis has been to measure and compare activity periods or active body temperatures (both indicators of time use), microhabitat (space), and diet (food). Continuous data are generally compared using standard statistical measures. Categorical data, such as prey types, are usually analyzed using niche breadths as measures of variability within a species and niche overlaps as measures of similarity in utilization patterns between species. Niche breadth is usually measured with Simpson's (1949) diversity index. Niche overlap is calculated with MacArthur and Levins's (1967) equation (see Pianka 1986). Pianka (1986) and others point out that niche overlap values can be difficult to interpret because high overlap (suggesting competition) could indicate convergence on a superabundant resource rather than competition for a rare resource. This has been demonstrated in one system (see Dunham 1980).

To facilitate discussion, I divide comparative studies into the following categories: (1) within a single desert, (2) between several deserts, and (3) analysis of habitat gradients and temporal stability. For the most part, these comparative studies address the question of how desert lizard communities are structured; indirect inference was used to speculate on the causes underlying the observed structure.

At the end of my review of comparative studies, I will comment on some particularly innovative analyses of comparative data that attempt to more directly determine the importance of competition.

Single-Desert Studies

Single-desert studies are designed to determine the relationships between sympatric species. They focus on determinants of alpha diversity. Ideally, these studies take place in a uniform habitat (but see Habitat Gradients and Temporal Stability). Because deserts are not uniform on a geographic scale, single-desert studies with multiple replications often provide insight into the mechanics of species interactions, within the perspective of variation attributable to abiotic factors including structural diversity of the habitat, local climate, and others.

Three species of terrestrial colubrid snakes, *Coluber constrictor*, *Masticophis taeniatus*, and *Pituophis melanoleucus*, comprised 98 percent by numbers and 96 percent by biomass of a snake community consisting of eight species in the Great Basin Desert of northern Utah (Brown and Parker 1982). Four years of data on diets, microhabitats, and activity revealed that the three species were clearly separated along a food axis (Figure 9.1), and niche complementarity (Schoener 1974a) was apparent between the food and place niche axes. The snakes overwinter in communal dens.

In a study of snakes as rodent predators, the diets of 5 of 20 species of Chihuahuan Desert snakes included significant proportions of rodents (Reynolds and Scott 1982). Between these five species differences existed in the species and relative frequencies of mammals taken, habitat utilization, and foraging. Such differences indicate niche separation in space and food.

Single-locality desert lizard studies are numerous, and range from detailed descriptions of the ecology of a single species (which I omit) to comparisons between all lizards at a given locality. In some instances, specific taxa (families: e.g., Pianka and Pianka 1976; Pianka and Huey 1978; genera: e.g., Mitchell 1979; Case 1979) are studied. Although many of these studies are short-term, some extend over several years. Because Pianka's (see Pianka 1986) studies have always involved numer-

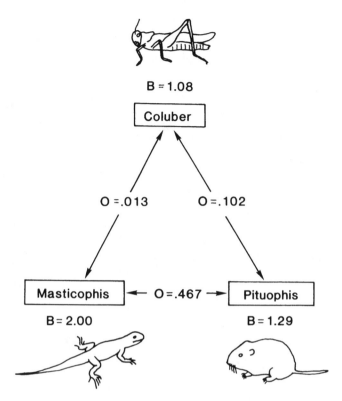

Figure 9.1. Quantitative analysis of the food niche of three sympatric colubrid snake species in a Great Basin cold desert shrub habitat. Relations of the three species indicated by niche breadth values (B) at each boxed name and niche overlap values (O) midway along arrows showing comparison of each species pair. Major prey group is depicted for each (redrawn from Brown and Parker 1982).

ous study sites, they will be considered later. I summarize single-desert studies in Table 9.1. Many studies contain either direct or indirect data on use of time, space, and food. It is notable that all studies demonstrate niche segregation on at least one axis. Many studies conclude that competition is responsible for the observed patterns of resource partitioning, although alternative explanations may be given. Species comparisons are generally based on pooled samples collected over a relatively short period, with the result being that neither time nor space are controlled as variables.

To determine the severity of problems caused by lack of control over space or time, I consider a 2-year study that adequately sampled six sympatric lizard species. During 1974 and 1975, lizards were studied in a Sonoran Desert habitat north of Phoenix, Arizona (see Vitt, Van Loben Sels, and Ohmart 1981 for description of the study area). Three species (*Urosaurus graciosus*, *Urosaurus ornatus*, and *Sceloporus magister*) are primarily arboreal and separate on the basis of a combination of perch

Table 9.1 Summary of studies on desert reptile communities indicating resource partitioning.

Desert Locality and Source	No. Species Studied[a]	Duration of Study	Resource Dimensions Examined[b]				Taxa Involved	Proposed Explanation for Apparent Partitioning[c]
			Macrohabitat	Microhabitat	Time	Food		
Great Basin Desert[d] Northern Utah (Brown & Parker 1982)	3 (1, 3)	4 yr	No	Yes	No	Yes	Snakes	Competition and differential effects of climatic factors.
Sonoran Desert[d] Baja California, Mexico (Case 1979)	2 (1, 1)	3 yr	Yes	Yes	No	Yes	Lizards	Interspecific competition.
Central Arizona (Vitt, Van Loben Sels, & Ohmart 1981)	3 (1, 2)	2 yr	Yes	Yes	No	Yes	Lizards	Tree size preference for two species; competition.
Mojave Desert[d] San Bernardino County, CA (Bury 1982)	8 (8, 3)	2 yr	…	…	…	…	Lizards, tortoises	Only collections to determine species composition.
Chihuahuan Desert Southeastern Arizona (Mitchell 1979)	4 (1, 1)	1 yr	Yes	Yes	Yes	Yes	Lizards	Habitat was most important, but higher than average overlaps suggested that these lizards form a guild.
Bolsón de Mapimí, Mexico (Barbault & Maury 1981; Maury & Barbault 1981; Maury 1981a,b)	11 (2, 7)	3 yr	Yes	Yes	Yes	Yes	Lizards	Patchy habitat structure and variable rains and productivity; competition of minimal importance.
Bolsón de Mapimí, Mexico (Barbault, Grenot, & Uribe 1978)	7 (5, 2)	2 mo	…	…	…	Yes	Lizards	Diet differences between species are attributed to microhabitat preferences and foraging mode.
Big Bend National Park, TX (Dunham 1980)	2 (2, 1)	4 yr	No	No	No	Yes	Lizards	This experimental study concludes that competition is severe during low resource periods but nonexistent at other times. When it occurs, the effect is asymmetrical.

Location (Reference)	Species	Duration					Taxa	Comments
Doña Ana County, N.M. (Whitford & Creusere 1977; Creusere & Whitford 1982)	12 (7, 3)	5 yr	Yes	Yes	Yes	....	Lizards	This long-term study demonstrated seasonal and yearly variation in lizard density and species composition at two study sites. It is concluded that rainfall and its effect on food availability explain the fluctuations.
South-central New Mexico (Medica 1967)	4 (1, 1)	2 yr	Yes	....	Yes	Yes	Lizards	Changes in distribution and density between a wet and dry year suggest food may be limiting some of the time. Interspecific competition is suggested.
Trans-Pecos area of southwest Texas (Milstead 1957, 1965)	4 (1, 1)	3 yr	Yes	....	....	Yes	Lizards	Interspecific competition. Ten years between studies revealed changes in population densities.
Sinai sand dunes								
Northern Sinai Peninsula (Werner 1982)	18 (17, 11)	....	Yes	....	....	....	Tortoises, lizards, snakes	A survey of the Sinai Peninsula with no real ecological data, no explanation.
Kalahari Desert[d]								
Northwest South Africa, southern Botswana (Pianka & Huey 1978; Huey et al. 1974)	2 (1, 1)	1 yr	Yes & no	Yes	No	Yes	Lizards	Competition based on character displacement in sympatry.
Northwest South Africa, southern Botswana (Pianka 1971)	23 (13, 5)	1 yr	Yes	Yes	Yes	....	Lizards	Deals primarily with lizard species density, concluding that lizard species density at a given site is similar to N. American desert but lower than Australian desert due to competition between birds and lizards.
Northwest South Africa, southern Botswana (Huey & Pianka 1977)	4 (1, 1)	> 1 yr	Yes	Yes	Yes	Yes	Lizards	Two species are only narrowly sympatric, either due to competition or adaptations to discontinuous aspects of the physical environment.
Northwest South Africa, southern Botswana (Pianka & Huey 1978)	7 (5, 1)	1 yr	Yes	Yes	Yes	Yes	Lizards	Competition, based on a comparison with Australian geckos.

Table 9.1 *Continued*

Desert Locality and Source	No. Species Studied[a]	Duration of Study	Resource Dimensions Examined[b]				Taxa Involved	Proposed Explanation for Apparent Partitioning[c]
			Macrohabitat	Microhabitat	Time	Food		
Sechura Desert								
Northern coastal Peru (Huey 1979)	4 (1, 1)	3 mo	Yes	Yes	No	Yes	Lizards	Uses parapatry and niche dimension complementarity to compare species, concluding that observed results, although consistent with a competition hypothesis, are also consistent with other explanations.
Arabian Desert								
Coastal lowlands of eastern Arabia (Arnold 1984)	23 (14, 5)	4 mo	Yes	Yes	Yes	Yes	Lizards	No conclusions drawn. There is good separation of species by niche dimensions based on the data.
Australian Desert[d]								
Most of the continent (Pianka 1969)	7 (1, 1)	16 mo	Yes	Yes	Yes	Yes	Lizards	Interspecific competition.
Most of the continent (Pianka 1972)	94 (23, 5)	?	Yes	. . .	. . .	. . .	Lizards	This is an attempt to understand present day distributions and habitat segregation of lizards in a historical context. It is concluded that speciation resulted from fluctuations of distinct habitats in time and space.

[a]The first no. in parentheses = no. of families, the second = no. of genera.

[b]Yes = differences; No = no differences.

[c]In some cases my determination of whether or not a study demonstrated niche segregation in a given category was subjective because the study did not directly address the issue.

[d]Pianka's multi-desert studies are not included here, but are summarized in Pianka (1986).

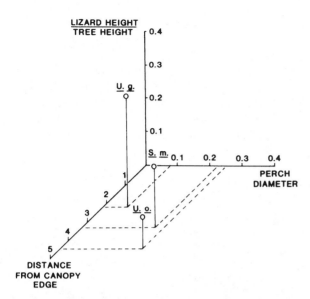

Figure 9.2. Foraging microhabitat comparisons for *Urosaurus graciosus* (*U. g.*), *Urosaurus ornatus* (*U. o.*). and *Sceloporus magister* (*S. m.*). Measurements in meters (redrawn from Vitt, Van Loben Sels, and Ohmart 1981).

height, distance from canopy edge, and relative position in trees (Figure 9.2). The two *Urosaurus* separate horizontally as well, on the basis of tree size, although narrow zones of syntopy exist. Differences in prey use between these species appear to result from differences in microhabitats. Three other diurnal species are abundant in this habitat: *Callisaurus draconoides*, *Uta stansburiana*, and *Cnemidophorus tigris*. These are terrestrial and consequently separate out from the group of arboreal species on a microhabitat axis. Due to temporal variation in rainfall, which translates into temporal variation in insect diversity and abundance (see Vitt, Van Loben Sels, and Ohmart 1981), diets of all lizards vary considerably from month to month within a season and within the same month between seasons (Vitt, Van Loben Sels, and Ohmart 1981; unpublished data). Niche breadths (Figures 9.3 and 9.4) and overlaps (Table 9.2) also vary within and between seasons. The arboreal species, which have high diet overlap (some of the overlap might be reduced if ants had been identified to species; see Vitt, Van Loben Sels, and Ohmart 1981), separate by microhabitat, whereas the three terrestrial species frequently occur in the same microhabitats but exhibit consistently lower overlaps in diets (Table 9.2). Even considering the temporal variation in diet niche breadth and overlap values, the data indicate that resources (food, and in the arboreal species, microhabitat) are partitioned during nearly all time periods. These results suggest that lack of control over time and space, given that large samples are taken, may not be a problem. The dynamic nature of niche breadth and overlap values is presumably a consequence of seasonal and yearly variation in resource levels associated with the effect of timing and intensity of rainfall on vegetation and, consequently, insect populations. Other studies demonstrate increased abundance and diversity of insects during relatively wet years (Dunham 1980). Problems interpreting overlap values still remain, however (see Dunham 1980). Indirect evidence suggests that resources might have been limiting during one year of the study on the six species described above. There was a reduction in reproductive output of females during 1975, which was dry compared to 1974 (Vitt, Van Loben Sels, and Ohmart 1978; Van Loben Sels and and Vitt 1984; Vitt and Van Loben Sels unpublished data). However, even this reproductive response fails to directly implicate competition as the structuring agent in this community because it does not necessarily indicate species interaction.

The two most similar (and most closely related) species, *U. graciosus* and *U. ornatus*, occupy microhabitats similar to those in central Arizona throughout their

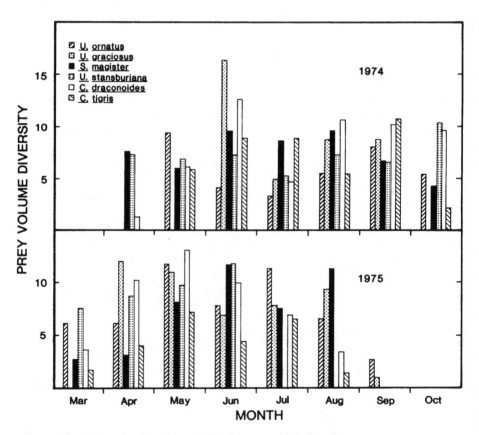

Figure 9.3. Seasonal and yearly variability in prey number diversity for six sympatric Sonoran Desert lizard species (*Urosaurus ornatus*, *Urosaurus graciosus*, *Sceloporus magister*, *Uta stansburiana*, *Callisaurus draconoides*, and *Cnemidophorus tigris* (from Vitt, Van Loben Sels, & Ohmart unpublished data).

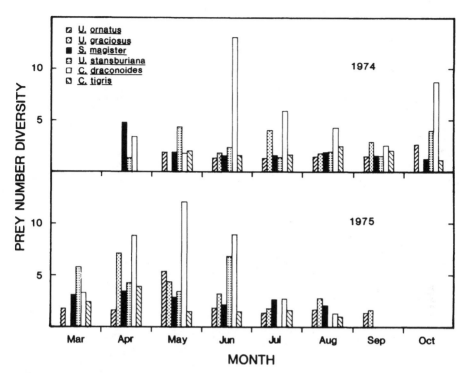

Figure 9.4. Seasonal and yearly variability in prey volume diversity for six sympatric Sonoran Desert lizard species (*Urosaurus ornatus*, *Urosaurus graciosus*, *Sceloporus magister*, *Uta stansburiana*, *Callisaurus draconoides*, and *Cnemidophorus tigris* (from Vitt, Van Loben Sels, & Ohmart unpublished data).

respective ranges, even in the absence of the other congener. The possibility exists that autecological requirements exclusive to each are sufficient to explain the horizontal aspect of their habitat distributions. *Urosaurus ornatus* is a relatively mesic-adapted species within the *Urosaurus* radiation. It was undoubtedly much more widespread historically, as suggested by its present fragmented geographic distribution (from Mexico to southern Wyoming), and invaded the northern part of its range from mainland Mexico. Its present distribution in the Sonoran Desert is restricted to riparian habitats associated with river courses and structurally diverse rocky habitats associated with isolated mountain ranges, suggesting that drying conditions restricted its distribution within the Sonoran Desert. The present distribution of *U. graciosus* includes the southern Mojave Desert and portions of the Sonoran Desert (including the Colorado Desert). Moreover, it is likely that *Urosaurus graciosus* invaded the Sonoran Desert from Baja California following divergence from the closely related species *U. microscutatus* in Baja (Murphy 1983). The Baja California *Urosaurus* radiation originated in southern Baja California following isolation of *Urosaurus* from mainland Mexico populations (see Murphy 1983 for a description

Table 9.2 Niche overlap in prey taxa by numerical and volumetric frequencies for six sympatric desert lizards, *Urosaurus ornatus*, *U. graciosus*, *Sceloporus magister*, *Uta stansburiana*, *Callisaurus draconoides*, and *Cnemidophorus tigris*.

Species and Month	U. ornatus #	U. ornatus Vol.	U. graciosus #	U. graciosus Vol.	S. magister #	S. magister Vol.	U. stansburiana #	U. stansburiana Vol.	C. draconoides #	C. draconoides Vol.	C. tigris #	C. tigris Vol.
U. ornatus												
April			—	—	—	—	—	—	—	—	—	—
May			—	—	0.94	0.20	0.86	0.47	0.31	0.53	0.03	0.07
June			1.00	0.36	1.00	0.75	0.97	0.57	0.52	0.21	0.02	0.15
July			0.92	0.25	1.00	0.85	0.99	0.74	0.64	0.90	0.16	0.41
Aug			0.99	0.64	0.98	0.19	0.99	0.50	0.29	0.60	0.11	0.07
Sept			0.98	0.73	1.00	0.69	1.00	0.50	0.09	0.12	0.08	0.23
U. graciosus												
April	0.82	0.11			—	—	—	—	—	—	—	—
May	0.80	0.09			—	—	—	—	—	—	—	—
June	0.98	0.86			0.99	0.40	0.96	0.30	0.53	0.49	0.02	0.12
July	0.99	0.81			0.93	0.31	0.91	0.15	0.62	0.17	0.15	0.28
Aug	0.97	0.74			0.98	0.19	0.99	0.47	0.28	0.20	0.11	0.07
Sept	0.97	0.06			0.97	0.49	0.97	0.27	0.10	0.28	0.09	0.33
S. magister												
April	0.96	0.44	0.76	0.34			0.88	0.35	0.66	0.04	0.05	0.32
May	0.50	0.35	0.54	0.34			0.70	0.56	0.00	0.33	0.01	0.12
June	0.98	0.26	0.96	0.28			0.96	0.47	0.52	0.37	0.02	0.18
July	0.87	0.32	0.87	0.26			0.99	0.65	0.64	0.84	0.16	0.52
Aug	0.98	0.25	0.96	0.32			0.98	0.60	0.28	0.31	0.16	0.66
Sept	—	—	—	—			0.99	0.52	0.11	0.11	0.11	0.42
U. stansburiana												
April	0.67	0.46	0.55	0.30	0.72	0.79			0.65	0.10	0.00	0.00
May	0.76	0.54	0.78	0.19	0.49	0.54			0.69	0.76	0.14	0.33
June	0.74	0.10	0.71	0.06	0.74	0.45			0.60	0.42	0.22	0.69
July	—	—	—	—	—	—			0.65	0.75	0.16	0.38
Aug	—	—	—	—	—	—			0.33	0.27	0.17	0.50
Sept	—	—	—	—	—	—			0.09	0.27	0.09	0.46

C. draconoides

April	0.11	0.10	0.14	0.20	0.11	0.36	0.19	0.59	0.00	0.00
May	0.51	0.48	0.55	0.27	0.70	0.54	0.73	0.51	0.01	0.12
June	0.44	0.15	0.44	0.13	0.45	0.22	0.85	0.18	0.06	0.30
July	0.42	0.19	0.42	0.15	0.45	0.30	—	—	0.72	0.50
Aug	0.06	0.06	0.06	0.08	0.20	0.11	—	—	0.90	0.38
Sept	—	—	—	—	—	—	—	—	0.98	0.60

C. tigris

April	0.08	0.81	0.02	0.03	0.15	0.34	0.07	0.42	0.11	0.05
May	0.03	0.67	0.01	0.07	0.01	0.27	0.51	0.53	0.49	0.27
June	0.01	0.20	0.01	0.04	0.05	0.22	0.52	0.26	0.74	0.37
July	0.05	0.06	0.05	0.05	0.20	0.67	—	—	0.91	0.42
Aug	0.01	0.01	0.01	0.03	0.15	0.05	—	—	1.00	0.94
Sept	—	—	—	—	—	—	—	—	—	—

SOURCE: Vitt, Van Loben Sels, & Ohmart (unpublished data).

NOTE: Data recalculated using Simpson's (1949) measure of niche breadth and MacArthur and Levins's (1967) measure of niche overlap. Data by month for 1974 appear in the upper right section of each matrix, with similar data for 1975 in the corresponding lower left section. Dashes indicate either samples that were too small for comparison, or missing data.

of "transgulfian vicariance"). Therefore, *U. graciosus* was a xeric-adapted species long before coming into contact with *U. ornatus*, and its arboreal habitat consisted of relatively small trees and bushes rather than the large trees (*Salix, Populus,* and large *Prosopis*) characteristic of major river systems in the Sonoran Desert.

A similar explanation is given for niche complementarity in *Phyllodactylus* geckos of the Sechura Desert of Peru (Huey 1979). Comparisons of niche dimensions among the four *Phyllodactylus* in sympatry and allopatry revealed that niche dimensions for each species were similar and independent of occurrence of other species. Species differences in niche dimensions in sympatry may simply reflect suites of autecological adaptations not caused by interspecific competition.

More compelling evidence that competition plays a role in structuring subsets of desert lizard communities derives from apparent ecological character displacement. Two species of subterranean skinks (*Typhlosaurus*) in the southern Kalahari Desert diverge in morphology and diet in areas of sympatry, suggesting ecological shift as the result of competition (Huey and Pianka 1974; Huey et al. 1974; but see Dunham, Smith, and Taylor 1979 for an alternative explanation).

A remaining question is whether patterns in lizard communities, given seasonal and yearly fluctuations, are for the most part stable over extended time periods. The only available data stem from a return trip by Pianka (in Pianka 1986) to one of his Australian study sites about ten years after he originally worked there. He found a striking similarity in species composition and relative abundances of lizard species and resource utilization patterns. Individual species had diets and microhabitat patterns similar to those of ten years earlier. Overall, diversity of prey eaten increased slightly and diversity of microhabitats declined. Diet breadth between the two time periods (see Table 10.6 in Pianka 1986) varied for each species within the same range of variation observed on a seasonal and yearly basis for desert lizards in central Arizona (Figures 9.3 and 9.4).

A major criticism of much of the earlier work on desert lizard communities is that resource availability, particularly food, was unknown. In large-scale community studies, direct measurement of resources would be logistically impossible. An alternative is to consider all of the resources used by the community as an approximation of resource availability (Pianka 1986) and compare resource utilization patterns of each species with that of the entire community. This electivity approach (Schoener 1974b) easily identifies specialists because their utilization patterns deviate significantly from the community resource spectrum. It also allows direct comparison of resource use by individual species with the community resource spectrum if the community resources are arranged in a ranked order by abundance (however, there are inherent problems with this approach; see Pianka 1986, and more recently, Morton and James 1988). Figure 9.5 illustrates this kind of comparison for Kalahari Desert lizards. Electivities are determined by dividing species utilization of a prey item by the relative abundance of that prey in the community resource spectrum. The results vary considerably from species to species but it is clear that only two species in this example elect to use the most abundant prey (only one exclusively) and that each species exhibits different electivities. The important finding is that each lizard species uses food resources nonrandomly.

Comparison between Deserts

Ideally, comparative studies should be repeated in similar habitats. Studies at a single site run the risk of producing results that are not representative of the questions of interest. In the event that habitat gradients exist between sites, replications can identify potential sources of variation between the issues of interest. Pianka (summarized in Pianka 1986) effectively used this approach as the rationale for gathering an enormous amount of data on lizards in three North American deserts, one African desert, and the Great Victorian Desert of Australia.

Initially, the expectation from comparisons of desert reptile communities between continents was that there would be overall similarities in number of species and broad patterns of community structure (Pianka 1986); in other words, communities in similar environments would be convergent. This was not the case. Pianka (1969a, 1972, 1986) found that the Australian desert contained many more species than other deserts (up to 42 sympatric species with twice as many genera as other deserts) with less niche overlap among species than in other deserts. These differences are summarized in Table 9.3 (see Pianka 1986 for the detailed data resulting in this summary). Within each desert lizard community, lizard species exhibited separation along at least one niche axis and niche complementarity was common (Pianka 1975, 1986).

Habitat Gradients and Temporal Stability

The vegetational and structural composition of habitats varies in both space and time. Habitat gradients can be rather subtle as one moves across flatland desert or quite abrupt where different habitats contact.

The use of numerous study sites within each desert has revealed interesting insight into the effect of habitat gradients on lizard communities. Four species of skinks (*Mabuya*) occur in the southern Kalahari (Huey and Pianka 1977). Two terrestrial species, *M. occidentalis* and *M. variegata punctulata*, are broadly sympatric, whereas two semiarboreal species, *M. spilogaster* and *M. striata sparsa*, are only narrowly sympatric even though each occurs in sympatry with the two terrestrial species. The two terrestrial species separate on the basis of microhabitat, time of activity in the summer, and prey taxa, whereas the two semiarboreal species are nearly identical to each other in all niche dimensions measured. The overall geographical distributions of the two arboreal species fall within two habitat zones, sand ridge and flatland. However, when examined on a finer scale, the occurrence of each arboreal species in both habitats suggests that habitat separation is not the only contributor to ecological separation. Intense competition in zones of sympatry resulting from identical microhabitat utilization is presumably also important (Huey and Pianka 1977). Using the same data, Schoener, Huey, and Pianka (1979) demonstrated a decrease in habitat niche breadth and an increase in food niche breadth for *Mabuya* in narrow sympatry, a result consistent with the compression hypothesis.

Deserts are intersected by arroyos and rivers, and interrupted by various-sized mountain ranges. Abrupt habitat gradients, containing variability in structural and vegetational characteristics, occur at these interfaces. Along these gradients, species composition and interactions may differ considerably from those in more typical and relatively more uniform desert flatland habitats. These transitional habitats may also experience greater temporal variation that could influence reptile communities.

A Chihuahuan Desert lizard community of 11 diurnal species consisted of sets of four to six species (Figure 9.6) unequally distributed along a habitat gradient (Maury and Barbault 1981). Most interesting is the observation that species in closely related taxa rarely occurred together. *Cophosaurus texanus* in cerro, piedmont, and bajada habitats was replaced by *Holbrookia maculata* in playa habitats, and *Holbrookia* was replaced by *Uma exsul* in dune habitats. The same was true to varying degrees for species of *Phrynosoma*, sceloporine lizards, and species of *Cnemidophorus*. Closely related species would be expected to use resources in a more similar way than would distantly related groups due to relative similarity in morphology, behavior, and foraging mode. Consequently, the potential for competition should be greater between species within each group. Habitat specificity alone may account for the apparent lack of interaction between *Uma*, *Holbrookia*, and *Cophosaurus*. All of the species of *Uma* are restricted to windblown sand patches regardless of the presence or absence of *Holbrookia* or *Cophosaurus*. Barbault and Maury (1981) concluded that competition between the lizard species was probably not the only factor, nor even the main factor, accounting for the organization of this desert lizard community. Autecological microhabitat preferences resulting from a patchy habitat and high variability of rains and productivity are used to explain the apparent structure.

Innovative Analyses

Partially as the result of the difficulties of interpreting data from comparative studies, even those conducted over relatively long periods, some rather innovative analyses have been used to support conclusions drawn by comparative studies. I already discussed the use of electivities in analyzing diet data; these made it much easier to compare species using the prey spectrum available. Three other quantitative

Figure 9.5. Patterns of food utilization among Kalahari lizards, based on the data pooled for all study sites. The upper panel illustrates the overall diet by volume of the entire lizard fauna, with prey categories ranked from those most used to those least used (a crude bioassay of food "availabilities"). The same ranking is followed in the lower panels, which show estimated electivities for each food resource state by each species of consumer. Prey categories from left to right: I = termites; Co = beetles; A = ants; G = grasshoppers and crickets; Lv = all insect larvae; U = miscellaneous arthropods, including unidentified insects; Sp = spiders; Sc = scorpions; V = all vertebrates; So = solpugids; Lp = butterflies and moths; H = bugs (Hemiptera and Homoptera); P = plant materials; W = wasps and other non-ant hymenopterans; D = flies; B = roaches; M = mantids and phasmids; N = adult neuropterans; E = insect eggs and pupae (redrawn from Pianka 1986).

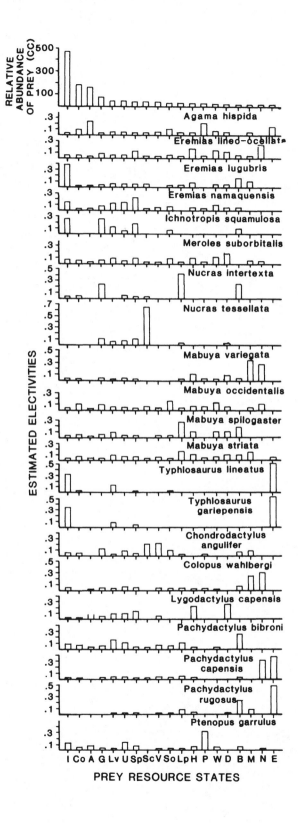

RELATIVE ABUNDANCE OF PREY (CC)

ESTIMATED ELECTIVITIES

Agama hispida
Eremias lined-ocellata
Eremias lugubris
Eremias namaquensis
Ichnotropis squamulosa
Meroles suborbitalis
Nucras intertexta
Nucras tessellata
Mabuya variegata
Mabuya occidentalis
Mabuya spilogaster
Mabuya striata
Typhlosaurus lineatus
Typhlosaurus gariepensis
Chondrodactylus angulifer
Colopus wahlbergi
Lygodactylus capensis
Pachydactylus bibroni
Pachydactylus capensis
Pachydactylus rugosus
Ptenopus garrulus

I Co A G Lv U SpScV So LpH P W D B M N E

PREY RESOURCE STATES

Table 9.3 Major differences and similarities between continental desert lizard systems.

	North America	Kalahari	Australia
Thermal climate	Colder, shorter growing season (especially in north)	Warmer	Warmer
Precipitation	Great Basin: distributed fairly evenly over year Mojave: winter precipitation Sonoran: bimodal annual rainfall, with peaks in both late summer and midwinter	Summer rains	Summer rain, variable annual precipitation
Higher taxa (number of lizard families)	5	5	5
Number of lizard genera	12	13	About 25
Species diversity			
Lizards	Low (4–11 species)	Medium (11–18 species)	High (15–42 species)
Snakes	High (4.5 species)	Low (2.2 species)	Medium (3.6 species)
Birds (all)	Low (7.8 species)	Medium (22.8 species)	High (28.3 species)
Insectivorous groundfeeding birds only	Low (about 2–3 species)	High (7.3 species)	Medium (4.8 species)
Small mammals	High (5.3 species)	. . .	Low (1.5 species)
Insect food resource diversity	High (8.7)	Low (4.4), dominated by termites	Intermediate (6.6)
Microhabitat resource diversity	Low (2.8)	High (6.8), variable	High (7.0)
Lizard niche breadth range and average			
Dietary	1.1–7.3 (4.4)	1.1–8.2 (3.9)	1.0–9.4 (3.8)
Microhabitat	1.0–3.9 (2.2)	1.0–6.0 (3.4)	1.0–6.4 (3.0)
Proportional utilization coefficients	Fit decaying exponential	Fit decaying exponential	Fit decaying exponential
Estimated electivities	Fit decaying exponential	Fit decaying exponential	Fit decaying exponential
Niche overlap			
Diet	Intermediate	Highest	Lowest
Microhabitat	All-or-none	Skewed toward low values	Skewed toward low values
Overall	Highest	Intermediate	Lowest
Guild structure	Little evident	Present	Conspicuous
Morphospace	Intermediate	Smallest	Largest

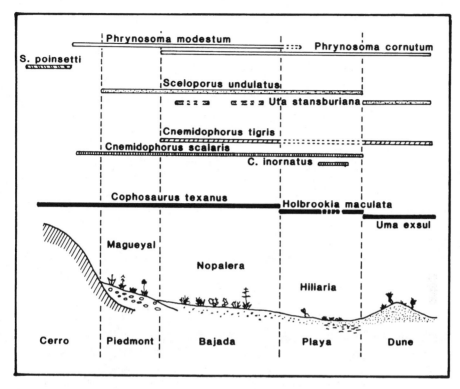

Figure 9.6. Lizard species distributions along the topographic gradient "cerro to dune" in the Chihuahuan Desert of Mexico (redrawn from Barbault and Maury 1981).

analyses have been used recently, morphometric analyses of lizard species in desert communities, computer simulations of "pseudocommunities," and the success of "alien" species introduced into communities in computer simulations.

Underlying morphological analyses is the notion that morphology reflects ecology and consequently analysis of morphology could, if carefully done, be used to make predictions about the ecology of given species (Ricklefs and Travis 1980). Morphometric analyses based on 10 morphological characteristics were performed by Pianka (1986) for the lizard faunas of North American deserts, the Kalahari, and the Australian desert. Within each system, species separated in morphospace in a manner consistent with microhabitat and dietary differences. Overall ecological similarity (microhabitat times dietary overlap) correlated negatively with morphological similarity. The significant correlations were weak, but species most dissimilar in morphology were also most dissimilar in niche characteristics. Likewise, the Australian ant specialist *Moloch horridus* and the North American ant specialist *Phrynosoma platyrhinos* were more similar to each other than to other members of their respective communities. Similar to niche analyses in comparative studies, morphological analyses do not provide direct evidence for the importance of competition in these communities. The results are, however, consistent with results obtained from niche analyses.

A particularly innovative analysis involves comparison of real communities with artificial communities constructed from randomized versions of the real communities (Sale 1974). Because Pianka (1986) has similar data for three different desert lizard communities, he has been able to construct pseudocommunities by taking random sets of species from the three-desert total. Four different kinds of pseudocommunities were constructed, ranging from relatively unrealistic ones with little structure, to realistic ones randomly interchanging observed resource utilization coefficients. Each real community was compared to 100 pseudocommunities of each type. Some of the original community's guild structure disappeared in all four of the treatments. The observation that randomized versions of the lizard communities lose structure, even with four different treatments, supports the notion that the real communities are not randomly assembled, but rather "possess substantial guild structure" (Pianka 1986, p. 116).

It is also possible to juggle niche relationships within a community to search for readjustments among species and to estimate how tightly species are packed. The introduction of computer-simulated alien species allows manipulation of guild structure by removing an individual species's contribution to the community resource matrix and replacing it with another similar species's resource spectrum (Pianka 1986). The results of this manipulation are quite encouraging in that for all three desert lizard communities, residents outperformed aliens in most substitutions. An example appears in Table 9.4. Performance was determined by comparing the resident's observed population density with the computer-generated population density of the alien after reaching equilibrium. Original substitutions were performed using Pianka's (1986) standard diet data categories. In a reanalysis, with expanded diet matrices (categories more finely broken down), residents outperformed aliens in 76.7 percent of the trials for the Kalahari and 75 percent of the trials for the Australian desert (Pianka 1986). The result of these simulations suggests that species interactions are important in maintaining structure within desert lizard communities.

Experimental Studies of Desert Reptiles

Given the controversy over the importance of competition in structuring communities and the problems of interpreting data from comparative studies, experimental manipulations of natural populations offer the opportunity to determine (1) whether competition is important at all, (2) how frequently it occurs, and (3) whether the effect is symmetrical. The logistics of large-scale community manipulations are overwhelming, but recent studies on the sympatric desert lizards *Urosaurus ornatus* and *Sceloporus merriami* provide interesting insights into these questions.

These two iguanid lizards occur in virtual syntopy in west Texas, both are insectivorous, and both are of near-equal body size (Dunham 1980). Two of six study plots were used as controls, *Sceloporus merriami* was removed from two plots, and *Urosaurus ornatus* was removed from the other two plots. A demonstration of competition would simply be an increase in population size of one species after the other

Table 9.4 Computer simulation of number of times that "residents" outperform "aliens," by species in the Kalahari.

Species	Total[a]	Number[b]	%
Agama hispida	90	64	71.1
Eremias lineo-ocellata	90	63	70.0
Eremias lugubris	72	37.5 (tie)	52.1
Eremias namaquensis	42	22	52.4
Mabuya occidentalis	90	56	62.2
Mabuya striata	56	43	76.8
Meroles suborbitalis	72	45	62.5
Chondrodactylus angulifer	90	67	74.4
Pachydactylus capensis	72	49	68.1
Ptenopus garrulus	90	53	58.9
Typhlosaurus lineatus	72	47.5 (tie)	66.0
Colopus wahlbergi	72	39.5 (tie)	54.9
Mabuya variegata	42	26	61.9
Mabuya spilogaster	6	4	66.7
Pachydactylus rugosus	42	22.5 (tie)	53.6
Ichnotropis squamulosa	2	1	50.0
Typhlosaurus gariepenis	30	20.5 (tie)	68.3
Nucras tessellata	6	4	66.7
Pachydactylus bibroni	20	12	60.0
Overall totals	1,056	677	64.1

SOURCE: Pianka (1986).
[a]Number of alien introductions.
[b]Number of times the alien was outperformed by the resident.

was removed, assuming that all else is equal. Presumably, the use of controls (no lizards removed), holding noncompetitive variables constant, and several repetitions reduces the probability that an erroneous result will be missed. The long-term nature of the study (4 years) increases the probability that a competitive effect will be detected if in fact competition is not continual.

Dunham (1980) first determined the relationship between rainfall and physical condition of individual lizards. Seasonal and annual variation in rainfall predictably influenced prey availability, which predictably influenced individual foraging success, body mass, total body lipids, and growth rates, among other things. These individual responses to resources also predictably influenced reproductive output. Clutch production in both species was reduced during dry (low resource) years (Dunham 1981). But during 1975 and 1977, the two driest years, the removal of *Sceloporus merriami* resulted in an increase in density of *Urosaurus ornatus*, in other words, a direct demonstration of competition.

The Historical Development of Desert Reptile Communities

The most interesting and complex question about desert reptile communities concerns the sequence of events that led to the community organization presently observed. A recent reexamination of Australian desert lizards provides a format with

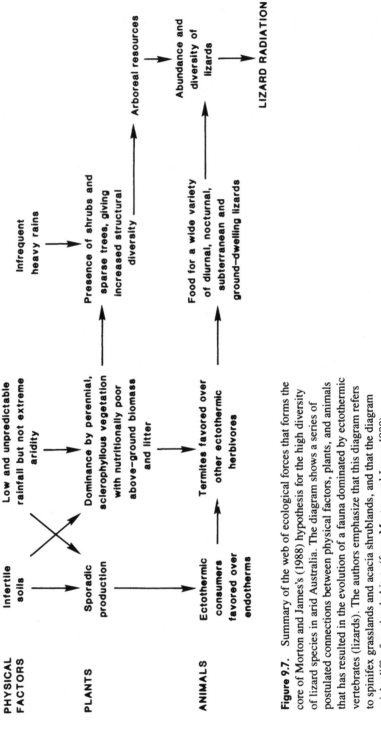

Figure 9.7. Summary of the web of ecological forces that forms the core of Morton and James's (1988) hypothesis for the high diversity of lizard species in arid Australia. The diagram shows a series of postulated connections between physical factors, plants, and animals that has resulted in the evolution of a fauna dominated by ectothermic vertebrates (lizards). The authors emphasize that this diagram refers to spinifex grasslands and acacia shrublands, and that the diagram might differ for other habitats (from Morton and James 1988).

which to address this question (Morton and James 1988). One of the key questions with respect to the Australian desert lizard fauna is why lizards have diversified yet birds and mammals are poorly represented. Morton and James (1988) point out that historically reptiles coexisted with a diversity of birds and mammals as arid zones in Australia developed, and, consequently, an answer to the question based entirely on geographical factors is unsupported. They suggest that "one pattern of resource availability—highly uncertain production" (Morton and James 1988, p. 242) best accounts for the high diversity and abundance of lizards in Australian deserts. Uncertain production constitutes a severe challenge to endothermic vertebrates with high and continual energetic demands and may explain the reduced number of endotherms, independent of the presence of reptiles. Morton and James's hypothesis relates soils and climate to the certainty of plant production and the plant types that dominate, which in turn influence the kinds of consumers in the system. This hypothesis is summarized in Figure 9.7. As the authors point out, key components of the hypothesis remain to be tested. Nevertheless, this integrative approach to the historical development of desert lizard communities promises to reorganize some of the current ideas. It is interesting that Pianka (1986, p. 26) also suggested the importance of variability in production: "Productivity alone may not promote diversity, but annual variability in production is perhaps more likely."

Factors Potentially Influencing Desert Reptile Communities

I have reviewed the various methods used to examine desert reptile communities and summarized some of the results. A majority of desert reptile community studies address relationships between species at a single point in evolutionary time and consequently shed little light on the historical development of each community. The following circumstantial evidence suggests the importance of competition in maintaining structure of desert reptile communities, at least at the local level:

1. Taken together, it is clear that within discrete reptile communities, species separate on the basis of time, food, or space.
2. Distribution of ecologically similar (and often taxonomically close) lizard species is often parapatric, with narrow zones of sympatry (if any). The distribution breaks can be horizontal (different substrate types) or vertical (different sized trees or perches).
3. When lizard communities are revisited after long periods, the structure remains similar to what it originally was (Pianka 1986).
4. Long-term comparative studies reveal variation in niche characteristics, but differences between species are maintained.
5. Experimental manipulations reveal that the removal of one competitor results in increased growth and abundance of the other.
6. Comparion of real communities with randomly generated pseudocommunities suggests that organization in the real communities is nonrandom.
7. Alien species introduced into structured communities in computer simulations do poorly compared to residents (Pianka 1986).

The results of Dunham's (1980) experimental study of desert lizard competition underscores the sporadic nature of competitive effects. Indeed, it is encouraging that, given the probable sporadic nature of species interactions, comparative studies have produced data consistent with the notion that competition maintains structure.

The species composition of given deserts and relationships between interacting species have been the primary focus of most desert reptile community studies. Few have dealt with historical events or aspects of the physical and biotic environment that resulted in differential speciation and distributional events and ultimately may determine the set of interacting species in a given area. Likewise, few studies have asked why other major taxa are poorly represented in deserts (but see Pianka 1986; Morton and James 1988). I have summarized some of the abiotic and biotic factors that may contribute to the structure of a given community in a direct or indirect way (Table 9.5). Some variables are amenable to experimental manipulations but many are not. Of overriding importance, if the evolution of desert reptile communities is to be understood, are the consideration of phylogenetic relationships, and dating of major diversification events (Morton and James 1988).

Suggested Future Research

Desert reptile communities are extremely complex, with each component species bringing a history of adaptations and constraints potentially independent of interactions it may face within its community at a given time. Understanding the dynamics of desert reptile communities requires consideration of the phylogenetic history of the reptiles as well as an appreciation for the geological and vegetational history of the desert. The lack of convergence in lizard communities from deserts of different continents can be attributed partially to history, but this should not negate the potential importance of ecological interactions at the local level (see Pianka 1986). The history of ecological factors during the development of arid regions appears to be a critical determinant of which taxa are excluded and which taxa diversify. The certainty of production is particularly important (Morton and James 1988). Ecological interactions should be easier to detect once the potential effects of history have been sorted out.

Desert reptiles constitute only part of desert communities, and their importance varies between deserts (Pianka 1986). Other vertebrates and, in some instances, invertebrates, may exert a significant impact on desert reptile communities, as competitors and as predators. Unfortunately little is known of these interactions.

Because of the complexity of desert reptile communities, potential avenues of research seem limited only by one's imagination. Long-term comparative studies, particularly across habitat gradients and in zones of sympatry between taxonomically and ecologically similar species, could contribute significantly to identifying determinants of resource partitioning. The importance of natural variation in species interactions could also be examined. For example, how stable are narrow zones of sympatry, like that reported in semiarboreal *Mabuya* (Huey and Pianka 1977)? How stable is the structure of a given reptile community? A careful consideration of the

Table 9.5 Summary of biotic and abiotic factors presumably affecting structure of desert reptile communities.

Factor	Presumed Effect
Historical	
Age or recency of desert	Older deserts should have more diverse faunas.
Source of reptile fauna	Deserts close to geographic areas with a history of a diverse reptile fauna (e.g., tropical forest) should have more diverse reptile communities than deserts distant or isolated from rich source areas.
History of climatic stability	Deserts with a long-term history of climatic stability should have more diverse reptile faunas than deserts periodically experiencing drastic climatic change *and* there should be more species of reptiles exhibiting special adaptations for desert life in deserts that have been climatically stable for long periods of time than in deserts that have not been climatically stable. Exactly the opposite is suggested for Australia (Morton and James 1988); climatic unpredictability, through its effect on certainty of productivity, provides an advantage for ectothermic vertebrates over endothermic ones and can result in high diversity.
Abiotic	
Latitude	Increasing species number with decreasing latitude, partially due to the effect of the source of the fauna.
Elevation	Increasing species number with decreasing elevation.
Temperature	Increasing species number with increasing temperature.
Moisture	Drier deserts should have fewer species—but this could be offset by the effect of moisture on certainty of productivity.
Soils	Poor soils would decrease productivity overall, but the effect on lizard species could be variable depending on interaction of soils with other factors.
Structural diversity of the habitat	Greater species number with increasing habitat diversity.
Biotic	
Habitat productivity	Effect might be indirect, through its influence on prey diversity and density.
Predictability of productivity	Less predictable productivity might exclude potential endothermic competitors.
Prey species richness	Increasing richness or diversity of desert reptiles.
Prey species diversity	Increasing richness or diversity of desert reptiles.
Prey abundance	Increasing richness or diversity of desert reptiles, particularly if predation influences population densities of reptiles.
Predators	Effect could be variable, direct on population densities (but not necessarily uniform across taxa), and indirect through its effect on competitive interactions, food availability, etc.
Competitors	Effect could be variable, indirect or direct, and may not be symmetrical. Also may be sporadic.

SOURCE: Modified from Vitt (1986).
NOTE: For each factor and presumed effect, it is assumed that all else is equal. The interaction of two or more factors could produce exactly the opposite result expected on the basis of any one factor alone, and in reality, all else is never equal!

history of productivity for each desert system could address some of the questions raised by Morton and James (1988) and place desert lizard communities in perspective within desert ecosystems.

Many relatively simple, but critical, questions regarding present-day organization of desert lizard communities can be addressed using field experiments. For example, are desert lizards resource limited, and if so how regularly? This could begin to be addressed simply by supplementing experimental plots with food to determine whether densities increase compared to control plots. Controlled removal experiments like the *Urosaurus ornatus–Sceloporus merriami* study (Dunham 1980) would yield more information on the influence of one species on another. The role of predation in desert lizard communities could also be examined with field experiments. One such study (Turner, Medica, and Smith 1974), has already demonstrated the effect of a predatory lizard, *Gambelia wislizenii*, on survivorship of *Uta stansburiana*. Experimental studies should be designed carefully to reduce the possibility that indirect (and usually undiscovered) effects produce the observed result.

Acknowledgments

I thank Janalee Caldwell for commenting on the manuscript. Financial support for collection of previously unpublished data stemmed from grants to R. D. Ohmart.

Bibliography

Arnold, E. N. 1984. Ecology of lowland lizards in the eastern United Arab Emirates. *Journal of the Zoological Society of London* 204: 329–354.

Arnold, S. J. 1972. Species densities of predators and their prey. *American Naturalist* 106: 220–236.

Barbault, R., C. Grenot, and Z. Uribe. 1978. Le Partage des resources alimentaires entre les especies de lezards du desert de Mapimi (Mexique). *Terre et Vie* 32: 135–150.

Barbault, R., and M. Maury. 1981. Ecological organization of a Chihuahuan Desert lizard community. *Oecologia* 51: 335–342.

Brown, W. S., and W. S. Parker. 1982. Niche dimensions and resource partitioning in a Great Basin Desert snake community. Pp. 59–81 in N. J. Scott, Jr. (ed.), *Herpetological Communities: A Symposium of the Society for the Study of Amphibians and Reptiles and the Herpetologists' League, August 1977*. U.S. Fish and Wildlife Service, Wildlife Research Report 13, Washington, D.C.

Bury, R. B. 1982. Structure and composition of Mojave Desert reptile communities determined with a removal method. Pp. 135–142 in N. J. Scott, Jr. (ed.), *Herpetological Communities: A Symposium of the Society for the Study of Amphibians and Reptiles and the Herpetologists' League, August 1977*. U.S. Fish and Wildlife Service, Wildlife Research Report 13, Washington, D.C.

Case, T. J. 1979. Character displacement and coevolution in some *Cnemidophorus* lizards. *Fortschritte der Zoologie* 25: 235–282.

Cody, M. L. 1974. *Competition and the Structure of Bird Communities*. Princeton, N. J.: Princeton University Press.

———. 1975. Towards a theory of continental species diversities: Bird distributions over Mediterranean habitat gradients. Pp. 214–257 in M. L. Cody and J. M. Diamond (eds.), *Ecology and Evolution of Communities.* Cambridge, Mass.: Belknap Press of Harvard University Press.

Connell, J. H. 1983. On the prevalence and relative importance of interspecific competition: evidence from field experiments. *American Naturalist* 122: 661–696.

Connor, E. F., and D. Simberloff. 1979. The assembly of species communities: chance or competition? *Ecology* 60: 1132–1140.

Cowles, R. D., and C. M. Bogert. 1944. A preliminary study of the thermal requirements of desert reptiles. *Bulletin of the American Museum of Natural History* 83: 265–296.

Creusere, M. F., and W. G. Whitford. 1982. Temporal and spatial resource partitioning in a Chihuahuan Desert lizard community. Pp. 121–128 in N. J. Scott, Jr. (ed.), *Herpetological Communities: A Symposium of the Society for the Study of Amphibians and Reptiles and the Herpetologists' League, August 1977.* U.S. Fish and Wildlife Service, Wildlife Research Report 13, Washington, D.C.

Dunham, A. E. 1980. An experimental study of interspecific competition between the iguanid lizards *Sceloporus merriami* and *Urosaurus ornatus. Ecological Monographs* 50: 309–330.

———. 1981. Populations in a fluctuating environment: the comparative population ecology of *Sceloporus merriami* and *Urosaurus ornatus. Miscellaneous Publications of the University of Michigan Museum of Zoology* 158: 1–62.

Dunham, A. E., G. R. Smith, and J. N. Taylor. 1979. Evidence for ecological character displacement in western North American catostomid fishes. *Evolution* 33: 877–896.

Huey, R. B. 1979. Parapatry and niche complementarity of Peruvian desert geckos (*Phyllodactylus*): the ambiguous role of competition. *Oecologia* 38: 249–259.

Huey, R. B., and E. R. Pianka. 1974. Ecological character displacement in a lizard. *American Zoologist* 14: 1127–1136.

———. 1977. Patterns of niche overlap among broadly sympatric versus narrowly sympatric Kalahari lizards (Scincidae: *Mabuya*). *Ecology* 58: 119–128.

———. 1981. Ecological consequences of foraging mode. *Ecology* 62: 991–999.

Huey, R. B., E. R. Pianka, M. E. Egan, and L. W. Coons. 1974. Ecological shifts in sympatry: Kalahari fossorial lizards (*Typhlosaurus*). *Ecology* 55: 304–316.

Huey, R. B., E. R. Pianka, and T. W. Schoener. 1983. *Lizard Ecology: Studies on a Model Organism.* Cambridge, Mass.: Harvard University Press.

MacArthur, R. H. 1965. Patterns of species diversity. *Biological Review* 40: 510–533.

———. 1972. *Geographical Ecology.* New York: Harper and Row.

MacArthur, R. H., and R. Levins. 1967. The limiting similarity, convergence and divergence of coexisting species. *American Naturalist* 101: 377–385.

Maury, M. E. 1981a. Food partition of lizard communities at the Bolson de Mapimi (Mexico). Pp. 119–142 in R. Barbault and G. Halffter (eds.), *Ecology of the Chihuahuan Desert: Organization of some Vertebrate Communities.* Instituto Ecologia Mexico.

———. 1981b. Variability of activity cycles in some species of lizards in the Bolson de Mapimi (Mexico). Pp. 101–118 in R. Barbault and G. Halffter (eds.), *Ecology of the Chihuahuan Desert: Organization of some Vertebrate Communities.* Instituto Ecologia Mexico.

Maury, M. E., and R. Barbault. 1981. The spatial organization of the lizard community of the Bolson de Mapimi (Mexico). Pp. 79–87 in R. Barbault and G. Halffter (eds.), *Ecology of the Chihuahuan Desert: Organization of some Vertebrate Communities.* Instituto Ecologia Mexico.

Medica, P. A. 1967. Food habits, habitat preference, reproduction, and diurnal activity in four sympatric species of whiptail lizards (*Cnemidophorus*) in south central New Mexico. *Bulletin of the Southern California Academy of Sciences* 66: 251–276.

Milstead, W. W. 1957. Some aspects of competition in natural populations of whiptail lizards (genus *Cnemidophorus*). *Texas Journal of Science* 9: 410–447.

———. 1965. Changes in competing populations of whiptail lizards (*Cnemidophorus*) in southwestern Texas. *American Midland Naturalist* 73: 75–80.

Mitchell, J. C. 1979. Ecology of southeastern Arizona whiptail lizards (*Cnemidophorus*: Teiidae): population densities, resource partitioning and niche overlap. *Canadian Journal of Zoology* 57: 1487–1499.

Morton, S. R., and C. D. James. 1988. The diversity and abundance of lizards in arid Australia: a new hypothesis. *American Naturalist* 132: 237–256.

Murphy, R. W. 1983. Paleobiogeography and genetic differentiation of the Baja California herpetofauna. *Occasional Papers of the California Academy of Sciences* 137: 1–48.

Pianka, E. R. 1966. Convexity, desert lizards, and spatial heterogeneity. *Ecology* 47: 1055–1059.

———. 1967. Lizard species diversity. *Ecology* 48: 333–351.

———. 1969a. Habitat specificity, speciation, and species density in Australian desert lizards. *Ecology* 50: 498–502.

———. 1969b. Sympatry of desert lizards (*Ctenotus*) in western Australian. *Ecology* 50: 1012–1030.

———. 1971. Lizard species density in the Kalahari Desert. *Ecology* 52: 1024–1029.

———. 1972. Zoogeography and speciation of Australian desert lizards: an ecological perspective. *Copeia* 1972: 127–145.

———. 1973. The structure of lizard communities. *Annual Review of Ecology and Systematics* 4: 53–74.

———. 1975. Niche relations of desert lizards. Pp. 292–314 in M. L. Cody and J. M. Diamond (eds.), *Ecology and Evolution of Communities*. Cambridge, Mass.: Belknap Press of Harvard University Press.

———. 1986. *Ecology and Natural History of Desert Lizards*. Princeton, N.J.: Princeton University Press.

Pianka, E. R., and R. B. Huey. 1978. Comparative ecology, resource utilization and niche segregation among gekkonid lizards in the southern Kalahari. *Copeia* 1978: 691–701.

Pianka, E. R., and H. D. Pianka. 1976. Comparative ecology of twelve species of nocturnal lizards (Gekkonidae) in the Western Australia desert. *Copeia* 1976: 125–142.

Pritchard, P. C. H. 1967. *Living Turtles of the World*. Hong Kong: T. F. H. Publ.

Quinn, J. F., and A. E. Dunham. 1983. On hypothesis testing in ecology and evolution. *American Naturalist* 122: 602–617.

Reynolds, R. P., and N. J. Scott. 1982. Use of a mammalian resource by a Chihuahuan snake community. Pp. 99–118 in N. J. Scott, Jr. (ed.), *Herpetological Communities: A Symposium of the Society for the Study of Amphibians and Reptiles and the Herpetologists' League, August 1977*. U.S. Fish and Wildlife Service, Wildlife Research Report 13, Washington, D.C.

Ricklefs, R. E., and J. Travis. 1980. A morphological approach to the study of avian community organization. *Auk* 97: 321–338.

Sale, P. F. 1974. Overlap in resource use, and interspecific competition. *Oecologia* 17: 245–256.

Schall, J. J., and E. R. Pianka. 1978. Geographical trends in numbers of species. *Science* 201: 679–686.

Schoener, T. W. 1974a. Resource partitioning in ecological communities. *Science* 185: 27–39.

―――. 1974b. Some methods for calculating competition coefficients from resource utilization spectra. *American Naturalist* 108: 332–340.

Schoener, T. W., R. B. Huey, and E. R. Pianka. 1979. A biogeographic extension of the compression hypothesis: competitors in narrow sympatry. *American Naturalist* 113: 295–298.

Simberloff, D. 1983. Competition theory, hypothesis testing, and other community ecological buzzwords. *American Naturalist* 122: 626–635.

Simpson, E. H. 1949. Measurement of diversity. *Nature* 163: 688.

Strong, D. R., Jr. 1983. Natural variability and the manifold mechanisms of ecological communities. *American Naturalist* 122: 636–660.

Turner, F. B., P. A. Medica, and D. D. Smith. 1974. Pp. 118–128 in *Reproduction and Survivorship of the Lizard,* Uta stansburiana, *and the Effects of Winter Rainfall, Density and Predation on these Processes.* U.S. International Biological Program Desert Biome Research Memo 74–26.

Van Loben Sels, R. C., and L. J. Vitt. 1984. Desert lizard reproduction: annual variation in *Urosaurus ornatus* (Iguanidae). *Canadian Journal of Zoology* 62: 1779–1787.

Vitt, L. J. 1986. Communities. Pp. 335–365 in R. A. Seigel, J. T. Collins, and S. S. Novak, (eds.), *Snakes: Ecology and Evolutionary Biology.* New York: Macmillan.

Vitt, L. J., and J. D. Congdon. 1978. Body shape, reproductive effort, and relative clutch mass in lizards: resolution of a paradox. *American Naturalist* 112: 596–608.

Vitt, L. J., R. D. Van Loben Sels, and R. D. Ohmart. 1978. Lizard reproduction: annual variation and environmental correlates in the iguanid lizard *Urosaurus graciosus.* *Herpetologica* 34: 241–253.

―――. 1981. Ecological relationships among arboreal desert lizards. *Ecology* 62: 398–410.

Werner, Y. L. 1982. Herpetofaunal survey of the Sinai Peninsula (1967–77), with emphasis on the Saharan sand community. Pp. 153–162 in N. J. Scott, Jr. (ed.), *Herpetological Communities: A Symposium of the Society for the Study of Amphibians and Reptiles and the Herpetologists' League, August 1977.* U.S. Fish and Wildlife Service, Wildlife Research Report 13, Washington, D.C.

Whitford, W. G., and F. M. Creusere. 1977. Seasonal and yearly fluctuations in Chihuahuan Desert lizard communities. *Herpetologica* 33: 54–65.

Wiens, J. A. 1977. On competition and variable environments. *American Scientist* 65: 590–597.

Woodbury, A. M. 1951. A snake den in Tooele County, Utah: Introduction—a ten-year study. *Herpetologica* 7: 4–14.

10   The Ecology of Desert Birds

John A. Wiens

Over the past 30 years, ecologists in general and avian ecologists in particular have argued over alternative views of community structuring. One perspective is that bird communities are little more than coincidental assemblages of species whose species-specific traits enable them to occupy the same environments and use similar resources. This "autecological" view has its historical roots in the individualistic concepts of plant communities championed by Gleason (1926) and Curtis (1959) and in Grinnell's (1917) notion of the ecological niche. At the opposite pole, communities are held to be tightly integrated collections of species whose coexistence is achieved by partitioning resources so as to minimize interspecific competition. This view is firmly rooted in Gause's (1934) competitive exclusion principle and Hutchinson's (1957) formulation of niche theory, and has been amply reinforced by the work of MacArthur and his colleagues (e.g., MacArthur 1972; Cody and Diamond 1975). The competition viewpoint dominated research and thinking in avian community ecology from the late 1950s through much of the 1970s, but recent disillusionment with that paradigm (e.g., Wiens 1977; Simberloff 1982) has been accompanied by calls for renewed consideration of the species-specific attributes of community members (Wiens 1983, in press; James et al. 1984; Andrewartha and Birch 1984).

What this means is that, had this chapter been written during the early 1970s, the focus would have been on resource partitioning patterns among coexisting species and their effects on local and regional species diversity, accompanied by some inferences about the scarcity of resources in the desert and the importance of interspecific competition in determining which species could coexist in communities. Now, however, a different and broader focus is warranted. I believe that, in order to understand how desert bird communities are put together, we must begin by considering the constraints that desert environments impose on birds and how birds respond to those constraints. The interaction of environmental constraints and species-specific responses determines which members of the species pool available to occupy a desert location may occur there (Figure 10.1). Biological interactions such as competition, predation, parasitism, or mutualism may then act as selective filters affecting the distribution and abundance of species and, ultimately, the composition and structure of communities (Figure 10.1). Past history, local disturbance, and chance may intercede at any of these steps to distort the patterns that might be expected on the basis of species' attributes and between-species interactions alone.

This view has the disadvantage of being much more complex and possibly murkier than the traditional, competition-based view, but it has the advantage of drawing attention to the diversity of factors involved in the assembly of bird communities in deserts and thereby leading to insights that are more likely to be correct. My treatment in this chapter will therefore follow the outline in Figure 10.1. Because the ways in which bird species respond to desert conditions are so important in determining their occurrence and dynamics, and because much more is known about these responses than about the magnitudes and effects of interactions among species, I will emphasize these responses. Much of what I review is not really about bird communities at all, but rather is about the ecology of birds in the desert.

How Do Birds Cope with Desert Environments?

Many desert animals escape the extremes of heat and dryness typical of most deserts by being nocturnal and/or using sheltered microhabitats where temperatures are lower and humidity greater (e.g., burrows). Not so birds. Although there are exceptions, such as some caprimulgids and owls, a few parrots, some desert shorebirds such as plovers (*Vanellus*) and coursers (*Rhinoptilus* and *Cursorius*), and sandgrouse of the genus *Nyctiperdix* (Maclean 1967, personal communication), most desert birds are strictly diurnal, and few use burrows to any extent. They are therefore directly exposed to the stresses of desert environments. The problems are even more severe for species such as larks, sandgrouse, plovers, coursers, thick-knees, bustards, pratincoles, or nighthawks that nest on the ground in exposed locations, for they must prevent overheating of the eggs and young and provide water as well as food to the nestlings (Dawson and Bartholomew 1968). In addition, because they are active endotherms, birds have high energy demands. Many species meet these demands by feeding extensively on seeds, which at times may be produced in abundance in deserts. Adoption of a granivorous diet, however, carries costs. Seed production usually depends on rainfall and, like rainfall, may be quite variable in time and space. What is a good location for breeding one year may not be suitable again for several years. Moreover, seeds contain far less preformed water than do insects, so many granivores are more dependent on sources of free water than are insectivores. The distribution of such granivorous birds is linked closely with water sources, but it is also affected by variations in food supplies. It is no wonder that the mobility of birds is often regarded as the key to their survival in desert environments (Evenari 1985; Williams and Calaby 1985; Brown 1986).

The adaptations that enable birds to live in deserts are not limited to mobility, however (Figure 10.1). Many of the features that we might consider adaptations for desert conditions are in fact really preadaptations—general attributes of avian biology that happen to compensate for the stresses of desert environments (Miller 1963; Maclean 1984). Indeed, conditions in many deserts may not be all that extreme or stressful from the viewpoint of a bird; we may tend to regard them as such only because we, as humans, have greater water conservation problems than most species (Mares et al. 1977).

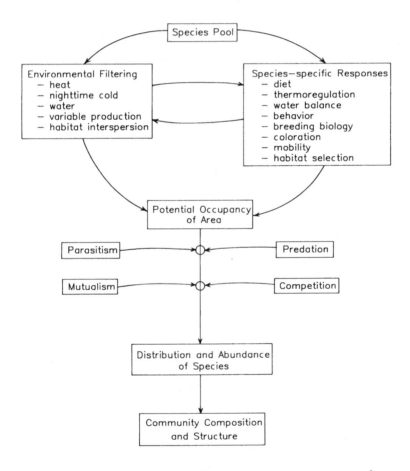

(All components are influenced by disturbance, history, and chance)

Figure 10.1.   A flow chart depicting the factors influencing the composition and structure of desert bird communities.

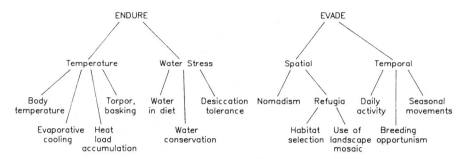

Figure 10.2. Components of tactics for enduring desert environmental stresses and for evading those stresses.

Not all birds live in deserts, however, so it is especially important to examine the attributes of those species that do regularly occur there. Following the approach of Solbrig and Orians (1977) and Low (1979), we may distinguish between features that enable species to *endure* desert conditions and those that allow them to *evade* the extremes (Figure 10.2). Bartholomew and Cade (1963), Dawson and Bartholomew (1968), Serventy (1971), Maclean (1974), Dawson (1976, 1981), and Louw and Seely (1982) review many of these features.

Endurance Tactics

Thermoregulation Most of the characteristics that enable birds to endure extremes of heat and dryness in deserts are physiological. In particular, because birds in general maintain a relatively high body temperature (about 40°C), they are not likely to begin to incur a heat load until ambient temperatures reach 39°–41°C. Many birds are also able to tolerate a limited amount of hyperthermia, accumulating a heat load by elevating body temperature by as much as 3°C for short periods of time (Dawson and Hudson 1970; Dawson 1976). Desert birds might be expected to have higher body temperatures than species in other habitats in order to minimize heat load problems even further, but apparently this is not so (Dawson and Schmidt-Nielsen 1964). Under laboratory conditions, desert birds, like nondesert species, are unable to survive for long periods at ambient temperatures above 45°–46°C, although in nature many desert birds encounter and are able to endure higher temperatures for periods of several hours each day (Weathers personal communication).

Between the point at which body temperature equals ambient temperature and the upper lethal temperature, birds must thermoregulate by losing heat. Some heat may be lost by radiant energy transfer via exposed surfaces such as legs or apteria under the wings. In some cases, such as the Ostrich (*Struthio camelus*), radiant heat gain may be reduced by erecting the body feathers, thereby increasing the thickness of the insulative coat (Crawford and Schmidt-Nielsen 1967; Louw, Belonje, and Coetzee 1969). Sandgrouse reduce radiant heat gain by means of a dense coat of brown underdown (Thomas 1984). Most often, however, thermoregulation is achieved primarily by evaporative cooling.

Birds use three avenues of evaporative cooling. For some desert species, such as the Budgerigar (*Melopsittacus undulatus*) and the Zebra Finch (*Poephila guttata*), a good deal of their evaporative cooling occurs through the skin (Smith and Suthers 1969; Bernstein 1971; Lasiewski, Bernstein, and Ohmart 1971). Although this mechanism has been studied in relatively few species, it may be widespread among birds (Webster, Campbell, and King 1985; Webster and King 1987). At somewhat higher temperatures (beginning at about 41°–44°C), panting is employed to enhance evaporation from respiratory surfaces. This is supplemented in some forms (e.g., some pigeons, owls, sandgrouse, and caprimulgids) by gular fluttering, a rapid vibration of the chin and throat areas that moves air across the moist surfaces of the buccal cavity and esophagus (Bartholomew, Lasiewski, and Crawford 1968; Dawson and Hudson 1970). These forms of evaporative cooling are quite effective in the dry desert air; a number of desert (as well as nondesert) species are able to dissipate all of their heat production by evaporative cooling at ambient temperatures of 43°–45°C (Serventy 1971; Dawson 1981). Using various thermoregulatory mechanisms, Ostriches are capable of maintaining a body temperature of more than 40°C for as long as 8 hours under ambient temperatures up to 50°C (Schmidt-Nielsen et al. 1969), while the Spotted Nightjar (*Caprimulgus guttatus*) of the Australian arid zone is able to prevent overheating at an ambient temperature of 52°C (Dawson and Fisher 1969). In some cases (such as the nightjar, as well as sandgrouse and some pigeons), unusually low metabolic rates contribute to reducing heat load problems (Dawson and Fisher 1969; Dawson and Bennett 1973; Maclean 1985), but a reduction in metabolism is far from universal among desert birds.

High temperatures are not the only thermal problem facing desert birds, however. Because desert air is so dry and clear, heat may be lost rapidly at night, and nighttime temperatures may be quite low. Birds are quite capable of metabolically thermoregulating under these nighttime conditions, but it may be energetically expensive to do so, especially for small birds. A few arid zone species, such as the Greater Roadrunner (*Geococcyx californianus*), may become moderately hypothermic at night, rewarming in the morning sun by basking (Ohmart and Lasiewski 1971). Other species, such as Common Poorwills (*Phalaenoptilus nuttallii*), White-backed Swallows (*Cheramoeca leucosternum*), and Crimson Chats (*Ephthianura tricolor*), may become torpid for varying lengths of time, especially if cold weather depresses the availability of their insect prey (Withers 1977; Serventy 1970; Ives 1973).

Water Economy   If heat is one great challenge of the desert, water scarcity is the other. Evaporative cooling may enable birds to withstand temperatures in excess of their own body temperature, but it comes at a cost of a considerable water loss. Because there are strong pressures on birds to minimize the load they must carry in flight, water storage is not an option available to flying species. Moreover, unlike some desert mammals, most birds make relatively little use of water produced by metabolism to offset their losses (but see Bartholomew 1972). The Zebra Finch, which is better at utilizing metabolic water than most passerines, is able to compensate for only half of its evaporative water loss in this way (Bartholomew and Cade 1963; Cade, Tobin, and Gold 1965). The magnitude of the problem is apparent from calculations that suggest that a Zebra Finch exposed to an ambient temperature of about 43°C for 5 hours will evaporate an amount of water equivalent to nearly 30 percent of its body weight (Calder 1964).

Birds are preadapted to reduce water losses in other ways, however. By excreting nitrogenous wastes as uric acid rather than urea, birds require less than a tenth the amount of water required by mammals to excrete a given amount of nitrogen (Dawson and Bartholomew 1968). The capacity of birds for renal concentration of electrolytes, on the other hand, is substantially less than that of some desert mammals. Desert species such as the Galah (*Cacatua roseicapilla*) and the Budgerigar, however, have greater renal concentration abilities than do some nondesert birds, and the osmotic permeability of the large intestine is also reduced (Skadhauge and Kristensen 1972; Skadhauge 1974). A few desert species, such as the Ostrich, roadrunners, the partridge *Ammoperdix heyi*, the Australian Dotterel (*Peltohyas australis*), the Australian Pratincole (*Stiltia isabella*), and the Double-banded Courser (*Rhinoptilus africanus*), possess well-developed nasal salt glands that enable them to lose ions with a small expenditure of water (Schmidt-Nielsen et al. 1963; Ohmart 1972; Maclean 1967, 1976b; Jesson and Maclean 1976), but such glands are absent in desert passerines (Rounsevell 1970; Dawson 1981). Water loss in excrement is generally less in desert-dwelling bird species than in related species occupying more mesic environments, and many desert species are able to restrict fecal water content even more when they are deprived of free water. Given unrestricted access to water, Zebra Finches and Budgerigars produce excrement containing 80 percent water by mass, but nondrinking individuals reduce this amount to 55–60 percent (Cade and Dybas 1962; Lee and Schmidt-Nielsen 1971). Some other species, such as Black-throated Sparrows (*Amphispiza bilineata*), Stark's Larks (*Spizocorys starki*), and Grey-backed Finch-Larks (*Eremopterix verticalis*), may produce feces containing as little as 30 percent water when denied free water (Smyth and Bartholomew 1966; Willoughby 1968). Much of the reduction in water loss under conditions of water deprivation, however, is due to a restriction of cutaneous water loss (Lee and Schmidt-Nielsen 1971).

It is doubtful that any desert birds can employ these mechanisms to permit them to exist on a seed diet for long periods without drinking, although on physiological grounds one might expect such water independence (evaporative water loss < metabolic water production) to be possible in some relatively large (>40 g) granivores. Under laboratory conditions, species such as Zebra Finches, Budgerigars, Black-throated Sparrows, Bourke's Parrots (*Noephema bourkii*), and Scaly-feathered Finches (*Sporopipes squamifrons*) are indeed able to reduce the rates of evaporative, fecal, and cloacal water loss enough to maintain a water balance with the gains from preformed water in their seed diet and from water of oxidation, at least under moderate temperatures (Serventy 1971; Dawson 1981). All of these species, however, are small (<30 g). This apparent paradox may be resolved as follows (R. E. MacMillen personal communication; see also Weathers 1981). The granivore species that are able to survive without free water have a higher lower critical temperature (T_{lc}) limit to the zone of thermoneutrality than do other species. Because the rate of evaporative water loss decreases with ambient temperature (T_a) below T_{lc} while the rate of metabolic water production increases, there will be a T_a at which losses are balanced by metabolic water production. Below this ambient temperature the species can achieve independence of free water. By shifting T_{1c} to higher temperatures, the value of T_a at which evaporative losses are balanced by metabolic production increases, permitting water independence at higher ambient temperatures. The cost of this strategy is greater energy expenditures and demands, but the benefit is greater metabolic water production that may permit independence from free water, at least under some ambient temperatures. Small birds are more likely to be successful at this strategy in arid regions than are large birds because their absolute energy demands per individual are less, even though their mass-relative rates of energy expenditure are greater. Thus, by elevating T_{1c}, some small granivores may increase the range of temperatures over which they can subsist on a seed diet independently of free water and thereby be able to move some distance from water. Despite such adaptations, however, water independence is still a temperature-dependent response.

How do desert birds normally obtain the water to meet their requirements? Many, of course, use free surface water. Accounts of the hordes of finches, parrots, or doves visiting desert water holes are frequent in the journals of early explorers or naturalists venturing into the desert (Serventy 1971). In their survey of the incidence of drinking by birds inhabiting the arid interior of Australia, however, Fisher, Lindgren, and Dawson (1972) found that 71 of 118 species (60 percent) visited water either sporadically or not at all on days when temperatures were more than 25°C. Most of these species were carnivorous or insectivorous; the species most dependent on drinking were small granivores and a few honeyeaters.

Diet clearly has a considerable effect on the reliance of birds on free drinking water. Insects contain a substantial amount of preformed water, whereas seeds in field conditions may contain anywhere from less than 5 percent to somewhat more than 20 percent water, depending upon humidity (Morton and MacMillen 1982). Because birds usually obtain seeds from exposed, dry microsites, water content is likely to be toward the lower end of this range. Few desert birds are totally granivorous, and those that are, such as doves and sandgrouse, are obligate drinkers (Maclean 1974, 1985). Most granivorous species feed upon insects opportunistically, especially if the availability of surface water is limited. Adoption of a granivorous diet clearly makes it difficult to maintain water balance without drinking water, but this cost may be countered by the greater availability and dependability of seed supplies in many deserts (Morton and MacMillen 1982). Many species in Australian deserts are largely granivorous (Fisher, Lindgren, and Dawson 1972), but only 25–30 percent of the species recorded in the Namib and Kalahari deserts of Africa are primarily granivorous (although most of the individuals in these avifaunas are seedeaters) (Maclean 1974). Thus, granivory is the prevalent dietary mode in the avifaunas of many deserts, although obligate granivory is relatively uncommon.

The problems that birds face in enduring heat and water scarcity in the desert may be heightened during breeding, when the adults must restrict their activity to the vicinity of the nest site and must provide shading and water for the young. Perhaps the extreme of this problem is confronted by sandgrouse (*Pterocles*), which nest in open, extremely arid situations. Because they are strictly granivorous, the water content of the diet fed to the chicks is low, and the adults must therefore provide them with free water to drink. The birds (primarily males) accomplish this by flying considerable distances to water sources, wetting specialized belly feathers, and carrying water back to the chicks in their plumage. Maclean (1970e) suggested that *Pterocles namaqua* in the Kalahari may fly up to 80 km to water (a round trip of 160 km), and Cade and Maclean (1967) determined that a male of this species can hold an average of 22 ml of water in its belly plumage. Maclean (1983, 1985) has reviewed this remarkable behavior of sandgrouse and the structural modifications that accompany it.

Evasion Tactics

The above features enable birds to endure the constraints imposed by desert environments. In combination with these features, many species also exhibit traits that permit them to evade or circumvent these conditions, at least to a degree (Figure 10.2). These tactics range in scale from short-term adjustments in activity or microhabitat use to long-term distributional shifts.

Activity The thermal regimes to which an individual is exposed may vary considerably between microhabitats in a location, and there is substantial diurnal variation in the temperatures of these microhabitats as well (Martindale 1983; Wiens 1985). Birds can therefore avoid thermal stress to some extent by shifting their use of microhabitats and changing their activity levels as ambient temperatures change. During summer in the Great Basin shrub deserts of North America, for example, nights may drop below freezing but midday temperatures may exceed 50°C a few centimeters above the ground in open areas. Sage Sparrows (*Amphispiza belli*) begin the day by singing from sunlit locations on the tops of shrubs and foraging in open areas between shrubs. As temperatures rise, they initially reduce their singing activity and shift foraging to shaded areas beneath shrubs. During the hottest period of the day, they become quiescent in the center of shrub canopies, where temperatures may be 10°–20°C cooler than in exposed locations near the ground. As temperatures fall later in the day, the activity sequence is reversed (J. A. Wiens and J. T. Rotenberry personal observations).

Similar temporal and spatial changes in activity patterns of birds have been observed in most desert settings. By reducing activity during the hottest times of the day, individuals reduce the rate of heat production and the elevation of body temperature that accompany vigorous activities, and thereby minimize the need to dissipate metabolic heat (Dawson and Bartholomew 1968; Phillips, Butler, and Sharp 1985). More generally, desert birds may reduce their heat generation and energy demands by engaging in less aggressive behavior than birds in other environments and forming longer pair bonds, reducing the need for vigorous courtship displays (Davies 1982).

Breeding Opportunism Several aspects of the breeding biology of birds may be adjusted to the environmental conditions of deserts. Perhaps most obvious is the

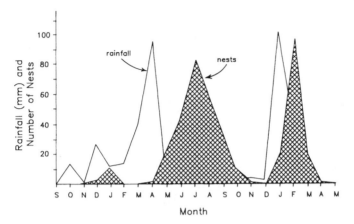

Figure 10.3. The relationship between rainfall and breeding activities (number of nests found) of several species of larks in the Kalahari sandveld (modified after Maclean 1970a).

ability of a number of desert species to initiate breeding opportunistically in response to unpredictable rainfall and the burst of production that often follows such events. The view that developed from studies during the 1950s and 1960s held that breeding activity is often suppressed by drought (Keast 1959), perhaps for several years, but that individuals may respond rapidly to drought-breaking rains, with courtship activity beginning even at the very sight of rainfall (Immelmann 1963). In arid environments, rainfall rather than photoperiod was proposed to be the primary zietgeber controlling breeding activity (Marshall 1959; Lofts and Murton 1968).

There is certainly some support for this view. In central and western Australia, Zebra Finches, Budgerigars, Bourke's Parrots, Black-faced Woodswallows (*Artamus cinereus*), and Grey Teal (*Anas gibberifrons*) have been reported to initiate breeding activity shortly after unseasonal or drought-breaking rains (Immelmann 1963; Frith 1967; Serventy 1971). Many of the species in the Kalahari sandveld breed opportunistically at any time of the year after rains; the several species of larks (Alaudidae) exemplify this opportunism especially well (Maclean 1970b,d) (Figure 10.3). In species such as Zebra Finches, the response to rainfall may be enhanced by the ability of individuals to breed at a very early age and to continue breeding for many successive broods, by the concentration of courtship, pair formation, and nest building into a very short interval, and by the group stimulation associated with breeding in large colonies (Serventy 1971).

The amount of rainfall necessary to release breeding activity varies depending on local conditions. In Ecuador, Marchant (1959) reported that breeding begins with rainfall in excess of 13 mm; the corresponding value in arid parts of southwestern Africa is perhaps 20 mm (Maclean 1976a). Given an adequate rainfall, some species can initiate breeding extremely rapidly. First eggs have been laid as little as 6 days (*Philetarius socius*), 7 days (*Malcorus pectoralis, Bradornis infuscatus, Pyrocephalus rubinus*), or 8 days (*Sporopipes squamifrons*) following rains (Maclean

1970d; Marchant 1959); for Zebra Finches, the lag period is perhaps 13 days (Immelmann 1963). Generally, insectivorous species breed more rapidly after rains and have a more protracted breeding period than do granivores (Maclean 1970d, 1976c; Nix 1976), perhaps because the food supply of the latter species takes longer to develop following rainfall.

It has also been suggested that, in association with opportunistic breeding and the capacity of at least some species to continue breeding for several successive broods, the duration of the breeding season in arid-zone birds should be greater than that of species in other habitats (Serventy 1971; Immelmann 1973). The analyses on which these suggestions were based, however, did not take differences in latitude into account. When Wyndham (1986) adjusted his measure of breeding season duration for the strong (and well-known) effects of latitude, he found that (a) there was no difference between sites at the same latitudes in Africa and Australia, although both had significantly longer mean breeding seasons than sites at the same latitudes elsewhere; (b) birds at the same latitudes in arid and nonarid portions of Australia had similar breeding durations; and (c) contrary to Immelmann's (1970) view, there was no evidence that the arid-zone birds of Australia are more dependent on irregular rainfall to initiate breeding than are those of arid Africa.

The failure of arid-zone birds as a group to match our expectation of a prolonged breeding season may reflect our misperception of the extent of breeding opportunism in these avifaunas. Although there are clear examples of such opportunistic breeding in response to rainfall at any time of year, it would be a mistake to believe that it is the norm among desert birds. In the Kalahari, resident raptors, Ostriches, and intra-African migrants are seasonal breeders, regardless of rainfall, while the Double-banded Courser breeds throughout the year independently of season or rainfall (Maclean 1970c). Many of the species occurring in the arid portions of Australia also follow a seasonal pattern of breeding (Maclean 1976b,c; Davies 1977; Wyndham 1983, 1986), and this is certainly true also of colder, high-latitude deserts in North America (J. A. Wiens and J. T. Rotenberry personal observations) and Patagonia (Maclean 1976a). When good conditions persist following unusually heavy rains, even "opportunistic" species may become seasonal breeders (Maclean 1976c). Where Zebra Finches have expanded into rural environments in southeastern Australia, for example, their breeding is unrelated to rainfall but coincides with flushes of ripening seeds in spring, midsummer, and autumn (Zann and Straw 1984).

The view that rainfall acts as a trigger to release opportunistic breeding in arid-zone birds is thus an oversimplification. Breeding opportunism may really be characteristic only of some nomadic species, and even then the pattern may be superimposed on a more basic breeding seasonality (Schodde 1982). The extent and duration of spring breeding may vary with the amount of rainfall in the previous seasons, being depressed if little or no rain occurs and massive if the rains are good. When breeding does occur nonseasonally in response to rainfall, it may largely be unsuccessful (Davies 1973, 1976).

Other Features of Breeding Biology Aspects of breeding other than timing and duration may also reflect ways of dealing with the environmental constraints of deserts. Nest sites may often be limiting, especially for species nesting in shrubs or trees (where microclimates may be cooler and predation rates lower). The colonial breeding habits of some desert species, such as Zebra Finches and Sociable Weavers (*Philetarius socius*), may have evolved as a result of the scarcity of nesting sites (Maclean 1974).

Species that nest on the ground in exposed locations have, in a sense, traded increased nest site availability against the increased climatic stresses and predation pressures to which they are subject. They often reduce heating of the nest by placing it on the side of a sheltering rock or clump of vegetation that is shaded during the hottest part of the day (Maclean 1974, 1976a). Nonetheless, conditions in such locations may be severe enough that sheltering the eggs or young requires the continuous presence of a bird at the nest. Perhaps as a consequence, both adults must share in parental care, and monogamous pairing is the rule in ground-nesting desert birds (Maclean 1974). Seedsnipes (*Thinocorus*) and the Australian Dotterel cover their eggs with nesting material when they are away from the nest, thereby sheltering them from predation and from overheating (Maclean 1969, 1973). Apparently no other desert birds have adopted this behavior.

Many ground-nesting desert species are also cryptically colored, often matching the soil color of the regions in which they breed (Serventy 1971). In the Kalahari, for example, soils are predominately either red or gray, and they are inhabited by red and gray larks, respectively (Maclean 1974). Background matching seems not so well developed in South American aridland species, although many species are cryptically patterned (Maclean 1976c). Not all desert birds are cryptic, however, the most notable exceptions being predominantly black species such as ravens (*Corvus*), several wheatears and chats (*Oenanthe*), and finch-larks (*Eremopterix*). In some cases, the black coloration may reduce energy expenditures required for thermoregulation at low temperatures (Hamilton and Heppner 1967; Lustick 1969; but see also Walsberg, Campbell, and King 1978), while the black-and-white plumage of some other species may disrupt their outline and make them more cryptic to predators (Serventy 1971).

Maclean (1976a) suggested that true desert species in the Namib and Kalahari frequently have small clutches, smaller than those of related species in more mesic areas, although the evidence that clutch sizes are really lower once latitudinal effects are considered is rather inconclusive. Adjustments in clutch sizes within a species in accordance with changes in environmental conditions are more apparent, however. In breeding seasons following good rains in the Chihuahuan Desert of New Mexico, for example, Cactus Wren (*Campylorhynchus brunneicapillus*) clutch sizes are larger than in years of lower rainfall (Marr and Raitt 1983), and similar variations in clutch sizes in relation to rainfall or drought have been noted in other desert species (Ricklefs 1965; Maclean 1974).

Movement   Perhaps the most conspicuous evasive tactic of desert birds is their movement. If conditions deteriorate in a particular locality, individuals may simply fly away. The degree to which this is an effective tactic, of course, depends on whether or not better conditions are available elsewhere. Many seasonal breeders migrate to more mesic environments during the nonbreeding seasons, often moving considerable distances. Other species are sedentary residents that may make local movements into more productive habitat patches but generally do not move far. Most fabled are the nomadic species, which undertake erratic, unpredictable, and large-scale movements in response to poor conditions in one area and/or good conditions in another (Davies 1984). In Australian species, there is a close correlation between movement patterns and the phenology of rainfall and vegetation growth (Nix 1976). Because the locations of surface water sources in many deserts do not change as much as the locations of food supplies do, nomadism is more likely to be related to unpredictability in food availability than to water scarcity (Davies 1984). Indeed, most nomadic species are largely or entirely granivorous. In the Kalahari, Maclean (1970d) calculated that 7 of the 10 nomadic species (70 percent) were seedeaters, whereas 27 of the 48 resident species (56 percent) were largely insectivorous. All of the nomadic species of larks in southern Africa are granivorous (Maclean 1970b). Although they were not derived from strictly nomadic species, Pulliam's (1985, 1986; Pulliam and Parker 1979) observations that finches wintering in desert grasslands in North America move farther south in years of low rainfall than in years of abundant seed crops also support this generalization.

Not all nomads are granivores, of course. Some raptors, especially those that prey chiefly on rodents that exhibit population outbreaks associated with rainfall and vegetation growth, may also be nomadic. In Australia, species such as Spotted Harriers (*Circus assimilis*), Fork-tailed Kites (*Milvus migrans*), and Letter-winged Kites (*Elanus scriptus*) are known to move over large areas to follow mouse plagues (Baker-Gabb 1983; Davies 1984).

Nomadism has often been considered to be a characteristic feature of desert birds. Keast (1960) estimated that 26–30 percent of the entire Australian avifauna are nomadic, whereas only 6 percent undertake true south-north migrations. Most of the nomadic species are found in the arid zone (Serventy 1971). However, such assessments were based primarily on anecdotal accounts or short studies at single locations. As more of these species have been studied intensively and as the results of banding operations have begun to appear, it has become clear that the extent of nomadism among Australian species (and probably those elsewhere in the world) has been overstated. Species such as Emus (*Dromaius novaehollandiae*), Black-tailed Native Hens (*Gallinula ventralis*), Grey Teal, and White-browed Woodswallows (*Artamus superciliosus*) in Australia or several larks and finch-larks in Africa are probably true nomads. For others, the situation is less clear. The Budgerigar, for example, has long been considered the epitome of a desert nomad, moving over large regions en masse to exploit food and water resources whenever and wherever they appear (e.g., Serventy 1971; Rowley 1975). Wyndham's (1980, 1983) studies, however, have shown that the species exhibits an underlying north-south pattern of movements, following the rainfall from the north in summer to the south in winter. Because the seasonal "migration" is made in short daily hops, it is not very evident, and the tendency for large numbers of birds to aggregate about water holes or ephemeral food concentrations enhances the appearance of nomadism. The Budgerigar appears to follow a movement strategy involving elements of both migration and nomadism, depending on environmental circumstances, and this pattern probably applies to many desert species.

Habitat selection   The mobility of birds gives them the ability not only to shift from one location to another as environmental conditions change seasonally or unpredictably, but to exercise a choice between habitats in landscape mosaics in the areas they do occupy. Although many desert areas are characterized by low overall rates of primary production (Noy-Meir 1973), there may be substantial variation in production levels between habitat patches in a local mosaic. In the northern Chihuahuan Desert in New Mexico, for example, Ludwig (1986, 1987) found that aboveground annual net primary production ranged from 30 to nearly 600 g/m^2. Average production over a 6-year period was greatest in the lower slopes of swales (over 400 g/m^2), while the upper slopes of basins, where precipitation runoff was rapid, had the lowest average production (less than 100 g/m^2). Annual variation in production was greatest in the large arroyos or wadis dissecting alluvial fans (30–456 g/m^2). Reichman (1984) has reported even greater magnitudes of spatial variation in seed abundances in desert soils.

There are thus substantial differences in potential food supplies (as well as shelter and availability of nest sites) between habitat and microhabitat patches in the landscapes of all but the most barren deserts. Birds actively respond to this variation in their selection of habitats to occupy. Because the most suitable patches are often small, gregariousness of individuals is favored; indeed, many desert species breed colonially and/or feed in flocks. These traits are especially evident among granivores, whose food supplies are often more concentrated in space and time than are those of insectivores or carnivores (Wiens and Johnston 1977). A partial exception to this generalization, however, is provided by the Dune Lark (*Mirafra erythrochlamys*) of the Namib, which is predominantly granivorous but forms only small flocks, probably because of low populations and sparse seed supplies (Cox 1983; Boyer 1988; G. L. Maclean personal communication).

The Composition of Desert Avifaunas

The species that comprise the avifaunas of desert locations are some subset of those species that, through an appropriate combination of the endurance and evasive tactics described above, are able to cope with the constraints of desert existence and are not excluded by biological interactions with other species or are not absent as a consequence of history, chance, or disturbance (Figure 10.1). Because many of these traits are possessed by birds in general, a great many species are potential occupants of desert environments. Viewing the situation from a physiological perspective, Bartholomew and Cade (1956) suggested that any bird that can satisfy its habitat requirements in the desert is a candidate for establishment there, because it is as likely to be as effective physiologically as many birds already occupying this environment. In desert scrub habitats in California, for example, perhaps 60 percent of the 40 species present are primarily associated with other habitat types but occur in the desert scrub as ecological "peripherals" of a sort (Dawson and Bartholomew 1968).

Table 10.1 Major bird taxonomic constituents of the arid zones of South America, Africa, and Asia.

Taxon	Distribution	% of Species	
		Arid	Non-arid
Burhinidae	Neotropical	100	0
Cursoriinae	Palaeotropical	52	48
Thinocoridae	South America	75	25
Pteroclididae	Africa and Asia	81	19
Furnariidae (*Geositta, Cinclodes, Upucerthia, Eremobius, Chilia*)	South America	60	40
Furnariidae (*Asthenes*)	South America	55	45
Tyrannidae (*Muscisaxicola*)	South America	64	36
Alaudidae	Africa and Asia	67	33
Turdidae (*Oenanthe, Cercomela*)	Africa and Asia	72	28
Mimidae	Neotropical	51	49

SOURCE: After Maclean (1976a).

Constituent Species

Generally, levels of speciation and endemism of birds are low in deserts (Serventy 1971), and most desert species are derived from avifaunas of neighboring regions or habitats. Moreau (1966) recognized only 25 "typically desert" species in the Sahara. There, as in the Kalahari, larks and wheatears are major elements of the desert avifauna (Maclean 1970a,b). Udvardy (1958) categorized only 29 species of North American birds as desert species; MacMahon's list (1979) contained 31 species. Deserts constitute 70 percent of the land area of Australia, yet Keast (1961) recognized only 17 of the 570 or so species known to breed in Australia (3 percent) as characteristic of the desert. Such categorizations, of course, depend on how one defines "desert" species, but the criteria are rarely mentioned. Williams and Calaby (1985) suggested that "about forty" species are largely restricted to desert areas in Australia (7 percent of the avifauna), while Schodde (1982) recognized 88 species (15 percent) as autochthonous to the Australian arid zone or at least distributionally centered in it. By any criteria, however, the number of bird species that are largely confined to deserts is small.

Certain taxa of birds contribute disproportionately to these desert avifauna, however. The stone-curlews (Burhinidae), sandgrouse (Pteroclididae), and seedsnipes (Thinocoridae) are largely arid-zone families (Table 10.1), and the extensive radiation of larks (Alaudidae) in the deserts of Africa and Asia (Maclean 1970a,b) has already been mentioned. In addition to the stone-curlews and seedsnipes, several other members of typically wading-bird taxa (e.g., pratincoles and coursers, Glareolidae; several species of *Charadrius*, *Vanellus*, and *Peltohyas* in the Charadriidae) have evolved to exploit desert environments effectively (Maclean 1973). Ostriches are typically found in deserts in Africa, as are Emus in Australia and, to a lesser extent, Rheas (*Rhea*) in South America. In Australia, three of the five species of Australian chats (Ephthianuridae) are primarily arid-zone occupants. A diversity of

Table 10.2 Bird species characteristic of four North American deserts.

Species	Sonoran	Mojave	Chihuahuan	Great Basin
Kestrel (*Falco sparverius*)		X		X
Scaled Quail (*Callipepla squamata*)	X		X	
Gambel's Quail (*Callipepla gambelli*)		X		
Mourning Dove (*Zenaida macroura*)	X	X	X	
White-winged Dove (*Zenaida asiatica*)	X	X		
Greater Roadrunner (*Geococcyx californicus*)	X	X	X	
Elf Owl (*Micrathene whitneyi*)	X		X	
Burrowing Owl (*Speotyto cunicularia*)		X		X
Lesser Nighthawk (*Chordeiles acutipennis*)	X	X	X	
Common Poorwill (*Phalaenoptilus nuttallii*)	X	X	X	
Gila Woodpecker (*Melanerpes uropygialis*)	X			
Ladder-backed Woodpecker (*Picoides scalaris*)	X	X	X	
Ash-throated Flycatcher (*Myiarchus cinerascens*)	X	X	X	
Common Raven (*Corvus corax*)	X	X	*C. cryptoleucus*	X
Verdin (*Auriparus flaviceps*)	X	X	X	
Cactus Wren (*Campylorhynchus brunneicapillus*)	X	X	X	
Mockingbird (*Mimus polyglottos*)	X	X	X	
Curve-billed Thrasher (*Toxostoma curvirostre*)	X	*T. lecontei*	X	
Crissal Thrasher (*Toxostoma dorsale*)	X	X	X	
Sage Thrasher (*Oreoscoptes montanus*)				X
Black-tailed Gnatcatcher (*Polioptila melanura*)	X	X	X	
Phainopepla (*Phainopepla nitens*)	X	X	X	
Loggerhead Shrike (*Lanius ludovicianus*)	X	X	X	
Black Phoebe (*Sayornis nigricans*)	X	X	*S. saya*	
Horned Lark (*Eremophila alpestris*)				X
Scott's Oriole (*Icterus parisorum*)	X	X		
Summer Tanager (*Piranga rubra*)	X	X		
Pyrrhuloxia (*Cardinalis sinuata*)	X		X	
House Finch (*Carpodacus mexicanus*)	X	X	X	
Sage Sparrow (*Amphispiza belli*)		X		X
Black-throated Sparrow (*Amphispiza bilineata*)	X	X	X	X
Brewer's Sparrow (*Spizella breweri*)				X

SOURCE: Modified from MacMahon (1979).

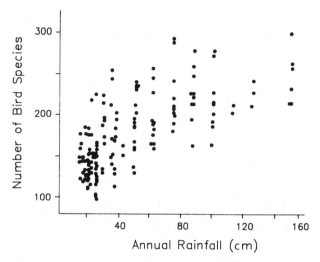

Figure 10.4. The relationship between the number of bird species occurring in 240 × 240 km grid blocks in Australia and mean annual precipitation (after Pianka and Schall 1981).

taxa contribute to the avifaunas of the major North American deserts (Table 10.2), although many members of most of these taxa occur in other environments as well.

Species Numbers, Guilds, and Densities

Although the species richness of many groups of plants and animals (e.g., succulents, lizards, scorpions, ants, tenebrionid beetles) is relatively high in deserts in comparison to other environments, that of birds is not. Not only are there few "true" desert species of birds, but the overall number of species occurring in a location is also relatively small. At a broad scale (58,000 km² blocks), species richness in Australia exhibits a strong positive correlation with mean annual precipitation (Pianka and Schall 1981). The relationship in fact appears to be curvilinear, richness falling especially rapidly in areas with less than 40 cm annual precipitation and being rather consistently low in the arid zone (<25 cm annual precipitation) (Figure 10.4). Interestingly, similar analyses of the North American avifauna (Cook 1969; Schall and Pianka 1978) indicate that species richness is *inversely* correlated with mean annual precipitation. The difference between continents may be due to the greater topographic and landscape diversity of the arid portions of western North America, which may increase bird species diversity at such broad scales of analysis. The avifaunas of different desert types within a continent exhibit only moderate levels of similarity in their bird community compositions (e.g., Table 10.2), the degree of similarity generally decreasing as a function of increasing distance between areas (MacMahon and Wagner 1985).

At the finer scale of local habitats or study plots, deserts usually contain only a few species. Comparisons between studies are difficult because different investigators have used different census methods (often none at all!) applied to areas of different sizes considered over periods of different durations (see Wiens 1989a). Still, the values give one some feeling for the number of species constituting local bird communities in deserts. In Western Australian deserts, for example, Pianka and Pianka (1970) recorded 11–35 species on eight study areas over a 16-month period. Keast (1959) listed 17 and 24 species present in desert spinifex and desert mulga habitats in central Australia. The *Atriplex*-dominated shrub deserts in northwestern New South Wales that I censused using 0.28 km^2 transect strips contained 2–9 species during the breeding periods in December, and 4–13 species during winter (May) (Wiens in press, personal observation). In the Sahel of Senegal, Morel (1968) found 13 species present in a 25-ha plot in the dry season (May); in the wet season (October), after palearctic migrants arrived, 38 species were recorded. Areas in the southern Chihuahuan Desert in Durango and Zacatecas, Mexico, averaged 25 breeding species in comparison with 14 species in Sonoran Desert study areas in Arizona and 6–9 species or 14–16 species in northern Chihuahuan Desert locations in New Mexico and Texas, respectively (Dixon 1959; Raitt and Maze 1968; Webster 1974). Weathers (1983) recorded 16–36 breeding species in his censuses in a variety of Sonoran Desert habitats in California. Sites in northern Great Basin shrub deserts, surveyed following the same procedures used in my Australian studies, supported 4–10 breeding species (J. A. Wiens and J. T. Rotenberry personal observations).

Another way to look at the species composition of a community is in terms of its "guild signature," the way species are distributed among functionally defined guilds. Guild analyses are sensitive to the ways in which guilds are defined, and these definitions are often a priori and arbitrary (Wiens 1989a). Morel (1968) classified the species occurring in the Senegal Sahel according to their food habits and migratory status (Table 10.3). During the dry season, no migrants were present and nearly half of the species were herbivorous (granivorous). In October, several species of migratory omnivores and insectivores were present and omnivores in general increased in frequency, but sedentary herbivores still dominated the community. In another study, Werger (1986) categorized the species breeding in several desert regions of the world according to their foraging position and diet (Table 10.3). By this accounting, carnivorous (mainly insectivorous) species predominated in all cases, arboreal forms being more frequent in the deserts of Western Australia and the Argentine Monte. There were few ground-feeding herbivorous species in the Monte, but elsewhere such species constituted 20–29 percent of the community. Arboreal-feeding herbivores were infrequent in all deserts. Overall, ground-feeding species dominated the avifaunas of the southern Kalahari and the Sonoran Desert, while arboreal-feeding forms dominated the Western Australian and Monte deserts. It is difficult to generalize from such comparisons; in order to do so, surveys from more areas are needed, and they must be analyzed by more objective guild categorizations.

Although few estimates of total densities of birds occupying desert habitats are available, it is apparent that they are low relative to those in more mesic environments (Wiens 1973). Rodin (1979) reported a density of 260 individuals/km^2 (I have adjusted all density estimates to this arbitrary scale) for a sandy desert site in Asian U.S.S.R., while Morel (1968) recorded densities at his Sahel site of 160 individuals/

Table 10.3 Percentages of bird species per guild category recorded in desert regions of the world.

	Senegal Sahel[a]		Southern Kalahari[b]	Western Australia[b]	Monte (Argentina)[b]	Sonoran Desert[b]
	May	Oct.				
Sedentary herbivores	46	32	. . .	. . .	. . .	. . .
Migratory herbivores	0	0	. . .	. . .	. . .	. . .
Sedentary omnivores	15	26	. . .	. . .	. . .	. . .
Migratory omnivores	0	11	. . .	. . .	. . .	. . .
Sedentary insectivores	39	18	. . .	. . .	. . .	. . .
Migratory insectivores	0	13	. . .	. . .	. . .	. . .
Ground herbivores	. . .	. . .	28	29	8	20
Ground carnivores	. . .	. . .	34	18	34	40
Arboreal herbivores	. . .	. . .	2	7	12	2
Arboreal carnivores	. . .	. . .	36	46	46	38

[a]SOURCE: Morel (1968).
[b]SOURCE: Werger (1986).

km^2 in May, increasing to 1,000 individuals/km^2 in October after the arrival of migrants. In the southern Chihuahuan Desert, breeding densities of 55–288 individuals/km^2 have been reported, while surveys in the Sonoran Desert of Arizona recorded an average of 210 individuals/km^2 (Webster 1974). In the northern Chihuahuan Desert of New Mexico and Texas, breeding densities of 42–88 individuals/km^2 (Raitt and Maze 1968) or 148–252 individuals/km^2 (Dixon 1959) have been reported. Weathers (1983) calculated average densities over the entire year for the habitats he surveyed in California Sonoran Desert; these values ranged from 263 to 743 individuals/km^2. There, densities were greatest in spring in alluvial plains, desert washes, and rocky slopes and uplands, in winter on the valley floor, and in fall and winter in the higher-elevation piñon-juniper and chaparral habitats. At several shrub desert locations in the northern Great Basin, we found densities of 176–461 individuals/km^2 (mean = 301 individuals/km^2) (J. A. Wiens and J. T. Rotenberry personal observations), whereas the breeding densities I recorded in Australian shrub deserts using the same procedures ranged from 102 to 219 individuals/km^2 (mean = 155 individuals/km^2). Wintering densities at the Australian sites were greater (mean = 259 individuals/km^2; range = 123–559). We did not systematically survey the North American sites during winter, but incidental observations indicated that there were few birds present. Thus, although many deserts contain more species and individuals during winter than during the summer breeding period as a consequence of the addition of migrants from other areas (drawn there, at least in good years, by the abundance of seeds), colder temperate deserts may lose species and individuals during the winter.

Noting that the Australian deserts generally lack the large numbers of granivorous rodents that characterize deserts in many other parts of the world, several investigators have suggested that the species richness of other granivores (ants and birds) should be greater there (Brown, Davidson, and Reichman 1979; Morton 1979, 1982, 1985; Morton and Davies 1983). In fact, in comparisons with North American desert locations of similar precipitation and productivity, Morton and Davidson (1988) found no differences in the abundance and species richness of seed-harvester ants on local plots. The limited quantitative data on birds suggest that, although species richness may be roughly equivalent in deserts of the two continents, densities are lower in Australia. Morton's (1985) comparisons of rates of removal of seeds from experimental baits in Australian, North American, and South American (Monte) deserts indicate that (a) mammals do indeed dominate consumption by granivores in North America but are of negligible importance in Australia and South America; (b) ants are the dominant granivores in the Australian desert sites; (c) removal rates by birds in Australia and North America are generally similar (although the experimental procedure probably underestimates their seed consumption); and (d) the overall intensity of seed predation by these granivores can be ranked, with North America as the highest, Australia the next highest, and South America the least. Preliminary observations indicate that seed abundance and size distributions are similar in Australian and North American deserts (S. R. Morton and F. Alexander personal communications), even though the deserts in Australia generally contain considerably more grass cover than most North American deserts (McCleary 1968). On the other hand, seed production and/or availability may be more variable in space and time and perhaps also lower in the Australian deserts. More detailed quantitative studies, extended to deserts in other continents (e.g., Mares and Rosenzweig 1978; Abramsky 1983), are needed to resolve this issue.

Ecological Convergence

Deserts in many parts of the world present similar challenges and constraints to the species occurring there, so one might expect there to be some degree of ecological convergence in both species characteristics and community attributes. On the other hand, the preceding analysis suggests that there may be some fundamental differences between deserts in different continents, which would act to thwart close convergences. At the species level, however, several examples of apparent convergence do exist. The sandgrouse of Africa and Asia and the seedsnipes of South America have both diverged from the basic arthropod diet of their relatives to become granivorous. Both depend on external sources of water and, although they differ in some respects, several species are remarkably similar in appearance, behavior, and ecology (Maclean 1968, 1969, 1985). The Western Meadowlark (*Sturnella neglecta*) of North America and the Brown Songlark (*Cinclorhamphus cruralis*) of Australia are remarkably similar in quantitative features of their morphology and life history attributes (Wiens in press, personal observation). Meadowlarks are grassland as well as desert inhabitants in North America, and their morphological and ecological convergence with the Yellow-throated Longclaw (*Macronyx croceus*), a steppe species of Africa, is well known. The Brewer's Sparrow (*Spizella breweri*) and Horned Lark (*Eremophila alpestris*) of North American shrub deserts closely match, respectively, the Southern Whiteface (*Aphelocephala leucopsis*) and Richard's Pipit (*Anthus novaeseelandiae*) of similar habitats in Australia in morphological features, although not in life history characteristics (Wiens, personal observation).

In their comparisons of the avifaunas of the Sonoran Desert of North America and the Monte of Argentina, on the other hand, Orians and Solbrig (1977) recognized several unrelated species pairs, for example, Verdins (*Auriparus flaviceps*) and Tufted Tit-spinetails (*Leptasthenura platensis*), Cactus Wrens and Short-billed Canasteros (*Asthenes baeri*), that were at least qualitatively similar in diet and foraging behavior but differed morphologically. A somewhat broader analysis of convergence has been made by Schluter (1986), who analyzed finch "communities" occurring in various habitat types in different parts of the world. His sample of desert localities was quite limited, but he found a suggestion (albeit nonsignificant) of unusual similarity in size and shape between finches in warm temperate deserts in California and Argentina, but not between finches occupying cold temperate deserts in Patagonia and Kazakhstan.

On a broader scale yet, I compared several attributes of bird communities breeding in quantitatively matched shrub desert sites in the northern Great Basin in North America and in the arid zone of northwestern New South Wales in Australia (Wiens in press). The vegetational structure of study plots in these two areas is remarkably similar, and the mean number of breeding species recorded in these plots (5.5 in Australia, 6.3 in North America) does not differ statistically. As noted above, however, total densities differ considerably between the continents.

To assess the possibilities of convergence between these communities in more detail, I categorized each species according to a number of ecological and life history traits, and then calculated the frequencies of these traits (adjusted for the relative densities of species on the survey plots) for each local assemblage. Most ecological and life history traits differ significantly in their occurrence in the Australian and North American assemblages (Wiens in press) (Table 10.4). On average, the Australian species have a much longer breeding period and more broods per year than the North American species, even though clutch sizes, overall, are quite similar. More Australian species breed in communal groups and/or form breeding colonies. The Australian species also feed in flocks during the breeding season more frequently than do the North American birds. The most impressive difference is in movement patterns: all of the North American species are typical migrants, whereas the Australian species are all either sedentary or nomadic. From what I have said previously in this chapter, one might expect many of these differences to be associated with a greater degree of granivory among the Australian species. Interestingly, however, there is no significant difference in general diet composition between the avifaunas as a whole (Wiens in press).

These differences in ecological and life history profiles of the communities point up the fact that, although the vegetation of these shrublands is quite similar in structure, the environments are vastly different in other ways that influence the birds. The Australian deserts, for example, are considerably warmer in winter than are the North American sites, and this may in part account for the differences in movement patterns. Perhaps most importantly, the spatial and temporal variance of food production may be much greater in Australia than in North America as a consequence of both greater edaphic variation and more variable precipitation (Wiens 1989a, in press; Morton and Davidson 1988). As Morton and Davidson observe,

Table 10.4   Mean values of various ecological/life history traits of bird species breeding in shrub desert habitats in Australia (N = 4 sites) and North America (N = 7 sites).

Trait	Mean Values		p^a
	Australia	North America	
Duration of nesting (days)	27.6	22.8	**
Breeding period (months)	5.9	3.1	***
Clutch size	3.4	3.3	ns
Broods per year	2.1	1.1	**
Mating system index[b]	1.1	1.0	ns
Communal breeding index[c]	1.5	1.0	***
Nest type index[d]	2.8	2.0	***
Nest height index[e]	2.0	1.9	ns
Breeding aggregation index[f]	2.6	1.0	***
Territory size index[g]	3.4	3.9	*
Diet index[h]	2.5	2.9	ns
Flock feeding index[i]	2.2	1.1	**
Migration index[j]	1.2	3.0	***

Source: Wiens (in press).
Note: Mean values obtained by multiplying the index value for each species by its relative density in the community; the ecological/life history trait means are therefore weighted on the basis of relative abundances of the species.
[a] ns = not significant; * = $p < .05$; ** = $p < .01$; *** = $p < .001$.
[b] 1 = monogamous; 2 = polygynous.
[c] 1 = no; 2 = yes.
[d] 1 = platform; 2 = cup; 3 = domed cup; 4 = cavity.
[e] 1 = ground; 2 = shrub; 3 = tree.
[f] 1 = pairs; 2 = group; 3 = small colonies (2–10 pairs); 4 = large colonies (>10 pairs).
[g] 1 = none; 2 = nest-centered; 3 = small; 4 = medium; 5 = large.
[h] 1 = granivorous; 2 = omnivorous; 3 = insectivorous (during breeding season).
[i] 1 = never; 2 = small flocks; 3 = medium flocks; 4 = large flocks (during breeding season).
[j] 1 = sedentary; 2 = nomadic; 3 = migratory.

infrequent, large cyclonic rainfalls can lead to production of a magnitude rarely seen in North American deserts, but these peaks are often separated by prolonged and widespread droughts. The absence of mountains in the Australian arid-zone landscape reinforces the climatic stress accompanying droughts, as there is no seasonal runoff to provide water and no nearby refuge of more mesic habitats like that characteristic of western North America. As a consequence, the Australian birds exhibit a variety of traits that are not seen in most North American shrub desert species.

Habitat Responses

How birds respond to habitat features has long been a focus of community investigations. This emphasis was strengthened by the work of MacArthur and his colleagues (e.g., MacArthur and MacArthur 1961; MacArthur, Recher, and Cody 1966) relating bird species diversity to a measure of vegetational profile structure, foliage height diversity. Pianka (1979, p. 322) has expressed a widely accepted view with particular reference to deserts by observing that more vegetationally diverse deserts "support a structurally complex vegetation which provides animals with ample opportunities for niche segregation involving differential use of microhabitats." The vegetational diversity, in this view, permits niche partitioning in habitat dimensions, allowing species that otherwise might compete to coexist, thereby increasing species richness. The relationship of habitat complexity to the alleviation of competition in bird communities is an untestable (or at least a frequently untested) proposition (Wiens 1983, 1989b), but there is little doubt that, over a broad range of habitats (say, deserts to forests) the correlation between vegetation height profile and bird species diversity does indeed hold. Within a narrower range of habitats, the relationship is less certain.

Attempts to relate bird community diversity in deserts to measures of vertical profile structure of the vegetation have generally failed. In the Sonoran Desert, for example, Tomoff (1974) found a clear relationship to the diversity of structural types or life forms of vegetation present in an area. He noted in particular the high proportion of bird species that nest in cavities, spinescent trees and shrubs, and cacti, and suggested that this may reflect strong predation pressure on nesting birds. Mares et al. (1977) reported similar observations for the Monte Desert in Argentina, as did Webster (1974) for the southern Chihuahuan Desert in Mexico. In an arid mountain range in South Australia, Shurcliff (1980) found that variations in bird species diversity between sites were closely associated with variations in vegetational life form diversity and plant species diversity. Avifaunal similarity between the sites was more closely related to the distance between them than to their similarity in vegetation structure, perhaps because most of the species exhibit fairly broad habitat amplitudes. In our studies of semiarid shrub deserts in the Great Basin of North America, we found that bird community richness increased with increasing structural and life form diversity and decreasing horizontal habitat patchiness (more even distribution) of the vegetation (Wiens and Rotenberry 1981). The diversity of desert bird communities clearly varies in relation to the complexity of the composition and structure of the vegetation, but in ways that are not easily measured with simple indices. Variations in predation pressure or food availability may be more important than competitive interactions in determining these patterns of habitat association. Careful studies of the factors underlying the habitat relationships of desert birds, however, have yet to be made.

Environmental Variation and Communities

In addition to acting as a strong selective force on life history attributes, the variability of desert environments may have fundamental effects on the distribution and abundance of species and thereby on community structure. Anecdotal accounts of the responses of birds to droughts or unseasonal deluges in some Australian deserts abound, as do reports of similar events in deserts elsewhere in the world (Serventy 1971). In a site in the Mojave Desert of North America, for example, LeConte's Thrashers (*Toxostoma lecontei*) were present in years of average or above-average precipitation, when insect food was abundant, but were absent during dry years (Miller and Stebbins 1964). George (1976) noted that an area in the Saharan desert of Morocco had a breeding population of 400 Spotted Sandgrouse (*Pterocles senegallus*) and 300–400 Desert Warblers (*Sylvia nana*) in one wet year but supported only 80 sandgrouse and 8 warblers in the following dry year (Wagner 1981). Very limited quantitative data available from long-term studies of several species in North American Sonoran Desert and desert grassland locations produce coefficients of variation of annual population estimates of 33–80 percent, only slightly greater than those for bird populations in more mesic environments (Wagner 1981).

In deserts in which rainfall is extremely variable and unpredictable in its occurrence and magnitude, resource availability to birds may vary between feast and famine, the drought-induced famines often lasting considerably longer than the feasts. As a consequence, the form and intensity of resource-based competition among species may vary. With a flush of resources following heavy rains, competitive pressures may be relaxed, and the community patterns of niche partitioning and minimal overlap expected on the basis of competition theory may disappear (Wiens 1977) (Figure 10.5). As resource levels drop in the absence of repeated (i.e., seasonal) rains, interspecific competition may intensify. Among birds, forms of interference competition may appear first, and such competition may produce community

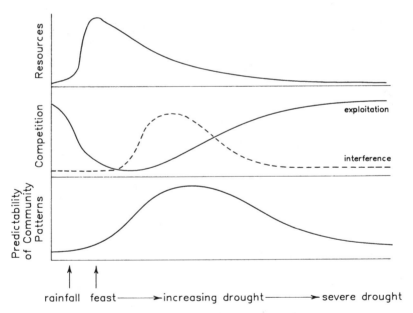

Figure 10.5. Hypothetical patterns of resource availability (top), the magnitude of exploitation or interference competition between species (middle), and the predictability or stability of patterns in community structure expected from competition theory (bottom) in relation to the occurrence of a rainfall that produces a flush of resources in the desert, followed by a prolonged drought.

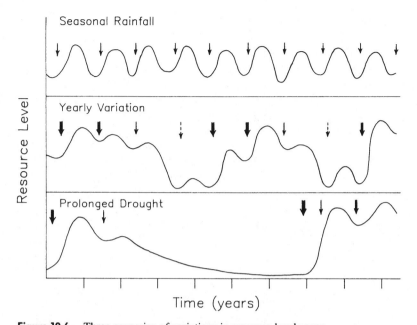

Figure 10.6. Three scenarios of variations in resource levels over several years in deserts. The top diagram depicts the pattern expected when rainfall is seasonal and relatively reliable from one year to the next. The middle diagram shows a seasonally dominated system in which there may be substantial yearly variation, producing (in this case) several years of abundant resources followed by shorter "crunch" periods in which resources are scarce. In the bottom panel, rainfall is less seasonal in its occurrence and is extremely episodic, periods of good rains being separated by several years of drought. The effects of these different scenarios on patterns of bird community structure are described in the text.

patterns most closely matching the predictions of competition theory. As the drought worsens, resource levels will continue to decline, and exploitation may replace interference as the dominant form of competition (Figure 10.5). If the drought becomes severe, the effects of competition may be overridden by more direct physiological limitation of individuals. Populations may initiate nomadic movements as resource levels drop, and different species are likely to emigrate at different resource availability levels. As a consequence of these factors, community membership may be in a constant state of flux, and the biological interactions that do occur may have relatively little effect on overall community patterns. Community structure, in the sense of repeatable, predictable patterns of guild organization, niche partitioning and overlap, and the like, disintegrates (Figure 10.5).

This sort of scenario is likely to be expressed only in the most unpredictable and variable desert environments (Figure 10.6), such as portions of central and western Australia. If rainfall (and pulses of production) follows a regular seasonal pattern with relatively small variations between years (as may occur in some Sonoran Desert locations), resource levels will be more predictable, community membership will be more stable, and the role of biological interactions such as competition in creating community patterns will be greater. In intermediate situations, there may be a basic seasonality to rainfall and production, but annual variation may be great. Several seasons or years of above-average rains may produce a flush of resources that persists for some time, and the drought intervals between these flushes may be relatively short (Figure 10.6). In this case, competitive effects on community structure may be most evident only during the intermittent "crunch" periods between times of relative resource abundance (Wiens 1977). These conditions may typify many desert regions.

These arguments are, of course, speculative. We have too little long-term empirical information to judge which of the above scenarios may apply to which deserts (or, indeed, whether the scenarios are at all correct). Clearly, however, the temporal variability of many desert environments, coupled with the mobility of birds, makes it difficult to detect clear and consistent patterns in the structure of desert bird communities, especially in short-term investigations.

Deserts are variable in space as well as in time, however, and this spatial variation also influences community dynamics. Because of the patterns of precipitation runoff, deserts may exhibit a relatively fine-scale interdigitation of patches of habitat that differ markedly in productivity (Ludwig 1987). Bird distributions may follow this production patchiness, leading to strongly aggregated dispersion patterns. Species richness and community structure may vary dramatically between elements of a local landscape mosaic. Rainfall events may also be patchy on a somewhat broader scale, one area receiving good rains while another a few kilometers away remains dry. This sort of patchiness is especially prevalent in deserts with considerable topographic relief, where convective thunderstorms are frequent but isolated (Rotenberry, Hinds, and Thorp 1976). Again, birds may respond to such events by shifting their local or regional distributions, producing substantial spatial variance in community patterns even in relatively uniform vegetation. At an even broader scale, movements of nomadic species from areas experiencing severe droughts may affect the composition of communities in areas some distance away that have received normal rainfall. In studies in northwestern Victoria, Baker-Gabb (1983) recorded considerable yearly fluctuations in the number of nomadic raptors present, even though the local climate remained much the same. Recher et al. (1983) found that the community dynamics they recorded in woodlands in southeastern Australia were influenced by influxes of birds from the drought-stricken interior more than a thousand kilometers away. Desert bird communities are extremely open to the effects of events elsewhere (Wiens 1989b).

Future Directions

We know a fair amount about the autecology of desert birds (the upper portions of Figure 10.1) but our knowledge of their population and community dynamics and the factors influencing these dynamics is quite sketchy. We have plenty of inferences in the literature about the occurrence or magnitude of interspecific competition between desert birds and its effect on community structure, for example, but direct empirical tests of these inferences are lacking. Other forms of biological interactions (Figure 10.1) have received even less direct study. Desert environments are different enough from those in other biome types that we cannot extrapolate findings from these other regions to deserts with any degree of confidence.

There is a clear need for both more extensive and more intensive studies. Several priorities for such studies seem to me to be especially clear:

- Populations and communities should be surveyed quantitatively, using a standardized methodology.
- To develop a firmer understanding of the similarities and differences between the many desert types and regions of the world, these surveys should be conducted in many areas.
- Because conditions in the desert vary so much between years, long-term studies are essential. Ideally, such studies should span at least one complete wet-dry cycle in an area, and such cycles may take several years. By conducting investigations of this sort in different desert settings, some appreciation for whether or not the scenarios of Figures 10.5 and 10.6 are realistic may emerge.
- The spatial patchiness of deserts also merits closer study, at several scales of resolution. At a fine scale, investigations might focus on the patterns of habitat patch selection by birds and the factors underlying this selection. Such studies should include careful measurements of resource levels. On a broader scale, examination of the linkages between different habitat types in a local or regional landscape mosaic might prove rewarding. A better understanding of the movement patterns and population dynamics of desert birds is essential.
- The relationship of desert birds to other groups of desert organisms should be given greater consideration. Some beginnings in this direction have been made in studies of granivore guilds (birds, ants, and rodents). The granivore studies should be expanded, and parallel investigations initiated with respect to other resource guilds, such as insectivores.

Deserts are special places, with a unique beauty. Too few ecologists appreciate this beauty, however, and not many are willing to spend their time there. There are few ecological studies of birds in the deserts of Asia, the Middle East, or Afghanistan, and even in relatively well-studied areas like Australia, North America, and the Namib much remains to be learned. To do serious ecological studies in deserts is difficult and often uncomfortable, but the rewards, in both science and aesthetics, are great.

Acknowledgments

My research in the deserts of North America and Australia has been supported by the National Science Foundation of the United States, the National Geographic Society, and a Fulbright Fellowship. Gordon Maclean, Richard MacMillen, and Wesley Weathers read an initial version of this chapter and offered many helpful suggestions.

Bibliography

Abramsky, Z. 1983. Experiments on seed predation by rodents and ants in the Israeli desert. *Oecologia* 57: 328–332.

Andrewartha, H. G., and L. C. Birch. 1984. *The Ecological Web: More on the Distribution and Abundance of Animals.* Chicago: University of Chicago Press.

Baker-Gabb, D. J. 1983. The breeding ecology of twelve species of diurnal raptor in northwestern Victoria. *Australian Wildlife Research* 10: 145–160.

Bartholomew, G. A. 1972. The water economy of seed-eating birds that survive without drinking. Pp. 237–254 in *Proceedings of the 15th International Ornithological Congress, Oxford.*

Bartholomew, G. A., and T. J. Cade. 1956. Water consumption of House Finches. *Condor* 58: 406–412.

———. 1963. The water economy of land birds. *Auk* 80: 504–539.

Bartholomew, G. A., R. C. Lasiewski, and E. C. Crawford. 1968. Patterns of panting and gular flutter in cormorants, pelicans, owls, and doves. *Condor* 70: 31–34.

Bernstein, M. H. 1971. Cutaneous water loss in small birds. *Condor* 73: 468–469.

Boyer, H. J. 1988. Breeding biology of the Dune Lark. *Ostrich* 59: 30–37.

Brown, J. H. 1986. The roles of vertebrates in desert ecosystems. Pp. 51–71 in W. G. Whitford (ed.), *Pattern and Process in Desert Ecosystems.* Albuquerque: University of New Mexico Press.

Brown, J. H., D. W. Davidson, and O. J. Reichman. 1979. An experimental study of competition between seed-eating desert rodents and ants. *American Zoologist* 19: 1129–1143.

Cade, T. J., and J. A. Dybas, Jr. 1962. Water economy of the Budgerygah. *Auk* 79: 345–364.

Cade, T. J., and G. L. Maclean. 1967. Transport of water by adult sandgrouse to their young. *Condor* 69: 323–343.

Cade, T. J., C. A. Tobin, and A. Gold. 1965. Water economy and metabolism of two estrilidine finches. *Physiological Zoology* 38: 9–33.

Calder, W. A. 1964. Gaseous metabolism and water relations of the Zebra Finch, *Taeniopygia castanotis. Physiological Zoology* 37: 400–413.

Cody, M. L., and J. M. Diamond, eds. 1975. *Ecology and Evolution of Communities.* Cambridge, Mass.: Harvard University Press.

Cook, R. E. 1969. Variation in species density in North American birds. *Systematic Zoology* 18: 63–84.

Cox, G. W. 1983. Foraging behaviour of the Dune Lark. *Ostrich* 54: 113–120.

Crawford, E. C., and K. Schmidt-Nielsen. 1967. Temperature regulation and evaporative cooling in the Ostrich. *American Journal of Physiology* 212: 347–353.

Curtis, J. T. 1959. *The Vegetation of Wisconsin.* Madison, Wis.: University of Wisconsin Press.

Davies, S. J. J. F. 1973. Environmental variables and the biology of native Australian animals in the mulga lands. *Tropical Grasslands* 7: 127–134.

———. 1976. Environmental variables and the biology of Australian arid zone birds. Pp. 481–488 in *Proceedings of the 16th International Ornithological Congress, Canberra.*

———. 1977. The timing of breeding by the Zebra Finch *Taeniopygia castanotis* at Mileura, Western Australia. *Ibis* 119: 20–23.

———. 1982. Behavioural adaptations of birds to environments where evaporation is high and water is in short supply. *Comparative Biochemistry and Physiology* 71(A): 557–566.

———. 1984. Nomadism as a response to desert conditions in Australia. *Journal of Arid Environments* 7: 183–195.

Dawson, W. R. 1976. Physiological and behavioural adjustments of birds to heat and aridity. Pp. 455–467 in *Proceedings of the 16th International Ornithological Congress, Canberra.*

———. 1981. Adjustments of Australian birds to thermal conditions and water scarcity in arid zones. Pp. 1651–1674 in A. Keast (ed.), *Ecological Biogeography of Australia.* The Hague: W. Junk.

Dawson, W. R., and G. A. Bartholomew. 1968. Temperature regulation and water economy of desert birds. Pp. 357–394 in G. W. Brown (ed.), *Desert Biology,* vol. 1. New York: Academic Press.

Dawson, W. R., and A. F. Bennett. 1973. Roles of metabolic level and temperature regulation in the adjustment of Western Plumed Pigeons (*Lophophaps ferruginea*) to desert conditions. *Comparative Biochemistry and Physiology* 44(A): 249–266.

Dawson, W. R., and C. D. Fisher. 1969. Responses to temperature by the Spotted Nightjar (*Eurostopodus guttatus*). *Condor* 71: 49–53.

Dawson, W. R., and J. W. Hudson. 1970. Birds. Pp. 223–310 in G. C. Whittow (ed.), *Comparative Physiology of Temperature Regulation,* vol. 1. New York: Academic Press.

Dawson, W. R., and K. Schmidt-Nielsen. 1964. Terrestrial animals in dry heat: desert birds. Pp. 481–492 in D. B. Dill (ed.), *Handbook of Physiology.* Washington, D.C.: American Physiological Society.

Dixon, K. L. 1959. Ecological and distributional relations of desert scrub birds in western Texas. *Condor* 61: 397–409.

Evenari, M. 1985. Adaptations of plants and animals to the desert environment. Pp. 79–92 in M. Evenari, I. Noy-Meir, and D. W. Goodall (eds.), *Hot Deserts and Arid Shrublands, A.* Amsterdam: Elsevier.

Fisher, C. D., E. Lindgren, and W. R. Dawson. 1972. Drinking patterns and behavior of Australian desert birds in relation to their ecology and abundance. *Condor* 74: 111–136.

Frith, H. J. 1967. *Waterfowl in Australia.* Sydney, Australia: Angus & Robertson.

Gause, G. F. 1934. *The Struggle for Existence.* Baltimore, Md.: Williams & Wilkins.

George, U. 1976. *The Deserts of This Earth.* New York: Harcourt Brace Jovanovich.

Gleason, H. A. 1926. The individualistic concept of the plant association. *Bulletin of the Torrey Botanical Club* 53: 1–20.

Grinnell, J. 1917. The niche-relationships of the California Thrasher. *Auk* 34: 427–433.

Hamilton, W. J. III, and F. Heppner. 1967. Radiant solar energy and the function of black homeotherm pigmentation: an hypothesis. *Science* 155: 196–197.

Hutchinson, G. E. 1957. Concluding remarks. *Cold Spring Harbor Symposia on Quantitative Biology* 22: 415–427.

Immelmann, K. 1963. Drought adaptations in Australian desert birds. Pp. 649–657 in *Proceedings of the 13th International Ornithological Congress.*

———. 1970. Environmental factors controlling reproduction in African and Australian birds—a comparison. *Ostrich* (Supplement) 8: 193–204.

———. 1973. Role of environment in reproduction as a source of "predictive" information. Pp. 121–147 in D. S. Farner (ed.), *Breeding Biology of Birds.* Washington, D.C.: National Academy of Sciences.

Ives, N. 1973. Overnight torpidity in Australian arid-country birds. *Emu* 73: 140.

James, F. C., R. F. Johnston, N. O. Wamer, G. J. Niemi, and W. J. Boecklen. 1984. The Grinnellian niche of the Wood Thrush. *American Naturalist* 124: 17–30.

Jesson, R. A., and G. L. Maclean. 1976. Salt glands in the neonatal Australian Pratincole. *Emu* 76: 227.

Keast, A. 1959. Australian birds: their zoogeography and adaptations to an arid continent. Pp. 89–114 in A. Keast, R. L. Crocker, and C. S. Christian (eds.), *Biogeography and Ecology in Australia.* The Hague: W. Junk.

———. 1960. Bird adaptations to aridity on the Australian continent. Pp. 373–375 in *Proceedings of the 12th International Ornithological Congress*.

———. 1961. Bird speciation in the Australian continent. *Bulletin of the Museum of Comparative Zoology, Harvard University* 123: 305–495.

Lasiewski, R. C., M. H. Bernstein, and R. D. Ohmart. 1971. Cutaneous water loss in the Roadrunner and Poor-will. *Condor* 72: 332–338.

Lee, P., and K. Schmidt-Nielsen. 1971. Respiratory and cutaneous evaporation in the Zebra Finch: effect on water balance. *American Journal of Physiology* 220: 1598–1605.

Lofts, B., and R. K. Murton. 1968. Photoperiodic and physiological adaptations regulating avian breeding cycles and their ecological significance. *Journal of Zoology* 155: 327–394.

Louw, G. N., P. C. Belonje, and H. J. Coetzee. 1969. Renal function, respiration, heart rate and thermoregulation in the Ostrich (*Struthio camelus*). *Scientific Papers of the Namib Desert Research Station* 42: 43–54.

Louw, G. N., and M. Seely. 1982. *Ecology of Desert Organisms*. London: Longman.

Low, W. A. 1979. Spatial and temporal distribution and behaviour. Pp. 769–795 in D. W. Goodall, R. A. Perry, and K. M. W. Howes (eds.), *Arid-Land Ecosystems: Structure, Functioning and Management*, vol. 1. Cambridge: Cambridge University Press.

Ludwig, J. A. 1986. Primary production variability in desert ecosystems. Pp. 5–17 in W. G. Whitford (ed.), *Pattern and Process in Desert Ecosystems*. Albuquerque, N.Mex.: University of New Mexico Press.

———. 1987. Primary productivity in arid lands: myths and realities. *Journal of Arid Environments* 13: 1–7.

Lustick, S. 1969. Bird energetics: effects of artificial radiation. *Science* 163: 387–389.

MacArthur, R. H. 1972. *Geographical Ecology*. New York: Harper and Row.

MacArthur, R. H., and J. W. MacArthur. 1961. On bird species diversity. *Ecology* 42: 594–598.

MacArthur, R. H., H. Recher, and M. Cody. 1966. On the relation between habitat selection and species diversity. *American Naturalist* 100: 319–332.

McCleary, J. A. 1968. The biology of desert plants. Pp. 141–194 in G. W. Brown (ed.), *Desert Biology*, vol. 1. New York:.

Maclean, G. L. 1967. The breeding biology and behaviour of the Double-banded Courser *Rhinoptilus africanus* (Temminck). *Ibis* 109: 556–569.

———. 1968. Field studies on the sandgrouse of the Kalahari desert. *Living Bird* 7: 209–235.

———. 1969. A study of seedsnipe in southern South America. *Living Bird* 8: 33–80.

———. 1970a. The biology of the larks (Alaudidae) of the Kalahari sandveld. *Zoologica Africana* 5: 7–39.

———. 1970b. Breeding behaviour of larks in the Kalahari sandveld. *Annals of the Natal Museum* 20: 381–401.

———. 1970c. The breeding biology and behaviour of the Double-banded Courser *Rhinoptilus africanus* (Temminck). In H. P. Riley (ed.), *Evolutionary Ecology*.

———. 1970d. The breeding seasons of birds in the south-western Kalahari. *Ostrich* (Supplement) 8: 179–192.

———. 1970e. Desert adaptations of sandgrouse. *African Wildlife* 24: 7–15.

———. 1973. A review of the Australian desert waders, *Stiltia* and *Peltohyas*. *Emu* 73: 61–70.

———. 1974. Arid-zone adaptations in southern African birds. *Cimbebasia* 3: 163–176.

———. 1976a. Arid-zone ornithology in Africa and South America. Pp. 468–480 in *Proceedings of the 16th International Ornithological Congress, Canberra*.

———. 1976b. A field study of the Australian Dotterel. *Emu* 76: 207–215.

————. 1976c. Rainfall and avian breeding seasons in north-western New South Wales in spring and summer 1974–75. *Emu* 76: 139–142.

————. 1983. Water transport by sandgrouse. *BioScience* 33: 365–369.

————. 1984. Arid-zone adaptations of waders (Aves: Charadrii). *South African Journal of Zoology* 19: 78–8i.

————. 1985. Sandgrouse: models of adaptive compromise. *South African Journal of Wildlife Research* 15: 1–6.

MacMahon, J. A. 1979. North American deserts: their floral and faunal components. Pp. 21–82 in D. W. Goodall, R. A. Perry, and K. M. W. Howes (eds.), *Arid-Land Ecosystems: Structure, Functioning and Management*, vol. 1. Cambridge: Cambridge University Press.

MacMahon, J. A., and F. H. Wagner. 1985. The Mojave, Sonoran and Chihuahuan deserts of North America. Pp. 105–202 in M. Evenari, I. Noy-Meir, and D. W. Goodall (eds.), *Hot Deserts and Arid Shrublands, A*. Amsterdam: Elsevier.

Marchant, S. 1959. The breeding season in S.W. Ecuador. *Ibis* 101: 137–152.

Mares, M. A., W. F. Blair, F. A. Enders, D. Greegor, A. C. Hulse, J. H. Hunt, D. Otte, R. D. Sage, and C. S. Tomoff. 1977. The strategies and community patterns of desert animals. Pp. 107–163 in G. H. Orians and O. T. Solbrig (eds.), *Convergent Evolution in Warm Deserts*. Stroudsburg, Pa.: Dowden, Hutchinson, & Ross.

Mares, M.A., and M. L. Rosenzweig. 1978. Granivory in North and South American deserts: rodents, birds, and ants. *Ecology* 59: 235–241.

Marr, T. G., and R. J. Raitt. 1983. Annual variations in patterns of reproduction of the Cactus Wren (*Campylorhynchus brunneicapillus*). *Southwestern Naturalist* 28: 149–156.

Marshall, A. J. 1959. Internal and environmental control of breeding. *Ibis* 101: 456–478.

Martindale, S. 1983. Foraging patterns of nesting Gila Woodpeckers. *Ecology* 64: 888–898.

Miller, A. H. 1963. Desert adaptations in birds. Pp. 666–674 in *Proceedings of the 13th International Ornithological Congress*.

Miller, A. H., and R. C. Stebbins. 1964. *The Lives of Desert Animals in Joshua Tree National Monument*. Berkeley, Calif.: University of California Press.

Moreau, R. E. 1966. *The Bird Faunas of Africa and its Islands*. New York: Academic Press.

Morel, G. 1968. Contribution à la synécologie des Oiseaux de Sahel sénégalais. *Memoires ORSTOM* 29: 1–179.

Morton, S. R. 1979. Diversity of desert-dwelling mammals: a comparison of Australia and North America. *Journal of Mammalogy* 60: 253–264.

————. 1982. Granivory in the Australian arid zone: diversity of harvester ants and structure of their communities. Pp. 257–262 in W. R. Barker and P. J. M. Greenslade (eds.), *Evolution of the Flora and Fauna of Arid Australia*. Frewville, South Australia: Peacock Publications.

————. 1985. Granivory in arid regions: comparison of Australia with North and South America. *Ecology* 66: 1859–1866.

Morton, S. R., and D. W. Davidson. 1988. Comparative structure of harvester ant communities in arid Australia and North America. *Ecological Monographs* 58: 19–38.

Morton, S. R., and P. H. Davies. 1983. Food of the Zebra Finch (*Poephila guttata*), and an examination of granivory in birds of the Australian arid zone. *Australian Journal of Ecology* 8: 235–243.

Morton, S. R., and R. E. MacMillen. 1982. Seeds as sources of preformed water for desert-dwelling granivores. *Journal of Arid Environments* 5: 61–67.

Nix, H. A. 1976. Environmental control of breeding, post-breeding dispersal and migration of birds in the Australian region. Pp. 272–305 in *Proceedings of the 16th International Ornithological Congress, Canberra*.

Noy-Meir, I. 1973. Desert ecosystems: environment and producers. *Annual Review of Ecology and Systematics* 4: 25–51.

Ohmart, R. D. 1972. Physiological and ecological observations concerning the salt-secreting nasal glands of the Roadrunner. *Comparative Biochemistry and Physiology* 43(A): 311–316.

Ohmart, R. D., and R. C. Lasiewski. 1971. Roadrunners: energy conservation by hypothermia and absorption of sunlight. *Science* 172: 67–69.

Orians, G. H., and O. T. Solbrig. 1977. Degree of convergence of ecosystem characteristics. Pp. 225–255 in G. H. Orians and O. T. Solbrig (eds.), *Convergent Evolution in Warm Deserts*. Stroudsburg, Pa.: Dowden, Hutchinson, & Ross.

Phillips, J. G., P. J. Butler, and P. J. Sharp. 1985. *Physiological Strategies in Avian Biology*. Glasgow, Scotland: Blackie & Son.

Pianka, E. R. 1979. Diversity and niche structure in desert communities. Pp. 321–341 in D. W. Goodall, R. A. Perry, and K. M. W. Howes (eds.), *Arid-Land Ecosystems: Structure, Functioning and Management*, vol. 1. Cambridge: Cambridge University Press.

Pianka, E. R., and J. J. Schall. 1981. Species densities of Australian vertebrates. Pp. 1675–1694 in A. Keast (ed.), *Ecological Biogeography of Australia*. The Hague: W. Junk.

Pianka, H. D., and E. R. Pianka. 1970. Bird censuses from desert localities in Western Australia. *Emu* 70: 17–22.

Pulliam, H. R. 1985. Foraging efficiency, resource partitioning, and the coexistence of sparrow species. *Ecology* 66: 1829–1836.

———. 1986. Niche expansion and contraction in a variable environment. *American Zoologist* 26: 71–79.

Pulliam, H. R., and T. A. Parker III. 1979. Population regulation of sparrows. *Fortschritte der Zoologie* 25: 137–147.

Raitt, R. J., and R. L. Maze. 1968. Densities and species composition of breeding birds of a creosotebush community in southern New Mexico. *Condor* 70: 193–205.

Recher, H. F., G. Gowing, R. Kavanagh, J. Shields, and W. Rohan-Jones. 1983. Birds, resources and time in a tablelands forest. *Proceedings of the Ecological Society of Australia* 12: 101–123.

Reichman, O. J. 1984. Spatial and temporal variation of seed distribution in Sonoran Desert soils. *Journal of Biogeography* 11: 1–11.

Ricklefs, R. E. 1965. Brood reduction in the Curve-billed Thrasher. *Condor* 67: 505–510.

Rodin, L. E. 1979. Productivity of desert communities in central Asia. Pp. 273–298 in D. W. Goodall, R. A. Perry, and K. M. W. Howes (eds.), *Arid-Land Ecosystems: Structure, Functioning and Management*, vol. 1. Cambridge: Cambridge University Press.

Rotenberry, J. T., W. T. Hinds, and J. M. Thorp. 1976. Microclimate patterns on the Arid Lands Ecology Reserve. *Northwest Science* 50: 122–130.

Rounsevell, D. 1970. Salt excretion in the Australian Pipit, *Anthus novaeseelandiae* (Aves: Motacillidae). *Australian Journal of Zoology* 18: 373–377.

Rowley, I. 1975. *Bird Life*. Sydney: Collins.

Schall, J. J., and E. R. Pianka. 1978. Geographical trends in numbers of species. *Science* 201: 679–686.

Schluter, D. 1986. Tests for similarity and convergence of finch communities. *Ecology* 67: 1073–1085.

Schmidt-Nielsen, K., A. Borut, P. Lee, and E. Crawford, Jr. 1963. Nasal salt excretion and the possible function of the cloaca in water conservation. *Science* 142: 1300–1301.

Schmidt-Nielsen, K., J. Kanwisher, R. C. Lasiewski, J. E. Cohn, and W. L. Bretz. 1969. Temperature regulation and respiration in the Ostrich. *Condor* 71: 341–352.

Schodde, R. 1982. Origin, adaptation and evolution of birds in arid Australia. Pp. 191–224 in W. R. Barker and P. J. M. Greenslade (eds.), *Evolution of the Flora and Fauna of Arid Australia*. Frewville, South Australia: Peacock Publications.

Serventy, D. L. 1970. Torpidity in Australian birds. *Emu* 70: 201–202.

———. 1971. Biology of desert birds. Pp. 287–339 in D. S. Farner and J. R. King (eds.), *Avian Biology*, vol. 1. New York: Academic Press.

Shurcliff, K. S. 1980. Vegetation and bird community characteristics in an Australian arid mountain range. *Journal of Arid Environments* 3: 331–348.

Simberloff, D. S. 1982. The status of competition theory in ecology. *Annales Zoologici Fennici* 19: 241–253.

Skadhauge, E. 1974. Cloacal resorption of salt and water in the Galah (*Cacatua roseicapilla*). *Journal of Physiology* (London) 240: 763–773.

Skadhauge, E., and K. Kristensen. 1972. An analogue computer simulation of cloacal resorption of salt and water from ureteral urine in birds. *Journal of Theoretical Biology* 35: 473–487.

Smith, R. M., and R. Suthers. 1969. Cutaneous water loss as a significant contribution to temperature regulation in heat stressed pigeons. *Physiologist* 12: 358.

Smyth, M., and G. A. Bartholomew. 1966. The water economy of the Black-throated Sparrow and the Rock Wren. *Condor* 68: 447–458.

Solbrig, O. T., and G. H. Orians. 1977. The adaptive characteristics of desert plants. *American Scientist* 65: 412–421.

Thomas, D. H. 1984. Sandgrouse as models of avian adaptations to deserts. *South African Journal of Zoology* 19: 113–120.

Tomoff, C. W. 1974. Avian species diversity in desert scrub. *Ecology* 55: 396–403.

Udvardy, M. D. F. 1958. Ecological and distributional analysis of North American birds. *Condor* 60: 50–66.

Wagner, F. H. 1981. Population dynamics. Pp. 125–168 in D. W. Goodall, R. A. Perry, and K. M. W. Howes (eds.), *Ecosystems: Structure, Functioning and Management*, vol. 2. Cambridge: Cambridge University Press.

Walsberg, G. E., G. S. Campbell, and J. R. King. 1978. Animal coat color and radiative heat gain: a re-evaluation. *Journal of Comparative Physiology* 126: 211–222.

Weathers, W. W. 1981. Physiological thermoregulation in heat-stressed birds: consequences of body size. *Physiological Zoology* 54: 345–361.

———. 1983. *Birds of Southern California's Deep Canyon*. Berkeley, Calif.: University of California Press.

Webster, J. D. 1974. The avifauna of the southern part of the Chihuahuan desert. Pp. 559–566 in R. H. Wauer and D. H. Riskind (eds.), *Transactions of the Symposium on the Biological Resources of the Chihuahuan Desert Region United States and Mexico*. Washington, D.C.: U.S. National Park Service.

Webster, M. D., G. S. Campbell, and J. R. King. 1985. Cutaneous resistance to water-vapor diffusion in pigeons and the role of the plumage. *Physiological Zoology* 58: 58–70.

Webster, M. D., and J. R. King. 1987. Temperature and humidity dynamics of cutaneous and respiratory evaporation in pigeons, *Columba livia. Journal of Comparative Physiology B* 157: 253–260.

Werger, M. J. A. 1986. The Karoo and southern Kalahari. Pp. 283–359 in M. Evenari, I. Noy-Meir, and D. W. Goodall (eds.), *Hot Deserts and Arid Shrublands, B*. Amsterdam: Elsevier.

Wiens, J. A. 1973. Pattern and process in grassland bird communities. *Ecological Monographs* 43: 237–270.

———. 1977. On competition and variable environments. *American Scientist* 65: 590–597.

———. 1983. Avian community ecology: an iconoclastic view. Pp. 355–403 in A. H. Brush and G. A. Clark, Jr. (eds.), *Perspectives in Ornithology.* Cambridge: Cambridge University Press.

———. 1985. Vertebrate responses to environmental patchiness in arid and semiarid ecosystems. Pp. 169–193 in S. T. A. Pickett and P. S. White (eds.), *The Ecology of Natural Disturbance and Patch Dynamics.* New York: Academic Press.

———. 1989a. *The Ecology of Bird Communities.* Vol 1: *Foundations and Patterns.* Cambridge: Cambridge University Press.

———. 1989b. *The Ecology of Bird Communities.* Vol. 2: *Processes and Variations.* Cambridge: Cambridge University Press.

———. In press. Ecological similarity of shrub-desert avifaunas of Australia and North America. *Ecology.*

Wiens, J. A., and R. F. Johnston. 1977. Adaptive correlates of granivory in birds. Pp. 301–340 in J. Pinowski and S. C. Kendeigh (eds.), *Granivorous Birds in Ecosystems.* Cambridge: Cambridge University Press.

Wiens, J. A., and J. T. Rotenberry. 1981. Habitat associations and community structure of birds in shrubsteppe environments. *Ecological Monographs* 50: 287–308.

Williams, O. B., and J. H. Calaby. 1985. The hot deserts of Australia. Pp. 269–312 in M. Evenari, I. Noy-Meir, and D. W. Goodall (eds.), *Hot Deserts and Arid Shrublands, A.* Amsterdam: Elsevier.

Willoughby, E. J. 1968. Water economy of the Stark's Lark and Grey-backed Finch-Lark from the Namib Desert of South West Africa. *Comparative Biochemistry and Physiology* 27: 723–745.

Withers, P. C. 1977. Respiration, metabolism and heat exchange of euthermic and torpid poorwills and hummingbirds. *Physiological Zoology* 50: 43–52.

Wyndham, E. 1980. Total body lipids of the Budgerigar, *Melopsittacus undulatus* (Psittaciformes: Platycercidae) in inland mid-eastern Australia. *Australian Journal of Zoology* 28: 239–247.

———. 1983. Movements and breeding seasons of the Budgerigar. *Emu* 82: 276–282.

———. 1986. Length of birds' breeding seasons. *American Naturalist* 128: 155–164.

Zann, R., and B. Straw. 1984. Feeding ecology and breeding of Zebra Finches in farmland in northern Victoria. *Australian Wildlife Research* 11: 533–552.

11 Desert Mammal Communities

O. J. Reichman

One of the major attractions of deserts is their relative simplicity. Anyone who has worked in deserts realizes, however, that the emphasis should be on the word relative rather than the idea of simplicity. Nevertheless, deserts do exist under regimes of low productivity, tend to be open, and are composed of discrete, measurable entities that lend themselves to manipulation and ecological investigations. Much has been learned about ecological processes, including the nature and structure of communities, from considerations of desert floras and faunas.

Because of the characteristics mentioned above, deserts have drawn the attention of ecologists, and therefore some features common to most deserts are relatively well known. For example, the physiological adaptations of desert plants and animals have attracted biologists specifically because of the extreme ambient conditions that can occur in deserts, especially in relation to aridity and temperature. Thus, investigators have utilized deserts to measure physiological parameters and to test hypotheses relating to adaptations. Also, certain groups of plants (e.g., cacti) and animals (lizards, snakes, some arthropods) are considered primarily desert organisms or exhibit impressive diversities in desert areas, and so biologists are drawn to deserts in their search for information about these taxa.

Even with many attributes to attract ecologists, most deserts, especially those outside North America, have not been extensively studied. Because the amount of research done in specific ecosystems is probably related to the number of universities or other research entities in the region, less research has been conducted in deserts than in some other types of habitats. A few research facilities are located in major deserts (e.g., the Desert Ecological Research Unit in the Namib Desert, and a number of universities in the American Southwest), but for the most part deserts are relatively underpopulated for the same reasons they are relatively simple—low productivity and a scarcity of water—and are, consequently, understudied. This, coupled with political exigencies, leaves a number of the world's deserts poorly known.

Several mammal taxa possess physiological and anatomical specializations for arid environments or have their distributions centered in deserts (e.g., heteromyids in North America, jirds and gerbils in the Old World, several genera of small dasyurid marsupials in Australia). Some desert mammal faunas are quite diverse. For example, almost two dozen species of small mammals occur in the Kalahari Desert (Nel 1978), and a similar number inhabit arid regions of central Australia (Morton and Baynes 1985). Nevertheless, mammals are not especially diverse in many true deserts nor are most desert mammals well known (Mares 1990). North American biologists and ecologists interested in mammals and community structure may not at first appreciate this circumstance because the heteromyid rodents of North America are among the best-known groups of vertebrates in the world and have contributed significantly to our understanding of community patterns in animals (Reichman and Brown 1983; Brown 1984), perhaps leaving the unwarranted impression that all desert mammal communities are as well known as those in North America.

In this chapter I will attempt to compare what is known about these North American desert mammals to information from several deserts in other parts of the world. I will begin by summarizing what is known about community structure in North American desert mammal communities. By community structure I mean regular patterns of traits (such as abundance, diversity, body size, microhabitat use) that represent adaptive responses of individuals to resources and other individuals (conspecific and heterospecific). Using the literature, I will then compare information from six other deserts to North American deserts. Three of the other deserts have been analyzed specifically with respect to community organization. The remaining three have insufficient data to make detailed analyses, but what is known about them can be compared to North American deserts to see if there are similarities in community structure. I will end with a review of contributions the study of desert mammals has made to community ecology, and a comment on how the studies have evolved.

North American Deserts

The vast majority of ecological research directed at understanding the organization of desert mammal communities has been conducted in North American deserts and has concentrated on the highly desert-adapted rodent family Heteromyidae. In absolute terms, however, what is known even in these deserts is a small fraction of what needs to be known. For example, only a handful of the more than 30 species of heteromyid rodents in North America have been studied in relation to their contributions to the communities in which they live. Nevertheless, they are the best known of all desert mammal systems, and they will be used as a gauge to analyze mammals from other deserts. There may also be pitfalls associated with using heteromyids as models for other desert systems, and these will be discussed.

Before beginning a discussion on the community organization of North American desert mammals, it is appropriate to consider the broader aspect of guild similarities among mammals of the four major North American deserts (the warm Chihuahuan, Sonoran, and Mojave deserts, and the cold Great Basin Desert). Root (1967) defines a guild as "a group of species that exploit the same class of environmental resources in a similar way. This term groups together species without regard to taxonomic position, that overlap significantly in their niche requirements." Brown (1973) applied a similar analysis to desert rodents, and in 1976 MacMahon broadened the approach to include all desert mammals except bats.

MacMahon (1976) first compared the similarity of species between four sites representing the four deserts of North America and several sites that are known to have been "good" desert sites in historical times but have developed into grasslands in the past few decades. Although none of the sites exhibited a high degree of similarity (<80 percent) in the number of species shared, the hot deserts share many more species between them than any of them share with the cold Great Basin Desert (MacMahon 1976).

MacMahon (1976) also analyzed trophic similarity between the desert sites (i.e., guild structure, using 12 functional guilds) rather than taxonomic similarity, with the result that several of the deserts exhibited significantly similar patterns. Even higher trophic similarity values were recorded when pairs of contiguous deserts were analyzed. As might be expected, comparisons between hot deserts yielded the highest similarities, comparisons between grasslands next, then hot deserts with grasslands, hot deserts with the cold desert, and finally, the cold desert with grasslands. From this analysis, MacMahon (1976) concluded that hot deserts, the cold desert, and desert grasslands exist as three functional categories of deserts in North America.

Barry Fox and J. H. Brown (personal communication) analyzed mammal guilds in North American deserts and discovered that there was a tendency for each guild to be filled with a representative before a second species of any one guild was "added." This assembly rule suggests an ordering of community residents which may be based on finer and finer subdivision of some resource or resources.

The discussion on mammal communities that follows will revolve around the desert representatives of the rodent family Heteromyidae, including the genera *Dipodomys*, *Microdipodops*, *Perognathus*, and *Chaetodipus* (two other genera, *Liomys* and *Heteromys*, belong to this family but are centered in the tropics of Central America; Hafner and Hafner 1983; Schmidly and Wilkins 1991). Heteromyids are abundant in North American deserts and their local diversity is related to seed productivity (as determined indirectly by precipitation; Brown 1973, 1975; Brown and Davidson 1977). They possess a suite of traits that reveal their desert affinities, including anatomical, physiological, and behavioral water conservation mechanisms (Schmidt-Nielsen 1964; MacMillen 1983), nocturnal activity, and excellent hearing and night vision (Reichman and Price 1991). Heteromyids possess external cheek pouches into which seeds are gathered during foraging for the return trip to the burrow, where many of the rodent species store extensive quantities of seeds (Smith and Reichman 1984; Reichman et al. 1985). In fact, a major unifying characteristic of heteromyids is their reliance on the relatively stable seed resources in the soil (although there are herbivorous specialists among the desert heteromyids; Kenagy 1972). As we shall later see, much of what makes sense about the community structure of desert mammals relates to specializations associated with the use of seeds.

Brown and Harney (1990) list several demographic traits, such as relatively small litters, long adult survival, and facultative reproductive seasons, which distinguish heteromyids from most other small rodents. Nevertheless, several granivorous species of rodents belonging to the family Cricetidae (e.g., *Peromyscus, Reithrodontomys*) frequently coexist with heteromyids and utilize many of the same resources. Thus, a complete analysis of North American desert rodent communities must ultimately include some of these taxa (e.g., Bowers, Thompson, and Brown 1987). There are nongranivorous desert rodents in North America (e.g., ground squirrels, woodrats, and grasshopper mice), but investigators have concentrated on the granivorous component of desert mammal communities when considering resource allocation schemes (but see, for example, Rebar and Conley 1983 and Munger, Jones, and Bowers 1983).

Because of their desert adaptations and the many complementary studies conducted, the rodents of North American deserts serve as an excellent model to analyze and test general ecological principles (Brown 1984; Price 1986). Among the thousands of articles that exist in the literature on various ecological aspects of these rodents, a number over the last two decades have sought to directly address the nature of community organization. Several excellent reviews of this topic have been or are about to be published (volume 7 of the Great Basin Naturalist Memoirs and the forthcoming volume 10 of the American Society of Mammalogists Special Publications are devoted entirely to desert rodents and especially heteromyids; see also the review by Kotler and Brown 1988) and I refer the reader to those articles rather than reiterating all their information here.

Several approaches have been employed to analyze the community characteristics of desert rodents in North America. Brown (1985) outlines the two approaches he and his collaborators have taken over the last 15 years. His early work relied on comparisons of geographical patterns to elucidate traits that might be responsible for the observed organization of desert rodent communities. Subsequently, he and others conducted large-scale manipulations in an attempt to focus on specific causal factors in community structure. Price (1986) analyzes several general approaches to studying communities and uses her experience with desert rodents as an example of how these approaches have served to increase knowledge about communities. She notes that most ecologists have used techniques similar to those of Brown and his colleagues, concentrating on patterns of morphology and resource use, or experimentally manipulating biotic factors in the community to unveil underlying patterns. She suggests an alternative, mechanistic approach involving analyses, at the level of the individual, of traits that might promote coexistence in some measurable fashion.

Before we can discuss what features might structure desert rodent communities, we must first be confident that these communities truly exhibit regular patterns of organization. For decades the apparently uniform array of body size ratios exhibited by heteromyid rodent communities has been hypothesized to represent an underlying structuring phenomenon in North American deserts. Beginning with the observations of Grinnell and Orr (1934) that species in desert rodent communities appeared to be arrayed in uniform "bores" (as in the bore of a firearm) and proceeding to the investigations of Brown (1973, 1975), the apparently even body mass ratios have lent credence to this proposition.

These patterns became accepted by many researchers and numerous subsequent investigators set about determining what factors molded the patterns that seemed so clear. In the late 1970s and early 1980s, however, the assumption of a discrete community organization among desert rodents came under scrutiny as part of an overall analysis of whether competition was an important force structuring communities, regardless of the taxa involved. These investigations (e.g., Connell 1975; Bedard 1976; Wiens 1977; Connor and Simberloff 1979; Strong, Szyska, and Simberloff 1979) questioned whether observed patterns were real (i.e., statistically valid), and some used null models to compare actual body mass ratios in a known community with randomly determined ratios to detect if there was some underlying, nonrandom structure to the target communities.

When this type of analysis was applied to North American desert rodent communities, the previously assumed patterns were verified. Specifically, as Bowers and Brown (1982) note, the proposition that desert rodent communities are assembled in a random pattern with respect to body size is unequivocally rejected (Figure 11.1). The authors found that species of granivorous desert rodents of similar body size (i.e., body mass ratios less than 1.5) coexist less frequently in the same community and overlap less in their geographic range than would be expected based on chance assemblages. Brown and Kurzius (1987) further analyzed geographical patterns among 29 species of granivorous desert rodents at 202 sites in the southwestern United States. They discovered that most species were found in less than 30 percent of the assemblages that occurred within their geographic range, and that even though most communities had only two to five sympatric species, 124 combinations were found across the 202 sites. Thus, the exact species composition at specific desert sites is highly variable but the assemblages still retain distinct arrays of body sizes. The results of these critical analyses have been beneficial to investigations of desert rodent communities and, in a much broader sense, to community studies in general (e.g., Diamond and Case 1986); North American desert rodent communities are now generally accepted as one series of communities that exhibit one form of community structure (Simberloff and Boecklen 1981; Brown and Bowers 1984), as revealed by their nonrandom pattern of body mass ratios.

Once we accept that there are consistent, distinct patterns of body sizes in granivorous desert rodent communities, the next obvious step is to search for some force or forces that have served as strong selection pressures molding the communities. It is possible that the observed patterns resulted from autecological, adaptive responses within species to traits of the environment (e.g., microhabitat characteristics, resource distributions), perhaps even in allopatry. Subsequently, members of the community could have meshed to yield the observed patterns of body sizes. This seems unlikely, however, as two or more species are likely to have converged on the same environmental trait (perhaps the dominant microhabitat or food resource)

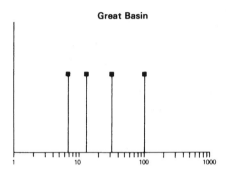

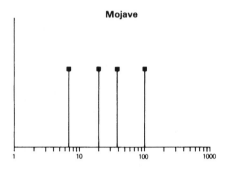

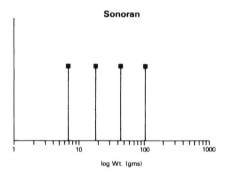

Figure 11.1.   Distribution of rodent body sizes (plotted on logarithmic scale so that equal spacing indicates equal ratios) for three deserts in North America. From smallest to largest, the species in the top panel are *Perognathus longimembris*, *Microdipodops megacephalus*, *Dipodomys merriami*, and *D. deserti*; middle panel: *P. longimembris*, *P. (Chaetodipus) formosus*, *D. merriami*, and *D. deserti*; bottom panel: *P. flavus*, *P. (Chaetodipus) penicillatus*, *D. merriami*, and *D. spectabilis*. All species are granivorous, and no data were available on relative densities (after Bowers and Brown 1982).

such that when species interacted ecologically, selection pressure to diverge in ecological space would occur. Thus, in the broadest sense, most investigators accept that some form of competition for a limited resource is responsible for the extant patterns. The two most obvious forms of competition are related to differences in foraging efficiency and the ability to avoid predators.

The structure of microhabitats in North American deserts appears to be the independent variable that most impinges on the foraging efficiencies of desert rodents. Numerous studies have illustrated the relationship between body size and microhabitat use (Rosenzweig and Winakur 1969; Rosenzweig 1973; Brown 1973; M'Closkey 1978, 1980, 1981; Price 1978a). Furthermore, microhabitat characteristics are directly related to such parameters as substrate particle size and seed patch densities and distributions (Reichman 1984; Price and Waser 1985; Price and Reichman 1987). Most studies confirm the generalization that larger, bipedal species tend to occur in open areas between bushes while smaller, quadrupedal species tend to be under bushes (Price 1978a; Bowers 1982, Brown 1988). Thompson (1982a,b, 1985) did not find as distinct a microhabitat segregation in a Mojave Desert rodent community, but did note that different species travelled different distances between shrubs, and thus differed in the amount of time they spent in the open versus under bushes. It is also possible that temperature regimes and radiant heat loss with respect to the presence or absence of cover may interact with body size. For example, small species might need to stay under bushes on cold nights to minimize heat loss while larger species could spend some time in more open areas (J. Fisher personal communication). Because so many aspects of body size and gait affect detection and efficient extraction of seeds, it is not surprising that morphology and microhabitat use are related.

Predators serve as the other major independent variable that may influence body size configurations in a community. Simple speed and erratic movements could deter some potential predators, but if all prey species adopted the same predator avoidance scheme it might be easier for predators to concentrate evolutionarily on the avoidance mechanisms. Conversely, natural selection would favor a range of escape mechanisms, which would minimize the opportunity for predators to concentrate on a single attack strategy (Schall and Pianka 1980). In this regard, different predator detection and avoidance traits are associated with various body sizes and gaits. For example, kangaroo rats have large eyes, excellent hearing, and are bipedal, traits likely to be beneficial in open areas, while small, quadrupedal mice might be more adept at scurrying through vegetation to avoid detection and capture.

As noted by Price and Brown (1983) and Price (1984), it is very difficult to assign actual costs and benefits to the use of various microhabitats or predator avoidance behaviors in a manner that allows precise analyses of how either of these selective forces mold body sizes and microhabitat use in desert rodent communities. Nevertheless, attempts have been made to investigate the roles each has played in the community structure of desert rodents. Simply because it is easier to manipulate resources than it is to manipulate predators, the former have received more scrutiny than the latter. Perhaps because of this asymmetry in available data, the proposition that resource acquisition is the major axis of specialization among desert granivorous rodents has held sway for a number of years (Price and Brown 1983). Recent research on the influence of predation on desert rodents (Kotler 1984a,b; Kotler et al. 1988; Brown et al. 1988), however, is documenting the obvious fact that this is an important factor structuring communities and much more research must be completed before the relative contribution of each can be ascertained.

While it would be redundant to review all that is known about North American desert mammals, a brief review of the available information is in order (readers are directed to the review articles mentioned earlier and Price and Brown 1983 for a more thorough presentation). Before we can consider the ways in which body size and microhabitat use might be related to differences in foraging, however, we must be satisfied that seeds are sufficiently limiting for natural selection to favor efficient foraging traits. That is, could scramble competition for seeds generate different solutions to the problem of securing the resource? Several lines of evidence suggest that seed availability does limit population densities and even the kinds of species present in communities (Abramsky 1978; Brown and Harney 1991). Seed additions positively affect rodent populations (Abramsky 1978; Brown et al. 1986), suggesting that seed availability limits these populations. The importance of seeds can also be shown indirectly by changes in seed densities in the absence of components of the granivore community. For example, Brown, Reichman, and Davidson (1979), Reichman (1979), and Brown et al. (1986) show that seed densities and spatial distributions change subsequent to removal of granivorous rodents or ants.

Interestingly, however, both taxa had to be removed to generate a dramatic increase in seed densities. This suggests that there was a density compensation on the part of the remaining taxon when the other was removed, which appears to have been the case (Brown and Davidson 1977; Munger and Brown 1981; Brown et al. 1986; Bowers, Thompson, and Brown 1987; but see Galindo 1986). The direct evidence of reciprocal density increases by one taxon in the absence of putative competitors provides strong evidence of competition for resources (although it is possible that by removing one competitor taxon the success of predators is altered in addition to or instead of altering resource availability; Holt 1984; Holt and Kotler 1987).

If we accept that desert granivores do compete for seeds, we can consider the direct and indirect ways that microhabitat characteristics mediate competition through rodent foraging efficiency. For example, Price and Heinz (1984) found that seed harvest rates decrease with soil particle size and increase with soil density, factors directly related to microhabitat (as is soil organic content, another factor likely to impinge on foraging efficiency; Price and Reichman 1987). Furthermore, Price and Heinz (1984) found that larger species of heteromyids have higher harvest rates than smaller species under the same soil conditions. The authors suggest that these differences in extraction efficiencies should generate species-specific microhabitat preferences. In a field test of microhabitat preference, Price and Waser (1985) found that major shifts in microhabitat use accompanied experimental manipulation of preferred seed patches such that heteromyids could be lured into inappropriate microhabitats with artificial seed arrangements, and Larsen (1986) found a similar phenomenon in a natural experiment. The density and configuration of vegetation in various microhabitats may also affect foraging efficiency in relation to body size and locomotory style by impeding larger, bipedal species, although this possibility has not been investigated.

Microhabitats also can indirectly affect foraging efficiency through their influence on the density and distribution of seed patches. Seeds are moved over the desert floor by wind and water and tend to sort out like pebbles in a stream (see also Seely this volume). Thus, small depressions and other wind shadows tend to accumulate seeds, while seed densities are lower and more uniform in areas that are topographically less diverse (Reichman 1984). This yields a pattern in which high density patches tend to be widely spaced while lower density patches (down to a single seed) are more uniformly dispersed. Because microhabitats differ in their microtopography, seed densities and distributions are related to microhabitat characteristics (Price and Reichman 1987).

Relying on such information, Price (1983) discussed how seed patchiness might lead to microhabitat specialization on the part of rodent foragers. Price and Waser (1985) further analyzed this proposition and offered experimental evidence that such specialization does occur. Kotler and Brown (1988) expand the discussion by analyzing several components of environmental heterogeneity, such as microhabitat, time, and spatial distribution of seeds, in relation to desert rodent coexistence.

Reichman (1981) presented a simple graphical model that relied on the relationship between gait efficiency (bipedal vs. quadrupedal) and seed distribution to suggest how known seed distributions in open areas and under vegetation cover might be subdivided between kangaroo rats and pocket mice. More recent studies have indicated that there probably are not differences in foraging efficiencies based on locomotory style (Thompson et al. 1980; Taylor, Heglund, and Maloiy 1981), but other authors still note that body size, independent of gait, can be an important correlate with efficient seed resource utilization (e.g, Harris 1984, 1986). Furthermore, a computer simulation model by Roberts and Reichman (manuscript) suggests that body size is implicated in some aspects of resource partitioning.

In the simulation we used data from the literature on seed distributions to construct a checkerboard desert in which seed patches were assigned to squares in relation to known spatial distributions (which exhibit highly skewed and kurtotic statistical distributions; Reichman 1984). Data from the literature on the costs of running, digging, and standing still, running speeds, time required to dig and pouch seeds, the average calories per seed, number of seeds within the digging range (depth) of the rodents, and number of seeds gathered per trip were incorporated into the simulation. Thus, all functions critical to the outcome were taken directly from known values in the literature or from mass-based functions generated from data in the literature. In particular, we used values for the costs of running from Thompson et al. (1980) to avoid the question of whether there is a plateau in energy costs associated with the locomotion of small bipeds (Thompson 1985).

Three body size/gait combinations, representing a 41-g kangaroo rat and a 29-g and an 11-g pocket mouse, were analyzed with the simulation. Each "species" was assigned an array of minimum patch densities (i.e., they would stop in any square with that density of seeds or greater) from 0 to 35 in increments of five seeds per patch (Figure 11.2). We averaged the time taken to obtain the minimum number of

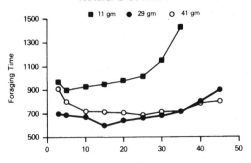

Natural Distribution

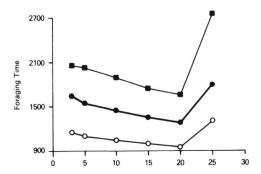

Normal Distribution

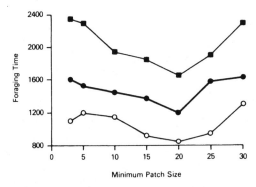

Uniform Distribution

Minimum Patch Size

Figure 11.2. Results from a computer simulation of the time required to secure enough seeds for 24 hours for three body sizes of heteromyids (representing two species of quadrupedal pocket mice and a larger, bipedal kangaroo rat) over a range of minimum patch sizes. The y-axis indicates the time in seconds to secure enough seeds for one day. The top panel is based on the distribution of seeds known to occur at a site in the Sonoran Desert while the other panels are based on hypothetical normal (middle) and uniform (bottom) distributions. See text for details.

seeds required for 24 hours, including all of the associated foraging costs, over 50 runs at each minimum patch density for each species. We took as the optimum patch density the value that yielded the least time necessary to obtain seeds for a 24-hour period, and these minima were compared between species. Results indicate that each species has a different patch density that minimizes the amount of time it takes to secure the daily requirement of seeds (Figure 11.2). Because the only differences between species were those related to their body sizes (i.e., seed distributions were essentially identical), the results suggest that some species-specific combination of morphological traits relating to foraging efficiency generates distinct patch use dynamics.

Significantly, when either uniform or normal distributions of seeds were incorporated into the simulation, rather than the highly skewed and kurtotic distribution known to actually occur in the Sonoran Desert, the three species showed virtually identical optima, although each required a different absolute amount of time to secure the minimum daily seed requirement (Figure 11.2). This suggests that it is the nature of the seed distribution (i.e., many low-density patches and few, widely spaced rich patches) which, in effect, "allows" subdivision of seed resources.

If kangaroo rats and pocket mice use microhabitats that differ in the spatial configuration of seeds, the species might differ in their "spatial memory." Rebar (1988) has recently discovered that kangaroo rats are better able to learn the spatial distribution of seeds than pocket mice. Furthermore, the author found that kangaroo rats use spatial cues for orientation while pocket mice tend to rely on specific response strategies (essentially "rules" for foraging patterns). She further suggests that these differences in foraging techniques are adaptive for the microhabitat and seed distributions encountered by the rodents in the field.

The discussion on microhabitat use so far has centered on indirect competition for resources. There is, however, evidence for direct competition through aggression. Blaustein and Risser (1976) noted that dominance among three species of kangaroo rats was positively correlated with body size. Frye (1983) found a similar pattern between two species of kangaroo rats and discussed how this might impinge on patterns of coexistence. Bowers, Thompson, and Brown (1987) discovered that removal of the largest kangaroo rat species from a complex community of desert rodents directly affected the microhabitat use, behavior, and success of a smaller species of kangaroo rat, and that this alteration reverberated throughout the community. Rebar and Conley (1983) discovered asymmetrical microhabitat shifts when either a kangaroo rat or grasshopper mouse was removed from experimental areas, and the authors allude to the possibility that interspecific aggression by the carnivorous mouse might have had a part in the asymmetry. While aggression is probably another manifestation of resource competition, it is a more direct form of resource allocation than those based on differences in foraging efficiency.

Several interesting characteristics that are associated with differences in body sizes among heteromyids are not obviously linked to microhabitat preferences but may promote coexistence among sympatric forms. For example, Rosenzweig and Sterner (1970) found that larger species could husk seeds more efficiently than smaller species (seed sizes used tended to be outside the range typically available to heteromyids; Reichman 1976; Price and Reichman 1987).

In another approach related to body size, MacMillen and Hinds (1983) and Hinds and MacMillen (1985) propose an intriguing model relating body size to water regulatory efficiency. The authors found that water use efficiency is inversely related to body size within pocket mice but that it is fixed across all body sizes of kangaroo rats at a level similar to the largest species of pocket mice. The authors suggest that as the Southwest became more arid over the last few thousand years, kangaroo rats remained relatively large and came to rely on diets that had high metabolic water content. Pocket mice, with a somewhat restricted locomotory style, had to rely on more eclectic diets. Thus, these quadrupeds retained their weight-based water use efficiencies and evolved reduced body sizes in extremely xeric environments. Concurrent with reduced body sizes, the smaller species also relied on winter torpor to avoid costly environmental extremes (Reichman and Brown 1979). It is possible that these different approaches to water conservation could promote coexistence among sympatric forms through specialization on specific seed types. Reichman (1975) found that sympatric kangaroo rats and pocket mice had quite similar diets, but other features of seed use that promote imbibition of water (such as seed storage; Frank 1988), could factor into patterns of coexistence related to water conservation mechanisms.

As noted earlier, much less work has been conducted on the effect of predators in shaping desert rodent communities. Nevertheless, several observed patterns and field manipulations indicate that predation is also an important force affecting desert rodent communities. For example, substrate and pelage color often match among desert rodents (Benson 1933; Dice and Blossom 1937; Lay 1974). Benson (1933) found dramatic differences in pelage coloration in adjacent populations of mice that inhabited black lava flows and nearby light sand dunes and ascribed the differences to selection for background matching enforced by predatory owls.

Another obvious pattern in heteromyid communities is that large species tend to be bipedal while small species are quadrupedal (with the small bipedal species of *Microdipodops* being the exceptions). Reichman (1981) ascribed this difference to foraging efficiency in relation to specific seed distributions, but Thompson et al. (1980) and Thompson (1985) have shown this to be unlikely and point to efficient predator escape as the probable selective force (Bartholomew and Caswell 1951; Bartholomew and Cary 1954; Thompson, 1985). Their argument is that bipedal, ricochetal locomotion, usually associated with kangaroo rats in open microhabitats, is effective for avoiding avian and terrestrial predators, while quadrupedal locomotion best serves the smaller species, which tend to occur where vegetation is more dense (Thompson 1982a,b, 1985).

Desert rodents, and especially the heteromyids, are noted for their inflated auditory bullae and associated hearing acuity (Webster and Webster 1975, 1980). The importance of this trait in avoiding predators was dramatically illustrated by Webster and Webster (1971), who compared the survivorship of deafened and intact animals in a wild population of kangaroo rats. Thirty-three percent of the intact animals disappeared one month after release, but 78 percent of the deafened individuals were not trapped within one month. While it is not possible to determine the exact proportion of the losses that were due to predation (as opposed to migration or other causes), the substantial difference in disappearance can reasonably be ascribed to differences in susceptibility to predators in the absence of functional hearing.

Another common observation of desert rodent communities is that the rodents frequently curtail or modify their aboveground activities during the bright phases of the moon (e.g., Kaufman and Kaufman 1982). Such avoidance of moonlight has been interpreted as an attempt to avoid being seen and caught by vigilant predators.

Observations of the avoidance of moonlight lead directly to recent experiments by Kotler and his colleagues in which simulated moonlight was manipulated using lanterns (Kotler 1984a,b, 1985). Trays of seeds were placed in microhabitats varying in the intensity of the simulated moonlight (and, presumably, in susceptibility to predation, as perceived by the rodents). Differences in the rates and amounts of seeds extracted indicated that the species of desert rodents involved altered their foraging behavior in response to the light conditions. More recently, Kotler et al. (1988) and Brown et al. (1988) analyzed the direct effect of predation by owls in flight cages on heteromyid foraging. They discovered that both pocket mice and kangaroo rats were behaviorally flexible and would shift to foraging under bushes in the presence of owls. Furthermore, the authors found that the large bipeds (kangaroo rats) were captured more frequently than the small quadrupedal pocket mice, but that there was no difference in predation rates on bipeds and quadrupeds of roughly similar sizes (*Dipodomys merriami* and *Chaetodipus baileyi*). These experiments represent the first attempts to determine the effects of predators on the relative success of desert rodent species that differ in body size and microhabitat use. Much more needs to be understood about the effects of predators, but these preliminary results are intriguing.

To date, in the face of the asymmetry in available information, the preponderance of data implicates the complex interactions between mechanisms of seed allocation and microhabitat use as the primary forces structuring granivorous desert rodent communities (Price and Brown 1983; Price 1984). While it may initially be efficient to accept the simplifying assumption that only this one factor significantly affects desert rodent community structure, this obviously is not the case. As more sophisticated theories and experiments accumulate evidence for the effects of both competition and predation, we can develop a more accurate view of how these factors interact.

From my own perspective, it seems that in North American deserts seed abundance and distribution serve as the cornerstone of community organization among rodents (especially heteromyids). The basic resource need for animals is energy and nutrition, and seeds are the primary persistent food resource in North American deserts. Thus, rodents moving into or evolving in these deserts would naturally gravitate toward this stable resource. The emerging granivores also would have to cope with those factors that impinged on their success in obtaining sufficient seeds to survive and reproduce. The two strongest limiting factors were probably seed availability and predator pressures. Seed availability is dependent on both absolute abundance and distribution, factors associated with productivity, abiotic physical characteristics of the habitat, and efficiency of acquisition. Selection for the latter component could drive rodent communities toward body size and microhabitat specialization as members of the evolving communities evolutionarily jockeyed for position along the seed resource axis. The same logic could be applied to predator pressures, generating anatomical specializations and microhabitat preferences that minimized the chances of succumbing to a predator.

As local communities of seed eaters were evolving, the relative importance of resource competition and predator pressures probably varied tremendously between areas. In a scenario where community makeup was being resolved, desert rodents from different areas might be subjected to different degrees of pressure in relatively competitor-free or predator-free habitats. Under these influences, community members might be driven off in distinctive directions (anatomically, behaviorally, physiologically) by the initial set of conditions encountered. Eventually the twin selective forces of efficient resource acquisition and predator avoidance would come to some equilibrium, but the initial nudge by one or the other of the forces might have contributed significantly to current patterns. Both the plants that produce seeds the rodents consume, and predators that eat the rodents, presumably involve responses to evolutionary changes in their consumers and their prey, respectively. Thus, selective pressures in a mature, relatively stable community should consist of continuous and subtle evolutionary thrusts and parries between community members.

The vast majority of North American desert rodent communities are probably in relatively stable phases at this time. There are, however, communities where competitors or predators are especially abundant or rare. An example occurs on islands in the Sea of Cortez. Several species of heteromyids inhabit islands that differ primarily in their competitor and predator components. If the current configuration of island inhabitants has been stable for several millennia, comparisons of behavior and use of microhabitats of target species could reveal the nature and degree of adaptation to the competitors and predators with which they have evolved. While there are many conceptual and logistical problems with research on the islands, preliminary results suggest that rodent populations have responded as expected evolutionarily to the different selection pressures in their environments. For example, individuals on predator-poor islands have shorter hind feet and are less "fearful" than individuals on a nearby predator-rich island (Reichman n.d.).

Desert Mammals Outside North America

Very little is known about the community structure of mammals from any deserts other than those in North America. While literature exists on various aspects of desert mammal biology (especially physiology) that might impinge on community structure, not enough work has been done in most deserts to allow even rudimentary analysis of community patterns. Mares (1991) presents the most comprehensive review to date of the literature on mammals in deserts around the world. His article, in a book on the biology of heteromyid rodents, compares what is known about the inhabitants of other deserts to North American heteromyids, and thus does not deal with mammals other than rodents. This, as it turns out, is not a major concern, as the huge majority of what is known about desert mammals relates to rodents (although see Schmidt-Nielsen 1964 for physiological and ecological adaptations of individual species of desert mammals other than rodents).

It is known that some desert "communities" are so simple, involving only one or two species, that analysis of their "structure" may initially appear trivial. For example, in some portions of the structurally simple Namib Desert a single species of fossorial insectivore (the Namib golden mole) and two species of gerbils coexist. The golden mole clearly has an exclusive niche and one might expect that in such a simple habitat the coexisting rodents would subdivide one or more critical resources in some obvious and straightforward manner. David Boyer, who has investigated this community (personal communication), has not yet found any basic differences in such obvious traits as diet, microhabitat use, or activity patterns. Thus, to date, even this simple community has defied "solution."

Some exquisite research has been done on the ecology of systems involving one or two species of mammals which could lend itself to analyses of community patterns were they extended to other features and other residents in the same community. This is not to demean the efforts made by researchers, but rather to point out the pioneering nature of their investigations and logistical difficulties encountered when attempting field work in distant deserts.

An example of such thorough studies is the work done by Taylor on heat management and water conservation mechanisms of several large ungulate species in arid regions of Africa. In a study of the gemsbok (*Oryx*), eland (*Taurotragus*), and gazelle (*Gazella*), Taylor (1969) investigated the ability of these animals to withstand extreme temperatures without seeking shelter. The gemsbok could maintain body temperatures approaching 45°C for up to 12 hours. While such high body temperatures would kill most mammals, the oryx apparently has circulatory specializations that allow it to tolerate these conditions. For example, the brain, the most temperature-sensitive organ in the body, is isolated from the elevated temperatures by a countercurrent system that shunts blood by nasal mucosa where evaporation on the surface of the mucosa cools the blood before it passes to the brain. Taylor and Lyman (1972) found a similar countercurrent system in Thompson's gazelle which, while running, maintained a brain temperature almost 4°C lower than temperatures measured in the nearby carotid artery. A variety of other adaptations, such as pale coloration and reduced metabolic rates at high temperatures, are also known to assist several large ungulates to cope with high ambient temperatures.

Unacceptable rates of water loss might accompany the high temperatures encountered by desert ungulates were it not for a variety of water conserving traits employed by them. For example, gemsbok often feed after sunset when dry vegetation imbibes moisture that condenses during the relatively cool nights (Taylor 1969). Furthermore, gemsbok excavate the belowground storage organs of plants, which are nutritious and contain significant amounts of water. Several physiological traits also limit water loss, including reduced metabolic rates, reductions in evaporative water loss (except under the extreme heat loads discussed above), and reduced respiratory rates. Taylor (1969) found that elands also consume moist vegetation and exhibit physiological traits that free them from the need to drink water.

Taylor (1968) and Taylor and Lyman (1967) discovered distinct adaptations to high temperatures in Grant's gazelle, which extends into the harsh northern deserts of Kenya, and Thompson's gazelle, which is restricted to less arid regions. Grant's gazelle exhibits higher rates of water loss than Thompson's gazelle, which seems paradoxical at first glance. The authors, however, found that at extremely high temperatures Grant's gazelle employs evaporative cooling to keep its body temperature within acceptable limits, while Thompson's gazelle maintains a higher body temperature and does not resort to evaporative cooling. Thus, even these two closely related gazelle species possess different physiological strategies for coping with extreme environmental conditions and these traits seem to be associated with habitat segregation across their range.

While Taylor's work was not designed to deal directly with the structure of communities of coexisting ungulates in African deserts, his results reveal several physiological specializations that might explain spatial patterns of distribution. For example, ungulate communities might be relatively diverse in less arid regions, incorporating over a dozen species. As habitats grade into more xeric areas, the less specialized species drop out, leaving only those that can tolerate the regional or periodic droughts. In the most extreme areas, such as the dunes of the Namib and Kalahari deserts, only a few specialists such as the gemsbok and springbok can survive. Within a specific habitat, sympatric forms might subdivide available food on the basis of moisture content, relying on distinctive foraging behaviors and physiological adaptations for ecological separation along the resource gradient. While this is mere speculation, it is a reasonable hypothesis on how physiological specialization might affect the community structure of a diverse ungulate fauna.

There are numerous studies on small mammals, especially rodents, which, like Taylor's work, concentrate on subsets of the information necessary to understand their community organization (see, for example, the chapters in Prakash and Ghosh 1975). While most of these investigations were not directed toward an analysis of mammal community structure per se, they contain elements known to be important in those desert mammal communities that are well known. For example, Christian (1979) analyzed the demographic patterns of three species of Namib Desert rodents and Streilein (1982a,b,c) studied features of habitat selection, population characteristics, and water balance in small mammal populations in the semiarid Caatinga of Brazil. In a few cases investigators have begun to analyze desert mammal faunas specifically with respect to community organization. For example, Morton (1979) compared Australian deserts to those in North America, and Meserve (1981a,b) carried out investigations of community organization in arid Chilean areas. Abramsky and his colleagues (Abramsky 1980; Abramsky and Sellah 1982; Abramsky and Rosenzweig 1984; Rosenzweig, Abramsky, and Brand 1984; Abramsky, Rosenzweig, and Brand 1985; Rosenzweig and Abramsky 1986) present features of rodent communities in the Negev Desert and apply sophisticated analyses to determine how the species relate to each other and their habitats. Nel (1978) presents extensive information about the microhabitat use, diets, and relative densities of a diverse small mammal fauna in the Kalahari Desert. The work of these few investigators and their colleagues goes right to the heart of the matter of community structure while also providing general information about desert mammals.

Examples from Specific Deserts

Even though so few details are known about the structure of most desert mammal communities, it is possible to look for those features that are known to indicate aspects of community organization in North American desert mammal faunas (e.g., uniform body size ratios, microhabitat segregation) among the mammals of other deserts. If similar patterns emerge, there is reason to suspect that there may be similar underlying selection pressures that mold those communities as they do in North American deserts. To make such comparisons I will rely on the work of those authors who have dealt specifically with the community organization of desert mammals (see above) and several other reports that present information about mammals trapped at specific desert locales. Only a few reports give complete information about the species present, their body masses, diets, microhabitat specializations, and relative densities, so I have had to piece together information from disparate sources. Therefore, details are imprecise and data presented are, no doubt, imbued with some errors pertaining to the specific sites discussed. Nevertheless, I used the best information available in the hope that it will provide a glimpse at what might be occurring in other deserts.

For each of six desert areas (in Chile, Afghanistan, Sudan, Israel, the Kalahari, and Australia) I will briefly describe the habitat characteristics and present, if available, information on the body size of the rodents, their relative densities, their trophic levels, and their microhabitat affinities. These data are summarized in a series of figures that will allow comparisons between deserts.

Chile Glanz (1977), Meserve and Glanz (1978), and Meserve (1981a,b) carried out a series of comparisons of niche characteristics of rodents in communities in California and Chile. Four numerically dominant species and three less common species were captured in a semiarid (mean annual rainfall = 127 mm) thorn scrub community in Parque Nacional Fray Jorge in Coquimbo Province, Chile (Meserve 1981a) (Figure 11.3). The area is characterized by drought-deciduous and evergreen shrubs, an herbaceous understory of forbs and grasses, and sandy open areas. Three of the four dominant species do not overlap in body size (Figure 11.3); the two species that are quite similar in size (*Akodon olivaceus* and *Phyllotus darwini*) prefer roughly similar microhabitats ("moderate to high cover"; Glanz 1977) but exhibit different diets. *Akodon olivaceus* maintains higher density minima and maxima than *P. darwini* (Figure 11.3).

In an extensive survey of habitat associations, Meserve (1981a) noted that the octodontid species *Octodon degus* exhibited an affinity for low cover and open ground while the other three, more closely related, dominant species (*Akodon olivaceus*, *A. longipilus*, and *Phyllotus darwini*) preferred areas with more cover. The four dominant sympatric species also exhibited a high degree of habitat overlap, but they represent three distinct trophic guilds (herbivore, insectivore, granivore) plus one species that is a generalist omnivore (Figure 11.3). The author proposes that trophic level is the resource axis along which the sympatric species are arrayed and notes that two members of the same feeding guild almost never co-occur as common species at this site or any other Chilean site studied so far. Thus, a tentative conclusion is that there does appear to be a degree of body size spacing in this community, and that minor microhabitat preferences and distinct dietary specializations promote the coexistence of the two species that are very similar in size. In a broader geographic pattern, Meserve and Glanz (1978) found rodent diversity showed a significant correlation to measures of precipitation and herbaceous cover (which were highly correlated with each other), a relationship that also exists for North American deserts (Brown 1973, 1975). These habitat measures showed a decreasing cline northward in Chile, grading into more xeric areas with lower rodent diversities.

Afghanistan   Gaisler (1975) presents ecological information about a number of rodent species near the town of Jalalabad in east-central Afghanistan. Although the five species that were studied intensively were captured in the Jalalabad Valle, a relatively moist area, they are characteristic of the surrounding desert habitat. The most striking feature of this community is the relatively large size of most of its constituents (Figure 11.4) compared to other deserts. There is a relatively large gap

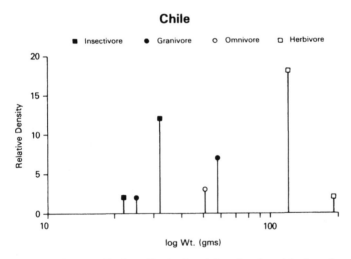

Figure 11.3.   Distribution of body sizes (plotted on logarithmic scale so that equal spacing indicates equal ratios) for a semiarid region (Parque Nacional Fray Jorge) in Chile. From smallest to largest, the species are *Marmosa elegans, Oryzomys longicaudatus, Akodon olivaceus, A. longipilis, Phyllotus darwini, Octodon degus,* and *Abrocoma bennetti.* The relative densities represent estimates of numbers per hectare (density data from P. Meserve personal communication; remaining data after Meserve 1981a, b).

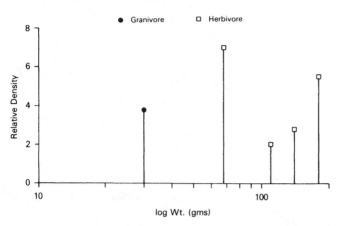

Figure 11.4. Distribution of body sizes (plotted on logarithmic scale so that equal spacing indicates equal ratios) for arid region near the Jalalabad Valle, Afghanistan. From smallest to largest, the species are *Calomyscus bailwardi*, *Meriones libycus*, *M. persicus*, *Tatera indica*, and *Nesokia indica*. Weights and relative densities were extrapolated from Gaisler (1975), who also provides some information on microhabitat use.

between the smallest species (*Calomyscus bailwardi*; 20 g) and the next largest species, *Meriones libycus*, but otherwise the body sizes seemed arrayed in fairly uniform ratios. The densities of the species vary only two- to threefold. Although Gaisler (1975) does not present information on the specific diets of the rodents, he implies that the four largest species are herbivores that will also consume fruits, seeds, and miscellaneous vegetation. Again by implication, the author indicates that there may be some habitat specialization between inhabitants of the flat areas and the rocky areas, but there is insufficient information to suggest specific microhabitat preferences. The largest species, *Nesokia indica*, is primarily a burrowing species, and thus inhabits a microhabitat distinct from the other four species.

While very little information pertinent to community structure is available for this assemblage, the data we have suggests that there is at least the possibility of some type of resource allocation that is reflected in the relative body sizes.

Sudan Four species of rodents occur in the deserts and semideserts near Khartoum, Sudan, while two species occur in the jebels (rocky islands) scattered throughout the region (Happold 1975) (Figure 11.5). There appear to be three distinct size classes in this sandy community, with the largest and smallest species being quite rare. The two intermediate species, *Jaculus jaculus* and *Gerbillus pyramidum*, are similar in body size and both exhibit bipedal locomotion. These species are similar in many characteristics, but *Jaculus jaculus* tends to run away from predators, is relatively "tame," and does not eat hard seeds or store food, while *G. pyramidum* hides from predators, is more aggressive than its cohabitant, and does eat and store hard seeds. While the rodents' sizes and microhabitat use appear to be similar, Happold (1975) indicates that they may have distinctive diets that could promote coexistence in the Sudan desert.

The two species that occur in the jebels are also very similar in size (*Gerbillus campestris*, 27 g; and *Acomys cahirinus*, 30 g). Although Happold (1975) does not dwell on their characteristics, the former is a biped while the latter is quadrupedal. This distinction may indicate major differences in foraging strategies like those noted for heteromyid rodents.

Israel Abramsky and his collaborators have begun to consider specific details of rodent communities in the deserts of Israel (Abramsky 1980, 1984; Abramsky and Sellah 1982; Rosenzweig, Abramsky, and Brand 1984; Abramsky, Rosenzweig, and Brand 1985). As many as five species of *Gerbillus*, two species of *Meriones*, and one species of *Jaculus* may coexist in some sandy areas of Israel (Rosenzweig, Abramsky, and Brand 1984; Abramsky, Rosenzweig, and Brand 1985). All five species of *Gerbillus* are granivorous and are arrayed across a relatively small range of body sizes (Figure 11.6). *Gerbillus nanus* and *G. henleyi* were too rare to be incorporated into the analyses, but Rosenzweig, Abramsky, and Brand (1984) analyzed species interactions in relation to habitat selection between the three remaining *Gerbillus* species. The authors determined that *G. pyramidum* and *G. allenbyi* have

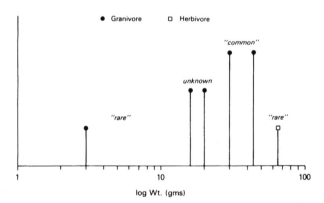

Figure 11.5. Distribution of body sizes (plotted on logarithmic scale so that equal spacing indicates equal ratios) for an arid region near Khartoum, Sudan. From smallest to largest, the species are *Gerbillus watersi*, *G. campestris*, *Acomys cahirinus*, *G. pyramidum*, *Jaculus jaculus*, and *Meriones crassus*. Happold (1975) notes only that two species were common and two were rare; the densities of the other two species are unknown (after Happold 1975).

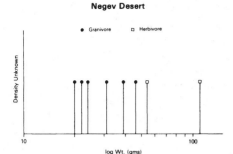

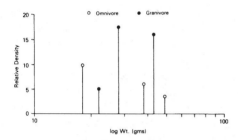

Figure 11.6.   Distribution of body sizes (plotted on logarithmic scale so that equal spacing indicates equal ratios) for two sites in the Negev Desert, Israel. The top panel is from near Holot Agur; from smallest to largest the species are *Gerbillus henleyi, G. gerbillus, G. nanus, G. allenbyi, G. pyramidum, Jaculus jaculus, Meriones crassus,* and *M. sacrementi.* The weights are from Rosenzweig, Abramsky, and Brand (1984); information on the relative densities of all species were not available for this site. The bottom panel represents rodents from several areas near the previously mentioned site. The species, from smallest to largest, are *G. dasyurus, G. gerbillus, G. allenbyi, Acomys cahirinus, G. pyramidum,* and *A. russatus.* Both species of *Acomys* and *G. dasyurus* occur in rocky habitats while the other species are confined to sandy substrates. Relative densities are averages for the number captured at those sites where all three species in either of the microhabitats were captured (after Abramsky, Rosenzweig, and Brand 1985).

a substantial impact on each other and are competing locally. The third common species, *G. gerbillus*, did not appear to be affected by any of the other species, but may have influenced the distribution of the rarer species. *Jaculus* was difficult to trap and was not included in the analyses. The two *Meriones* species are much larger than the granivores and occupy a different trophic level (both are herbivores) (Figure 11.6). In addition, one species of *Meriones* is twice as large as the other.

In another investigation across several sites in the Sinai, Abramsky, Rosenzweig, and Brand (1985) obtained information on the microhabitat use, diets, and densities of six broadly sympatric rodent species (other species occurred in the area, but the authors included only the common forms in their study). Three species of granivorous gerbils (*G. pyramidum*, *G. gerbillus*, and *G. allenbyi*) coexisted in a sandy microhabitat, and were somewhat similar in body size (although their densities differed; Figure 11.6). Three other species, *G. dasyurus*, *Acomys cahirinus*, and *A. russatus* were sympatric in rocky microhabitats. The body masses of these three species were more uniformly arrayed than the three species in the sandy microhabitats. While comparing a regression method and a distribution method for detecting habitat selection, the authors identified specific independent variables with which each species was associated. Their analysis revealed that *G. allenbyi* was the least adapted to open sandy dunes. Of the remaining two gerbils, *G. gerbillus* was found in the poorest habitats, with *G. pyramidum* choosing areas in between the other two species in terms of productivity. The information on the natural history of the species inhabiting the rocky microhabitats was insufficient to allow thorough analysis.

In addition to analyses of existing patterns, Abramsky (1980, 1984) and Abramsky and Sellah (1982) conducted manipulation experiments along the coastal sand dunes of Israel designed to determine the effect of two similar species (*Meriones tristrami* and *Gerbillus allenbyi*) on each other. While these species differ significantly in size, they tend to inhabit similar microhabitats and eat similar seed species (even though *Meriones* consumes more vegetation than *Gerbillus*; Abramsky 1980). The smaller *G. allenbyi* is an obligate sand dweller while its cohabitant has more general requirements that include sandy areas. The authors found that where *M. tristrami* occurred in the absence of *G. allenbyi*, the former includes sand dunes in its range, but where *G. allenbyi* was present in the sand dunes, the larger species seemed to be displaced to soil types other than sand.

When the sand-inhabiting *G. allenbyi* was removed from exclusion plots, Abramsky and Sellah (1982) anticipated that *M. tristrami* would move into the vacated habitat type. This did not occur (even though *M. tristrami* occurs in sand dunes in nearby areas), and the authors suggest that this reflects past competitive interactions that have been settled evolutionarily by inherited habitat preferences in sympatric populations of the two species.

Several recent investigations of desert rodent systems in Israel have dealt with broad questions about community organization and habitat selection. Some of the details are in the papers cited above, but two other articles offer intriguing hypotheses. In one analysis, Abramsky and Rosenzweig (1984) tested the proposition that rodents might be specializing on specific proportions of different microhabitats rather than retaining exclusive control over a single microhabitat. Their hypothesis is based on Tilman's (1982) proposal that many species of plants can coexist on a limited resource base by specializing on different ratios of nutrients. Abramsky and Rosenzweig (1984) treat microhabitats as nutrients and find that desert rodent species diversity matches the prediction that increases in productivity across a geographic range will initially yield increased species diversity, but that the diversity will rapidly fall off as productivity continues to increase. The decline is hypothesized to be the result of a decline in one or more particular "nutrients" (in this case, microhabitats) even in the face of a general increase in productivity. While their results are consistent with the predictions, the authors note that other hypotheses concerning patterns of diversity generate the same predictions. While the authors' proposal to equate microhabitats to nutrients is intriguing, significantly more needs to be learned about the subsets of habitats available to desert rodents, and their ratios at specific locations, before the hypothesis can be properly evaluated.

In a somewhat related analysis, Rosenzweig and Abramsky (1986) use information about two species of coexisting desert rodents to test their hypothesis concerning centrifugal community organization. The authors suggest that two similar species might share the same preferred resource (e.g., one particular microhabitat) while each also specializes on a different, less preferred resource. The biological basis of this structuring force is that, while the ideal primary resource is the same for each participating species, each is differentially adapted to tolerate deprivation of one resource. They model the isolegs ("lines of equal optimal behavior drawn in a state space") of the proposed arrangement and show that the distinguishing feature of centrifugally organized communities is that the isolegs have negative slopes. This has two important and novel implications for communities. One is that the niche breadth of a species should decrease when a competitor is removed, freeing the primary resource for the remaining species which can then end its reliance on the secondary resource. Secondly, competition will be a constant force as the participating species continually attempt to monopolize the primary, favored, shared resource. This, in turn, means that competition will be a constant factor in their coexistence and will never be resolved evolutionarily.

Kalahari Desert   One of the most complex assemblages of desert rodents that has been studied occurs in the Kalahari Desert along the Nossob River (Nel and Rautenbach 1975; Nel 1978). Here, 16 species of rodents inhabit several microhabitats (low and high dunes, riverbed, and plateaus) along with two species of small insectivores. Thirteen species of rodents occur exclusively or occasionally in the low dunes but three of these, a large porcupine, the large spring hare, and a fossorial molerat, are sufficiently distinct to exclude in the following analysis, leaving 10 species that are sympatric in the same microhabitat. The body sizes represented in the low dune microhabitat range from the tiny *Mus minutoides* (4.7 g) to the large, colonial *Parotomys brantsii* (80 g; Figure 11.7).

The community inhabitants are arrayed across a range of body masses, with several species clustering around the 30–45 g category and around the 65–80 g category (Figure 11.7). Details presented by Nel (1978), however, indicate that ecological features seem to distinguish each member within these clusters. For exam-

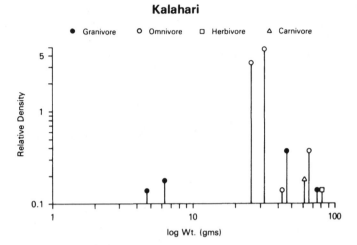

Figure 11.7. Distribution of body sizes (plotted on logarithmic scale so that equal spacing indicates equal ratios) for rodents captured in the Low Dune microhabitat in the Kalahari Desert. The species, from smallest to largest, are *Mus minutoides*, *Dendromus melanotis*, *Gerbillurus paeba*, *Rhabdomys pumilo*, *Aethomys namaquensis*, *Desmodillus auricularis*, *Zelotomys woosnami*, *Tatera brantsii*, *Thallomys paedulcus*, and *Parotomys brantsii*. Relative densities (plotted on a logarithmic scale) represent relative numbers trapped over several trapping periods; detailed microhabitat information was not presented (after Nel 1978).

ple, in the 30–45 g grouping, *Rhabdomys* is diurnal and forages primarily in the bushes while the very similar-sized *Aethomys* and *Desmodillus* are separated on the basis of their diets (the former being omnivorous, the latter granivorous). In the 60–80 g cluster, *Zelotomys* feeds entirely on flesh. *Tatera* is granivorous, as is the slightly larger *Thallomys*, but the latter forages and nests in trees; the large *Parotomys* is diurnal, colonial, and herbivorous.

While these qualitative ecological characteristics have not been subjected to rigorous analyses to detect the degree of body mass separation and the actual mechanisms promoting coexistence, there is clearly an appearance of ecological separation nested within a fairly uniform array of body masses (Figure 11.7). This impressive desert mammal fauna deserves further attention both because of its diversity and because there is enough already known about it compared to other deserts outside North America that detailed questions can now be asked.

Australia   Morton (1979, 1985) and Morton and Baynes (1985) have presented substantial amounts of information on Australian desert mammals and compared them to their North American counterparts. Certainly one of the most striking discoveries Morton and Baynes (1985) made is that an average of approximately 40 percent of the rodent and small marsupial species have disappeared from several desert localities in Australia where they were known to exist less than a century ago (the disappearances are local; the species still occur at other sites). This is a striking biological phenomenon in itself, but it also makes community pattern analyses difficult because much of the evolutionary past is not represented in communities that exist today. Fortunately, the authors present information on past conditions, garnered from old records and skeletal remains, and these data will be used to look at several desert sites that exhibit different levels of diversity.

Historically, small marsupials outnumbered rodents at most of the desert sites presented by Morton and Baynes (1985). For example, at the Gibson Desert, Great Sandy, and Alice Springs sites the marsupial/rodent species ratios were 7:6, 14:5, and 13:7 respectively. Virtually all of the marsupials are insectivores, although a few are carnivores (Morton 1979). Most of the rodents are omnivorous or herbivorous and only a few are granivorous, unlike most of the North American desert rodents (Morton 1979; Morton and Baynes 1985).

The body mass ratios of the rodents appear to be fairly evenly spaced at the three sites mentioned above (these sites were chosen because of their relatively diverse rodent faunas) (Morton and Baynes 1985) (Figure 11.8). At all three sites there are two or three species that cluster around 25 g, but in each case the rodents appear to have diets that do not completely overlap. For example, at Alice Springs, *Leggadina forresti* (22 g), *Pseudomys desertor* (25 g), and *Notomys alexis* (29 g) all consume seeds, but the smallest species tends toward omnivory, the intermediate species toward herbivory, and the largest species toward true granivory (Morton 1979; Morton and Baynes 1985). Little information is available on the microhabitat preferences of these species, so there is the possibility of segregation along this axis as well.

Small insectivorous marsupials (primarily dasyurids) in the arid regions of Australia exhibit a phenomenal diversity (four times the insectivore diversity found in North American deserts; Morton 1979, 1982). Fox (1982) and Morton (1982) provide excellent analyses of the characteristics of these communities. Morton (1982) notes that even though many details are lacking, it appears that the communities are strongly affected by both short-term and long-term variability in the food supply, which may generate patterns of periodic local extinctions. Fox (1982) conducted thorough studies on the body size relationships among dasyurids (primarily across broad areas of mesic eastern Australia) and he concludes that ratios of body mass are uniform between species, but that the actual ratios are lower than occur in other mammal communities. Furthermore, the patterns do not appear to be based on chance associations of the range of body sizes "available" for the assemblages.

Australian ecologists are rapidly generating data on the mammalian communities of the region. The unique phylogenetic history of the mammals in Australia and their spectacular diversity in some arid regions should provide more exciting insights into desert mammal community structure in the near future.

Obviously, the data presented for these desert areas vary in how much they reveal about discrete patterns of community structure. A few, such as those from India, the Sudan, and Afghanistan, provide only the slimmest evidence that their desert rodent communities might be organized similarly to North American deserts. The results from Chile are more complete and reveal specific community patterns and forces

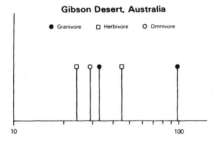

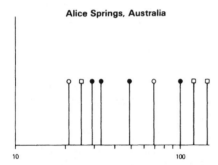

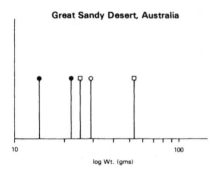

log Wt. (gms)

Figure 11.8. Distribution of body sizes (from Watts and Aslin 1981;
plotted on logarithmic scale so that equal spacing indicates equal
ratios) for rodents at three sites in Australia. The species, from
smallest to largest, at the Gibson Desert site: *Pseudomys
hermannburgensis, P. desertor, Notomys alexis, P. praeconis,
N. longicaudatus,* and *Rattus tunheyi*; Alice Springs:
*P. hermannburgensis, Leggadina forresti, P. desertor, N. alexis,
P. fieldi, Zyzomys pedunculatus, N. longicaudatus, R. tunheyi,* and
Leporillus apicalis; Great Sandy Desert: *P. hermannburgensis,
L. forresti, P. desertor, N. alexis,* and *P. nanus*. No information was
available on microhabitat preferences. Several of the species included
have disappeared in the last century, along with a number of
coexisting, insectivorous, dasyurid marsupials (distribution data after
Morton and Baynes 1985).

that might yield those patterns. The research from the Kalahari Desert, Israel, and Australia is even more thorough and provides important information for a general understanding of communities in addition to what has been uncovered specifically about mammals. The net result is that while each of these communities retains its own distinct characteristics, all exhibit patterns of body mass ratios that suggest that underlying selection forces have molded the extant community. Many of the details of how the allocation of resources (e.g., food resources, predator escape behaviors) has been resolved remain unknown but there is sufficient information to warrant further investigation of these communities.

Discussion

The information on the deserts discussed above (even the better-known representatives) provides only a sliver of insight into their mammal communities compared to what is known about North American desert mammal communities (Genoways and Brown 1991). Even though there is a tremendous amount still to learn about North American heteromyid communities, this ecological group, and the literature it has engendered, has come to serve as the model for analyses of other desert mammal systems (Brown 1985).

As valuable as the data on heteromyids are, however, it is possible that the wealth of information on them has done something of a disservice to researchers studying other desert mammal systems. As it turns out, most other deserts do not appear to resemble North American deserts, especially in the degree of granivory exhibited by their inhabitants. Much of the value of what is known about North American desert systems revolves around the fact that the diverse community is composed primarily of granivores. Other species co-occur (e.g., herbivorous ground squirrels and woodrats, and insectivorous grasshopper mice), but investigations have centered on the granivores, including distantly related taxa (e.g., birds and ants; Brown, Reichman, and Davidson 1979; Brown et al. 1986). Specific features of the seed resource (size, nutritional quality, distribution, production and abundance, and storability) serve as the centerpiece of adaptation to North American deserts for many taxa.

Seeds, which are abundant in North American deserts and are very persistent compared to ephemeral resources like vegetation and insects, play a crucial role in the organization of desert mammal communities. North American deserts are actually quite productive, so it was reasonable to assume that seeds would be even more valuable as a persistent resource in drier, less productive deserts and thereby serve as the focus of adaptation in those deserts as well. For a variety of ecological and historical reasons, however, this may not be the case, as the work of Mares and Rosenzweig (1978), Nel (1978), Morton (1979, 1985), Morton and Baynes (1985), and Mares (1991) indicates. Certain taxa of desert mammals outside North America are granivorous (e.g., jirds and gerbils), but many other genera are insectivorous, herbivorous, or omnivorous. The gradual realization that granivory may not play the central role in other deserts can be seen in the change in understanding from Morton's 1979 paper to the interpretations of Morton and Baynes (1985). Morton, who received some of his training in North America, went back to his native Australia to apply what he had learned to the deserts there, only to find that circumstances were significantly different.

While results from the study of North American desert mammal communities may have misdirected similar studies in other deserts, the overriding importance of the North American systems is their contribution to a more general understanding of community structure independent of the taxa involved. For example, because of the substantial amount of information about the apparent uniform array of body masses at many different locales (e.g., Brown and Kurzius 1987), desert rodents were an important component of the development of the null model approach to community structure analysis (Bowers and Brown 1982). In fact, they remain one of the best-documented and most-accepted examples of a nonrandom pattern in community organization.

Furthermore, several approaches to analyzing community structure have developed around studies of North American desert rodents so their impact has spread beyond specific details to considerations of the most efficacious approaches to community analysis. For example, Price (1986) reviews several approaches to the investigation of communities and uses data from desert rodents to compare their strengths and weaknesses. Null models of community organization have been applied to determine whether North American desert rodent communities are organized randomly and the results have been beneficial to studies of these communities (they are not structured randomly) and to community ecology in general (Bowers and Brown 1982). Even more specifically, Rosenzweig and Abramsky (1986) relied on information from desert rodent communities in Israel to analyze general concepts of centrifugal community organization and resource specialization and point out that the generality of this concept must be tested on other communities.

Finally, many studies have concentrated on the specific details on the axes along which desert rodents might, or actually do, subdivide resources. These have included seed size, spatial distribution and energy content, aggression or intimidation between individual rodents, and predator avoidance adaptations. Even though deserts are unique and relatively simple habitats, certain points that have been learned about how resources (food, space, predator avoidance schemes) are divided among coexisting species should be applicable to other systems where competition is important.

Several ecological and behavioral traits of desert mammals are still virtually unknown and might add information important to a quest for an understanding of communities. For example, all interactions between desert mammals, their resources, and their predators are probably filtered through their parasite loads. That is, most animals are parasitized throughout their lives and these parasites can either reduce the success of their hosts or completely change host behavior in ways that might impinge on community relationships. While some of the basics about water metabolism in desert animals are known, the complex relationships between water, nutrition, food storage, and torpor as alternative or mixed strategies of coping with the desert environment remain to be investigated. We are also just beginning to learn about social interactions of desert mammals that may affect communities (e.g., Jones 1984; Randall 1984, 1987).

In addition to these specific areas of investigation, there are several approaches that I think will be especially effective in revealing underlying traits of desert mammal communities (actually, of many communities). The first is to use geographical patterns to test a priori predictions about community structure (e.g., Abramsky and Rosenzweig 1984; Brown and Kurzius 1987). Another productive approach would be to locate areas that possess unique combinations of mammals and study the selection pressures that affect them (such as the islands in the Sea of Cortez). And finally, the use of clever, rigorously conducted field manipulations may be the most effective means of analyzing community properties.

As important as desert mammals, especially rodents, have been as models for analyses of community patterns, they have also served as a model of how community studies are initiated and evolve. I do not propose that this is how such studies should evolve, but rather that this is one way they do evolve. Specifically, desert rodent research began with autecological investigations of single species of heteromyids, seemingly the most desert-adapted group of rodents, and their "niche"—where they lived and what they did for a living (although there were some early considerations of community properties in the work of Benson 1933, Dice and Blossom 1937, and Hall 1946). Desert mammals were especially important as tools for physiological studies because of adaptations to the potentially harsh environment (that is, environments that seemed harsh to humans). Schmidt-Nielsen (1964) and his many collaborators contributed significantly to this tend, as did MacMillen (1964).

Then, like small expanding spheres of interest, these studies evolved to include two-species systems (frequently one bipedal species and one quadrupedal species; e.g., Rosenzweig 1973; Frye and Rosenzweig 1980; Reichman and Oberstein 1977) and multispecies systems (Brown 1973, 1975). Comparisons of geographical patterns (Brown 1973, 1975) and manipulations of microhabitats (Rosenzweig 1973) or seed resources in the field (Price 1978a,b; Wondolleck 1978; Frye and Rosenzweig 1980; Harris 1984) and lab (Reichman and Oberstein 1977; Trombulak and Kenagy 1980) were most often analyzed and the results incorporated into ideas concerning community patterns. As the spheres of attention expanded, other taxa and other parameters were incorporated. For example, Rebar and Conley (1983) investigated the interaction between a kangaroo rat species and a cricetid rodent, the grasshopper mouse, and found asymmetrical increases in response to reciprocal removals. Eventually even more distantly related taxa, such as granivorous birds and ants, were brought into the analyses of desert granivore guilds (Pulliam and Brand 1975; Brown, Reichman, and Davidson 1979; Davidson et al. 1985; Brown et al. 1986).

More recently investigations into the effect of predation have appeared (Kotler 1984a,b, 1985; Thompson 1985; Brown et al. 1988; Kotler et al. 1988). These studies have taken on the difficult task of manipulating the risk (or perceived risk) of predation to rodent populations to determine how this factor affects rodent behavior and success in relation to body size and gait, and microhabitat use.

As new layers of information about desert granivore communities were added, claims were staked and schools of thought developed which tended to concentrate on specific aspects of the granivore community. At times, the debates over the relative importance of one aspect over another became acrimonious, even though most of the investigators recognized that no one simple answer to questions concerning community organization would be forthcoming from their investigations. Presentations at meetings and articles in the literature record actual or implied accusations of one author ignoring the points of view of another.

I suggest that the development of studies on North American desert granivore systems actually follows from two natural phenomena associated with basic research and not from schemes to promote one view or another. The first is that individual investigators tend to study systems that they find interesting and convenient (because of proximity or training). Secondly, new studies follow naturally from previous ones in the pursuit of further details—this is the working definition of original research. Thus, the newest research is something that previous investigators have not gotten around to doing or have not thought to do. Initial attempts to investigate any system usually begin with the most obvious and tractable components. Therefore, with the desert granivore system it was appropriate initially to concentrate on one, and then two, and then several species of the highly desert-adapted species of heteromyids. Surely investigators tackling these interactions were aware that other taxa of mammals, and even granivorous birds and ants, were important constituents of the community, but the initial work required the unrealistic, simplifying assumptions associated with working with a subset of the larger granivore community. The simplifications made in early investigations or models represent, in most cases, the best good-faith efforts that were possible at the time, rather than passive or active neglect of alternative ideas or a lack of appreciation for the true complexity of the community. My basic point is that, at least in the granivore communities, many of the questions are obvious, but it takes squadrons of investigators and many years to think of ways to answer them and then actually carry out the research effort.

In conjunction with the expanding sphere of information about granivores came increased sophistication in the concepts and analyses of community structure. These new ideas, heaped on top of the burgeoning mound of information about desert granivores, have brought us to the point of reflection and reevaluation represented by the reviews of Price and Brown (1983), Price (1986), the article by Kotler and Brown (1988), and the Mares, Brown and Harney, and Reichman and Price chapters in the forthcoming book on heteromyids (Genoways and Brown 1991).

The acrimony exhibited during the evolution of studies on North American granivore communities is probably typical of the way scientific investigations develop. Younger scientists, using information gathered earlier, and applying newfound concepts and theories, impinge on the investments of earlier workers. Possessiveness, envy, and, increasingly, competition for resources (jobs and research grants) may be natural parts of the cultural landscape. Perhaps competition structures the community of desert granivore scientists just as it does desert granivore communities.

Acknowledgments

All ecologists are deeply indebted to those investigators who have gathered and analyzed data from deserts around the world. In addition, I am personally grateful for the responses of Mike Bowers, Jim Brown, Joel Brown, Burt Kotler, Mike Mares, Peter Meserve, and Mike Rosenzweig when asked for assistance or information. I also thank J. U. M. Jarvis, who served as my host on a visit to Africa (and the myriad other scientists in Africa who shared their insights) where I first began to appreciate that not all deserts are like the Sonoran Desert.

Bibliography

Abramsky, Z. 1978. Small mammal community ecology: changes in species diversity in response to manipulated productivity. *Oecologia* 34: 113–123.

——. 1980. Ecological similarity of *Gerbillus allenbyi* and *Meriones tristrami*. *Journal of Arid Environments* 3: 153–160.

——. 1984. Population biology of *Gerbillus allenbyi* in northern Israel. *Mammalia* 48: 197–206.

Abramsky, Z., and M. L. Rosenzweig. 1984. Tilman's predicted productivity-diversity relationships shown by desert rodents. *Nature* 309: 150–151.

Abramsky, Z., M. L. Rosenzweig, and S. Brand. 1985. Habitat selection of Israel desert rodents: comparison of a traditional and a new method of analysis. *Oikos* 45: 79–88.

Abramsky, Z., and C. Sellah. 1982. Competition and the role of habitat selection in *Gerbillus allenbyi* and *Meriones tristrami*: a removal experiment. *Ecology* 63: 1242–1247.

Bartholomew, G. A., and G. Cary. 1954. Locomotion in pocket mice. *Journal of Mammalogy* 35: 386–392.

Bartholomew, G. A., and H. Caswell. 1951. Locomotion in kangaroo rats and its adaptive significance. *Journal of Mammalogy* 32: 155–169.

Bedard, J. 1976. Coexistence, coevolution, and convergent evolution in seabird communities: a comment. *Ecology* 57: 177–184.

Benson, S. 1933. Concealing coloration among some desert rodents of the southwest United States. *University of California Publications in Zoology* 40: 1–70.

Blaustein, A., and A. Risser. 1976. Interspecific interactions between three species of kangaroo rats (*Dipodomys*). *Animal Behaviour* 24: 381–385.

Bowers, M. A. 1982. Foraging behavior in heteromyid rodents: field evidence for resource partitioning. *Journal of Mammalogy* 63: 361–367.

Bowers, M. A., and J. H. Brown. 1982. Body size and coexistence in desert rodents: chance or community structure? *Ecology* 63: 391–400.

Bowers, M. A., D. B. Thompson, and J. H. Brown. 1987. Foraging and microhabitat use in desert rodents: the role of a dominant competitor. *Oecologia* 72: 77–82.

Brown, J. H. 1973. Species diversity of seed-eating rodents in sand dune habitats. *Ecology* 54: 775–787.

——. 1975. Geographical ecology of desert rodents. Pp. 315–341 in M. L. Cody and J. Diamond (eds.) *Ecology and Evolution of Communities*. Cambridge, Mass.: Harvard University Press.

——. 1984. Desert rodents: a model system. *Acta Zoologica Fennica* 172: 45–49.

——. 1985. Organization of North American desert rodent associations: insights from geographic comparisons and perturbation experiments. *Australian Mammalogy* 8: 131–136.

Brown, J. H., and M. A. Bowers. 1984. Patterns and processes in three guilds of terrestrial vertebrates. Pp. 282–296 in D. Strong, D. Simberloff, L. Abele, and A. Thistle (eds.),

Ecological Communities: Conceptual Issues and the Evidence. Princeton, N.J.: Princeton University Press.

Brown, J. H., and D. W. Davidson. 1977. Competition between seed-eating rodents and ants in desert ecosystems. *Science* 196: 880–882.

Brown, J. H., D. W. Davidson, J. C. Munger, and R. S. Inouye. 1986. Experimental community ecology: the desert granivore system. Pp. 41–62 in J. Diamond and T. J. Case (eds.), *Community Ecology.* New York: Harper and Row.

Brown, J. H., and B. Harney. 1990. Population and community ecology of heteromyid rodents in temperate habitats. In H. Genoways and J. H. Brown (eds.), *Biology of Heteromyids.* American Society of Mammalogists Special Publication #10. In press.

Brown, J. H., and M. Kurzius. 1987. Composition of desert rodent faunas: combinations of coexisting species. *Annales Zoologici Fennici* 24: 227–237.

Brown, J. H., O. J. Reichman, and D. W. Davidson. 1979. Granivory in desert ecosystems. *Annual Review of Ecology and Systematics* 10: 201–227.

Brown, J. S. 1988. Patch use as an indicator of habitat preference, predation risk, and competition. *Behavioral Ecology and Sociobiology* 22: 37–47.

Brown, J. S., B. P. Kotler, R. J. Smith, and W. O. Wirtz II. 1988. The effects of owl predation on the foraging behavior of heteromyid rodents. *Oecologia* 76: 408–415.

Christian, D. 1979. Physiological correlates of demographic patterns in three Namib Desert rodents. *Physiological Zoology* 52: 329–339.

Connell, J. H. 1975. Some mechanisms producing structure in natural communities: a model and evidence from field experiments. Pp. 460–491 in M. L. Cody and J. Diamond (eds.), *Ecology and Evolution of Communities.* Cambridge, Mass.: Harvard University Press.

Conner, E. F., and D. Simberloff. 1979. The assembly of species communities: chance or competition? *Ecology* 60: 1132–1140.

Davidson, D. W., J. H. Brown, and R. S. Inouye. 1980. Competition and the structure of granivore communities. *BioScience* 30: 233–238.

Davidson, D. W., D. A. Sampson, and R. S. Inouye. 1985. Granivory in the Chihuahuan Desert: interactions within and between trophic levels. *Ecology* 66: 486–502.

Diamond, J., and T. J. Case, eds. 1986. *Community Ecology.* New York: Harper and Row.

Dice, L. R., and P. M. Blossom. 1937. Pp. 1–129 in *Studies of Mammalian Ecology in Southwestern North America with Special Attention to the Colors of Desert Mammals.* Carnegie Institute of Washington, Publication #485.

Fox, B. 1982. A review of dasyurid ecology and speculation on the role of limiting similarity in community organization. Pp. 97–116 in M. Archer (ed.), *Carnivorous Marsupials.* Royal Society of New South Wales, NSW, Australia.

Frank, C. 1988. The influence of moisture content on heteromyid rodent seed preferences. *Journal of Mammalogy* 69: 353–357.

Frye, R. 1983. Experimental field evidence of interspecific aggression between two species of kangaroo rat (*Dipodomys*). *Oecologia* 59: 74–78.

Frye, R., and M. L. Rosenzweig. 1980. Clump size selection: a field test with two species of *Dipodomys. Oecologia* 47: 323–327.

Gaisler, J. 1975. Comparative ecological notes on Afghan rodents. Pp. 59–74 in I. Prakash and G. Ghosh (eds.), *Rodents in Desert Environments.* The Hague: W. Junk.

Galindo, C. 1986. Do desert rodent populations increase when ants are removed? *Ecology* 67: 1422–1423.

Genoways, H., and J. H. Brown, eds. 1991. *Biology of Heteromyids.* American Society of Mammalogists Special Publication #10. In press.

Glanz, W. E. 1977. Comparative ecology of small mammal communities in California and Chile. Ph.D. dissertation, University of California, Berkeley.

Grinnell, J., and R. T. Orr. 1934. Systematic review of the Californicus group of the rodent genus *Peromyscus*. *Journal of Mammalogy* 15: 210–217.

Hafner, J. C., and M. S. Hafner. 1983. Evolutionary relationships of heteromyid rodents. *Great Basin Naturalist Memoirs* 7: 3–29.

Hall, E. R. 1946. *Mammals of Nevada*. Berkeley, Calif.: University of California Press, 710 pp.

Happold, D. C. D. 1975. The ecology of rodents in the northern Sudan. Pp. 15–46 in I. Prakash and G. Ghosh (eds.), *Rodents in Desert Environments*. The Hague: W. Junk.

Harris, J. 1984. Experimental analysis of desert rodent foraging ecology. *Ecology* 65: 1578–1584.

———. 1986. Microhabitat segregation in two desert rodent species: the relation of prey availability to diet. *Oecologia* 68: 417–421.

Hinds, D. S., and R. MacMillen. 1985. Scaling of energy metabolism and evaporative water loss in heteromyid rodents. *Physiological Zoology* 58: 282–298.

Holt, R. D. 1984. Spatial heterogeneity, indirect interactions, and the coexistence of prey species. *American Naturalist* 124: 377–406.

Holt, R. D., and B. P. Kotler. 1987. Short-term apparent competition. *American Naturalist* 130: 412–430.

Jones, W. T. 1984. Natal philopatry in bannertailed kangaroo rats. *Behavioral Ecology and Sociobiology* 15: 151–155.

Kaufman, D. W., and G. Kaufman. 1982. Effect of moonlight on activity and microhabitat use by Ord's kangaroo rat (*Dipodomys ordii*). *Journal of Mammalogy* 63: 309–312.

Kenagy, G. J. 1972. Saltbush leaves: excision of hypersaline tissue by a kangaroo rat. *Science* 178: 1094–1096.

Kotler, B. 1984a. Effects of illumination on the rate of resource harvesting in a community of desert rodents. *American Midland Naturalist* 111: 383–389.

———. 1984b. Predation risk and the structure of desert rodent communities. *Ecology* 65: 689–701.

———. 1985. Owl predation on desert rodents which differ in morphology and behavior. *Journal of Mammalogy* 66: 824–828.

Kotler, B., and J. S. Brown. 1988. Environmental heterogeneity and the coexistence of desert rodents. *Annual Review of Ecology and Systematics* 19: 281–308.

Kotler, B., J. S. Brown, R. J. Smith, and W. O. Wirtz II. 1988. The effects of morphology and body size on rates of owl predation on desert rodents. *Oikos* 53: 145–152.

Larsen, E. 1986. Competitive release and microhabitat use among coexisting rodents: a natural experiment. *Oecologia* 69: 231–237.

Lay, D. M. 1974. Differential predation of gerbils (*Meriones*) by the little owl, *Athene brahma*. *Journal of Mammalogy* 55: 608–614.

M'Closkey, R. T. 1978. Niche separation and assembly in four species of Sonoran Desert rodents. *American Naturalist* 112: 683–694.

———. 1980. Spatial patterns in types of seeds collected by four species of heteromyid rodents. *Ecology* 61: 486–489.

———. 1981. Microhabitat use in coexisting desert rodents: the role of population density. *Oecologia* 50: 310–315.

MacMahon, J. A. 1976. Species and guild similarity of North American desert mammal faunas: a functional analysis of communities. Pp. 133–148 in David A. Goodall (ed.), *Evolution of Desert Biota*. Austin, Tex.: University of Texas Press.

MacMillen, R. E. 1964. Pp. 1–66 in *Population Ecology, Water Relations, and Social Behavior of a Southern California Semidesert Rodent Fauna.* University of California Publications in Zoology, No. 71.

———. 1983. Adaptive physiology of heteromyid rodents. *Great Basin Naturalist Memoirs* 7: 65–76.

MacMillen, R. E., and D. S. Hinds. 1983. Water regulatory efficiency in heteromyid rodents: a model and its application. *Ecology* 64: 152–164.

Mares, M. A. 1991. Heteromyids and their ecological counterparts: A pandesertic view of rodent ecology and evolution. In H. Genoways and J. H. Brown (eds.), *Biology of Heteromyids.* American Society of Mammalogists Special Publication #10. In press.

Mares, M. A., and M. L. Rosenzweig. 1978. Granivory in North and South American deserts: rodents, birds and ants. *Ecology* 59: 235–241.

Meserve, P. 1981a. Resource partitioning in a Chilean semi-arid small mammal community. *Journal of Animal Ecology* 50: 745–757.

———. 1981b. Trophic relationships among small mammals in a Chilean semiarid thorn scrub community. *Journal of Mammalogy* 62: 304–314.

Meserve, P., and W. E. Glanz. 1978. Geographic ecology of small mammals in the northern Chilean arid zone. *Journal of Biogeography* 5: 135–148.

Morton, S. R. 1979. Diversity of desert-dwelling mammals: a comparison of Australia and North America. *Journal of Mammalogy* 60: 253–264.

———. 1982. Dasyurid marsupials of the Australian arid zone: an ecological review. Pp. 117–130 in David Archer (ed.), *Carnivorous Marsupials.* Royal Zoological Society of New South Wales, NSW, Australia.

———. 1985. Granivory in arid regions: comparison of Australia with North and South America. *Ecology* 66: 1859–1866.

Morton, S. R., and A. Baynes. 1985. Small mammal assemblages in arid Australia: a reappraisal. *Australian Mammalogy* 8: 159–169.

Munger, J. C., and J. H. Brown. 1981. Competition in desert rodents: an experiment with semipermeable membranes. *Science* 211: 510–512.

Munger, J. C., T. Jones, and M. A. Bowers. 1983. Desert rodent populations: factors affecting abundance, distribution, and genetic structure. *Great Basin Naturalist Memoirs* 7: 91–116.

Nel, J. A. J. 1978. Habitat heterogeneity and changes in small mammal community structure and resource utilization in the southern Kalahari. *Bulletin of the Carnegie Museum of Natural History* 6: 118–131.

Nel, J. A. J., and I. L. Rautenbach. 1975. Habitat use and community structure of rodents in the southern Kalahari. *Mammalia* 39: 9–29.

Prakash, I., and P. K. Ghosh, eds. 1975. *Rodents in Desert Environments.* The Hague: W. Junk.

Price, M. V. 1978a. The role of microhabitat in structuring desert rodent communities. *Ecology* 59: 910–921.

———. 1978b. Seed dispersion preferences of coexisting desert rodent species. *Journal of Mammalogy* 59: 624–626.

———. 1983. Ecological consequences of body size: a model of patch choice in desert rodents. *Oecologia* 59: 384–392.

———. 1984. Microhabitat use in rodent communities: predator avoidance or foraging economics? *Netherlands Journal of Zoology* 34: 63–80.

———. 1986. Structure of desert rodent communities: a critical review of questions and approaches. *American Zoologist* 26: 39–49.

Price, M. V., and J. H. Brown. 1983. Patterns of morphology and resource use in North American desert rodent communities. *Great Basin Naturalist Memoirs* 7: 117–134.

Price, M. V., and K. Heinz. 1984. Effects of body size, seed density, and soil characteristics on rates of seed harvest by heteromyid rodents. *Oecologia* 61: 420–425.

Price, M. V., and O. J. Reichman. 1987. Spatial and temporal heterogeneity in Sonoran Desert soil seed pools, and implications for heteromyid rodent foraging. *Ecology* 68: 1797–1811.

Price, M. V., and N. Waser. 1985. Microhabitat use by heteromyid rodents: effects of artificial patches. *Ecology* 66: 211–219.

Pulliam, R., and M. Brand. 1975. The production and utilization of seeds in plains grassland of southeastern Arizona. *Ecology* 56: 1158–1166.

Randall, J. 1984. Territorial defense and advertisement by footdrumming in bannertail kangaroo rats (*Dipodomys spectabilis*) at high and low population densities. *Behavioral Ecology and Sociobiology* 16: 11–20.

———. 1987. Sandbathing as a territorial scent-mark in the bannertail kangaroo rat, *Dipodomys spectabilis. Animal Behaviour* 35: 426–434.

Rebar, C. 1988. Comparative learning abilities and foraging strategies of *Dipodomys merriami* and *Chaetodipus intermedius*. Ph.D. dissertation, Kansas State University, Manhattan.

Rebar, C., and W. Conley. 1983. Interactions in microhabitat use between *Dipodomys ordii* and *Onychomys leucogaster. Ecology* 64: 984–988.

Reichman, O. J. 1975. Relationships of desert rodent diets to available resources. *Journal of Mammalogy* 56: 731–751.

———. 1976. Relationships between dimensions, weights, volumes, and calories of some Sonoran Desert seeds. *Southwestern Naturalist* 20: 573–575.

———. 1979. Desert granivore foraging and its impact on seed densities and distributions. *Ecology* 60: 1085–1092.

———. 1981. Factors influencing foraging in desert rodents. Pp. 195–213 in A. Kamil and T. Sargent (eds.), *Foraging Behavior: Ecological, Ethological, and Psychological Approaches*. New York: Garland-STMP Press, 534 pp.

———. 1984. Spatial and temporal variation in seed distributions in desert soils. *Journal of Biogeography* 11: 1–11.

Reichman, O. J., N.d. Microhabitat use and foraging behavior of pocket mice on predator-rich and predator-poor islands. In preparation.

Reichman, O. J., and J. H. Brown. 1979. The use of torpor by *Perognathus amplus* in relation to resource distribution. *Journal of Mammalogy* 60: 550–555.

———. 1983. Opening remarks: Symposium on the biology of desert rodents. *Great Basin Naturalist Memoirs* 7: 1–2.

Reichman, O. J., A. Fattaey, and K. Fattaey. 1986. Management of sterile and mouldy seeds by a desert rodent. *Animal Behaviour* 34: 221–225.

Reichman, O. J., and D. Oberstein. 1977. Selection of seed distribution types by *Dipodomys merriami* and *Perognathus amplus. Ecology* 58: 636–643.

Reichman, O. J., and M. V. Price. 1991. Ecological aspects of heteromyid foraging. In H. Genoways and J. H. Brown (eds.), *Biology of Heteromyids*. American Society of Mammalogists Special Publication #10. In press.

Reichman, O. J., D. Wicklow, and C. Rebar. 1985. Ecological and mycological characteristics of caches in the mounds of *Dipodomys spectabilis. Journal of Mammalogy* 66: 643–651.

Roberts, E., and O. J. Reichman. Manuscript. Computer simulation of the effects of resource distribution on desert rodent community structure.

Root, R. B. 1967. The niche exploitation pattern of the blue-gray gnatcatcher. *Ecological Monographs* 37: 317–350.

Rosenzweig, M. L. 1973. Habitat selection experiments with a pair of coexisting heteromyid species. *Ecology* 58: 636–643.

Rosenzweig, M. L., and Z. Abramsky. 1986. Centrifugal community organization. *Oikos* 46: 339–348.

Rosenzweig, M. L., Z. Abramsky, and S. Brand. 1984. Estimating species interactions in heterogeneous environments. *Oikos* 43: 329–340.

Rosenzweig, M. L., and P. Sterner. 1970. Population ecology of desert rodent communities: body size and seed husking as a basis for heteromyid coexistence. *Ecology* 51: 217–224.

Rosenzweig, M. L., and J. Winakur. 1969. Population ecology of desert rodent communities: habitats and environmental complexity. *Ecology* 50: 558–571.

Schall, J. J., and E. R. Pianka. 1980. Evolution of escape behavior diversity. *American Naturalist* 115: 551–566.

Schmidly, D. J., and K. T. Wilkens. 1991. Zoogeography. In H. Genoways and J. H. Brown (eds.), *Biology of Heteromyids*. American Society of Mammalogists Special Publication #10. In press.

Schmidt-Nielsen, K. 1964. *Desert Animals: Physiological Problems of Heat and Water.* Oxford, England: Clarendon Press, 277. pp.

Simberloff, D., and W. Boecklen. 1981. Santa Rosalia reconsidered: size ratios and competition. *Evolution* 35: 1206–1228.

Smith, C. C., and O. J. Reichman. 1984. The evolution of food caching behavior by birds and mammals. *Annual Review of Ecology and Systematics* 15: 329–351.

Streilein, K. E. 1982a. Ecology of small mammals in the semi-arid Brazilian Caatinga. II. Water relations. *Annals of the Carnegie Museum* 51: 109–126.

———. 1982b. The ecology of small mammals in the semi-arid Brazilian Caatinga. III. Reproductive biology and population ecology. *Annals of the Carnegie Museum* 51: 251–269.

———. 1982c. The ecology of small mammals in the semi-arid Brazilian Caatinga. IV. Habitat selection. *Annals of the Carnegie Museum* 51: 331–343.

Strong, D. R., L. A. Szyska, and D. Simberloff. 1979. Tests of community-wide character displacement against null hypotheses. *Evolution* 33: 897–913.

Taylor, C. R. 1968. Hygroscopic food: a source of water for desert antelopes? *Nature* 219: 181–182.

———. 1969. Metabolism, respiratory changes, and water balance of an antelope, the eland. *American Journal of Physiology* 217: 317–320.

Taylor, C. R., N. C. Heglund, and G. M. O. Maloiy. 1981. Energetics and mechanics of terrestrial locomotion: I. Metabolic energy consumption as a function of speed and body size in birds and mammals. *Journal of Experimental Biology* 97: 1–21.

Taylor, C. R., and C. P. Lyman. 1967. A comparative study of the environmental physiology of an East African antelope, the eland, and a Hereford steer. *Physiological Zoology* 40: 280–295.

———. 1972. Heat storage in running antelopes: independence of brain and body temperatures. *American Journal of Physiology* 222: 114–117.

Thompson, S. D. 1982a. Microhabitat utilization and foraging behavior of bipedal and quadrupedal heteromyid rodents. *Ecology* 63: 1303–1312.

———. 1982b. Structure and species composition of desert heteromyid species assemblages: effects of a simple habitat manipulation. *Ecology* 63: 1313–1321.

————. 1985. Bipedal hopping and seed-dispersion selection by heteromyid rodents: the role of locomotion energetics. *Ecology* 66: 220–229.

Thompson, S. D., R. E. MacMillen, E. Burke, and C. R. Taylor. 1980. The energetic cost of bipedal hopping in small mammals. *Nature* 287: 223–224.

Tilman, D. 1982. *Resource Competition and Community Structure*. Princeton, N.J.: Princeton University Press Monographs in Population Biology #17.

Trombulak, S. C., and G. J. Kenagy. 1980. Effects of seed distribution and competitors on seed harvesting efficiency in heteromyid rodents. *Oecologia* 44: 342–346.

Watts, C. H., and H. J. Aslin. 1981. *The Rodents of Australia*. Sydney, Australia: Angus and Robertson.

Webster, D. B., and M. Webster. 1971. Adaptive value of hearing and vision in kangaroo rat predator avoidance. *Brain, Behavior and Evolution* 4: 310–322.

————. 1975. Auditory systems of Heteromyidae: functional morphology and evolution of the middle ear. *Journal of Morphology* 146: 343–376.

————. 1980. Morphological adaptations of the ear in the rodent family Heteromyidae. *American Zoologist* 20: 247–254.

Wiens, J. A. 1977. On competition and variable environments. *American Scientist* 65: 590–597.

Wondolleck, J. T. 1978. Forage-area separation and overlap in heteromyid rodents. *Journal of Mammalogy* 59: 510–518.

12 Sand Dune Communities

Mary K. Seely

Sand dunes are often perceived as the archetypal desert environment, harsh and lifeless. However, the dune environment may be less severe than that afforded by saline soils or stony deserts and may offer the potential for biotic assemblages to develop. It is the relatively favorable water relations of sandy soils, under low rainfall conditions typical of deserts, that are responsible for their advantageous status.

Desert dunes, composed mainly of quartz sand grains (Bagnold 1954), are fairly similar throughout the world (Figure 12.1). Where sand is abundant, winds are strong, and rainfall is scant, mobile sands are shaped into dunes with characteristic features (McKee 1979). Vegetation may grow on the dune base and plinth (slope); the mobile crest, slip face, and avalanche base are vegetationless. Dunes may be separated by interdune areas, which may be relatively free of mobile sand. Desert dunes may cover square meters or hundreds of square kilometers, and attain heights from a few meters to a maximum of about 300 m.

This chapter is organized around communities living on mobile sand dunes. I first describe relevant physical characteristics and then use selected examples from various sand dune communities to illustrate response patterns of components of the biota to their physical environment and the processes that are thought to be involved in the responses. Only those concepts of desert communities relevant to dunes are considered. Literature dealing specifically with sand dunes is limited, however, and most information about mobile sands is contained in more general desert works. To extract those principles of community organization applicable to the dune component of the desert habitat, I relied extensively on several excellent, recent reviews of specific deserts (Walter and Box 1983; MacMahon and Wagner 1985; le Houérou 1986; Orshan 1986; Shmida, Evenari, and Noy-Meir 1986).

This chapter focuses on mobile sand dunes with at least part of their surfaces unvegetated. This excludes consideration of those sandy soils in deserts that are not shaped into mobile sand dunes. Many examples will be taken from the Namib dunes, with which I am most familiar. In the Namib, fog, a relatively predictable moisture source (Pietruszka and Seely 1985), appears to have facilitated evolution of a relatively rich and unusual desert biota (Seely 1978). Consequently, abiotic and biotic interrelationships and processes may be more obvious than they are in dunes supporting a more depauperate fauna and flora.

In general, my thesis is that the species composition of biotas inhabiting mobile sand dunes, and the abundance and distribution of dune organisms, are largely determined by species response to abiotic factors. Although biotic processes such as competition, predation, and mutualism do occur, available evidence usually in the absence of controlled experimentation suggests that they are relatively unimportant in establishing community patterns. Consequently, I will emphasize the importance of physical processes in understanding sand dune communities.

Physical Characteristics of Active Sand Dunes

Dune sand grains range from 0.01 to 2.00 mm in diameter (Bagnold 1954; Foth and Turk 1972), with mobile dune sand mostly falling between 0.15–0.27 mm (Lancaster 1981). In general, mean grain size decreases from dune base and plinth to crest and slip face (Lancaster 1981) (Figure 12.2). In the absence of organic matter and salts, properties of dune sands are determined mainly by physical characteristics (Quézel 1971; Robinson and Seely 1980).

With rain, sandy substrata offer more favorable moisture conditions for plant growth than do finer-grained soils (Noy-Meir 1973). Little or no runoff, deeper and more rapid penetration, smaller capillary forces resulting in lower evaporation and a mulching effect, and lower field capacity and wilting point coefficients allow more plant cover and productivity (Leistner 1967; Zohary 1973; Orshan 1986). Figure 12.2 indicates depth of water penetration into a typical Namib sand dune. In Central Asia, vegetationless dunes accumulate fresh water lenses at less than 100 mm annual rainfall, which may reflect storage of half of the rain in a suspended water table; dunes with vegetation accumulate little or no water (Walter and Box 1983). Quézel (1971) reported a water content of 4 percent by weight for Sahara dune sand. At 750 mm depth in Central Asian dunes, water content is 1–3 percent (Walter and Box 1983). In the Kalahari, Leistner (1967) noted moisture in loose sand at 300 mm depth in crest and slope regions. Field capacity of fine slip face sand is 6.4 percent in the Namib, while that of coarser plinth sand is only 2.5 percent (Robinson and Seely 1980). A wilting point for Central Asian dune sand was given as less than 1 percent (Walter and Box 1983), while 1.4 percent (slip face) and 1.2 percent (plinth) were measured in the Namib (W. J. Fölscher personal communication 1979).

Wind has a dominant influence on all mobile sand dunes (Seely 1984). Threshold wind velocity for sand movement is 4.5 m/second (Bagnold 1954). Differential deflation and deposition produces a mosaic of compacted and uncompacted sand across

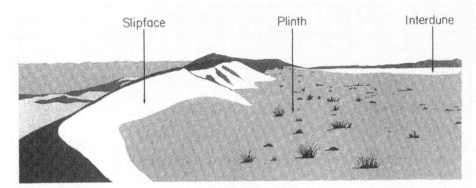

Figure 12.1. Profile of the leeward side of a typical linear dune.

a single dune (Figure 12.2). Water content of sandy soil and wind action are interrelated; sand stability and compaction directly affect water infiltration, whereas moisture content of sand influences the ability of wind to move sand. In the longer term, particularly in desert dunes with very low rainfall, wind may move vast amounts of sand (Livingstone 1986), thereby drying out surface layers. Therefore, even though the amount of available water in sandy soil is greater than that in finer-grained soils, the upper several meters of sand on a dune may be very dry.

Sand dunes provide a modified climate for organisms living near or below the surface (Figure 12.3). On the surface, temperature, humidity, and wind speed are primarily affected by orientation of inclined planes of the dune to solar radiation and wind direction (Koch 1962). Where clumps of vegetation occur, small islands of modified conditions (e.g., Heatwole and Muir 1979; Larmuth 1979) can be exploited by some organisms (Seely, de Vos, and Louw 1977). Below the sand surface, amplitude of variation of temperature and humidity is reduced (Edney, Haynes, and Gibo 1974); momentary effects of wind are reduced or absent. Circadian temperature variations disappear at 300 mm below surface and annual variations at 1,000 mm. At depths greater than about 100 mm, temperatures usually are moderate, so that a thermal refuge is available for animals able to retreat below the sand surface (Seely, Roberts, and Mitchell 1988).

Nutrients for plant growth may be secondary limiting factors in dunes. Levels of available nutrients in mobile sands are lower than those in stable sands or other desert soils (Bowers 1982; Walter and Box 1983), and well below levels thought necessary for normal growth (Bowers 1982). In general, however, sand dune environments are controlled by interactions of wind, mobile sand, and available soil moisture, and these combined effects determine most response patterns of plants and animals.

Desert Sand Dune Vegetation

Patterns of Plant Assemblages on Mobile Sand Dunes

The number of plant species found on active dunes is usually low (Table 12.1). Nature of the vegetation is generally predictable, although actual composition of the flora varies with dune field. Distribution of plant species within a dune field is primarily related to their differential tolerances for mobile sand conditions and varying soil moisture, as is demonstrated, for example, by plant assemblages across the main Namib sand sea (Yeaton 1988) (Figure 12.4). Within a desert region, the number of species in common between dune fields may be relatively high, even where the annual rainfall differs (Table 12.2).

Although relatively coarse-grained sandy soils of dunes are more hospitable to plant growth than are finer-grained soils elsewhere in deserts, species richness is not always greater on dunes. The dune flora of the North American Gran Desierto in Sonora has only 75 species, compared to 238 in nearby non-dune areas (Simmons

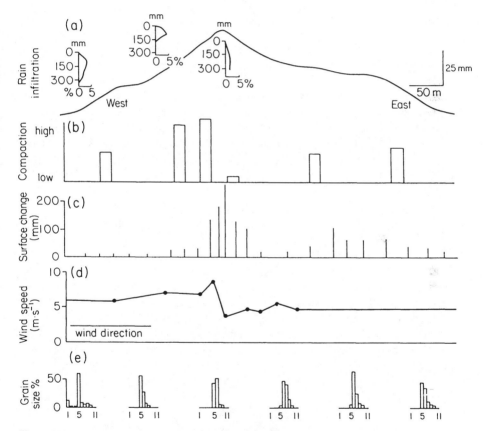

Figure 12.2. Physical characteristics of a Namib dune important to the biota (West = windward; East = leeward): *a*, depth of water infiltration (percent by mass) into sand on the windward dune base (left), upper plinth (center), and crest (right) after 20 mm rainfall; *b*, relative degree of compaction; *c*, amount of sand movement indicated by mean weekly surface change; *d*, typical variation of wind speed across a dune; *e*, grain size distribution (redrawn from Coineau et al. 1982; Louw & Seely 1982; & Livingstone 1986, 1987).

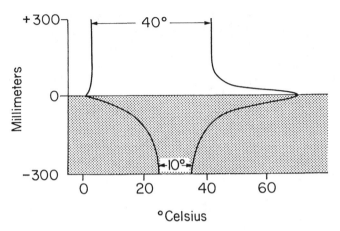

Figure 12.3. Range of temperatures above and below the sand
surface typical of the Namib dune sea.

1966; Felger 1980). Lack of habitat diversity and rigors of moving sand may limit
species richness on North American desert dunes (Bowers 1982) and are probably
important elsewhere. In the central Namib dune sea, eight plant species are confined
to sand dunes and one species to interdunes, whereas over 250 species grow on the
nearby gravel plains (Robinson 1976). In the southern Kalahari, by contrast, Leistner
(1967) found 240 plant species on sandy soils and 72 species elsewhere. These
generalizations may be confounded, however, by variation in size between areas
compared and lack of uniformity of definition of sand dunes. The question of species
richness in sands versus other substrata should be reviewed in relation to relative
harshness of conditions obtaining.

   In North American dunes, 9 percent of the flora is endemic (Bowers 1982),
compared to 56 percent in the Kara Kum sand desert in Central Asia (Walter and
Box 1983). The Kara Kum is much older and occupies a larger and more contiguous
area. Of 10 common species on mobile dunes of the central Namib (Robinson and
Seely 1980), 7 are endemic. The antiquity of the Namib (Ward, Seely, and Lancaster
1983) may provide a partial explanation for the high proportion of endemic species.
The relatively high proportion of endemic species in some dunes may reflect selection pressures by moving sand (Bowers 1982) as well as regional factors such as age,
size, or isolation of the dune field. In the Middle East, desert dune vegetation has a
close affinity to coastal dune vegetation and may be derived from littoral ancestors
(Zohary 1973). Several coastal/inland dune species pairs found in western North
America are thought to have evolved in this direction (Bowers 1982).

   On sand, number of species is partly dependent on stability of the substratum
(Table 12.3). On an individual dune, the number of species usually decreases from
more stable base to mobile crest (Figure 12.5). On extremely mobile sand near the
dune crest, a single species may dominate. In Old World deserts, this species, in
many instances, is one of the perennial grasses of the genera *Stipagrostis* or *Aristida*
in the tribe Aristideae (de Winter 1965). In the Sinai, *Stipagrostis scoparia* is the

only species that grows on very mobile dune crests (Danin 1983). In the Namib, *S. sabulicola* occupies this position (Louw and Seely 1980; Yeaton 1988), and in the Kalahari it is *S. amabilis* (but see below) (Leistner 1967). In the Kara Kum, *Aristida karelinii* occupies mobile sand together with two trees and a shrub (Walter and Box 1983). In Australia, the dune cane-grass *Zygochloa* sp. colonizes dune crests (Mabbutt 1984).

Sand stability also influences types of plants growing on dunes, but there are marked regional variations. Succulents are not found on dunes in the Kalahari (Leistner 1967) and are uncommon on active dunes in North America (Bowers 1982). In the Namib, however, the woody-stemmed leaf succulent *Trianthema hereroensis* (Aizoaceae) (Seely, de Vos, and Louw 1977) is codominant in the coastal half of the main sand sea. In southern Kalahari regions with an annual rainfall of less than 200 mm, dune crests are the only sand habitat capable of supporting trees and shrubs in appreciable numbers. In the Kara Kum, trees occur as dominants on sand dunes. They are not limited to groundwater sources and may reach heights of up to 9 m (Walter and Box 1983). In North America, trees are uncommon on active dunes (Bowers 1982) and, in the Namib, *Acacia erioloba* is found only occasionally at the base of dunes on the higher rainfall, inland margin of the main dune sea.

Occurrence of annuals or perennials on sand dunes also appears to be determined by mobility of sandy substratum. Dominant plants on dunes are usually perennials, although annual individuals may numerically exceed perennials after episodic rainfall. However, immediately after unusually high rainfall in the Namib, annuals were uncommon or absent from upper dune slopes, whereas perennial species germinated profusely (Seely and Louw 1980). In the Gran Desierto, Sonora, although 65 percent of the flora is ephemeral (Felger 1980), on 510 m of line transects on high and low dunes only eight species of perennials and no annuals were encountered. On dune slopes of the Kalahari (Leistner 1967) and of the Namib (Seely and Louw 1980), perennial grasses of the genera *Cladoraphis, Eragrostis*, and *Stipagrostis* are common dominants. Three or four different perennial species in different localities are

Table 12.1 Number of plant species reported from mobile sand dune habitats.

Desert	No. of Species	Dunes (source)
Chihuahuan	60	White Sands (Reid 1980 in Bowers 1982)
Sonoran	97	Algodones Dunes (WESTEC Services 1977)
Sonoran	75	Gran Desierto (Felger 1980)
Great Basin	53	Pink Coral Sands (Castle 1954)
Kara Kum	19	Sand desert (Walter & Box 1983)
Middle East	24	Important on mobile dunes and interdune flats (Zohary 1973)
Sahara	9–18	Northwestern Sahara: Grand Erg Occidental, Erg er Raoui, Erg Iguidi, Erg Chech (Pierre 1958)
Sahara	1	Large areas of Grand Erg Oriental (<50 mm annual rainfall) (Wagner & Graetz 1981)
Kalahari	240	Southwestern dunes (Leistner 1967)
Namib	20	Main sand sea (Robinson 1976)

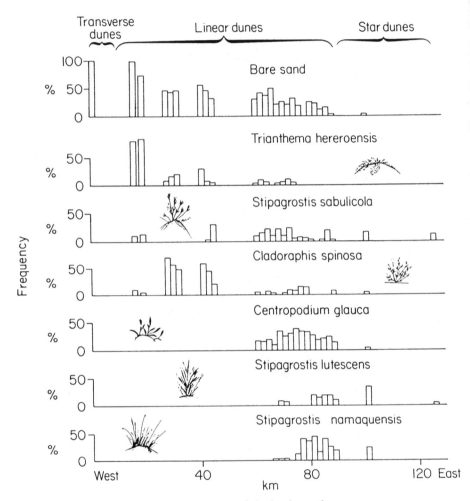

Figure 12.4. Percent frequency of occurrence of six dominant plant species in 5 by 10 m contiguous quadrats over single dunes located across the northern edge of the main Namib sand sea. Dune types change from transverse (crescentic) at the coast to linear further inland to star and multicyclic (network) along the eastern edge of the field (Barnard 1973), in response to changes in wind speed and direction (Lancaster, Lancaster, & Seely 1984). Climate varies from cool, foggy coastal conditions (mean annual rainfall < 25 mm) to warmer, more continental conditions 140 km inland (mean annual rainfall 70–100 mm); seasonal temperature variation is minimal (Lancaster, Lancaster, & Seely 1984). Bare sand refers to the percent of quadrats without any vegetation.

dominant on mobile sand dunes throughout the Middle East (Orshan 1986). In the Sahara, the perennial shrubs *Cornulaca monocantha* (Quézel 1971), *Calligonum comosum*, and *Stipagrostis pungens* characterize mobile sands, whereas vegetation of sand flats is composed of psammophilic annuals such as *Danthonia forskalii* and *Plantago ciliata* (Grenot 1974). The shift in importance from annuals to perennials as sandy substrata become more mobile is well illustrated by life-form spectra Walter and Box (1983) compiled for three degrees of sand stability in the Kara Kum (Table 12.3). Dominance of perennial over annual species in mobile sand results from ability of perennials to withstand moving sand and to use poor but reliable resources more efficiently (Shmida, Evenari, and Noy-Meir 1986), for example, low levels of stored soil moisture or fog.

On mobile dunes, perennial plants cover only a small fraction of surface sand yet contribute a major portion of biomass. In the Namib, two perennial species covered

Table 12.2 Species richness and community coefficients of plants growing in different sandy areas in the Sahara.

Area (source)	No. of Species	CC[a]
Northern Sahara (le Houérou 1986)		
Arid zone (100–400 mm; drifting sands)	18	
		40
Desert zone (25–100 mm; sand dunes and veils)	12	
Northwestern Sahara (Pierre 1958)		
Humid erg	18	
		67
Semihumid erg	12	30
Dry erg		38
	9	

[a]CC = community coefficient = (2 × no. of species in common/sum of species in each habitat) × 100.

Table 12.3 Percentage of species on sand of three degrees of mobility in the Kara Kum.

Species Class	Sand Hollows	Immobile Sand	Mobile Sand
Trees and shrubs	11.9	10.2	26.3
Semishrubs and dwarf shrubs	1.0	2.3	10.5
Perennial herbs	8.9	14.8	21.1
Annual herbs	70.3	67.1	31.6
Fungi (large)	6.9	4.5	10.5
Moss	1.0	1.1	
Total no. species	101	88	19

SOURCE: Walter & Box (1983).

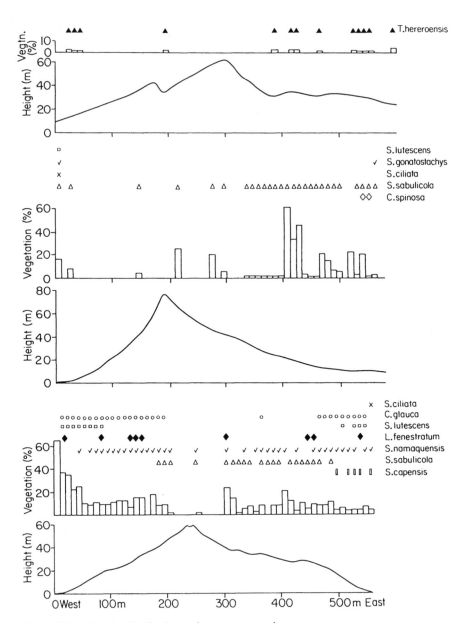

Figure 12.5. Species distribution and percent vegetation cover over three linear dunes situated *a*, 16 km, *b*, 56 km, and *c*, 88 km inland from the coast in the main Namib sand sea (*Centropodium glauca, Cladoraphis spinosa, Limeum fenestratum, Sesamum capensis, Stipagrostis ciliata, S. gonatostachys, S. lutescens, S.* cf. *namaguensis, S. sabulicola, Trianthema hereroensis*).

less than 1 percent of plinth area, yet constituted 59 percent of plant and associated detritus biomass in a dry period (Seely and Louw 1980).

Generally, grasses, particularly C_4 grasses (Smith and Nobel 1986), are more abundant under summer than under winter rainfall regimes in deserts (Louw and Seely 1982; Vogel, Fuls, and Danin 1986). This pattern appears to hold true on mobile dunes. Although Leistner (1967) found perennial grasses on undisturbed dunes in the Kalahari, woody shrubs and annual species grew on similar dunes subjected to heavy grazing by domestic stock. In the Sahara, a few shrubs and only one grass species remain on degraded sandy substrata or are active in stabilizing mobile sand dunes (le Houérou 1986). A long history of grazing by domestic animals may partially explain greater prevalence of dicotyledonous plants instead of grasses in Sahara dunes when compared with undisturbed sand dunes of southern Africa.

In addition to living vegetation, plant litter plays an important role in mobile dunes and its biomass may far exceed aboveground living biomass (Hadley and Szarek 1981). Discarded organic matter, including leaves, seeds, and flower parts, is distributed by the wind and accumulates as detritus on and at the base of slip faces and around plants (Brinck 1956; Robinson and Seely 1980; Seely and Louw 1980). Within the Kara Kum dunes, a constant change of currently active plant groups leads to production of organic matter throughout the year (Walter and Box 1983). In the Namib, perennial dune grasses flower and produce seed seasonally (e.g., Louw and Seely 1980), whereas a succulent reproduces throughout the year (Seely, de Vos, and Louw 1977), ensuring continuous renewal of detritus. Thus, detritus, although very patchy in distribution (Figure 12.6), represents a largely aseasonal, relatively persistent (Figure 12.7) form of organic matter in mobile dunes.

In summary, plant species richness on sandy substrata is usually relatively low, and decreases with increasing mobility of sand. Within a desert area, mobile dunes support fewer species than sand sheets or hollows, and crests of individual dunes support fewer species than dune base or plinth. Perennial plants, in particular grasses, comprise an important portion of plant assemblages of mobile sand dunes. Detritus constitutes a large proportion of the organic biomass. In subsequent sections, I discuss the importance of abiotic interactions and the relative lack of importance of biotic processes in determining observed vegetation patterns.

Autecological Responses of Plants to Mobile Sand Dunes

Similarity of vegetation patterns on mobile sand dunes in different regions suggests that similar autecological responses have evolved in plants from different origins. Instability of sandy substratum is the main dune characteristic to which the majority of convergent plant adaptations are oriented.

Instability of sand surface affects growth plasticity of some plant species. For example, in the Namib superficial portions of *Trianthema hereroensis* (Aizoaceae) and *Acanthosicyos horrida* (Cucurbitaceae) continue growing throughout the year while sand accumulates around and covers the plant base (Seely, de Vos, and Louw 1977; personal observation). Many grass species of the genus *Stipagrostis* are stimulated to grow by sand smothering them. Because of the short life span of its roots,

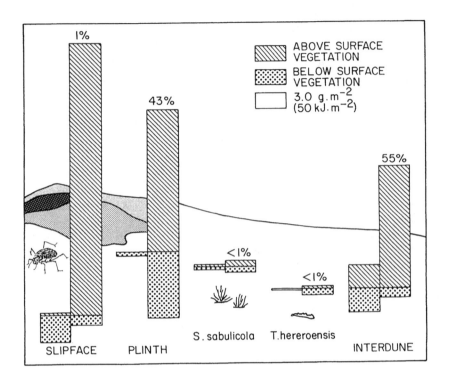

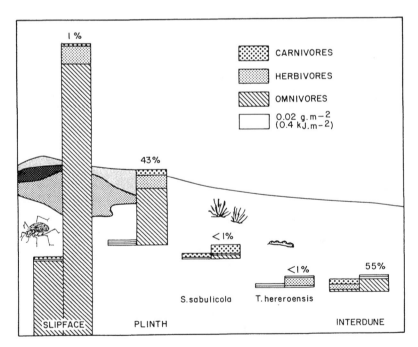

S. scoparia only grows where it is continuously covered by new sand, and dies if sand cover is removed by wind or new sand is not deposited (Danin 1983). When roots of young individuals of *S. sabulicola* are excavated by wind to depths of about 300 mm, aerial parts fall onto the sand, often inducing sand accumulation, initiating adventitious roots, and reestablishing active growth. Most dune plants in North America cope with burial more successfully than they do with excavation (Bowers 1982).

Roots of plants living in mobile sand usually grow in one of two patterns. Shrubs and other dicotyledons often have long tap roots. Some long roots may exploit deep water reserves only after upper sand layers have dried (Orshan 1986). The second pattern is exemplified by perennial grasses, which often have profuse growth of lateral and adventitious roots, not penetrating the dune further than a few meters, but, nevertheless, anchoring the plant. In the case of *Aristida karelinii* in the Kara Kum, 40 to 350 adventitious roots grow radially up to 10 to 15 m in length. Most roots lie between 150 and 600 mm below the surface, but may penetrate to a depth of 2,200 mm (Walter and Box 1983). Similar dense growths of adventitious roots occur in Namib dune perennial grasses. Adventitious roots also provide a method of vegetative reproduction, an advantage in deserts where low rainfall and high sand mobility hinder germination. In addition to their profuse growth of subsurface adventitious roots, *Stipagrostis sabulicola*, *S.* cf. *namaguensis*, *S. lutescens*, and *Cladoraphis spinosa* have an extensive system of lateral surface roots, in some cases reaching out more than 20 m. For *S. sabulicola*, and possibly some other *Stipagrostis* species in the Namib, these roots are involved in uptake of fog water (Louw and Seely 1980). Extensive lateral root development is uncommon in North American dune species; when it occurs, it is more important for plant stabilization than for taking up water (Bowers 1982).

Sand instability may cause unusual growth patterns in dune plants. *Stipagrostis pungens*, growing in mobile sand at its southern limits on the edge of the Sahel, forms a collapsing tussock with wiry stems that creep along the ground or even climb trees (Monod 1986). Bowers (1982) mentions several examples of unusual gigantism, elongation, or configuration of plants growing on mobile sands in North America.

The sand dune habitat also appears to favor "multi-season" growth in facultative annual/perennial species with indeterminate growth patterns. In the Namib, for example, *Stipagrostis gonatostachys* grows in interdune valleys as an annual species, but on the lower plinth it is a multiseason species that may live 4 or 5 years. Greater water availability at the base of dunes relative to interdunes may cause this difference in growth pattern. Growth patterns of *Stipagrostis ciliata* also vary from annual to perennial with varying depth of sand. Flowering phenology of some species is

Figure 12.6. Patchy distribution of vegetation (plants and detritus) and animals on the three main habitats (and associated with the dominant perennial plant species) of a linear dune during a dry (left) and wet (right) period in the Namib dunes. Figures above bars indicate percent of the dune environment occupied by that habitat (redrawn from Seely & Louw 1980).

affected by sand mobility: some plants grow and flower throughout the year in mobile sands, whereas they flower only in spring in semistable sands (Danin 1983).

Not only growth, but also germination and establishment are affected by mobility and moisture content of sand. In the Namib, germination can occur on all areas of a dune except the slip face, where water retention is lowest (Figure 12.2) and seeds are deeply buried. In the Kalahari, dune crests may be colonized by species that produce abundant, heavy seeds that germinate rapidly. Several prostrate annuals, and short-lived perennials with aerial parts offering little wind resistance and with underground organs that can survive partial exposure, fall into this category (Leistner 1967). In the Negev, *Stipagrostis scoparia* germinates abundantly only after several consecutive days of rain (Danin 1983). In the Namib, abundant germination of perennial dune grasses and *Trianthema hereroensis* has been recorded only following unusually high rainfall (>100 mm in one season) (Seely and Louw 1980), although some germination may occur between 15 and 20 mm rainfall, most frequently in lower dune hollows. Although germination may occur over much of the sand dune surface, establishment of reproductive plants is less widespread, for example, *S. sabulicola* and *T. hereroensis* in the Namib (Figure 12.8). Differential establishment and growth of these two species are thought to relate to soil moisture patterns, which in turn are related to differences in sand compaction, and to distinct methods of uptake of fog water used by the two species (Seely, de Vos, and Louw 1977; Louw and Seely 1980).

In summary, response of dune plants to mobile sand dunes consists of plasticity of growth patterns, above and below sand, and of plasticity of seasonality and duration of reproduction. Germination and establishment are also controlled by wind-induced variations in sand stability.

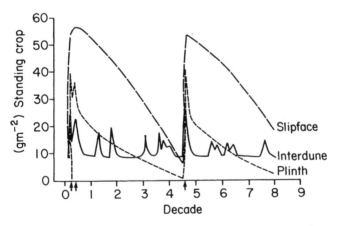

Figure 12.7. Variable persistence of vegetation and detritus in three main habitats of the Namib dunes. The hypothetical curves were drawn from measurements of standing biomass before and after rains and from the long-term precipitation record. Arrows indicate major rainfall events (redrawn from Louw & Seely 1982).

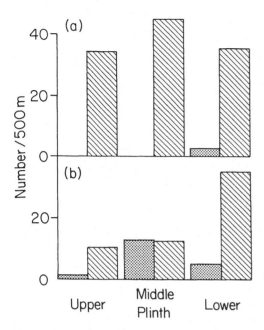

Figure 12.8. Mean number of individuals of *a*, *Trianthema hereroensis*, and *b*, *Stipagrostis sabulicola* established and growing during a dry period (left bars) and recently germinated after > 100 mm rain (right bars) on the upper, middle, and lower plinth of a linear dune in the Namib (mean number/500 m over 8 km of dune).

Desert Sand Dune Animals

Patterns of Animal Assemblages on Mobile Sand Dunes

Larger (e.g., ungulates) and more mobile (e.g., birds) desert animals often are not restricted to particular substrata (Seely and Griffin 1986), but many surface-active arthropods (Crawford and Seely 1987; Polis and McCormick 1986; Polis and Yamashita this volume) and reptiles (Grenot 1976) living in sand dunes are unable to live on more compact substrata. Pierre (1958) recorded 26 orders of insects in the northwest Sahara; representatives of 12 orders occur only in sand dunes. Mobile dune faunas are limited to a few taxa, within which a number of species may be present (Koch 1962) (Table 12.4). Dune-adapted species may occur in a variety of sandy habitats and, in the northwest of Sahara, various sand accumulations have a number of species in common (Pierre 1958) (Table 12.5). More similar substrata have a greater number of species in common than do less similar substrata (Table 12.6). Of 41 tenebrionids recorded on mobile dunes of the central Namib, 60 percent of species and 10 percent of genera are endemic to the central Namib dune area (Holm and Scholtz 1980). Tenebrionid species involved all belong to ancestrally wingless lineages (Koch 1962), a condition which probably contributed to the high level of endemicity.

In the Namib, 1 of 5 mammal species, 3 of 10 reptile species, and 15 of 48 invertebrate species occur only on dunes and avoid nearby interdunes (Robinson and Seely 1980). One reptile and four invertebrate species live only on interdunes. Although some species are habitat specific, there is a high overall community coefficient (CC; CC = [2 × number of species in common/sum of species in each habitat] × 100) of similarity between dune and interdune faunas (CC = 51). Pit trapping of surface-active arthropods for a short period gave a similar CC of 50 (Crawford and Seely 1987). Species richness on more stable interdunes is lower than on mobile sands. In a dry year, animals in interdunes contribute less biomass (g/m^2) than those of slip faces, but more than those of dune slopes. In a wet period, biomass of the interdune fauna was lower than that of the other two habitats (Seely and Louw 1980).

In the Namib, species richness of surface-active arthropods varies across the main sand sea (Figure 12.9). Number of species appears to be influenced by dune height, ratio of vegetated to unvegetated surface, quantity of detritus on slip faces,

Table 12.4 Number of species of insects and arachnids living in the dunes of the Sahara and the Namib.

	No. of Species	
	Sahara[a]	Namib[b]
Insects		
Coleoptera	123	50
Hymenoptera	35	9
Neuroptera	21	2
Diptera	8	5
Heteroptera	8	2
Mantodea	4	1
Homoptera	3	6[c]
Thysanura	5	9[d]
Lepidoptera	2	. . .
Orthoptera	2	7
Isoptera	1	2
Collembola	. . .	e
Total	212	93
Percent tenebrionids	27	34
Arachnids		
Scorpions	. . .	3
Spiders	. . .	14
Solpugids	. . .	8
Mites	. . .	10[f]

[a]SOURCE: Pierre (1958).
[b]SOURCE: Holm & Scholtz (1980).
[c]Curtis (1985).
[d]Watson & Irish (1988).
[e]Present but number of species not known.
[f]Coineau & Seely (1983).

Table 12.5 Comparison of insect community coefficients between sandy habitats in the Sahara.

Habitat Number and Habitat	Habitat Number									
	2	3	4	5	6	7	8	9	10	11
Humid dunes										
1. Gr Erg Occidental	73	51	34	12	42	14	13	16	17	50
2. Erg er Raoui		52	30	14	44	17	15	42	20	46
Semi-humid dunes										
3. Erg Iguidi (North)			71	32	41	10	8	39	10	44
4. Erg Iguidi (South)				50	36	7	6	35	7	31
Dry dunes										
5. Erg Chech					18	12	8	24	11	13
6. Groups of isolated dunes						27	23	23	42	53
Sand on rocks										
7. Plains							59	9	48	20
8. Hills								8	50	17
9. Sand sheets									17	34
Nebkas										
10. On plains										35
11. In watercourses										

SOURCE: Tables XVIII, XIX, XX in Pierre (1958).
NOTE: Community coefficient = (2 × no. of species in common/sum of species in each habitat) × 100.

Table 12.6 Comparison of community coefficients of total number of tenebrionid beetles and ants on dunes and sandy extradune substrates (Sahara) and on dunes and less sandy extradune substrates (Namib).

	All Sandy Accumulations (Sahara)		Sandy and Less Sandy Habitats (Namib)	
	N	CC[a]	N	CC[a]
Tenebrionids				
Dunes/watercourses	31/18	45	33/31	19
Dunes/plains	31/19	40	33/10	5
Watercourses/plains	18/19	54	31/10	20
Ants				
Dunes/plains	. . .	. . .	8/33	24

SOURCE: Table XVIII in Pierre (1958); Holm & Scholtz (1980); Wharton & Seely (1982); Marsh (1986).
[a]CC = community coefficient = (2 × no. of species in common/sum of species in each habitat) × 100.

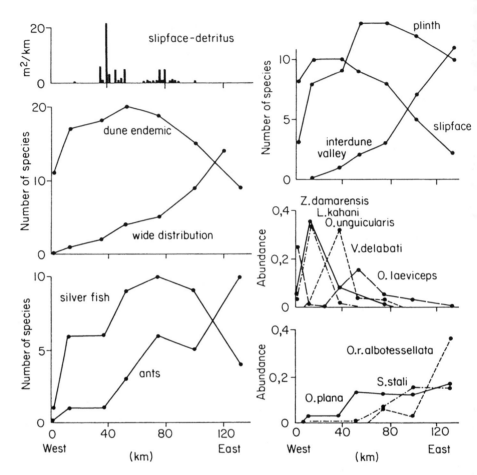

Figure 12.9. Occurrence across the central Namib dune gradient of *a*, detritus (m²/km slip face); *b*, widespread and dune endemic tenebrionid beetles; *c*, thysanurans and ants; *d*, tenebrionid species inhabiting slip face, plinth, and interdune habitats; *e*, individuals of tenebrionid species living on the dune slip face; *f*, individuals of tenebrionid species living on the dune plinth. Abundance = proportion of total number of individual tenebrionids captured at each site (*Lepidochora kahani, Onymacris laeviceps, O. plana, O. rugatipennis albotessellata, O. unguicularis, Stips stali, Vernayella delabati, Zophosis damarensis*).

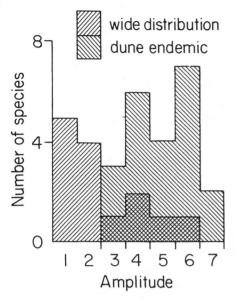

Figure 12.10. Amplitude of distribution (number of sites across central Namib gradient from which species were captured) of endemic and widespread tenebrionid beetles.

and amount of fog, in other words, autecological adaptations to particular conditions as they vary from coast to inland. Tenebrionids restricted to dunes have a wider geographical distribution within the sand sea than those also occurring on nondune substrata (Figure 12.10).

On a single mobile dune, degree of stability and compaction of sand affects distribution of species, for example, tenebrionid beetles (Figure 12.11), thysanura (Figure 12.12), and spiders (Figure 12.13). Some species occur over an entire dune whereas others do not. Dune vegetation may support a permanently associated (satellite) fauna, or attract a wide-ranging (errant) fauna (Pierre 1958; Koch 1961), which varies with location on a dune.

Carnivorous species are often abundant in deserts (Cloudsley-Thompson 1974). Holm and Scholtz (1980) found the invertebrate fauna of a Namib dune to be composed almost equally of carnivores (35 percent) and detritivores or omnivores (36 percent), with slightly fewer herbivorous species (28 percent). On mobile sand dunes of the Sahara, herbivores are the most important component of the invertebrate fauna (Pierre 1958). Although Crawford and Seely (1987) trapped almost equal numbers of detritivore and carnivore species on sandy Namib substrata, detritivore abundance was more than four times greater than that of carnivores; herbivores were absent. Detritivore species were more numerous in the main dune system, where one-third of species were tenebrionids. In the northern Namib, proportions of carnivorous and detritivorous species differed widely between adjacent habitats, suggesting that detritus processing and carnivory may proceed at habitat-specific

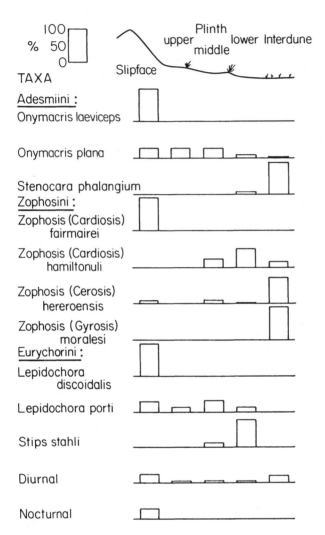

Figure 12.11. Distribution of 10 species of tenebrionid beetles in three tribes on the slip face, plinth, and interdune of a linear dune in the central Namib. Bars indicate percent of total catch for the species and percent of all beetles ($N = 1,917$) captured diurnally or nocturnally (lower two rows).

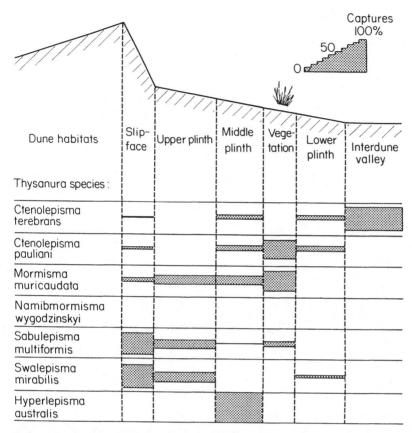

Figure 12.12. Distribution of seven species of Lepismatidae on a linear dune in the central Namib as a percentage weighted capture (abundance/activity index). (*Namibmormisma muricaudata* includes *N. setosa* data) (redrawn from Watson & Irish 1988).

rates. In more mesic dune habitats, where rain had fallen, the proportions of carnivore species and individuals were greater than in drier habitats. In contrast, using excavation and censusing techniques, Seely and Louw (1980) found the ratio of carnivores to detritivores and omnivores decreased from 1:1.23 to 1:7.44 from a dry to a wet period. Detritivorous species were dominant on least stable sands.

Omnivorous feeding habits are adopted by many desert species and dune species are no exception. Dune carnivores are mainly trophic generalists eating a variety of prey taxa (e.g., Seely and Louw 1980; Polis, Sissom, and McCormick 1981). Many detritivores are omnivores; for example, tenebrionid beetles will consume flesh, fruits, or any water-rich food. The endemic Namib dune lark, *Mirafra erythrochlamys*, spends two-thirds of its time consuming seed and one-third consuming insects associated with dune vegetation (Cox 1983). Gerbils on the Namib dune also consume many insects and spiders in addition to seeds and other plant material

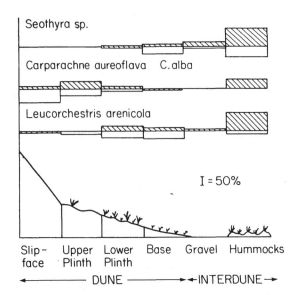

Figure 12.13. Distribution of nests of three genera of heteropodid spiders on Namib sand dunes. Solid bars above the line are from a single, well-vegetated central dune site; open bars below the line are a composite from three sites across the main sand sea (redrawn from J. R. Henschel unpublished data).

(Boyer 1987; J. R. Henschel personal communication 1988). Rapid burying of seeds by mobile sand appears to limit the potential for granivory on active dunes.

Nocturnal activity is a common adaptation of desert animals for escaping hostile environmental conditions and perhaps predation. In North America the majority of desert tenebrionids are considered nocturnal (Thomas 1983). On mobile sand dunes in the Namib, however, Koch (1962) reported that more sand-restricted tenebrionid species were diurnal than were nocturnal, and Robinson and Seely (1980) noted that 5 of 8 reptile species and 23 of 41 invertebrate species (12 of 23 tenebrionid species) were mainly diurnal. Watson and Irish (1988) found 6 of 7 dune thysanuran species active day and night and the seventh active only nocturnally. Crawford and Seely (1987) captured almost equal numbers of nocturnal and diurnal surface-active arthropod species in sandy Namib habitats. Koch (1961) noted that diurnal dune tenebrionids tended to be somewhat larger than extradune species. The large number of conspicuous diurnal tenebrionids suggests that predation on surface-active adult beetles is not an important factor determining the abundance and other aspects of their ecology (Seely 1985).

Many tenebrionids are long lived; for example, *Onymacris plana* were retrapped in Namib dunes after 500 days and *Pimelia angulata* from the Sahara were maintained in captivity for 497 to 918 days (Niethammer 1971). In the presence of an assured, but patchy and not very nutritious food source such as detritus, and apparent high mortality of immature forms (personal observation), selection appears to favor long-lived adult forms. These characteristics fit in with a suite of life history traits summarized as bet hedging (Stearns 1977), characteristics that are found in a number of desert organisms.

In summary, on mobile sand dunes the fauna usually constitutes a small fraction of the standing biomass. In the Namib dunes, the fauna represented slightly less than 0.2 percent of the total biomass during a dry and a wet year (Seely and Louw 1980). However, when compared with some other systems, for example, the 0.02 percent represented by the fauna of a tropical forest (Odum et al. 1970), the value for dunes may be viewed in perspective. The number of groups able to live in sand dunes is limited by the physical problems associated with sand, but, in some instances at least, speciation has been extensive. Sand compaction affects the distribution of animals in mobile sand dunes, but in a manner very different to that of plants. For detritivores consuming windblown detritus, the unstable slip face is the preferred habitat. In some groups the biomass per unit area is greater on the slip face than in any other dune habitat, resulting in a patchy distribution. Carnivorous species are numerous on mobile dunes and species with omnivorous feeding habits and diurnal activity patterns are common.

Autecological Responses of Animals to Mobile Sand Dunes

Animals in deserts must contend with various extreme environmental factors, for example, high temperatures and low and unpredictable rainfall and productivity (Louw and Seely 1982). As a result, only a few animal groups live exclusively in deserts, and many of these animals live only in restricted habitats or sets of conditions. In parallel with plants in arid conditions, sand dunes constitute a more hospitable habitat than some finer-grained soils for those animals able to take advantage of sand's special conditions. Small species that retreat daily from desert extremes to a refuge within the sand are particularly well suited to dune life.

Wind, through its effect on compaction and moisture content of sand, is the single most important physical influence on animals living in the dune environment. The degree of compaction and stability, and perhaps grain size (Coineau et al. 1982), dictate whether animals can "swim" or burrow through the sand (Table 12.7) and the rate at which they can move. The slip face is the predominant habitat where animals live that swim through very mobile sand (e.g., Coineau et al. 1982; Mitchell et al. 1987). Loose sand accumulated around the base of some plants may also present suitable conditions for swimming. Species that depend on burrow construction to penetrate the sand are, predictably, found more often on the lower plinth or base of a dune or beneath plants surrounded by more compacted sand. Major exceptions are the spiders, predominantly heteropodids in the Namib dunes (Lawrence 1965; J. R. Henschel personal communication 1988), that construct silk-lined tubes with trap doors on any part of a dune (Figure 12.13). In the Namib, many invertebrates and a few vertebrates, for example, a hare, a sand snake, and a chameleon, do not escape beneath the surface at all but take refuge from desert extremes on or beneath plants. These three vertebrates live in, but are not restricted to, sand dunes.

Most animals occupy the upper meter of sand; hence any moisture stored deeper

Table 12.7 Number of burrowing and
swimming animals in the dunes of the central
Namib.

Animals	Swimming	Burrowing
Mammals	1	2
Reptiles	4	3
Spiders	1	2
Scorpions	0	2
Solpugids	0	3
Tenebrionids	18	5

Source: Robinson & Seely (1980).

in the sand is not directly of benefit to either swimming or burrowing animals. When the upper meter is dry, vapor pressure differences between an animal and its surrounding sand are less favorable beneath the sand than they are above the surface for most of the diel cycle (Seely and Mitchell 1987). Continued use of a semipermanent burrow may, however, lead to an increase of moisture in surrounding sand. Moreover, on the lower plinth and dune base, where most burrowing takes place, the sand in the upper meter is more stable and may be more moist (e.g., Edney, Haynes, and Gibo 1974; Robinson and Seely 1980).

In the Namib, soil moisture is thought to control tenebrionid beetle populations through its effect on larval development. Although many species of dune tenebrionids reproduce throughout the year (Seely 1983), mortality of the immature forms appears to be very high. In years of exceptional rainfall, the long-lived adults may expand their distribution ranges but do not appear to increase feeding or reproduction rates. Nevertheless, adult populations increase by several orders of magnitude within the year (e.g., Wharton and Seely 1982), apparently the result of reduced mortality of larvae under moist soil conditions. Thomas (1979), studying non-dune tenebrionid beetles in North America, came to a similar conclusion concerning the importance of soil moisture. These observations also agree with those of Ayyad and Ghabbour (1986) who comment that precipitation is more important to invertebrate populations than is food. Although the exact mechanism by which water affects development is unknown, tenebrionid larvae take up moisture from unsaturated atmospheres (Coutchié and Crowe 1979) and perish in sand of low moisture content (personal observation). Soil moisture may have direct effects on populations of other taxa (e.g., thysanura) which also take up moisture from unsaturated amospheres (Edney 1971), although this has not been documented.

During the day, in sparsely vegetated or vegetationless habitats, animals on the surface are subjected to the full effects of solar radiation, and lethal temperatures can occur (Edney, Haynes, and Gibo 1974; Robinson and Seely 1980). Despite this, in the Namib at least half the species of reptiles and arthropods on mobile sand dunes forage diurnally (Robinson and Seely 1980; Crawford and Seely 1987). These species show temperature-dependent, unimodal, or bimodal activity patterns that change seasonally (Holm and Edney 1973; Holm and Scholtz 1980; Robinson and

Seely 1980) or from day to day (Seely et al. 1988). This temperature-dependent behavior is not unique to mobile sand dunes, nor are various morphological, behavioral, or physiological adaptations to high temperatures. However, the opportunity to escape at will to a better thermal niche below the sand is unique to dunes (Louw and Seely 1982). Some tenebrionids in the Sahara respond to near-lethal temperatures by digging into the sand (Ayyad and Ghabbour 1986), as do representatives of the same tribes in the Namib. The permanent availability and proximity of this refuge is thought to have led to the unusually high body temperatures (Hamilton 1973) selected by several species of tenebrionids when active on mobile Namib sand dunes (Seely, Roberts, and Mitchell 1988).

Most animal species restricted to mobile sand dunes, and all those not intimately associated with dune vegetation, spend more than half of their time inactive beneath the sand surface, and some individuals may emerge for less than an hour per week (e.g., scorpions; Bradley 1988 and Polis 1980; *Angolosaurus skoogi*; Mitchell et al. 1987). Moreover, timing and duration of surface activity can be highly variable.

Within the constraints imposed by potentially lethal diurnal temperatures, wind is a major determinant of the diurnal surface activity of many Namib dune species, through its effect on the distribution and availability of detritus (Seely 1984). Detritus is an important source of food for Namib tenebrionid beetles (Seely 1983), several lizards (Robinson and Cunningham 1978; Mitchell et al. 1987), thysanurans (Watson and Irish 1988), and termites, as it is for soil fauna of the Kara Kum (Walter and Box 1983). Many Namib dune tenebrionid beetles forage on detritus on the slip face only when winds are strong (Seely 1983), perhaps to take advantage of newly imported quality components. Herbivorous insects are not subject to the same constraints. Strong wind inhibits the activity of other Namib insects, such as the dune ant, *Camponotus detritus*, which feeds mainly on honeydew from scale insects associated with perennial vegetation (Curtis 1985), and two apterous dune dung beetles.

Wind has a direct effect on the water relations of all desert animals in that, at low water vapor pressure, increased wind speed causes increased desiccation stress. In addition, in the Namib, wind provides water for adult tenebrionids and several reptile species as it blows fog water into the dune environment at night or in the early morning. Many swimming, but no burrowing, species take direct advantage of this relatively reliable source of free water (Seely 1979). Many of the species forage diurnally but emerge to use fog moisture at night. Although Koch (1961) suggested that fog-dampened detritus (Tschinkel 1973) could be a source of water for Namib detritivores, none has been observed foraging when the detritus is damp. Foraging on dew-dampened detritus has been observed in the Negev (Broza 1979). Although windblown detritus is a ubiquitous food source in most deserts, the evolution of an extensive detritus-based assemblage of surface-active arthropods on mobile sand dunes appears to have taken place only in the Namib. Fog, providing a reliable source of moisture to the detritivores, may be the unique factor.

In summary, wind is the most important factor controlling the response of dune animals to their habitats, indirectly through its effect on substratum compaction and directly through its effect on surface foraging and fog uptake activity. Dune animals can swim through uncompacted sand and burrow in more stable sandy substrata.

Some dune-dwelling forms spend all their time beneath the sand surface, for example, termites and insect larvae, whereas others spend the majority of their time apparently inactive there, surfacing for relatively brief periods of foraging, drinking, or mating.

Biotic Interactions on Sand Dunes

Plant-Plant Interactions

Competition for water and interactions during succession as a dune surface becomes stabilized appear to be the two main possible manners in which plants influence each other on sand dunes, as space is not limiting. The former does not appear to be important in North America, where most dune species are deeply rooted and abundant water is stored at depth (Bowers 1982). Similarly, Walter and Box (1983) calculated that water was not limiting to plants on mobile sands in the Kara Kum. In the southern Kalahari, however, Leistner (1967) suggested that competition for water may occur on the more stable, lower parts of the dunes.

Abiotic factors, the interaction of wind and sand, are probably the most important influences on plant succession on sand dunes. Succession is a continuous process and may be reinitiated when long dry periods lead to death of established plants, when strong winds excavate plants, or when the sand surface is reactivated after grazing and trampling by stock or game. Plant succession usually proceeds by one species occupying mobile sand and at least partially stabilizing it. For example, in the Negev, after *Stipagrostis scoparia* stabilizes sand, and dies because of the short life span of the roots, it is replaced by *Artemisia monosperma* (Danin 1983). This latter species and *Convolvulus lanatus* are semishrubs possessing tap roots with corky bark which protects them from desiccation. Similarly, in the roots of *Cornulaca monacantha* xylem and phloem are distributed throughout the root, which can survive even though several meters of root are exposed (Danin 1983). After establishment these semishrubs can tolerate some continuing sand movement. In the Kara Kum, after *Aristida karelinii* stabilizes the sand, it is replaced by *Heliotropium arguzioides* and *Tournefortia sogdiana* (Boraginaceae) (Walter and Box 1983). Whereas clear sequences of species that pioneer, stabilize, and eventually may cover dunes occur in many Old World deserts, this sequence is often not apparent in the less extensive sandy regions of the New World (MacMahon and Wagner 1985). These distinctions may disappear with further study. Moreover, in both North American (Bowers 1982) and Namib dunes, alien weeds have not succeeded in entering the natural succession in mobile sand.

Two alternative explanations for apparent succession in mobile sands may apply; these differences may relate to differences in total rainfall between areas. Either the nurse plant syndrome may give the appearance of succession, or static zonation in response to sand compaction and available moisture may appear to be succession (e.g., Figure 12.8). For succession to occur, new plants must germinate or existing plants must reproduce vegetatively to replace pioneer species. In extreme desert conditions (<50 mm rainfall), germination is rare (e.g., Seely and Louw 1980), precluding initiation of succession in this way. In the less extreme rainfall regime of the Kalahari, however, Leistner (1967) mentions that germination may occur preferentially in areas protected by plant growth, establishing conditions for succession. In all examples of implied succession in dune sands, stabilization of the sand by one species which is then colonized by a second species is suggested in accord with the facilitation model of Connell and Slayter (1977).

Plant-Animal Interactions

Living Plants Most plant-animal interactions on sand dunes occur in areas of habitat substantially modified by plants. In sparsely vegetated dunes, individual plants represent islands where sand is often cooler, moister, more compact, or richer in nutrients compared to surrounding bare surface. For example, in the coastal portion of the Namib, fog water that precipitates on sparse vegetation falls to the sand. Shading and partial sand stabilization by the plant reduce water loss so that sand is moister beneath plants than in surrounding barren sand. A second effect of fog-water drip is to wash windblown salt off the plant onto the sand below. Salt solidifies sand under some plants, providing a substratum of which burrowing animals take advantage to prevent burrow collapse. The vegetative structure of plants modifies the thermal environment available to small animals (e.g., Larmuth 1979) above and below the sand surface, providing an additional thermal niche, which allows animals not restricted to dunes to occupy the habitat, at least temporarily.

In the relatively fragile environment of mobile sand dunes, animals may have a major effect on the plant's environment. Particularly noticeable is the process of destabilization of sand, including destruction of the biological crust (Danin 1983) from trampling and overgrazing by domestic stock (Walter and Box 1983; Monod 1986) or game. If the impact is severe, abiotic conditions may prevent reestablishment of the vegetation destroyed.

Animals may also induce modifications of the plant environment for other animals. Gall formation, stimulated by gelechiid moths, occurs commonly on two perennial *Stipagrostis* species in the central Namib dunes (R. L. Pretorius personal communication 1983). Empty galls provide a modified environment not available for insects elsewhere in mobile dune sands. The overall effect of galls on these plants is not known. Scale insects are commonly found on three species of perennial plants in the central Namib; the ant *Camponotus detritus*, restricted to the dune sands, derives a major part of its nutrition from these insects (Curtis 1985). Extent of impact on plants by these scale insects is unknown. These and many similar interactions are not necessarily restricted to mobile sand dunes. However, because many interactions are species specific and because there are so few plant species in the mobile sand dune environment, impact of a particular interaction upon the plant community may be greater than in more heterogeneous environments.

Except in the presence of domestic stock, herbivory does not usually account for consumption of a large proportion of available aboveground plant biomass (Heal and Maclean 1975). On the other hand, many animals found on mobile sand dunes would not be there were it not for plants. On Sahara dunes (Pierre 1958), two plant species alone maintain insects in a dry dune field. On semihumid dunes, three plants are hosts for insects and, on humid dunes, nine plant species play an important ecological role. Although number of plant species is limited, continual growth and long-term reproduction of many dune perennials assures a food source for herbivores for most of the year. Importance of belowground herbivory (e.g., Andersen 1987) in mobile dunes is unknown.

Although plants are sparsely distributed on mobile sand dunes, they serve as a focus for herbivores as well as many detritivores and carnivores (e.g., Seely, de Vos, and Louw 1977). Many predators, such as therevid fly larvae, neuropteran larvae, and golden moles, concentrate their foraging activity under dune plants (personal observation). We have no evidence, however, for density-dependent control of one species of plant or animal by another. Indeed, there are no data to suggest that biotic interactions of any type affect the community structure of plants in mobile dune habitats.

Organic detritus   Plant litter and discarded animal products, including carcasses, constitute organic detritus, deposited by wind action around plants and on slip faces. Further wind and sand action buries accumulated detritus and reversing winds subsequently uncover and redistribute this material. Continually produced, concentrated, and redistributed, detritus is almost continually present as a food source on mobile sand dunes.

Microbial and fungal decomposition of detritus is retarded by low moisture levels. As a consequence, macrodetritivores appear to play an important role in nutrient cycling in deserts (Crawford and Taylor 1984), especially relative to other, more mesic systems. However, dominance of macrodetritivores in the surface-active arthropod assemblage may be a characteristic peculiar to mobile sand dunes (Seely and Louw 1980; Crawford and Seely 1987).

Vast amounts of detritus accumulate on a dune slip face, particularly at the avalanche base (Seely and Louw 1980). In the Namib, a number of species of tenebrionid beetles (Koch 1961), thysanurans (Watson and Irish 1988), and lizards (Robinson and Cummingham 1978; Pietruszka et al. 1986) consume detritus on the surface of slip faces; the termite *Psammotermes allocerus* eats detritus after it has been buried by shifting sand (personal observation). Each of these groups is selective to some degree. The termites appear to consume most of the cellulose fraction, leaving seeds and insect chitin. One lacertid lizard, *Aporosaura anchietae*, consumes seeds of *Trianthema hereroensis* and perhaps insects from the detritus (Robinson and Cunningham 1978). Various species of tenebrionid beetles select different parts of detritus: insect fragments, seeds, and flower parts (S. A. Hanrahan personal communication 1988). As mentioned previously, although seeds are a nutritious component of detritus, sand movement appears to limit potential for strict granivory in active dunes.

Animal-Animal Interactions

Predation, competition, mutualism, and effects of pathogens and disease are interactions between animals most commonly invoked as important determinants of community structure. Modification of the environment by one animal for another is an interaction less well recognized. Specific examples of the latter phenomenon in mobile sand dunes are use of sand patches solidified by oryx urine for initiation of burrowing by gerbils in an otherwise unstable substratum (Seely 1977), use of abandoned gerbil burrows by some spiders (J. R. Henschel personal communication 1988), and use of an oryx carcass as a nest by *Camponotus detritus* (A. C. Marsh personal communication 1988).

Several attributes of desert faunas imply the importance of predation, including a high proportion of carnivores (Cloudsley-Thompson 1974; Wagner and Graetz 1981; MacMahon and Wagner 1985), and frequent occurrence of cryptic coloration (Wagner and Graetz 1981). Importance of predation in mobile sand dunes of the Namib is suggested by the high proportion of carnivores in the fauna (Holm and Scholtz 1980; Crawford and Seely 1987); however, this is contradicted by the presence of many potential prey animals that are diurnally active and conspicuous (Koch 1961).

On a mobile dune system, predators may be resident, migratory, or transitory. Because of various constraints on the biota occupying the unstable dune substratum, few large animals are able to live there permanently. Larger predators, however, may occupy richer bordering areas, for example, riverine systems, and intermittently enter the sand dune area. If mobile dunes are not large, potential predators might exert a continual pressure on dune fauna, for example, owls on rodents. In larger dune fields, only predators themselves living on or within the mobile substratum have a continuous effect. Although seven mammal, eight bird, two reptile, three arachnid, and a number of invertebrate species could be predators on dune tenebrionids in mobile dunes of the Namib, environmental constraints are thought to be more important than predation as selective factors operating on adult beetles (Seely 1985). To date, there is suggestive (e.g., J. R. Henschel personal communication 1988; Polis 1988) but no unequivocal evidence that predation structures faunal assemblages of mobile sand dune environments.

Inferential evidence for competition (e.g., niche segregation, resource partitioning, and character displacement) is to be found in even the most arid environments (Wagner and Graetz 1981; MacMahon and Wagner 1985), and dunes are no exception. However, despite examples of species packing and niche segregation of tenebrionid beetles (Hamilton 1971; Holm and Edney 1973; Holm and Scholtz 1980), there is no definitive evidence for competition among these organisms on mobile sand dunes. Moreover, populations of tenebrionids, the groups most often cited from mobile dunes, appear to be continually depressed by abiotic rather than biotic factors (Thomas 1979). In contrast is evidence which Kotler and Brown (1988) have obtained from dune-dwelling rodents that suggests competition is important. Beneath the sand surface there are also larvae and vertebrate species, particularly where the fauna is concentrated under scattered individual plants.

In mobile sand dunes, many substratum-restricted species depend on plant detritus as a major food source (e.g., Koch 1961). Although steadily renewed (Louw and Seely 1982) with intermittent massive inputs (Seely and Louw 1980), detritus is distributed patchily across dunes (Figure 12.6), as is vegetation. Under such conditions of resource availability, probability of interspecific competition may be low (e.g., Price 1984). Thus, even though dunes may present a more moderate environment for plants and animals than surrounding finer-grained substrata, potential for interspecific resource or interference competition there is no greater than for other extreme desert environments.

Conclusions and Future Research

Mobile sand dunes present the paradoxical situation of an environment hostile to most forms of life yet able to support a relatively rich biota derived from a limited group of organisms. Comparatively large grain size and instability of the substratum combine to provide an environment for both plants and animals that is less harsh under low rainfall conditions than are finer grained, more compact substrata.

Mobile sand dunes occur most commonly in low rainfall desert areas where precipitation is irregular and unpredictable. However, perennial plants dominate on mobile dunes, so that effects of transient perturbations, such as that resulting from desultory rainfall, are damped (Figure 12.7). Unless man intervenes by allowing overgrazing or otherwise totally destroying the habitat, mobile sand dunes are relatively stable, in terms of both persistence and resilience of the vegetative and the detrital trophic base. Transient, often unpredictable events, such as unusually high rainfall or unusually hot or cold temperatures, are important in deserts and may be particularly important in mobile sand dunes where the number of interacting species of plants and animals is lower than on other desert substrata. Study of the impact of such occasional events, either natural or experimental, on various combinations of plant and animal interactions in mobile sand dunes could provide an important method of testing the significance of these interactions generally.

Physical characteristics of mobile sand dunes are similar throughout the world. This similarity, considered with the sometimes high numbers of individuals and stability of the biota, renders sand dunes ideal models for comparative studies of many aspects of desert community ecology. In particular, controlled, manipulative experiments in sand dunes are essential to sort out many of the questions posed. Such an approach has been conspicuously lacking in most studies of mobile sand dunes to date.

Most of the principles concerning assemblages of plants on mobile sand dunes have been derived from comparisons between dune fields. However, the various studies involved have been carried out by numerous workers with differing aims. All current predictions could be tested by more controlled use of the available array of mobile desert sand dunes. For example, evidence concerning importance of competition and succession on mobile dunes should be investigated. Relative importance of sand mobility and moisture availability on flowering phenology is another question to which such an approach lends itself.

Much research on mobile sand dunes has focused on tenebrionid beetles (e.g., Pierre 1958; Koch 1962), as beetles are abundant, relatively rich in species, relatively large, active diurnally, and thus conspicuous. Although adults have received most attention and have served well for behavioral and ecophysiological research, it appears that larvae represent the critical life history stage in terms of population dynamics. Predation too may be more prevalent on the larval stage. Moreover, recent work on taxonomy of other well-adapted but less conspicuous animals living on and in mobile sand dunes has indicated that they too are rich in species. Whereas some of these other taxa are also detritivores (e.g., thysanurans; Irish and Mendes 1988), others are predators in the dune system (e.g., neuropteran larvae and heteropodid spiders). Work with some of these groups should prove rewarding for community studies.

Most research on dune animals has focused, to date, on their activities on the surface despite the fact that they spend most of their time beneath the sand. Future work will have to focus more on this area of study, though it is admittedly less tractable (Seely, Mitchell, and Louw 1985).

Mobile sand dunes support some very distinctive assemblages of organisms which, in terms of their density-independent interactions with the environment and each other, are organized in a predictable manner. Although biotic interactions commonly occur, none has been demonstrated to play a dominant role in organizing the community. Instead, biotic interactions on mobile sand dunes appear to take place within the overall framework of strong abiotic influence on the fauna, flora, and environment. Only further (particularly experimental) research exploiting the unique desert dune habitat, however, will allow us to confirm or negate current views.

Acknowledgments

I wish to thank J. R. Henschel, A. C. Marsh, D. Mitchell, and J. D. Ward for valuable comments and discussion, the Transvaal Museum and the Foundation for Research and Development of the Council for Scientific and Industrial Research for financial support, and the Department of Agriculture and Nature Conservation of Namibia for facilities and permission to work in the Namib Desert dunes.

Bibliography

Andersen, D. C. 1987. Below-ground herbivory in natural communities: a review emphasizing fossorial animals. *Quarterly Review of Biology* 62: 261–286.

Ayyad, M. A., and S. I. Ghabbour. 1986. Hot deserts of Egypt and the Sudan. Pp. 149–202 in M. Evenari, I. Noy-Meir, and D. W. Goodall (eds.), *Ecosystems of the World*. Vol. 12(B): *Hot Deserts and Arid Shrublands, B*. Amsterdam: Elsevier.

Bagnold, R. A. 1954. *The Physics of Blown Sand and Desert Dunes*. London: Methuen.

Barnard, W. S. 1973. Duinformasies in die Sentrale Namib. *Tegnikon*: 2–13.

Bowers, J. E. 1982. The plant ecology of inland dunes in western North America. *Journal of Arid Environments* 5: 199–220.

Boyer, D. C. 1987. Effect of rodents on plant recruitment and production in the dune area of the Namib Desert. M.S. thesis, University of Natal, Pietermaritzburg, Republic of South Africa.

Bradley, R. 1988. The influence of weather and biotic factors on the behaviour of the scorpion (*Paruroctonus utahensis*). *Journal of Animal Ecology* 57: 533–551.

Brinck, P. 1956. The food factor in animal desert life. Pp. 120–137 in K. G. Wingstrand (ed.), *Bertil Hanstrom: Zoological Papers in Honour of His 65th Birthday.* Stockholm: Lund Zoological Institute.

Broza, M. 1979. Dew, fog and hygroscopic food as a source of water for desert arthropods. *Journal of Arid Environments* 2: 43–49.

Castle, E. 1954. The vegetation and its relationship to the dune soils of Kane County, Utah. M.S. thesis, Brigham Young University, Provo, Utah.

Cloudsley-Thompson, J. 1974. *Deserts and Grasslands.* London: Aldus Books.

———. 1979. Adaptive functions of the colours of desert animals. *Journal of Arid Environments* 2: 95–104.

Coineau, Y., N. Lancaster, R. Prodon, and M. K. Seely. 1982. Burrowing habits and substrate selection in ultrapsammophilous tenebrionid beetles of the Namib Desert. *Vie et Milieu* 32: 125–131.

Coineau, Y., and M. K. Seely. 1983. Mise en évidence d'un peuplement de microarthropodes dans les sables fins des dunes du Namib central. Pp. 652–654 in Ph. Lebrun, H. M. André, A. De Medts, C. Grégoire-Wibo, and G. Wauthy (eds.), *Proceedings of the VIII International Colloquium of Soil Zoology.* Ottignies-Louvain-la-Neuve, Belgium: Dieu-Brichart.

Connell, J. H., and R. O. Slayter. 1977. Mechanisms of succession in natural communities and their role in community stability and organization. *American Naturalist* 111: 1119–1144.

Coutchié, P. A., and J. H. Crowe. 1979. Transport of water vapor by tenebrionid beetles. I. Kinetics. *Physiological Zoology* 52: 67–87.

Cox, G. W. 1983. Foraging behaviour of the dune lark. *Ostrich* 54: 113–120.

Crawford, C. S. 1981. *Biology of Desert Invertebrates.* Berlin: Springer-Verlag.

Crawford, C. S., and M. K. Seely. 1987. Assemblages of surface-active arthropods in the Namib dunefield and associated habitats. *Revue de Zoologie Africaine* 101: 397–421.

Crawford, C. S., and E. C. Taylor. 1984. Decomposition in arid environments: role of the detritivore gut. *South African Journal of Science* 80: 170–176.

Curtis, B. A. 1985. The dietary spectrum of the Namib Desert dune ant *Camponotus detritus.* *Insectes Sociaux* 32: 78–85.

Danin, A. 1983. *Desert Vegetation of Israel and Sinai.* Jerusalem: Cana Publishing House.

De Winter, B. 1965. The South African Stipeae and Aristideae (Gramineae). *Bothalia* 8: 201–404.

Edney, E. B. 1971. Some aspects of water balance in tenebrionid beetles and a thysanuran from the Namib Desert of South Africa. *Physiological Zoology* 44: 61–76.

Edney, E. B., S. Haynes, and D. Gibo. 1974. Distribution and activity of the desert cockroach *Arenivaga investigata* (Polyphagidae) in relation to microclimate. *Ecology* 55: 420–427.

Felger, R. S. 1980. Vegetation and flora of the Gran Desierto, Sonora, Mexico. *Desert Plants* 2: 87–114.

Foth, H. D., and L. M. Turk. 1972. *Fundamentals of Soil Science.* New York: John Wiley & Sons.

Grenot, C. J. 1974. Physical and vegetational aspects of the Sahara Desert. Pp. 103–164 in G. W. Brown, Jr. (ed.), *Desert Biology,* vol. 2. New York: Academic Press.

———. 1976. P. 323 in *Ecophysiologie du Lézard Saharien Uromastix acanthinurus Bell, 1825 (Agamidae herbivore).* Publication of the Zoology Laboratory Ecole Normale Supérieure, Paris, No. 7.

Hadley, N. F., and S. R. Szarek. 1981. Productivity of desert ecosystems. *BioScience* 31: 747–753.

Hamilton, W. J. III. 1971. Competition and thermoregulatory behavior of the Namib Desert tenebrionid beetle genus *Cardiosis*. *Ecology* 52: 810–822.

———. 1973. *Life's Color Code*. New York: McGraw-Hill.

Heal, O. W., and S. F. Maclean, Jr. 1975. Comparative productivity in ecosystems—secondary productivity. Pp. 89–108 in W. H. von Dobben and R. H. Lowe-McConnell (eds.), *Unifying Concepts in Ecology*. The Hague: W. Junk.

Heatwole, H., and R. Muir. 1979. Thermal microclimates in the pre-Saharan steppe of Tunisia. *Journal of Arid Environments* 2: 119–136.

Holm, E., and E. B. Edney. 1973. Daily activity of Namib Desert arthropods in relation to climate. *Ecology* 54: 45–56.

Holm, E., and C. H. Scholtz. 1980. Structure and pattern of the Namib Desert dune ecosystem at Gobabeb. *Madoqua* 12: 3–39.

Irish, J., and L. F. Mendes. 1988. New genera and species of ultrapsammophilous Namib Desert Lepismatidae (Thysanura). *Madoqua* 15(4): 275–284.

Koch, C. 1961. Some aspects of abundant life in the vegetationless sand of the Namib Desert dunes. *Journal of SWA Scientific Society* 15: 8–34, 77–92.

———. 1962. The Tenebrionidae of southern Africa XXXI. Comprehensive notes on the tenebrionid fauna of the Namib Desert. *Annals of the Transvaal Museum* 24: 61–106.

Kotler, B. P., and J. S. Brown. 1988. Environmental heterogeneity and the coexistence of desert rodents. *Annual Review of Ecology and Systematics* 19: 281–307.

Lancaster, J., N. Lancaster, and M. K. Seely. 1984. Climate of the central Namib Desert. *Madoqua* 14: 5–61.

Lancaster, N. 1981. Grain size characteristics of Namib Desert linear dunes. *Sedimentology* 28: 115–122.

Larmuth, J. 1979. Aspects of plant habit as a thermal refuge for desert insects. *Journal of Arid Environments* 2: 323–327.

Lawrence, R. F. 1965. Pp. 1–12 in *New and Little Known Arachnida from the Namib Desert, S.W. Africa*. Scientific Papers of the Namib Desert Research Station No. 27.

Le Houérou, H. N. 1986. The desert and arid zones of northern Africa. Pp. 101–147 in M. Evenari, I. Noy-Meir, and D. W. Goodall (eds.), *Ecosystems of the World*. Vol. 12B: *Hot Deserts and Arid Shrublands, B*. Amsterdam: Elsevier.

Leistner, O. A. 1967. *The Plant Ecology of the Southern Kalahari*. Botanical Survey of South Africa Memoir No. 38, Pretoria.

Livingstone, I. 1986. Geomorphological significance of wind flow patterns over a Namib linear dune. Pp. 97–112 in W. G. Nickling (ed.), *Aeolian Geomorphology*. Boston: Allen & Unwin.

———. 1987. Using the response diagram to recognize zones of aeolian activity: a note on evidence from a Namib dune. *Journal of Arid Environments* 13: 25–30.

Louw, G. N., and M. K. Seely. 1980. Exploitation of fog water by a perennial Namib dune grass, *Stipagrostis sabulicola*. *South African Journal of Science* 76: 38–39.

———. 1982. *Ecology of Desert Organisms*. London: Longman.

Mabbutt, J. A. 1984. The desert physiographic setting and its ecological significance. Pp. 87–109 in H. G. Cogger and E. E. Cameron (eds.), *Arid Australia*. Sydney, New South Wales: Surrey Beatty and Sons.

McKee, E. D. 1979. Introduction to a study of global sand seas. Pp. 1–19 in E. D. McKee (ed.), *A Study of Global Sand Seas*. U. S. Geological Survey Professional Paper 1052.

MacMahon, J. A., and F. H. Wagner. 1985. The Mohave, Sonoran and Chihuahuan deserts of North America. Pp. 105–202 in M. Evenari, I. Noy-Meir, and D. W. Goodall (eds.),

Ecosystems of the World. Vol. 12A: *Hot Deserts and Arid Shrublands, A*. Amsterdam: Elsevier.

Marsh, A. C. 1986. Checklist, biological notes and distribution of ants in the central Namib Desert. *Madoqua* 14: 333–344.

Mitchell, D., M. K. Seely, C. S. Roberts, R. D. Pietruszka, E. McClain, M. Griffin, and R. I. Yeaton. 1987. On the biology of the lizard *Angolosaurus skoogi* in the Namib Desert. *Madoqua* 15(3): 201–216.

Monod, Th. 1986. The Sahel zone north of the equator. Pp. 203–243 in M. Evenari, I. Noy-Meir, and D. W. Goodall (eds.), *Ecosystems of the World*. Vol. 12(B): *Hot Deserts and Arid Shrublands, B*. Amsterdam: Elsevier.

Niethammer, G. 1971. Die Fauna der Sahara. Pp. 499–603 in H. Schiffers (ed.), *Die Sahara und ihre Randgebiete*, vol. 1. München: Weltforum Verlag.

Noy-Meir, I. 1973. Desert ecosystems: environment and producers. *Annual Review of Ecology and Systematics* 4: 25–51.

Odum, H. T., W. Abbott, R. K. Selander, F. B. Golley, and R. F. Wilson. 1970. Estimates of chlorophyll and biomass of the Tabonuco Forest of Puerto Rico. Pp. I-3–I-19 in H. T. Odum (ed.), *A Tropical Rain Forest*, vol. 3. U.S. Atomic Energy Commission.

Orshan, G. 1986. The deserts of the Middle East. Pp. 1–28 in M. Evenari, I. Noy-Meir, and D. W. Goodall (eds.), *Ecosystems of the World*. Vol. 12(B): *Hot Deserts and Arid Shrublands, B*. Amsterdam: Elsevier.

Pierre, F. 1958. Pp. 7–332 in *Ecologie et Peuplement Entomologique des Sables Vifs du Sahara Nord-Occidental*. Publications du centre de recherches Sahariennes série biologie No. 1.

Pietruszka, R. D., S. A. Hanrahan, D. Mitchell, and M. K. Seely. 1986. Lizard herbivory in a sand dune environment: the diet of *Angolosaurus skoogi*. *Oecologia* 70: 587–591.

Pietruszka, R. D., and M. K. Seely. 1985. Predictability of two moisture sources in the Namib Desert. *South African Journal of Science* 81: 682–685.

Polis, G. A. 1980. Seasonal and age specific variation in the surface activity of a population of desert scorpions in relation to environmental factors. *Journal of Animal Ecology* 49: 1–18.

———. 1988. Foraging and evolutionary reponses of desert scorpions to harsh environmental periods of food stress. *Journal of Arid Environments* 14: 123–134.

Polis, G. A., and S. J. McCormick. 1986. Patterns of resource use and age structure among species of desert scorpions. *Journal of Animal Ecology* 55: 59–73.

Polis, G. A., W. D. Sissom, and S. J. McCormick. 1981. Predators of scorpions: field data and a review. *Journal of Arid Environments* 4: 309–326.

Price, P. W. 1984. Alternative paradigms in community ecology. Pp. 353–383 in P. W. Price, C. N. Slobodchikoff, and W. S. Gaud (eds.), *A New Ecology: Novel Approaches to Interactive Systems*. New York: John Wiley & Sons.

Quézel, P. 1971. Die Pflanzenwelt. 1. Teil. Flora und Vegetation der Sahara. Pp. 429–475 in H. Schiffers (ed.), *Die Sahara und ihre Randgebiete*, vol. 1. München: Weltforum Verlag.

Robinson, E. R. 1976. Phytosociology of the Namib Desert Park, South West Africa. M.S. thesis, University of Natal, Pietermaritzburg, Republic of South Africa.

Robinson, M. D., and A. B. Cunningham. 1978. Comparative diet of two Namib desert sand lizards (Lacertidae). *Madoqua* 11: 41–53.

Robinson, M. D., and M. K. Seely. 1980. Physical and biotic environments of the southern Namib dune ecosystem. *Journal of Arid Environments* 3: 183–203.

Seely, M. K. 1977. Sand solidified by gemsbok urine as selected burrow sites by gerbils. *Zoologica Africana* 12: 247–249.

————. 1978. The Namib dune desert: an unusual ecosystem. *Journal of Arid Environments* 1: 117–128.

————. 1979. Irregular fog as a water source for desert dune beetles. *Oecologia* 42: 213–227.

————. 1983. Effective use of the desert dune environment as illustrated by the Namib tenebrionids. Pp. 357–368 in Ph. Lebrun, H. M. André, A. De Medts, C. Grégoire-Wibo, and G. Wauthy (eds.), *Proceedings of the VIII International Colloquium of Soil Zoology.* Ottignies-Louvain-la-Neuve, Belgium: Dieu-Brichart.

————. 1984. The Namib's place among deserts of the world. *South African Journal of Science* 80: 155–158.

————. 1985. Predation and environment as selective forces in the Namib Desert. Pp. 161–165 in E. S. Vrba (ed.), *Species and Speciation.* Transvaal Museum Monograph No. 4. Pretoria: Transvaal Museum.

Seely, M. K., M. P. de Vos, and G. N. Louw. 1977. Fog imbibition, satellite fauna and unusual leaf structure in the Namib dune plant *Trianthema hereroensis. South African Journal of Science* 73: 160–172.

Seely, M. K., and M. Griffin. 1986. Animals of the Namib Desert: Interactions with their physical environment. *Revue de Zoologie Africaine* 100: 47–61.

Seely, M. K., and G. N. Louw. 1980. First approximation of the effects of rainfall on the ecology and energetics of a Namib Desert dune ecosystem. *Journal of Arid Environments* 3: 25–54.

Seely, M. K., and D. Mitchell. 1987. Is the subsurface environment of the Namib Desert dunes a thermal haven for chthonic beetles? *South African Journal of Zoology* 22: 57–61.

Seely, M. K., D. Mitchell, and G. N. Louw. 1985. A field technique using iridium-192 for measuring subsurface depths in free-ranging Namib Desert beetles. *South African Journal of Science* 81: 686–688.

Seely, M. K., D. Mitchell, C. S. Roberts, and E. McClain. 1988. Microclimate and activity of the lizard *Angolosaurus skoogi* on a dune slip face. *South African Journal of Zoology* 23(2): 92–102.

Seely, M. K., C. S. Roberts, and D. Mitchell. 1988. High body temperatures of Namib dune tenebrionids—why? *Journal of Arid Environments* 14: 135–143.

Shmida, A. 1985. Biogeography of the desert flora. Pp. 23–77 in M. Evenari, I. Noy-Meir, and D. W. Goodall (eds.), *Ecosystems of the World.* Vol. 12(A): *Hot Deserts and Arid Shrublands, A.* Amsterdam: Elsevier.

Shmida, A., M. Evenari, and I. Noy-Meir. 1986. Hot desert ecosystems: An integrated view. Pp. 379–387 in M. Evenari, I. Noy-Meir, and D. W. Goodall (eds.), *Ecosystems of the World.* Vol. 12(B): *Hot Deserts and Arid Shrublands, B.* Amsterdam: Elsevier.

Simmons, N. M. 1966. Flora of the Cabeza Prieta Game Range. *Journal of the Arizona Academy of Sciences* 4: 93–104.

Smith, S. D., and P. S. Nobel. 1986. Deserts. Pp. 13–62 in N. R. Baker and S. P. Long (eds.), *Photosynthesis in Contrasting Environments.* Amsterdam: Elsevier.

Stearns, S. C. 1977. The evolution of life history traits. *Annual Review of Ecology and Systematics* 8: 145–171.

Thomas, D. B., Jr. 1979. Patterns in the abundance of some tenebrionid beetles in the Mojave Desert. *Environmental Entomology* 8: 568–574.

————. 1983. Tenebrionid beetle diversity and habitat complexity in the Eastern Mojave Desert. *Coleopterists Bulletin* 37: 135–147.

Tschinkel, W. R. 1973. The sorption of water vapor by windborne plant debris in the Namib Desert. *Madoqua,* Series II, 2: 21–24.

Vogel, J. C., A. Fuls, and A. Danin. 1986. Geographical and environmental distribution of C₃ and C₄ grasses in the Sinai, Negev, and Judean deserts. *Oecologia* 70: 258–265.

Wagner, F. H., and R. D. Graetz. 1981. Animal-animal interactions. Pp. 51–83 in D. W. Goodall, R. A. Perry, and K. M. W. Howes (eds.), *Arid-Land Ecosystems: Structure, Functioning and Management*, vol. 2. Cambridge: Cambridge University Press.

Walter, H., and E. O. Box. 1983. The Karakum desert, an example of a well-studied eubiome. Pp. 105–159 in N. E. West (ed.), *Ecosystems of the World*. Vol. 5: *Temperate Deserts and Semi-Deserts*. Amsterdam: Elsevier.

Ward, J. D., M. K. Seely, and N. Lancaster. 1983. On the antiquity of the Namib. *South African Journal of Science* 79: 175–183.

Watson, R. T., and J. Irish. 1988. Lepismatidae (Thysanura: Insecta) of the Namib Desert sand dunes. *Madoqua* 15(4): 285–293.

WESTEC Services. 1977. Survey of sensitive plants of the Algodones Dunes. Unpublished report prepared for U.S. Department of the Interior, Bureau of Land Management, California Desert District Office, Riverside, California.

Wharton, R. A., and M. Seely. 1982. Species composition of and biological notes on Tenebrionidae of the lower Kuiseb River and adjacent gravel plain. *Madoqua* 13: 5–25.

Yeaton, R. I. 1988. Structure and function of the Namib dune grasslands: characteristics of the environmental gradients and species distributions. *Journal of Ecology* 76: 744–758.

Zohary, M. 1973. *Geobotanical Foundations of the Middle East*. 2 vols. Stuttgart: Gustav Fischer Verlag.

Food Webs in Desert Communities: Complexity via Diversity and Omnivory

13

Gary A. Polis

Deserts are often considered to be simple ecosystems (Noy-Meir 1974, 1981; Seely and Louw 1980; Wallwork 1982; Whitford 1986). Water is limiting and soils are nutrient-poor. Consequently, productivity is low and deserts appear to support a less diverse biota than most nonarid habitats. Nevertheless, we are still largely ignorant of how deserts function as ecosystems. In spite of the great deal of information we have accumulated, the inherent complexity of this relatively simple system has blocked our efforts to model or understand the structure and dynamics of deserts (Noy-Meir 1981). Most descriptions of desert ecosystems have focused on energy, nutrient, and water cycles. Few studies have attempted to describe desert communities by analyzing the feeding relationships between member species. The trophic linkage between consumers and their food is usually delineated by "food webs"— schematic descriptions of feeding by consumers.

Several approaches have been used to analyze food webs (see DeAngelis, Post, and Sugihara 1983; May 1986; Lawton 1989). One approach uses models based on stability analysis (e.g., see Pimm 1982; DeAngelis, Post, and Sugihara 1983; May 1983b, 1986). The results of these models are complex and beyond the scope of this chapter. However, stable food webs are predicted to be relatively simple, short and with few trophic levels, and exhibit little omnivory or looping (Pimm 1982) (see below). A second approach analyzes empirical food webs to determine regularities in their properties (e.g., Cohen 1978, 1988; Briand 1983; Pimm 1982; Yodzis 1984; Cohen, Briand, and Newman 1986). Food webs used for this approach were compiled by Cohen (1978), Briand (1983), Cohen, Briand, and Newman (1986), and Schoenly, Beaver, and Heumier (1991). Cohen, Briand, and Newman (1986) published a catalog of 113 webs and Schoenly, Beaver, and Heumier (1991) compiled 95 insect-oriented webs. Pimm (1982) argues that the empirically derived patterns are consistent with and validate the predictions of the dynamics models above (Pimm 1982; Pimm and Rice 1987). Some of these empirical patterns are presented in Table 13.1.

In this chapter, I present the trophic interactions within a desert community located in California. The food web is analyzed explicitly to evaluate the patterns in Table 13.1. The patterns of this community are quite different from the regularities abstracted from published food webs. Most obviously, this desert web is *much more complex*. This result indicates that the practice of abstracting empirical regularities yields an inaccurate and artifactual view of trophic interactions within communities. I end the chapter by generalizing the patterns from the California desert web to other webs.

General Problems in the Analysis of Empirical Food Webs

In this section, I present five substantial problems associated with the catalogs of published food webs (see also Glasser 1983; May 1983b; Taylor 1984; Paine 1988; Sprules and Bowerman 1988; Lawton 1989). These problems make the catalogs totally inadequate for the types of analyses that have been conducted. I contend that the practice of abstracting empirical regularities from these incomplete webs yields a highly inaccurate, tremendously oversimplified view of the trophic interactions within a community. In the next section, I illustrate these problems by presenting the web from one California desert community.

Inadequate Representation of Species Diversity

The major problem with published food webs is that the number of species in the analyzed communities is much smaller than the number that exist in real communities. Three types of food webs were compiled in the catalogs of webs (Cohen 1978). Sink and source webs describe subsets of the feeding interactions within a community. The community web is the only complete web, representing all feeding interactions between all members of the trophically linked assemblage of species. Consequently, this is the most difficult web to construct and is the rarest type of published web. Unfortunately, community webs are the most useful for analysis; some theoreticians (May 1983b) now demand their use. The vast majority (all?) of cataloged webs are not community webs. Most (all?) authors either simply ignored unfamiliar species and concentrated on taxa in their field of expertise, or lumped unfamiliar species into higher categories.

The practice of lumping species into higher categories is a severe problem. Cohen (1978) labeled these categories "kinds of organisms." Briand (1983, p. 253) states that "A 'kind of organism' (interchangeable henceforth with the term 'species') may be an individual species, or a stage in the life cycle of a size class within a single species, or it may be a collection of functionally or taxonomically related species."

Table 13.1   A partial summary of "features observed in real food webs"
as suggested by food web theorists.

Food Web Features and Source

Chain lengths are limited to "typically three or four" trophic levels. (iii)[a]
   (Pimm 1982; Briand 1983; Cohen, Briand, & Newman 1986)
Omnivores are statistically "rare." (iv)[a]
   (Pimm 1982; Yodzis 1984)
Omnivores feed on adjacent trophic levels. (v)[a]
   (Pimm 1982)
Insects and their parasitoids are exceptions to the second and third patterns, above. (vi)[a]
   (Pimm 1982; Hawkins & Lawton 1987)
Webs are usually compartmentalized between but not within habitats. (viii)[a]
   (Pimm 1982)
Loops are rare or nonexistent and do not conform to "biological reality." (i)[a]
   (Gallopin 1972; Cohen 1978; Pimm 1982; Pimm & Rice 1987; Lawton & Warren 1988)
The ratio of prey species to predator species is <1.0 (0.86–0.88). (x)[a]
   (Cohen 1978; Briand & Cohen 1984)
The proportion of species of top predators to all species in a community averages 0.29.
   (Briand & Cohen 1984)
Species interact directly (as predator or prey) with only 2–5 other species.
   (Cohen 1978; Cohen, Newman, & Briand 1985)

SOURCE: These purported empirical generalizations were derived by food web theorists
from the catalog of published food webs assembled by Briand & Cohen. All quoted
phrases are from Chapter 10 in Pimm (1982) (see especially Table 10.1).
[a]Roman numerals refer to pattern number in Table 10.1 in Pimm (1982).

"Kinds" are also called "trophic species" (Briand and Cohen 1984) and "species"
(Cohen and Newman 1986).

Kinds of organisms include "basic food," "benthos," "other carnivores" (matrix
1 in Briand 1983), "algae," "plankton," "birds" (matrix 9), "phytoplankton," "ice
invertebrates," "zooplankton," "fish" (matrix 21), and "herbs," "trees and bushes,"
"insects," "spiders," "soil insects and mites," and "parasites" (matrix 27). In total,
only 28.7 percent of the "kinds" in Briand's webs are real species (average ± s.d.
= 26.3 ± 23.5 percent); nine matrices are without any real species. The "kinds of
organism" simplification has been criticized by Glasser (1983), May (1983b), Taylor
(1984), Lawton (1989), and by Cohen (1978) himself in a self-critique (but see
Sugihara, Schoenly, and Trombla 1989).

The lumping problem is particularly evident in the consideration of certain taxa.
Plants, arthropods, parasites, and organisms that live in the soil or benthos are rarely
considered more than superficially. Pimm (1982, p. 168) postulates that there is a
"widespread antipathy among ecologists towards plant and invertebrate taxonomy"
resulting in the gross lumping of species at these taxonomic levels. Further, Taylor
(1984) notes that invertebrates are often analyzed in much less detail than vertebrates,
thus obscuring complexity and web structure. The incomplete presentation of these
important taxa is a serious flaw in present web analysis. In particular, arthropods
(insects and arachnids) play a central position in structuring desert and other terres-

trial communities. The approximately 800,000 identified species of insects represent about 89 percent of all identified animal species combined (Pearse et al. 1987; it is estimated that 5–50 million species of insects exist; May 1988).

Parasitoid Hymenoptera and Diptera by themselves are a diverse and important component of food webs. Askew (1971) estimates that more than 10 percent of all animal species are parasitoids. Other parasites (protozoa, helminths, and Acari) form a diverse and speciose group and feed on almost all organisms. They form another step in the flow of energy and can greatly influence population dynamics and community structure (May 1983a).

Soil organisms and their interactions are usually ignored or greatly simplified, although they are essential to understanding food web structure (Cousins 1980; Odum and Biever 1984; Rich 1984). Detritivory and subterranean herbivory form one of the major pathways of energy flow in terrestrial communities and are particularly important in deserts (see below). Detritus is a universal and major component of all food webs simply because all organisms die and all animals defecate.

Both the tactics of ignoring and lumping species produce the depauperate published webs compiled by theorists. For example, the number of species in Cohen, Briand, and Newman's (1986) catalog ranged from 3 to 48, with the average web describing a trophic community of 16.9 species. All communities have far more species.

This is illustrated by enumerating the species from the sandy desert within the Coachella Valley (CV) (Riverside County, California): 174 species of vascular plants, 138 species of vertebrates, more than 55 species of arachnids, and an unknown (but great) number of insects, Acari, and nematodes. The number of insect species is estimated to be from 2,000 to more than 3,000. I have identified 123 families of insects. A still-incomplete survey of insect diversity in the adjacent University of California Deep Canyon Desert Research Preserve has identified 24 orders, 308 families, and more than 2,540 species (Frommer 1986). Intensive work reveals high diversity in some taxa. For example, 147 species of Bombyliidae (bee flies) were recorded in Deep Canyon (Tabet and Hall 1984; J. Hall personal communication 1986). Timberlake recorded more than 500 species of bees within 2 miles of Palm Springs (C. Michener personal communication 1988). The beetles in a nearby (85 km distant) sand system (Palen Dunes) include 31 families, 120 genera, and 142 species (Andrews, Hardy, and Giuliani 1979).

Inadequate Dietary Information

Another major weakness of the cataloged webs is the inadequate information on the diet of most species. The number of prey items recorded is usually a function of the amount of time and effort devoted to observation. A "yield/effort" curve (Cohen 1978) is illustrated by analyzing the diet of the scorpion *Paruroctonus mesaensis* as a function of the number of cumulative nights data were collected in the field (Figure 13.1). The number of prey species tabulated continues to increase with observation time. The 100th prey species was recorded on the 181st survey night; an asymptote was never reached in 5 years and more than 2,000 person-hours of field time.

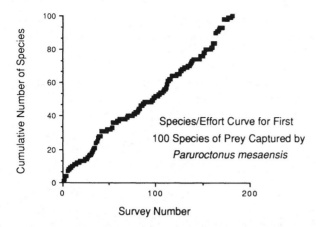

Figure 13.1. Number of prey species recorded for the scorpion *Paruroctonus mesaensis* in the Coachella Valley, Riverside County, California, as a function of number of surveys performed (yield/effort curve).

The shape of this yield/effort curve is likely typical of studies of many other species of consumers. This suggests that a great amount of effort must be devoted to approach the complete diet of even one species. It is unlikely that such an effort has been devoted to the majority of species in the cataloged webs. The diet of a few focal organisms may be known with relative certainty. However, the amount of effort and time needed to determine the diet of just the numerically dominant species is astronomical. Thus a food web containing all species still would be an inadequate description of the trophic relations in a community unless diets were known with more confidence.

Such inadequacy is manifested in the analysis of published webs. For example, cataloged webs show a high proportion (28.5 percent; Briand and Cohen 1984; 46.5 percent; Schoenly, Beaver, and Heumier 1991) of top predators (consumers without predators) (Briand and Cohen 1984). It is highly unlikely that one-fourth of the animals on this planet are free of predators. (I suspect that much fewer than 1 percent of species do not suffer predators sometimes during their lives; see the desert food web in Figure 13.7). Such a figure results from grossly incomplete data on predators and prey. For example, in the catalog of 113 webs, 57 chains were of length one; in other words, 57 herbivores were top predators with no recorded predators (Cohen, Briand, and Newman 1986)! Top predators in this catalog include, for example, spiders, mites, midges, mosquitos, bees, weevils, fish larvae, blackbirds, shrews, and moles, all of which inevitably have their own predators.

Temporal and Spatial Variation

Temporal and spatial differences in feeding ecology are undoubtedly universal. It is well known that the abundance and distribution of individual species, species composition, and diet change in time and space. A rich ecological literature recognizes that such differences are key factors shaping community patterns (see Wiens 1976; Caswell 1978; Seely, Vitt, Wiens, and Wisdom this volume). However, it is logistically difficult to quantify spatial and temporal variation. Consequently, food webs are usually either restricted to a particular season and study site or generalized to represent some community average (Cohen 1978). In either case, much of their richness is lost. Temporal and spatial differences in the diets of organisms living in the CV illustrate such changes.

With few exceptions, published analyses of annual food habits of CV species noted great differences through time. Several factors contribute to temporal differences. Normal changes in prey composition and abundance dictate which foods are available. Prey in the CV (and elsewhere) exhibit three general phenologies: pulsed (short but intense population eruptions lasting a few days or weeks), seasonal (present for 2–4 months), and annual (available throughout the year) (Polis 1979). Feeding on prey from all three phenologies produces diet changes over time. For example, diet similarity among adults of the same species (the scorpion *Paruroctonus mesaensis*) is a function of the time between the months compared: diet (prey species) overlap varied from 0.73 between adjacent months to 0.0 (i.e., no species eaten in common) between periods separated by five months (1 month [mean ± s.d.]: 0.73 ± 0.23; 2 months: 0.58 ± 0.32; 3 months: 0.52 ± 0.29; 4 months: 0.35 ± 0.05; 5 months: 0.0).

The sudden appearance of pulsed foods often is reflected in dramatic changes in diet. In the CV and other deserts, mass emergence of alate termites after some rains is accompanied by many predators (temporarily) concentrating on termites (Table 13.2). With the exception of adult *Dinothrombium pandorae* (Tevis and Newell 1962), all 31 of these predators feed on other prey when termites are not available. For example, after heavy rains in 1976 in the CV, the diet of *Messor pergandei* included 89.7 percent *Amitermes wheeleri* termites; normally arthropods form 2–10 percent of their diet (Gordon 1978).

Further, most vertebrates seasonally change between plant and animal foods. Granivorous birds and rodents primarily eat seeds during most months but normally feed on insects and spiders when arthropods become abundant (Bent 1958; Brown, Reichman, and Davidson 1979; Brown 1986; Wiens this volume). This switch provides protein during the spring breeding season for developing nestlings and allows permanent residency for some desert birds (Welty 1962; Brown 1986). In the CV, 8 of 13 primarily granivorous bird species are reported to eat arthropods (Appendix 13.1). Alternatively, many omnivorous or primarily arthropodivorous birds and rodents in deserts consume significant quantities of seed or fruit when arthropods are unavailable (see below; Brown, Reichman, and Davidson 1979; Brown 1986). Many carnivorous mammals in the CV seasonally become frugivorous when fruit appears, for example, coyotes, kit foxes, and grasshopper mice (Appendix 13.1; see below).

Diet is rarely the same throughout a species's geographical range (Welty 1962; Wiens and Rotenberry 1981; McCormick and Polis 1986). Spatial differences in the composition and abundance of prey constrain the consumer to eat different foods in different places. For example, the kangaroo rat *Dipodomys merriami* was studied in the deserts of southern California (Soholt 1973) and Arizona (Reichman 1975). Filaree (*Erodium cicutarium*) seeds formed 87 percent of the diet biomass in Arizona but only 9.5 percent in California. Vitt (this volume) shows that differences in diet for the same species of lizards at different sites approximate the magnitude of differences observed between different species of lizard.

In general, similarity in diet between populations of the same species should decrease monotonically, even over relatively short distances. For example, McCormick and Polis (1986) quantified differences in feeding ecology of *P. mesaensis* at two sites 65 km apart within the CV. Predator size, prey size, feeding rate, prey composition, and proportion of the same prey in the diet each were highly significantly different. Overall similarity in prey species was 83 percent.

Age Structure

Age-related changes in food choice and predators are normal phenomena not well incorporated into analyses of food webs. A population is composed of a complex of age groups that vary in their interactions with the environment (Werner and Gilliam 1984; Persson and Ebenmann 1988; Polis 1988a). Many species occur as a series of discrete age/size classes, each exhibiting significant differences in resource use, predators, and competitors (Polis 1984a; Werner and Gilliam 1984).

Age classes often use different foods, thereby greatly expanding the diet of a species. Pimm and Rice (1987) refer to such changes as "life history omnivory." "Ontogenetic diet shifts" are characteristic of those species that undergo metamorphosis during growth (Werner and Gilliam 1984), for example, holometabolic insects and anurans. Appendix 13.2 lists 27 insect families living in the CV whose juvenile diet (live arthropods) differs radically from their adult diet (plants).

For species that grow slowly through a wide size range, diet changes more gradually as prey size increases with predator size (e.g., reptiles, arachnids; Polis 1984a, 1988a). Scorpions are used to illustrate. Young are miniature adults in appearance. However, prey use, foraging behavior, home range, microhabitat use, temporal (seasonal and diel) patterns, and intra- and interspecific interactions change greatly during growth (Polis 1979, 1984a; Polis and McCormick 1987). For example, growing *P. mesaensis* increase 60 to 80 times in mass; instar 2 scorpions eat prey that average 5 mm in length, whereas adults use prey averaging three times larger, two-thirds of which are different species. In fact, differences in body sizes and resource use for the three age classes of *P. mesaensis* are equivalent to or greater than the differences between most real "biological species" (Polis 1984a). This magnitude of change is typical of other wide size range predators (e.g., most invertebrate phyla, hemimetabolous insects, larvae of holometabolous insects, arachnids, fish, and reptiles; Polis 1984a). Exceptions include birds and mammals.

Predators also change during growth. Juveniles fall prey to species too small to capture adults. Such developmental "escapes" are a common feature of all communities. For example, snakes eat eggs and newborn (but not adults) of carnivorous

Table 13.2 Partial list of prey taxa and predators that eat
termites, tenebrionids, and/or scorpions in the Coachella
Valley, Riverside County, California.

Taxon	Prey[a]
Termite species	
Amitermes coachellae Light	
A. emersoni Light	
A. minimus Light	
A. snyderi Light	
A. wheeleri (Desneux)	
Gnathamitermes perplexus (Banks)	
Heterotermes aureus (Snyder)	
Incisitermes minor (Hagen)	
Paraneotermes simplicicornis Light	
Reticulitermes tibialis Banks	
Tenebrionid species	
Araeoshizus sp.	
Batulius sp.	
Chilometopon sp.	
Cryptoglossa sp.	
Edrodes ventricosus	
Eleodes (two species)	
Eupsophulus castaneus	
Eussatus muricatus	
Notobius puberulus	
Philolitha opimas	
Stibia sp.	
Telabis sp.	
Class Arachnida	
O. Acari	
Dinothrombium pandorae	Ter
O. Araneae	
Aptostichus[b]	Ter, Ten, Sc
Latrodectus hesperus[b]	Ter, Ten, Sc
Statoda fulva[b]	Ter, Ten
S. grossa[b]	Ter, Ten
Pellenes sp.	Ter, Ten
Psilochorus sp.[b]	Ter, Ten
Diguetia mohavea[b]	Ter, Ten
O. Scorpionida	
Paruroctonus mesaensis[b]	Ter, Ten, Sc
P. luteolus[b]	Ter, Ten, Sc
Vaejovis confusus[b]	Ter, Ten, Sc
Hadrurus arizonensis[b]	Ter, Ten, Sc
O. Solpugida	
Eremochelis sp.[b]	Ter, Ten, Sc
Eremorhax sp.[b]	Ter, Ten, Sc

Taxon	Prey[a]
Class Insecta	
F. Asilidae	
Proctacanthella sp.[b]	Ter
F. Formicidae	
Pogonomyrmex californicus[b]	Ter, Ten
P. rugosus[b]	Sc
(Vero)Messor pergandei[b]	Ter, Ten
Myrmecocystus flaviceps[b]	Ter, Ten
F. Myrmeleontidae	
Antlion larvae (many species)[b]	Ter
F. Mantidae	
Stagmomantis californicus[b]	Ter, Ten
Class Reptilia	
F. Iguanidae	
Callisaurus draconoides	Ter, Ten, Sc
Cnemidophorus tigris	Ter, Ten, Sc
Coleonyx variegatus	Ten, Sc
Phrynosoma m'calli	Ter
P. platyrhinos	Ter, Ten
Uma inornata	Ter, Ten
Uta stansburiana	Ter, Ten, Sc
F. Leptotyphlopidae	
Leptotyphlops humilis[b]	Ter
F. Colubridae	
Chionactis occipitalis	Sc
Class Aves	
Loggerhead Shrike	Ter, Ten, Sc
Western Tanager	Ter, Ten
Western Meadowlark	Ter, Ten, Sc
Common Raven	Sc
American Kestrel	Ten, Sc
Red-tailed Hawk	Ten
Burrowing Owl	Ten, Sc
Western Screech-Owl	Sc
Great Horned Owl	Sc
Barn Owl	Sc
Roadrunner	Ten, Sc
Class Mammalia	
Antrozous pallidus (pallid bat)	Ter, Ten, Sc
Onychomys torridus (grasshopper mouse)	Ten, Sc
Vulpes macrotus (kit fox)	Ter, Ten, Sc
Canis latrans (coyote)	Sc

[a]Ter = termites; Ten = Tenebrionids; Sc = scorpions. Termite prey were predominantly alate (reproductive) termites, although ants consumed both workers and alates.

[b]Members of this taxon consumed by scorpions.

birds (see Figure 13.7) and desert tortoises (*Gopherus agassizi*; Barrows 1979); adults of small scorpion species eat juveniles of larger species. Alternatively, adults may be eaten by predators that do not or cannot eat small juveniles. Thus, vertebrate predators (e.g., owls, kit foxes) in the CV eat adult scorpions but ignore juveniles (Polis, Sissom, and McCormick 1981).

In summary, age/size-specific differences in predators, prey, and competitors are the norm (Polis 1984a, 1988a). Such differences may be major determinants of population dynamics and community structure. Unfortunately, ecologists and theoreticians have largely ignored the richness that age/size-specific analyses contribute to community and food web structure (but see Pimm and Rice 1987; Persson and Ebenmann 1988). Usually either only the diet of adults is considered or the diets of all age classes are combined. Age is recognized in only 22 of 875 kinds in Cohen's (1978) catalog and in 3 of 422 in Briand's (1983).

Looping

Looping is a feeding interaction in which A eats B, B eats C, and either B (mutual predation) or C (three-species loop) eats A. Cannibalism is a "self loop" (A eats A; Gallopin 1972). Food web theorists dismiss loops as "unreasonable structures" (Pimm 1982, p. 70; see also Gallopin 1972, p. 266; and Cohen 1978, p. 56). Pimm (1982, p. 67) summarizes from the catalog of webs: "I know of no cases in the real world with loops." He has modified this view to include loops in aquatic age-structured species (Pimm and Rice 1987), but still maintains that loops are rare in terrestrial systems.

Cannibalism (Polis 1981) and mutual predation (Polis, Myers, and Holt 1989) are taxonomically widespread interactions. Cannibalism was reported in more than 1,300 species and is a key factor in dynamics of many populations (reviewed by Polis 1981). Cannibalistic loops are frequent in the CV (see Figures 13.2–13.7). Ontogenetic reversal of predation among age-structured species is the most common form of mutual predation: juvenile A are eaten by B but adult A eat B (and/or juvenile B). This is observed among CV spiders, scorpions, solpugids, and predaceous insects, and among lizards, snakes, and birds (see Figures 13.5–13.7). For example, gopher snakes (*Pituophis*) eat eggs and young of burrowing owls (Coulombe 1971), whereas adult burrowing owls eat young gopher snakes (Bond 1942).

Two examples from the CV show that mutual predation can occur independent of age structure. First, black widow spiders (*Latrodectus hesperus*) catch three species of scorpions by using web silk to pull them up off the ground (Polis, Sissom, and McCormick 1981); black widows traveling on the ground are captured by these same scorpions (Polis and McCormick 1986b) (Figure 13.5). Second, CV ants (*Pogonomyrmex californicus, Myrmecocystus flaviceps,* and *Messor pergandei*) regularly include each other in their diets (Ryti and Case 1988). Killing and predation of winged reproductives (after swarming) and workers (during territorial battles) are regular interactions among ants and other social insects (Brian 1983; Polis, Myers, and Holt 1989).

The Coachella Valley: Trophic Relations Summed into Food Webs

The Coachella Valley is located in Riverside County, California (116°37' W, 33°54' N). It encompasses about 780 km² and spans an elevational cline from 320 m in the northwest to sea level in the southeast. Winters are mild, and summers, hot and dry. Air temperature in July annually exceeds 40°C and temperatures of more than 50°C have been recorded (Edney, Haynes, and Gibo 1974; Polis 1988b). It is a low-elevation rain shadow desert with an average annual rainfall at Deep Canyon Field Station of 116 mm and 75 mm at Indio. Extremes in rainfall range from 34 mm in 1961 to 301 mm in 1976.

Data Base

The sand dune/intergrading sand flat habitat of the CV was chosen to define the biological community. This community is well studied primarily due to the presence of the University of California Deep Canyon Desert Research Preserve and Station. Appendix 13.1 identifies vertebrate species in the CV, and Appendix 13.2, major arthropod taxa. Beginning in 1969, W. Mayhew surveyed the vertebrate fauna of sand dunes and sand flats for the Deep Canyon Transect Study (Mayhew 1981). This work defined the sand dune/sand flat habitat and established a list of 138 species of reptiles, birds, and mammals. Although Mayhew records mountain lions and badgers in the area historically, they are no longer present and hence not included in the food web. Furthermore, only the 56 birds that actually nest in the area are considered here. These birds represent 58 percent of the 97 species reported as residents for various periods of the year. Additional information on the vertebrates in sandy areas of the CV was obtained from Weathers (1983) and Ryan (1968).

Invertebrates of the community are less well known. Much knowledge of them is the result of my long-term research (since 1973) in the CV (e.g., Polis 1979; Polis and McCormick 1986a,b, 1987; McCormick and Polis 1986). Information on diet and species composition was obtained from more than 4,300 trap days for insects and from 15 years and more than 4,000 hours of fieldwork. Taxonomic lists of the arthropods of the CV are obtained from this fieldwork and from catalogs of insects (Frommer 1986; Wheeler and Wheeler 1973; Hawks 1982; Tabet and Hall 1984) and arachnids (Polis and McCormick 1986b).

Plants of the Deep Canyon Transect were surveyed by Zabriskie (1980). Only the plants that are widely distributed in the 0 to 245 m range of elevation in Zabriskie's list were considered to be members of the flora of the sand community.

Trophic relationships (food and predators) were determined from the literature and my work. I read about 820 papers to collect these data. Published data from studies in the CV were used with priority; however, I was forced to supplement these data with information from other regions, some quite near (e.g., the Palen Dunes) and some much further (e.g., the Chihuahuan Desert). When I could not find specific information on a particular taxon, I was forced to use more general sources (e.g., field guides, such as Bent's [1932–58] *Life Histories of North American Birds*). I could not find the diets (from desert areas) of 18 bird and one rodent species (see Appendix 13.1). Finally, interviews with scientists conducting research in the CV provided information on species identity and diet. Due to space limitations, I cannot include references to every trophic link depicted; however, important references are included. Interested readers can write for more information.

A series of representative subwebs depict trophic relations in the CV sand community. Subwebs proceed from plants and detritus to various secondary consumers. Subwebs are connected so that organisms in one web consume (or are consumed by) organisms in the next. Webs are incomplete as there are far too many species to include and adequate diet data are unavailable for many species. I thus concentrate on well-studied, focal species whose trophic interactions are relatively well known. Webs include only species that live in the CV and only interactions for which evidence exists. Similar complexity is expected for other, less-known species. Thus, the complexity in these webs is an understatement of what actually exists.

Consumers are classified in terms of resource specialization (i.e., number of species eaten within one group, e.g., plants) or trophic specialization (i.e., number of different types eaten, e.g., plants, detritus, arthropods) (Levine 1980). Species vary from resource specialists that eat a few species of the same resource type to trophic generalists (omnivores) that feed on several food types. Closed-loop omnivory is a special case of omnivory in which A eats both B and C but B also eats C (Sprules and Bowerman 1988).

Plant-Herbivore Trophic Relations

Herbivory describes feeding on several types of plant products: leaves, wood, roots, tubers, sap, nectar, seeds (granivory), and fruit (frugivory). In deserts, below-ground herbivory can be important because a large fraction of plant biomass is subterranean. Interactions between surface and subsurface herbivores are complex and involve both competition and facilitation (Seastedt, Ramundo, and Hayes 1988). This section focuses on above-ground herbivory. Below-ground herbivory is discussed in the following section.

Desert herbivores include nematodes, mites, insects, reptiles, birds, and mammals. I cannot list and discuss the herbivores of the 174 species of vascular plants found in the CV. Rather, I discuss broad groups of herbivores with the hope of conveying some of the complexity in the plant-herbivore link.

A wide variety of arthropods feed on desert plants (Orians et al. 1977; Powell and Hogue 1979; Crawford 1981; Wisdom this volume; Appendix 13.2 lists 74 families of CV arthropods that are herbivorous sometime in their lives). Illustrating with CV plants, more than 60 species of insects are reported from creosote (*Larrea tridentata*; Schultz, Otte, and Enders 1977), more than 200 species of insects eat mesquite (*Prosopis glandulosa*), and more than 89 species feed on ragweed (*Ambrosia dumosa*) (Wisdom this volume).

Feeding habits are known for many herbivorous arthropod families that live in the CV. Many beetles, thrips, bees, and wasps feed on flowers. Flies, wasps, bees, butterflies, and moths eat nectar. Pollen is a source of food for beetles, wasps, and bees. Grasshoppers and crickets, bugs, aphids, leafhoppers, larvae of beetles and Lepidoptera, and many other groups eat leaves. A diversity of arthropods also use plant exudates. A few primarily granivorous arthropods live in the CV sand community, for example, lygaeid bugs, bruchid weevils, and ants. It is likely that many generalized herbivorous arthropods eat seeds incidentally, but this is unconfirmed because of inadequate knowledge of diet. For example, seeds formed 11.1 percent of the diet of the camel cricket *Macrobaenetes valgum* in the CV (McCormick and Polis unpublished data).

Both resource specialists and generalists eat plants in the CV. Examples of specialists include the grasshopper *Bootettix punctatus* (eats only leaves of creosote; Mispagel 1978) and some gall midges (Cecidomyidae) that occur only on saltbush (*Atriplex*) (Hawkins and Goeden 1984). Insects on cactus usually do not attack other plants and generalists do not attack cactus (Mann 1969). (Damage caused by insects feeding on cactus facilitates the growth of several specialist and generalist phytophagous fungi, e.g., *Gloesporium lunatum.*)

Resource generalists apparently are more common in deserts as compared to specialists (Orians et al. 1977; Otte and Joern 1977; Crawford 1981, 1986). Examples of CV generalists include the lacebug *Corythucha morerelli* (15 species of plants; Silverman and Goeden 1979) and the camel cricket *M. valgum* (16 species of plants; McCormick and Polis unpublished data). *Messor pergandei*, a CV harvester ant, was recorded to collect 97 species of seed (Gordon 1978).

Most herbivorous arthropods are trophic specialists feeding on plants all their lives, for example, hemimetabolous insects whose nymphs eat the same foods as adults (Orthoptera, Hemiptera, some Homoptera, Thysanoptera). This is also true for some holometabolous insect herbivores (N.B., larvae and adult may feed on different parts), for example, curculionid, chrysomelid, scarabid, and buprestid beetles. It is not true for many other holometabolous insects. The larvae of several families (e.g., bombyliid flies, meloid beetles, pompilid, mutillid, and tiphiid wasps) are parasitic or predaceous on other arthropods, whereas adults feed on plants (Ferguson 1962; Andrews, Hardy, and Giuliani 1979; Powell and Hogue 1979; Wasbauer and Kimsey 1985) (Appendix 13.2). Nor is it true for trophically flexible species; for example, CV harvester ants take more than 40 categories of foods including seeds, parts (flowers, stems) from five plant species, spiders, and insects from at least six orders including four ant species (Ryti and Case 1988). *Macrobaenetes valgum* eats plant detritus (14.8 percent of its diet), animal detritus (40.7 percent), and even conspecifics (less than 1 percent). (Many herbivorous insects, especially Orthoptera, are normally cannibalistic; Polis 1984b, Walter 1987.)

Most (16 of 18) mammal species from the CV are reported to eat plant tissue (the two bats did not; Appendix 13.1). Plants (fruit) constituted 0.2–4.1 percent of the diet of the largest mammal, the coyote (Johnson and Hansen 1979). This is the smallest plant component for any of the 16 mammals. Morell (1972) reported that more than 50 percent of the scats of the desert kit fox contained plant material. The two rabbits and the gopher are apparently the only trophic specialists; however, they are resource generalists primarily eating stems and/or leaves of many plant species.

The omnivorous whitetail antelope squirrel fed on a seasonally changing proportion of plant material (10–60 percent foliage, 20–50 percent seeds, 62–95 percent total plants by volume; Bradley 1968). Rodents (*Dipodomys, Perognathus, Peromyscus*) fed on seeds and plant parts (Bradley 1968; Bradley and Mauer 1973; Meserve 1976). The ratio of seeds to plant parts (and arthropod prey) varied seasonally. In total, 15 mammal species are reported to eat seeds (only bats and the gopher do not). Nine species regularly consumed more than 50 percent seeds in their diet.

None of the mammals are specialists on particular plant species. The closest to being a resource specialist is the kangaroo rat *Dipodomys merriami*. Soholt's (1973) study at nearby (110 km distant) Joshua Tree National Monument indicates 17 different diet items for *D. merriami*; however, only 4 were present in stomachs (N = 172) by more than 0.5 percent mass. In contrast, *D. merriami* (N = 1,054) in Arizona regularly eat 12 species (Reichman 1975). All other herbivores living in the CV are resource generalists; for example, pocket mice (*Perognathus formosus*) feed on 27 plant species (French et al. 1974), and whitetail antelope squirrels feed on 24 species (Bradley 1968).

With the exception of the gopher and the two rabbits, all plant-eating mammals are trophic generalists that included arthropods in their diet (Appendix 13.1), for example, 1 percent (Soholt 1973) and 17 percent (Reichman 1975) for *D. merriami*, and 2–35 percent for the whitetail antelope squirrel (Bradley 1968) (this species also eats vertebrates; see below).

Many birds (34 of 56 species) in the sand community are herbivorous, feeding on many different plant parts (seeds, nectar, flowers, and fruit) (Walsberg 1975; Weathers 1983) (Appendix 13.1). Frugivorous birds are common in the CV. Fruits of *Lycium*, cactus, and mistletoe are consumed by such birds as the Cactus Wren, Phainopepla, Verdin, Anna's Hummingbird, and doves. Some birds include fruit as a minor component of an omnivorous diet (e.g., roadrunner, Scott's Oriole, Western Tanager, Western Bluebird, Warbling Vireo, Bewick's Wren). Granivory is also common in CV birds: 22 of 56 were reported with seeds in their diet (13 are primarily granivorous; Appendix 13.1). Many insectivorous birds consume seeds while foraging (e.g., 2.5 percent seeds in the diet of the predaceous Loggerhead Shrike; Miller 1931). Many of the primarily insectivorous birds eat significant quantities of seed when insects are scarce (Wiens this volume; Brown, Reichman, and Davidson 1979; Brown 1986).

None of the herbivorous birds are resource specialists on particular plant species. In fact, trophic specialists are rare. Of the 34 plant-eating birds, only 5 are not recorded to eat arthropods. For example, 4.8 percent of the prey mass eaten by Gambel's Quail in the CV is the harvester ant *M. pergandei* (Goldstein and Nagy 1985). This ant, a competitor for seeds, is an important seasonal source of water for quail. In general, fruit- and seed-eating birds regularly consume insects associated with their plant foods (Polis, Myers, and Holt 1989).

Two species of CV reptiles are primarily herbivorous, the desert tortoise (Burge and Bradley 1976) and the desert iguana (Minnich and Shoemaker 1970). Both are resource generalists eating a wide variety of plants (17–40 species for the tortoise) and plant parts (stems, flowers, buds, fruit). Only the tortoise is a trophic specialist.

The diet of the desert iguana contains 1–5 percent arthropods. Five of the nine other lizards (but none of the 10 snakes) eat a minor portion of plants (e.g., *Lycium* berries; see Appendix 13.1).

Detritus and Soil Biota

Detritus is a broad term applied to nonliving organic matter originating from a living organism. It may originate from plants (e.g., wood, senescent leaves, seed capsules, flower parts, and roots [especially rhizodeposition]) or animals (feces, urine, secretions, molted skin or fur, or dead animals). Dead vertebrates, referred to as carrion, are considered in the next section.

Most primary productivity within a community flows directly or indirectly through the detrital component of the food web. The idea of a trophic pyramid of herbivores feeding on plants and forming food of carnivores is an incorrect simplification because these two links of the food chain process only one-third to one-half the total energy (Macfayden 1963; Odum and Biever 1984). The percentage of net primary production eaten by herbivores varies from 1 percent to 50 percent; the rest enters the detrital system (Odum and Biever 1984). Macfayden (1963) suggests that 30–73 percent of the energy flow through all microarthropods is via detritivores. Living plants constitute only 6–8 percent of the total biomass of vegetation in the Namib desert; detritus forms 92–94 percent (Seely and Louw 1980). Nevertheless, Cousins (1980) is one of the few explicitly to incorporate detritus into food web analysis (see also Odum and Biever 1984). He disputes placing autotrophs alone at the basal position of food webs; rather, herbivory and detritivory should be considered equally important as links in a "trophic continuum." Energy, produced by autotrophs, is recycled and made available to other consumers by detritivores.

This is particularly so in desert ecosystems where the main energy flow often proceeds directly from autotrophs to detritivores (Evenari 1981; Wallwork 1982; Seely and Louw 1980; see Crawford, Zak and Freckman, Seely this volume). The plant-herbivore-carnivore link forms 12 to less than 33 percent of the fate of plant production in deserts; the remainder goes through the soil and the detrital food chain.

Organisms of wide diversity live within desert soils (Wallwork 1982; El-Kifl and Ghabbour 1984; Crawford, Zak and Freckman this volume). Many taxa of microorganisms (fungi, yeast, bacteria, protozoa) decompose detritus in desert soils (Vollmer, Au, and Bamburg 1977; Ghabbour et al. 1980; Venkateswarlu and Rao 1981). Several families of nematodes and mites are adapted to and widely distributed within deserts (see references below). Termites, Collembola, Thysanura, burrowing cockroaches, tenebrionid larvae, some ants, millipedes, and isopods are some of the more important of the many detritivorous soil arthropods (Crawford 1979, 1981, this volume). All are common in the CV.

Although some species in these taxa degrade organic material, others are facultative or obligate herbivores on belowground plant parts (Crawford 1979, 1981, 1986). Roots and tubers represent a rich food source in deserts (Ludwig 1977). Over 50 percent of net primary production is commonly allocated to belowground plant parts (Andersen 1987). For example, in Russian deserts, up to 80–90 percent (average =

65 percent) of the plant biomass is belowground (Rodin and Bazilevich 1964). Species from seven orders of insects, mites, nematodes, and some rodents have adopted belowground herbivory as their primary feeding mode (Andersen 1987).

Soil-inhabiting organisms are quite abundant in deserts. Nematode biomasses of $1-20$ g/m^2 normally occur (Freckman, Mankau, and Ferris 1975; Zak and Freckman this volume). Detritivorous arthropods represent 37–93 percent (average = 73 percent) of all individual macroarthropods in four deserts analyzed by Crawford (this volume). In particular, termites are abundant in deserts and their biomass is often an order of magnitude greater than that of any other animal taxon (Whitford 1986; see Crawford, MacKay, Polis and Yamashita this volume). Wallwork (1982) emphasized the importance of termites in desert food webs: they fix atmospheric nitrogen, consume large quantities of detritus, change the physical and chemical characteristics of soil, recycle nutrients within their colonies via tropholaxis and cannibalism, and ultimately release their nutrients to a diversity of predators.

The abundance and diversity of soil fauna vary greatly as a function of time and microhabitat (Freckman and Mankau 1977; Vollmer, Au, and Bamburg 1977; Franco, Edney, and McBrayer 1979; Whitford 1986; see Crawford, Zak and Freckman this volume). Marked successional and seasonal changes occur. Whitford (1986) presents substantially different food webs for buried versus surface litter. Microbial populations vary with soil depth, soil moisture, and stability of substrate (populations are 200 times more abundant in stabilized versus unstabilized sand dunes; Venkateswarlu and Rao 1981). Such spatial and temporal variation produces dynamic patterns of energy and nutrient flow.

A rich food web based on detritus and underground plant parts exists within desert soils (e.g., El-Kifl and Ghabbour 1984; Whitford 1986; see Crawford, Zak and Freckman this volume). Freckman and Mankau (1977) identified several distinct trophic roles of nematodes: herbivores and plant parasites, microbial feeders, fungal feeders, omnivores, omnivore-predators, and parasites. Several feeding groups of mites live in desert soils (Santos, Phillips, and Whitford 1981). In the Mojave, Franco, Edney, and McBrayer (1979) recognized four trophic groups of soil mites within the Prostigmata: fungivores and detritivores (three families), phytophages (one family), parasitic (one family), and predaceous (eight families). Predatory mites can be particularly abundant; Wallwork (1972) found 1:1 ratios (non-predatory:predatory mites) in litter in nearby Joshua Tree National Monument. These mites eat nematodes, Collembola, and other mites. The large number and diversity of predatory mites led Edney et al. (1974) to postulate the possibility of two or more predator trophic levels in the decomposer web. Wallwork (1972, 1982) and Santos, Phillips, and Whitford (1981) suggested that decomposer pathways in the soil fauna were regulated by mites that prey on nematodes that feed on microorganisms.

Soil trophic interactions become even more complex with the inclusion of macroarthropods. Almost all larger detritivores not only eat detritus, but also feed on microorganisms (bacteria, fungi, protozoa) feeding and living in the detritus (Janzen 1977). For example, in the CV, the burrowing cockroach *Arenivaga investigata* feeds belowground on living and decaying plant roots sheathed by mycorrhizae; fungal hyphae are very numerous in the crop (Hawks and Farley 1973). Further, many arthropod detritivores (e.g., cockroaches, tenebrionids, millipedes) are host to cellulolytic gut symbionts that degrade plant detritus (Crawford and Taylor 1984). Although such symbionts represent a separate energetic "trophic level," they are usually not included in food web analysis. Finally, several desert insects eat detritus directly or are predaceous on nematodes and microarthropods, for example, in the CV, larvae of (asilid, bombyliid, and therevid) flies and (staphylinid and clerid) beetles (Edney et al. 1974; Powell and Hogue 1979).

Interactions in desert soils are linked to the entire community. Many surface-dwelling desert organisms (and most arthropods) either spend part of their lives in the soil (as larvae, e.g., tenebrionids) or feed on macroarthropod taxa that live permanently or temporarily under the ground (e.g., Pianka and Parker 1972; Vitt and Ohmart 1977; Mitchell 1979; see Ghilarov 1964, 1968). For example, Polis (1979) reports that 46 percent of the prey individuals of *P. mesaensis* in the CV live in the soil as larvae. *A. investigata* forms 11.4 percent by frequency, and 22.9 percent by weight of this scorpion's diet. Tenebrionids and termites are particularly important conduits of energy flow from below to aboveground, where they are subject to intense predation by a diverse group of arthropod and vertebrate predators (Deligne, Quennedey, and Blum 1981; Wallwork 1982) (Table 13.2 lists known predators from the CV). Such predation by surface dwellers on soil insects exports much of the energy recycled by detritivores and links the soil subweb to that which occurs above the surface. Thus, even if consumers (herbivores and detritivores) operate in distinct microhabitats, energy flowing further into the community merges into the bodies of predators common to both consumers (Odum and Biever 1984).

Although the above studies outline trophic interactions within detritus and soil, no study has explicitly analyzed these interactions in the CV. The great majority of the studies cited above were conducted in deserts (the Mojave and Chihuahuan) geographically adjacent to the CV. I combined information from these studies (especially Whitford 1986 and Zak and Freckman this volume) with my field data to construct a soil/detritus food web (Figure 13.2) that should describe trophic interactions within the CV. Within this subweb, note the complexity (even with extensive lumping), loops, omnivory common at nonadjacent trophic levels, closed-loop omnivory, and food chains of four to five links. Also note the links between belowground consumers and their aboveground predators.

Finally, the complex subweb associated with a special class of detritus (animal feces) should be included. For example, Schoenly (1983) found 30 species of arthropods in mammal dung from the Chihuahuan Desert. Coprovores, predators, and omnivores which eat dung and other arthropods were abundant. A clear succession of species occurs, with termites eventually finishing the decomposition process. Most species in all trophic classes fed on foods other than dung, thus connecting this web to the rest of the desert community.

Carrion Feeders

The decomposition of carrion results from the cumulative action of microorganisms, necrophagous insects, and some vertebrates. A rich carrion fauna occurs

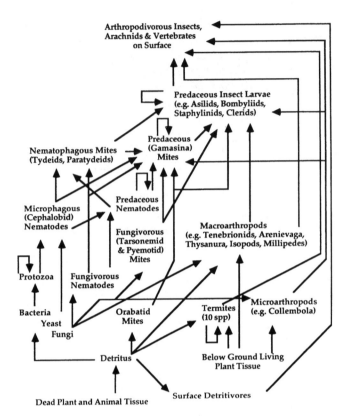

Figure 13.2. Trophic interactions within sandy soils in the Coachella Valley, Riverside County, California. The identity of a few of the important detritivorous species is as follows: Isopoda: *Venizillo arizonicus*; Collembola: Entomobryididae; Thysanura: *Leucolepisa arenaria*, *Mirolepisma deserticola*; Gryllacrididae: *Macrobaenetes valgum*; Blattidae: *Arenivaga investigata*; Isoptera: five species of *Amitermes*, and five other species; Tenebrionidae: 14 species. Not all trophic links are represented (e.g., for tenebrionids and termites). An arrow returning to a taxon indicates cannibalism. See Appendix 13.2 for identity of other detritivorous arthropods.

in deserts (Crawford this volume) and feeding interactions are complex. Many (28 to more than 500) species can be involved. Some species specialize on particular tissue, others do not. Trophic specialists and generalists occur. Species composition and diversity change through time.

Although carrion decomposition has not been studied in the CV, studies from the Chihuahuan Desert illustrate carrion use (McKinnerney 1978; Schoenly and Reid 1983). In particular, McKinnerney analyzed the fauna associated with carrion of two rabbit species common to both deserts. Overall, 63 arthropod and 4 vertebrate species were identified. Many families of Diptera and Hymenoptera and some families of Orthoptera, Coleoptera, and Lepidoptera consume carrion directly. Coyotes and Great Horned Owls also fed on rabbit carrion. Lizards, skunks, spiders, solpugids, reduviid assassin bugs, asilid robber flies, and carabid and staphylinid beetles were all generalist predators on necrophagous insects. Some predators specialized; for example, tachinid flies and braconid wasps specifically attacked dipteran larvae; larvae of clerid beetles attacked those of dermestid beetles. Hister beetles ate both fly and beetle larvae. Ants, silphid beetles, and Opiliones were omnivorous, eating both carrion and larvae; some larvae were those of other predators (forming loops via mutual predation).

Other trophic interactions occur. For example, some insects and spiders are attacked by parasitoids. Vertebrates not only eat carrion but also eggs and larvae of necrophagous insects. Carrion feeders also consume microorganisms within the carrion (Janzen 1977). Further, many carrion species are well-known cannibals, for example, callophorid fly larvae (see references in Polis 1981).

As with other detritus, much of the energy from carrion is exported to the rest of the community. McKinnerney (1978) noted that most organisms associated with carrion were opportunistic; for example, spiders, solpugids, Opiliones, ants, asilids, staphylinid beetles, reduviids, and the vertebrates not only eat insects associated with carrion (or carrion itself) but also prey on other species. In turn, all these species are eaten by other predators.

Arthropod Parasitoids

Parasitoids occur in several families of flies, wasps, and beetles (see Appendix 13.2). Parasitoids deposit eggs in or on an arthropod host; the developing larva feeds on and causes the death of the host (in contrast to parasites in general). Adults almost always feed on other foods (usually of plant origin). The trophic relations involving parasitoids are generally quite complex (Askew 1971; Price 1975; Pimm 1982; Hawkins and Goeden 1984; Polis and Yamashita this volume). Hawkins and Lawton (1987) estimate that each species of insect herbivore is host to 5–10 species of parasitoids.

A few studies detail parasitoid-host relationships in the CV sand community. Hawkins and Goeden (1984) studied insects associated with saltbush (*Atriplex*) galls. The system is complex with 67 species (40 common ones), at least five trophic links, and extensive omnivory. Gall-forming midges (3 species), parasitoid Hymenoptera (26 species), predators (4 species), and inquilines (7 species) interact within galls (Figure 13.3). The midges are trophic specialists on plants and either resource specialists on *Atriplex* or generalist gall formers on five other plant species. Most parasitoids are primary, attacking only midge larvae or inquilines; seven of these also feed on gall tissue. Two species are facultative hyperparasitoids. One, *Torymus capillaceus*, feeds on gall tissue, gall midges, and primary parasitoids. The other, a eupelmid, feeds on midges, primary parasitoids, and *T. capillaceus*. The top predator, a clerid beetle (*Phyllobaenus*), occurs in 10 percent of galls and feeds on at least 17 species from all animal groups. (This clerid also parasitizes 6 percent of *Diguetia* spider egg cases on *Atriplex*; see Figure 13.5.) Hawkins and Goeden argue that an omnivorous strategy of entomophagy and phytophagy is adaptive to these and other parasitoids because it provides more potential food, growth to a larger size, and potential competitive dominance (e.g., for *T. capillaceus*).

The trophic relations of *Photopsis* (an abundant mutillid wasp in the CV sand community) are diagrammed in Figure 13.4 (from Ferguson 1962 and my unpublished data). Females oviposit in the cells of developing larvae and the *Photopsis* larva consumes the entire host. They are not host specific and parasitize several species of hymenopteran larva, and are hyperparasitoid on parasitoids of these larva (i.e., other Hymenoptera, and immature stages of [stylopid, meloiid, and rhipiphorid] beetles and [bombyliid] flies). Some hyperparasitized Hymenoptera (e.g., sphecid wasps) also may be parasitoids on spiders (this is likely but not established). Up to 37 percent are destroyed when *Photopsis* larvae themselves are parasitized by

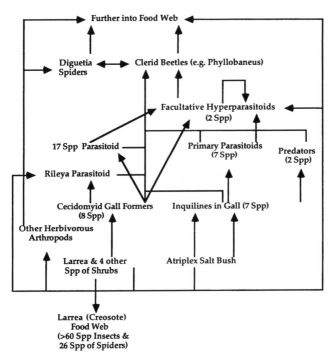

Figure 13.3. Trophic interactions within galls on saltbush (*Atriplex canescens*) within the Coachella Valley, Riverside County, California. In total, 67 species interact within galls formed by Cecidomyidae (midge) larvae. (Many of these species are also involved in the subweb centering on creosote [*Larrea divaricata*; see Shultz, Otte, and Enders 1977].) Interactions of some species are not fully represented (e.g., *Diguetia*, *Phyllobaenus*). An arrow returning to a taxon indicates cannibalism (Modified from Hawkins & Goeden 1984.)

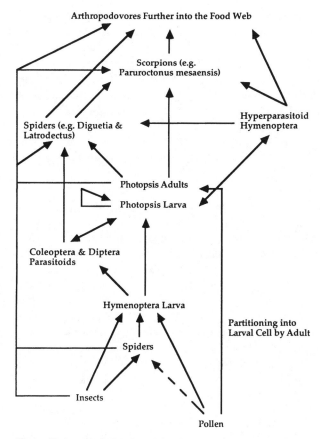

Figure 13.4. Trophic interactions involving parasitoid mutillid wasps in the genus *Photopsis* within the Coachella Valley, Riverside County, California. Interactions of some species are not fully represented (e.g., scorpions, spiders). A double-headed arrow indicates looping via mutual predation.

some of the same parasitoids (e.g., sphecids) that fall host to *Photopsis*. This is an example of looping via mutual parasitism and tertiary parasitism. Some *Photopsis* larvae are also host to bombyliid (e.g., *Bembix*) and stylopid parasitoids. Further, *Photopsis* larvae are also cannibalized (and self regulated?; Ferguson 1962) by conspecifics. Adults are herbivores on nectar, pollen, and flowers but also eat ground-nesting Hymenoptera. Adults frequently are prey to many predaceous arthropods (e.g., *Diguetia*, *Latrodectus*, *Steatoda*, and other spiders and *Paruroctonus* scorpions; see Figures 13.5 and 13.6).

These parasitoid subwebs are characterized by high omnivory on many trophic levels, closed-loop omnivory, frequent looping (via cannibalism, mutual predation, and three-species loops), and many trophic levels. Further, key species export energy from this subweb when they act as predators or prey in the rest of the community.

Many more parasitoids eat eggs and larvae of CV insects. For example, eggs of the grasshopper *Bootettix punctatus* occasionally suffer high mortality rates from ovipositing *Mythicomia* bombyliid flies (Mispagel 1978). Other dipteran parasitoids (Bombyliidae, Sarcophagidae, Tachinidae) develop within eggs and immature stages of Orthoptera, Coleoptera, and Hymenoptera. Several hymenopteran parasitoids are common in the CV, for example, tiphiid, mutillid, sphecid, ichneumonid, and Chalcidoidea wasps. These develop in Orthoptera, Neuroptera, Lepidoptera, Hymenoptera, and Diptera. Eggs of the common CV tiphiid *Brachycistus* develop on grubs of scarab beetles (Wasbauer 1973). Larvae of red velvet mites (*Dinothrombium pandorae*) are external parasitoids of CV grasshoppers (Tevis and Newell 1962) (adults are termite predators).

Spiders are hosts to parasitoids (pompilid, sphecid, and ichneumonid wasps as well as various dipterans). Adult wasps partition nests with captured spiders; developing wasp larvae eat the moribund spiders. Adults are usually nectar eaters. Wasps vary greatly in specificity, from resource specialists on particular families to generalists attacking several families (Wasbauer and Kimsey 1985; Austin 1985). Many pompilids (> 11 species) occur in the CV (Wasbauer and Kimsey 1985). The abundant *Aporus hirsutus* (and the less common *Psorthaspis planata*) feed trapdoor spiders (*Aptostichus*, Ctenizidae) to their larvae; adults drink sugar secretions from aphids and nectar from at least 10 plant species. Some CV pompilids are hyperparasitoids (e.g., *Evagetes mohave*). Spiders are also beset by a diversity of egg parasitoids and predators (Austin 1985; Polis and Yamashita this volume). Further, kleptoparasitic insects (particularly Drosophiloidea flies) and spiders parasitize spiders by robbing captured prey (Sivinski 1985; Polis personal observation).

Parasites

Parasites feed on their host over a long period of time; consequently, they do not cause the immediate (or, usually, the ultimate) death of their host. Although seldom represented in food webs, they constitute an extra trophic level; some ectoparasites form yet another "gratis" level when they themselves host parasites (Marshall 1981). Parasites are not well studied in natural communities. Of animals living in the CV, only the parasites of coyotes, lizards, and scorpions have been examined in any detail. Otherwise, there is little information on the hundreds of species of endo- and ectoparasites that infest animals in the sand community. The following information is likely characteristic of these unstudied taxa.

Telford (1971) identified protozoan and helminth endoparasites in 10 CV lizards. These lizards were infected by several protozoan parasites (average = 7.8 species; range = 3–10) including flagellates, ciliates, amoebas, sporozoans, and haemogregarines. Helminth parasites (average = 1.4 species; range = 0–5) included nematodes, cestodes, and Acanthocephala. In addition, each lizard was infested with an unknown number of mite species. Indeed, every lizard sustains its own food web, a community of parasites.

Gier, Kruckenberg, and Marler (1978) discussed coyote parasites. External parasites include mange mites, ticks (seasonal), lice (rare), and *Pulex* fleas (on all individuals). Adult fleas feed on the host's blood; flea larvae develop in the burrow and feed on organic debris, particularly adult flea feces. An internal parasite, the tapeworm *Taenia pisiformes*, occurs in 60–95 percent of all stomachs. The prime intermediate host of the tapeworm is the cottontail rabbit; the coyote infestation rate was correlated with the percentage of rabbit in the diet.

It is likely that most (all?) of the free-living animals in the CV harbor one or more parasite species. For example, examination of 1,525 birds (from 112 species) from deserts and other areas in the southwestern United States showed that 23.4 percent had blood parasites (Woods and Herman 1943; Welty 1962 lists the diverse parasite fauna of birds). Inspection of CV spiders and insects almost always reveals mite infestation. Many genera of spiders in the CV were reported with nematode parasites (Poinar 1985). CV scorpions support nematodes and eight species of (pterygosomid) mites (McCormick and Polis 1990).

Arthropod Predators

Arthropods are one of the most important conduits of energy flow in desert webs. Most consumed primary productivity in deserts is utilized by arthropods rather than vertebrates (Seely and Louw 1980). These arthropods, in turn, are eaten by a host of predators, the vast majority of which are other arthropods (Crawford 1981). Many predaceous arthropods are dense and may play important community roles (Polis and Yamashita this volume); in the CV these include species of scorpions, solpugids, spiders, mantids, antlions, robberflies, small carabid beetles, and facultatively predaceous ants. A great variety of arthropods are predaceous sometime during their lifehistory. In the CV, 8 families of mites, more than 23 families of arachnids, and at least 21 families of insects are predaceous both as juveniles and adults (Appendix 13.2). The complex life cycle of holometabolous insects often results in different feeding habits between life stages. At least 27 families of Coleoptera (e.g., Cleridae, Meloidae), Diptera (e.g., Tachinidae), and Hymenoptera (e.g., Tiphiidae) are trophic generalists: they are predaceous as larvae and herbivorous as adults (Appendix 13.2). For example, *Pherocera* (Therevidae) fly larvae, in addition to being cannibalistic, are predators on beetle, fly, and moth larvae in sandy CV soils, and adults are nectar feeders (Irwin 1971). Further, some parasitic Hymenoptera (e.g., Sphecidae, Pompilidae) function as predators: adults eat plant material (pollen, nectar) but capture live prey; these prey are eaten by developing larvae. Finally, some omnivorous taxa (e.g., ants, camel crickets, the gryllacridid *Stenopelmatus coahuilensis*) are occasionally but regularly predaceous.

Some of these predators are uncommon; others (mites and insects) are common only in some periods. However, many are quite dense and likely play an important role in the community (see Polis and Yamashita this volume). Particularly abundant arthropod predators in the CV include the scorpion *P. mesaensis*; several solpugids (e.g., *Chanbria coachella*); the spiders *Diguetia mohavea*, *Latrodectus hesperus*, *Steatoda fulva*, *S. grossa*, *Dictyna* spp., and several salticids (e.g., *Habronatus* spp.); antlions (e.g., *Brachynemurus*); mantids (e.g., *Stagmomantis californicus*); several robberflies (e.g., *Efferia*, *Proctacanthella*); small carabid beetles (e.g.,

Calosoma, Tetragonoderus); and facultatively predaceous ants (*M. pergandei, Pogonomyrmex* spp.).

Most are resource generalists (Polis and Yamashita this volume). For example, *P. mesaensis* is recorded to eat more than 125 prey species; *D. mohavea*, more than 70 species; and *L. hesperus*, 35 species (see below). In fact, some scorpions and spiders are neither true trophic specialists nor obligate predators: species of both scavenge dead arthropods, and some spiderlings are aerial plankton feeders, eating pollen and fungal spores trapped by their web (see references in Polis and Yamashita this volume). Facultative predators (e.g., the omnivores above) are trophic generalists that eat plant material, detritus, dead arthropods, and live prey.

A few predaceous arthropods tend to specialize. Adult velvet mites (*D. pandorae*) feed almost exclusively on termites. *Mimetus* spiders (Mimetidae) primarily prey on other spiders. Many parasitoids have preferred host species; others, however, are more generalized (see Figures 13.3 and 13.4).

Trophic interactions within predatory arthropods are complex. The generalized diet of most species is established by predator/prey size relationships: they catch whatever (smaller) prey they can subdue. Consequently, most smaller arthropods are potential prey, and predators commonly eat from all trophic levels including other predators (Polis and Yamashita this volume). For example, the diet of six CV predaceous arthropods averages 51.5 percent other predaceous arthropods (Table 13.3). Such predation is particularly common in deserts because predators form such a high proportion of desert arthropods (Crawford, Polis and Yamashita this volume). Age structure also promotes predator-predator feeding interactions. During their normal growth through a wide size range, predaceous arthropods are potential prey to larger predators. All these factors produce extensive cannibalism (Polis 1981), intraguild predation (Polis, Myers, and Holt 1989), looping, and mutual predation during ontogenetic reversals and closed-loop omnivory (e.g., see Figures 13.5 and 13.6).

Predaceous arthropods are particularly important in desert food webs. The greatest proportion of consumed primary productivity in deserts is utilized by herbivorous and detritivorous arthropods rather than vertebrates (Seely and Louw 1980 and included references). These arthropods, in turn, are eaten by a host of arthropodivores, the vast majority of which are other arthropods (Hadley 1980; Crawford 1981). Thus, predaceous arthropods are one of the most important conduits of energy flow in desert food webs.

It is impossible to construct a food web representing all CV predaceous arthropods. Therefore, I will present webs centered around three common species that I have studied extensively: the scorpion *P. mesaensis* (see earlier references) and the spiders *D. mohavea* and *L. hesperus* (Polis and McCormick 1986b; Polis and Sculteure unpublished data; Nuessly and Goeden 1984 also report the diet of *D. mohavea* in the CV).

Figure 13.5 diagrams trophic relations centered on *D. mohavea* and *L. hesperus*. Spiderlings of *D. mohavea* develop in egg sacs deposited within a retreat protected by the mother until her death. Eggs and spiderlings are then eaten by species that invade the retreat (Figure 13.5). Predaceous invaders include spiders (nine families,

Table 13.3 Diet classification of some representative predators of arthropods in the Coachella Valley, Riverside County, California.

Taxon[a]	% Arthropod Taxa in Diet		
	Predaceous & Parasitoid	Herbivorous & Detritivorous	N
Arachnids			
Hadrurus arizonensis[b]	53	47	15
Paruroctonus luteolus[b]	60	40	10
P. mesaensis[b]	47	53	126
Vaejovis confusus[b]	50	50	12
Diguetia mohavea[c]	45	55	71
Latrodectus hesperus[c]	54	46	35
Arachnid average ± s.d.	51.5 ± 5.4	48.5 ± 5.4	47.3
Lizards			
Callisaurus draconoides	45	55	22
Cnemidophorus tigris	46	54	15
Gambelia wislizenii	33	67	15
Phrynosoma platyrhinos	28	72	18
Uta stansburiana	39	61	28
Birds			
Blue-gray Gnatcatcher	36	64	69
Burrowing Owl	36	64	14
Loggerhead Shrike	35	65	17
Roadrunner	35	65	23
Mammals			
Ammospermophilus leucurus	71	29	7
Antrozous pallidus	31	69	16
Onychomys torridus	40	60	15
Pipistrellus hesperis	32	68	22
Vulpes macrotus	67	33	6
Vertebrate average ± s.d.	41.0 ± 12.9	59.0 ± 12.9	20.5

[a]A taxon is the designated unit in which the diet was classified by the author. It varies from species to families to orders.
[b]Scorpion species.
[c]Spider species.

>14 species), solpugids, mites, mantispids, chrysopids, and the clerid *Phyllobaenus* (from Figure 13.3). In general, eggs and spiderlings within egg sacs of most spider species are attacked by members of these taxa and of several wasp and fly families, and by other spiders (Austin 1985). Sibling cannibalism is also frequent (Polis 1981) and occurs among *D. mohavea* spiderlings. At least three trophic levels occur in the *D. mohavea* retreat: the clerid likely eats other egg predators and is itself host to a pteromalid wasp parasitoid.

Adult *Diguetia* prey on more than 70 species including 14 families of predatory insects (*Photopsis* and the same invading *Phyllobaenus* clerid are prey) and 8 spider species, including the same species of invading salticids and the araneophagous spider *Mimetus*. One-third of all *D. mohavea* webs included remains of salticid prey.

Adult *D. mohavea* are fed upon by *Mimetus*, *P. mesaensis*, birds, and a parasitoid pompilid wasp. *D. mohavea* is involved in at least four cases of looping mutual predation.

Predators constitute 54 percent of the diet of *L. hesperus* (Table 13.3). It falls prey to at least seven predators, including three species which it eats (mutual predators are *Steatoda grossa*, *S. fulva*, *P. mesaensis*, and other *L. hesperus*). The blue mud wasp (*Chalybian californicum*, Sphecidae) specializes on *Latrodectus* and other theridiid spiders (e.g., *Steatoda*) (Wasbauer and Kimsey 1985). *Tastiotenia festiva* (Pompilidae) is a hyperparasitoid preying upon both cached theridiid spiders (e.g., *Latrodectus* and *Steatoda*) and the developing wasp.

Paruroctonus mesaensis is one of the largest (adults = 2–2.5 g) and most dense (1,500–3,500/ha) arthropod predators in the CV. Its population biomass (g/ha) is the

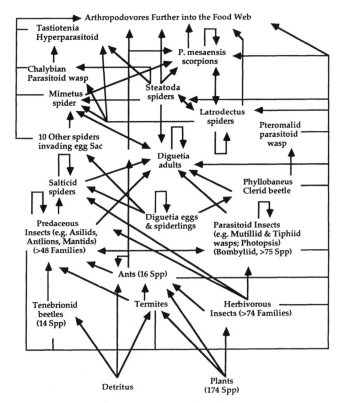

Figure 13.5. Trophic interactions above the soil surface involving a few of the predaceous arthropods living within the Coachella Valley, Riverside County, California. This subweb is focused around the spiders *Diguetia mohavea* and *Latrodectus hesperus*. An arrow returning to a taxon indicates cannibalism. No vertebrates are represented.

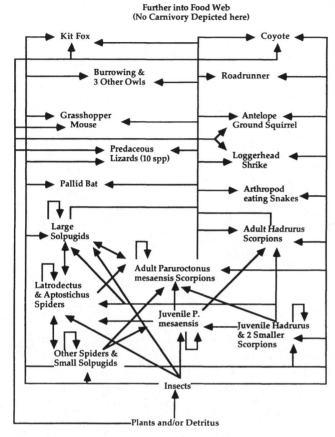

Figure 13.6. Trophic interactions centered around the prey and predators of the scorpion *Paruroctonus mesaensis*. An arrow returning to a taxon indicates cannibalism. A double-headed arrow indicates looping via mutual predation. Interactions of some species are not fully represented (e.g., spiders and insects) and carnivory on vertebrates is not depicted (see Figure 13.7).

greatest of any CV predator (vertebrate or invertebrate). Ontogenetic shifts during growth partially explain the existence of trophic interactions with more than 125 species of prey, including 47 percent other predators (Table 13.3) and mutual predation with at least 10 species (5 solpugids, 3 scorpions, 2 spiders; young *P. mesaensis* are eaten by the same species eaten by adults). Trophic relationships of *P. mesaensis* are presented in Figures 13.5 and 13.6.

Note the complexity of these webs: looping via mutual predation and cannibalism is frequent; (closed-loop) omnivory is the norm; omnivorous predators feed on herbivores, detritivores, predators, and predators of predators. Consequently, chain lengths are long even without including parasitoids or loops; for example, detritus–

Table 13.4 Feeding categories of vertebrates resident in the Coachella Valley, Riverside County, California.

Feeding Category	Reptiles			Birds			Mammals			All Vertebrates		
	1[oa]	2[ob]	Total	1[oa]	2[ob]	Total	1[oa]	2[ob]	Total	1[oa]	2[ob]	Total
Granivory	0	0	0	14	0	14	8	0	8	22	0	22
%	0	0	0	25	0	25	44	0	44	23	0	23
Herbivory	2	5	7	5	15	20	5	1	6	12	21	33
%	10	24	33	9	27	34	28	6	33	12	22	34
Arthropodivory	11	4	15	34	15	49	9	5	14	54	24	78
%	52	19	71	61	27	88	50	28	78	58	25	83
Carnivory	9	2	11	12	2	14	3	1	4	24	5	29
%	43	10	52	21	4	25	17	6	22	25	5	31
Total # species		21			56			18			95	

[a]Primary category: food class is a major (>90%) component of diet. Some omnivorous species (two reptiles, nine birds, seven mammals) belong to two (each >33% of diet) or three (each >20% of diet) primary categories.
[b]Secondary category: food <10% of diet.

termite–*Messor* ants–antlion–*Latrodectus–Steatoda–Mimetus–P. mesaensis*–Eremobatid solpugids–*Hadrurus* scorpion–(vertebrate subweb).

I strongly suspect that the trophic interactions depicted among these species are representative of the hundreds of other arthropod predators in the CV. Omnivory (due to age structure, opportunism, and the generally catholic diets of these arthropods), combined with a high diversity of insect and arachnid predators, necessarily creates the observed complexity. Complexity increases even further when we consider vertebrate predators of these arthropods.

Arthropodivorous Vertebrates

Arthropodivory is the consumption of arthropods. In contrast to the more familiar (and less cumbersome) word *insectivory*, arthropodivory means the consumption of all types of arthropods (insects, arachnids, myriapods, and terrestrial Crustacea). The majority of vertebrates (83 percent of the 95 species) in the CV include arthropods in their diet (Table 13.4; Appendix 13.1). Over half (58 percent) are primarily arthropodivorous, including 52 percent of the reptiles, 61 percent of the birds, and 50 percent of the mammals. The majority (80 percent) of vertebrates that are primarily carnivorous on other vertebrates (20 of 25 species) also feed on arthropods. Two-thirds of the primarily herbivorous/granivorous vertebrates (24 of 36) eat arthropods, at times in large quantities (e.g., 88–97 percent of the seasonal diet of the Sage Sparrow in the Great Basin Desert; Wiens and Rotenberry 1979). In total, 71 percent of all reptile species, 88 percent of the birds, and 78 percent of the mammals primarily or secondarily eat insects and/or arachnids. Only 17 species are not reported to eat arthropods.

Arthropodivorous vertebrates are usually resource generalists that include trophically distinct arthropods in their diet, for example, arachnid and insect predators, detritivores, and herbivores. Of 36 arthropodivorous birds whose diet is detailed sufficiently, 58 percent eat spiders in addition to insects. Seven of the 10 lizards eat spiders and 4, scorpions. Spiders are eaten by 3 of the 14 arthropod-eating mammals; scorpions, by 5. Predaceous arthropods form 28–71 percent (average = 41 percent) of the diet of the CV vertebrate arthropodivores listed in Table 13.3.

Vertebrates that eat arthropods also tend to be trophic generalists. Most (28 of 55 species = 51 percent) primary arthropodivores eat plants and 59 percent of the total 79 vertebrates that eat arthropods also eat plants. Of these 79 species, 32 percent are also carnivorous on other vertebrates.

A few arthropodivorous vertebrates tend to specialize. Ants form 89 percent by frequency (56 percent by volume) of the prey of the horned lizard *Phrynosoma platyrhinos* (Pianka and Parker 1975). However, these lizards eat 17 other categories of prey (including spiders and solpugids), and 20–50 percent of the diet of some individuals consists of beetles. The worm snake (*Leptotyphlops humilis*) mainly eats termites and ants. There are no other examples of vertebrates specializing on certain taxa of arthropods.

The trophic and resource generalization exhibited by arthropodivorous vertebrates is illustrated in the food web focused around the predators of *P. mesaensis* (Figure 13.6). All the birds, reptiles, and mammals that eat *P. mesaensis* also eat (predaceous, detritivorous, and herbivorous) insects and even plant material (44 percent of the 16 predators). Many (81 percent) of the vertebrate predators of *P. mesaensis* are also carnivorous on other vertebrates (Figure 13.7) (3 of the 11 arthropod predators of *P. mesaensis* also eat vertebrates).

Carnivorous Vertebrates

Carnivorous vertebrates kill and eat other vertebrates. Many CV vertebrates are carnivores: 24 of the 95 species are primarily carnivorous and 5 others prey on vertebrates at least occasionally (Table 13.4). Reptiles as a group are the most carnivorous (9 primary and 2 secondary out of 21 species = 52 percent; 8 of 10 snake species are primary carnivores). Carnivores form about the same proportion of the bird (12 primary and 2 secondary of 56 species = 25 percent) and mammal (3, 1 of 18 species = 22 percent) faunas.

All carnivores are resource generalists preying on many species of vertebrates. Most (19 of 24 primary carnivores = 79 percent) are trophic generalists that eat arthropods (71 percent) and/or plants (33 percent). For example, coyotes eat other mammals (12 of the 19 species in the CV: rabbits, rodents, gophers, whitetail antelope squirrels, and even kit foxes and other coyotes), birds (including eggs and nestlings, e.g., Roadrunners, doves, quail), snakes (e.g., gopher and kingsnakes), lizards (e.g., horned lizards), young tortoises, arachnids (scorpions), insects, and fruit (Johnson and Hansen 1979; Polis unpublished data) (Figure 13.7). Foods other than vertebrate prey are purposefully hunted; for example, coyotes excavate entire

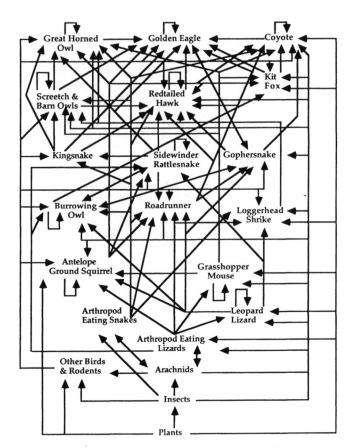

Figure 13.7. Trophic interactions involving a few of 96 vertebrates
that are resident in the Coachella Valley, Riverside County,
California. No top predator exists in this subweb or within the
Coachella Valley. An arrow returning to a taxon indicates
cannibalism. A double-headed arrow indicates looping via mutual
predation. The bottom of this subweb is simplified from the
preceding five subwebs (Figures 13.2–13.6).

ant nests in the CV (Ryti and Case 1986). Great Horned Owls consistently eat wolf
spiders, tarantulas, centipedes, and Orthoptera in addition to rodents, squirrels,
lizards, snakes, and other horned owls (Ohlendorf 1971; Johnegard 1988).

Many carnivores (33 percent) are reported to feed on carrion in addition to live
prey; these include sidewinder rattlesnakes, ravens, Golden Eagles, horned owls,
Red-tailed Hawks, coyotes, kit foxes, and whitetail antelope squirrels. Such scavenging means that these primary carnivores also include in their diets both microorganisms (Janzen 1977) and the rich arthropod fauna that use decaying carcasses,
for example, the horned owl (Ohlendorf 1971).

A food web focusing on many of the carnivores in the CV is presented in Figure 13.7. Note the frequency of omnivory (often on non-adjacent trophic levels) and closed loop omnivory. Most carnivores (9 of 16 species in Figure 13.7) are reported as cannibalistic. There are three cases of mutual predation. Each occurs because, although adult A eats B, the egg or nestling stages of A are eaten by B. For example, many snakes are nest predators (e.g., in the CV: whipsnake, sidewinder, gopher snake, and rosy boa). These snakes eat eggs and nestlings of species (e.g., the Burrowing Owl) whose adults are predators upon the same snakes (e.g., gopher snakes, sidewinder rattlesnakes).

Finally, note that only the Golden Eagle approaches the status of top predator (i.e., a species without predators). However, even the Golden Eagle may not be a true top predator: at other locations, gopher snakes eat Golden Eagle eggs; parasitic trichomoniasis and sibling fratricide/cannibalism cause nestling mortality (30 percent and 8 percent, respectively) (Olendoroff 1976; Palmer 1988). Thus, here is a community and food web with potentially no top predators.

Food Webs Compared

Sandy areas in the Coachella Valley are the habitat for about 175 species of plants, 100 vertebrates, and thousands of arthropods, parasites, and soil microorganisms. These species form a community trophically connected into a single food web that describes energy flow. In this section, I abstract trends from the CV food web and compare these with other (desert) food webs and with the "empirical generalizations" presented by food web theorists.

The Coachella Valley Food Web

Although I have presented only a few components of the CV food web, I summarize several general trends. An inspection of the subwebs (Figures 13.2–13.7) indicates the following:

1. Each subweb is complex. Complexity originates from the large number of interactive species, age structure, and the high degree of omnivory shown by many species. A web representing all species would increase complexity even more.
2. Long average chain lengths are characteristic. Even excluding looping and parasites, lengths of 6 to more than 11 links are the norm. Chains are lengthened primarily by trophic interactions within the soil biota and/or arthropods. Shorter chains exist (e.g., plant–rabbit–Golden Eagle) but these are less common because vertebrate consumers form a relatively small proportion of all species when arthropods and soil biota are not ignored.
3. Age structure is central to understanding the CV food web. Two growth-related factors greatly increase complexity. First, growth allows or requires individuals to change their diets ("life history omnivory"), sometimes gradually

(e.g., small and large conspecifics [reptiles, arachnids] eat different-sized prey), and sometimes radically (e.g., holometabolous insects that are predators as larvae and herbivores as adults). Second, potential predators change as a function of size. Younger stages are eaten by species that cannot eat adults.

4. Omnivory (feeding from several different trophic levels) is the norm. With the exception of the few species of strict herbivores (primarily insects), most species feed on several trophic levels. Food is frequently from non-adjacent trophic levels. "Closed-loop omnivory" is very common.

5. Looping during cannibalism and mutual predation is common among (and between) vertebrates and invertebrates. Cannibalism inevitably results when larger individuals eat smaller or younger conspecifics. Mutual predation results primarily from ontogenetic reversals of size relations (adult B eats juvenile A but B is eaten by the now-larger adult A). However, other factors also produce mutual predation (e.g., group predation among ants).

6. Top predators are rare or nonexistent. The largest predators in the CV (coyotes, kit foxes, horned owls, and Golden Eagles) suffer the fewest predators but each is the reported prey of other species.

7. Species composition and abundance change through time and space. Such variation, almost universal among species in the CV, adds a dynamic component to the web and increases omnivory.

8. Species from different microhabitats and times are connected trophically to one other. At first inspection, there appear to be several distinct compartments or "energy channels" within the CV, for example, diurnal versus nocturnal species, shrub versus ground species, surface versus subsurface species. However, extensive crossover exists among these potential compartments. Diel activity patterns of many CV species change from day to night as a function of temperature; nocturnal species eat diurnal prey (e.g., scorpions prey on honeybees, robberflies, and mantids). Plants and detritus are eaten during all time periods by species that live both above and below the surface. Many desert arthropods feed in the soil as larvae but export energy when they become surface-dwelling adults (e.g., tenebrionids). "Different channel omnivores" link communities together because they feed on prey from different subwebs (Moore, Walter, and Hunt 1988). Consumers eat food types rather than specializing on particular energy channels or trophic levels. Such decompartmentalization further increases complexity.

Other Desert Food Webs

The Coachella Valley is not unique in its diversity or properties of its food web. Other deserts are characterized by hundreds of thousands of species (see each chapter in this book). The same suites of plants, herbivores, detritivores, soil biota, parasites, parasitoids, arthropodivores, and carnivores are apparently present in all deserts. In particular, the universal existence of diverse assemblages of predaceous arthropods (Polis and Yamashita this volume) and soil organisms (Zak and Freckman this volume) must greatly contribute to a trophic complexity similar to that observed in the CV.

Omnivory is normal among consumers from deserts (Noy-Meir 1974; Orians et al. 1977; Seely and Louw 1980; Holm and Scholtz 1980; Hadley and Szarek 1981; Bradley 1983; Brown 1986). Desert granivores regularly supplement seeds with arthropods, not only for protein but also to balance water budgets (Brown, Reichman, and Davidson 1979; Brown 1986; Wiens this volume). Arthropodivores eat arthropods that are herbivores, detritivores, parasitoids, predators, and predators of predators (Orians et al. 1977; see any feeding summary); many also eat plants (Brown 1986). Noy-Meir (1974) hypothesizes that no true carnivores exist in the desert, only strongly flexible omnivores that prefer meat when they can get it (see also Brown 1986). He argues that almost all desert mammals, birds, and reptiles occasionally eat plants.

In general, low and unpredictable productivity resulting in chronic and seasonal food stress may promote flexibility in feeding. Noy-Meir (1974, p. 201) states that the "response of many animals, both herbivorous and carnivorous, to the uncertainty of food and water supply in arid environments has been dietary flexibility and opportunism. This is likely to cause difficulties in the quantitative analysis of energy transfer between species and trophic levels in the desert ecosystem. Instead of simple food chains or pyramids it may be necessary to consider rather complex food webs with many cross-links and shunts, changing in space and time." The same thoughts (omnivory and complex webs) are expressed by Hadley (1980) and Hadley and Szarek (1981).

Foraging theory provides one general explanation for such omnivory. Theory predicts and empirical studies almost universally show that diets are expanded either during periods of food scarcity or opportunistically to include temporarily available high-quality foods (such as fruit) (see references in Bailey and Polis 1987). Such dietary changes should be particularly frequent in deserts where prey availability changes by one to two orders of magnitude during the year (Noy-Meir 1974; Seely and Louw 1980; see Seely, Vitt, and Wiens this volume; for the CV see Polis 1988a).

To my knowledge, four desert food webs have been defined in the literature. Bradley (1983) presents a source web focused on the predators of camel crickets in sandy areas of the Chihuahuan Desert in New Mexico. Predators of the camel cricket (scorpions, solpugids, Burrowing Owls, grasshopper mice, and pallid bats) are extremely omnivorous (only two of the 27 possible species pairs are noninteractive, i.e., not linked as predator or prey). Closed-loop omnivory occurs for every predator. Looping is common: six cases of mutual predation occur and six of eight species are cannibalistic. Finally, at least 18 three-species loops exist (A eats B, B eats C, but C eats A).

Hadley and Szarek (1981) outline a source web based on the grasshopper *Trimerotropis*. It eats a variety of plants and is eaten by eight kinds of organisms. The web is quite incomplete but still contains closed loop omnivory and looping via mutual predation.

Whitford (1986) summarizes trophic interactions among soil detritivores in the Chihuahuan Desert. He simplifies the subweb by lumping species into broad trophic groups. He emphasizes that this web continually changes with time. Omnivores are present, but not in the majority. Closed-loop omnivory occurs but he does not illustrate looping via cannibalism or mutual predation.

Finally, the food web of the Namib Desert dune system has been compiled by Seely and Louw (1980) and Holm and Scholtz (1980) (see also Seely this volume). This is the only community web for a desert and discusses 136 species. However, these species tend to be lumped into broad trophic groups. Detritus is the most important energy source for Namib dune primary consumers (mainly tenebrionid beetles [33 species] and Thysanura; microbial decomposition is almost absent). Herbivory is relatively unimportant. Insects are eaten by a variety of arachnid and vertebrate arthropodivores. The authors comment on the relatively high diversity of arachnid predators (23 species; actually, >35 species exist; J. Henschel personal communication 1988). Further, 12 insects (not including omnivorous ants and parasitoid wasps) are predaceous. Reptiles and birds eat both herbivorous and predaceous arthropods and vertebrates. Vertebrates are eaten by two top predators, brown hyenas and jackals.

The authors note several trends. First, the Namib web is characterized by marked temporal and spatial variation in the distribution and abundance of species. Second, above- and belowground components are closely linked. Soil detritivores feed on detritus produced by surface-dwelling plants and animals. Surface and subsurface detritivores are the direct or indirect energy sources for all secondary consumers in this community. For example, the energy recycled by subsurface detritivores is exported to the surface when they become surface dwellers (e.g., post-metamorphic tenebrionids).

Third, Namib species are quite omnivorous. A few primary consumers eat arthropods in addition to plants or detritus; for example, the two gerbils frequently eat insects and spiders in addition to seeds, and two grasshoppers supplement herbivory with arthropodivory (and cannibalism). Some mutual predation and looping occur among the arachnids; for example *Leucorchestris* spiders eat conspecific and heterospecific spiders, scorpions, solpugids, and even geckos; these same geckos, spiders, scorpions, and solpugids eat *Leucorchestris* (Henschel in press). Almost all carnivorous species are omnivorous; for example, hyenas and lizards regularly eat plant matter. All carnivores also eat arthropods (insects from all trophic groups and arachnids). Such omnivory promotes extensive closed-loop omnivory.

Comparison with Published "Empirical Generalizations"

Analyses by theorists have produced a series of generalizations derived from catalogs of published food webs (e.g., Cohen 1978, 1988; Cohen and Briand 1984; Cohen, Briand, and Newman 1986; Pimm 1982; Pimm and Rice 1987; Briand 1983; Yodzis 1984). These generalizations (Table 13.1) are entering accepted ecological literature (see May 1986, 1988; Lawton and Warren 1988; Lawton 1989). For the following reasons, the food web of the CV offers little support for these patterns:

1. The number of interactive species in the CV food web is orders of magnitude greater than the average number (17.8: Briand 1983; 16.7: Cohen, Briand, and Newman 1986; 24.3: Schoenly, Beaver, and Heumier 1991) from the cataloged webs analyzed by theorists. In fact, Briand and Cohen's most speciose web contains only 48 species, a diversity less than that of each of

the following CV taxa: plants, nematodes, mites, arachnids, bees, beetles, bombyliid flies, and birds.

2. CV chain lengths average more links than 2.86 (Briand 1983), 2.71 (Cohen, Briand, and Newman 1986), or 2.89 (Schoenly, Beaver, and Heumier 1991).

3. Omnivory at adjacent and nonadjacent trophic levels is frequent in the CV web but "rare" in cataloged webs (Pimm 1982; Yodzis 1984). In cataloged webs, 22 percent (Schoenly's catalog) to 27 percent (Briand and Cohen's catalog) of all "kinds" are omnivorous; a much higher fraction of species in the CV web are omnivorous. Adequate diet data, not lumping arthropods, and the inclusion of age structure partially explain the ubiquity of omnivory in the CV. The long chains observed in the CV may be allowed because energy to the top consumers comes from many (lower) trophic levels in addition to adjacent upper levels.

4. Loops, "unreasonable" to modelers and purported to be "very rare in terrestrial" food webs (Pimm 1982, personal communication 1988) are neither rare nor abnormal in the CV.

5. Animals in the CV use many more foods and suffer many more predators than those in the cataloged webs. For example, Cohen (1978) and Schoenly, Beaver, and Heumier (1991) calculated the number of predators on each prey species (means = 3.2, 2.88 respectively) and the number of prey species per predator (2.5, 2.35). Overall, the species in the catalog of webs interact directly with an average of 3.2–4.6 other species (Cohen, Newman, and Briand 1985). Inspection of the diet and predators of CV species makes obvious that CV parameters are an order of magnitude greater. High values exist first because almost all CV consumers are resource generalists eating many species (Table 13.3) (published diets range from 15 to > 125 items; a few arthropod herbivores and some parasites are exceptions). Second, individual species are eaten by tens (Table 13.2) to hundreds (e.g., mesquite) of species (Figures 13.2–13.7).

6. Top predators form 28.5 percent (Briand and Cohen 1984) to 46.5 percent (Schoenly, Beaver, and Heumier 1991) of the kinds of organisms in cataloged webs but are rare or nonexistent in the CV food web. This great discrepancy is undoubtedly due to the inadequacy of diet information in cataloged webs, or because these webs only focus on a limited subset of a trophically linked community, or both.

7. Data from the CV web pose great difficulty in accepting Briand and Cohen's (1984) empirical observation that prey/predator ratios are less than 1.0 (0.88: Briand and Cohen 1984; 0.64: Schoenly, Beaver, and Heumier 1991); in other words, that the number of organisms heading rows (prey) in food web matrices is less than the number heading columns (predators).

The ratio in the CV and other real communities should be more than 1.0. As all heterotrophs must obtain food, every animal should head a column. Rows (prey) include plants and detritus in addition to all animals except those with no predators (i.e., top predators). Let x be the number of animal species in the web that are intermediate predators (they are both predator and prey). Then the total number of species of prey is x + the number of plant species, which is 174 in the CV; the number of predator species is x + the number of top predators, which is 0 or 1 in the CV. If there are more species of plants than top predators, then the ratio of prey to predators will always be greater than 1.0. Few (no?) real communities will have more top predators than autotrophs. The appearance that top predators are more speciose is an artifact discussed earlier. It appears that the lumping of species into kinds of organisms obliterates the actual relationship between prey and predator numbers, primarily because more species of plants are lumped than the (easily recognized) animals that are top predators.

8. Factors 1–7 make the CV web much more complex than cataloged webs. For example, the number of trophic links in cataloged webs varies from 31 (Cohen, Briand, and Newman 1986) to 43 (Schoenly, Beaver, and Heumier 1991); only 2 of Cohen, Briand, and Newman's (1986) 113 webs had more than 100 links. The average CV *subweb* (Figures 13.2–13.7) has 54.7 links and the carnivore subweb alone has 107.

9. The CV web provokes questioning of the utility of the concept of "trophic level." A trophic level is defined as a set of organisms with a number of food chain links in common between them and the primary producers. The nearly universal presence of omnivory and age structure makes this concept non-operational. What trophic level should we assign granivores, arthropodivores, or carnivores that seasonally, ontogenetically, or opportunistically eat all trophic levels of arthropods in addition to plant material and (for carnivores) vertebrates? Looping further blurs the concept of a trophic level. If A eats B but B eats A, is B on the first, third, or (after another loop) fifth trophic level (ad infinitum)? I am not alone in criticizing this concept (see Gallopin 1972; Rigler 1975; Cousins 1980; Levine 1980; Lawton 1989).

10. Two patterns from the catalog of published webs (the fourth and fifth items from Table 13.1) are confirmed with data from the CV web. First, separate compartments were not found to exist within one habitat (the sand dunes and flats of the CV). Second, it is true that the analysis of insects and other arthropods adds tremendous complexity. Arthropod-dominated systems do create more omnivory than those dominated by vertebrates. However, very few communities on this earth are not dominated (in both the number of individuals and species) by arthropods (Hawkins and Lawton 1987; May 1988). So not lumping arthropods, the most speciose taxon on this planet, should increase the complexity of any web.

Overall, a general lack of agreement exists between patterns from the CV and those from the catalog of published webs. Is the CV food web unique or are the cataloged webs so simplified that they have lost realism? The fact that cataloged webs depict so few species, absurdly low ratios of predators on prey and prey eaten by predators, so few links, so little omnivory, a veritable absence of looping, and such a high proportion of top predators argues strongly that cataloged webs do not adequately describe real biological communities. Taylor (1984), Paine (1988), and Lawton (1989) reach similar conclusions.

Table 13.5 General factors that promote omnivory in the food webs in the Coachella Valley (Riverside County, California) and elsewhere.

Life history omnivory is extremely common and widespread in aquatic, marine, and terrestrial habitats. Such diet shifts during ontogeny may be gradual (with growth) or abrupt (with metamorphosis).

Predators disregard the feeding history of prey (different-channel omnivory). The existence of multiple trophic levels within arthropods and within the soil biota causes consumers of these groups to feed on a diversity of trophic levels. For example, arthropodivores eat arthropods that are herbivores, detritivores, parasitoids, predators, and predators of predators.

Opportunistic feeding on abundant resources is commonly reported among consumers. Granivorous birds and rodents primarily eat seeds during most months but normally feed on insects and spiders when arthropods become abundant in spring. Many carnivorous mammals become frugivorous when fruits seasonally appear.

Foraging theory predicts and empirical studies almost universally show that diet reflects food availability and quality. Diets are expanded during periods of food scarcity.

Cannibalism and intraguild predation of heterospecifics from the same trophic level regularly occur among all trophic groups (herbivores, detritivores, and predators).

Arthropods, parasitoids, and gall fauna exhibit complex feeding relations.

Consumers of food in which other consumers live regularly eat these other consumers:

1. Scavengers not only eat carrion but also the microbes and various trophic groups of arthropods that live within these foods.
2. Frugivores and granivores commonly eat insects associated with fruit and seeds.
3. Detritivores eat detritus, microbes, and (often) smaller detritivores.

Self Critique and Prospectus

Several issues need to be addressed:

1. Are desert food webs (a)typical of other webs?

Deserts may differ in two main ways from other systems. First, deserts are often considered to be relatively simple ecosystems characterized by low productivity and species richness (Noy-Meir 1974; Seely and Louw 1980; Wallwork 1982; Whitford 1986; see Polis this volume). Should such depauperate communities translate into relatively simple food webs? If so, the complexity of the CV web is much less than that of more speciose systems.

Second, are desert consumers markedly more omnivorous than other consumers? It is clearly impossible to answer this question with rigor. I approach this issue by indicating that the features that promote omnivory in the CV food web are present in other systems (Table 13.5). Several authors make clear that a high level of omnivory is not restricted to deserts. Menge and Sutherland (1987) cite several studies showing that omnivory is rather frequent in some terrestrial communities and characterizes normal feeding relationships in aquatic and marine environments. Price (1975) maintains that omnivory is a normal feeding strategy throughout the animal kingdom. Walter (1987, p. 228) argues that opportunistic omnivory "appears to be the common feeding behavior" in soil microarthropod populations. He also produces evidence to show that omnivory is common among terrestrial vertebrates and inverte-brates. Moore, Walter, and Hunt (1988) provide strong evidence that omnivory is one of the most frequent and dynamically important trophic links among a diverse array of soil arthropods, protozoans, and nematodes in detrital food webs. Sprules and Bowerman (1988) concluded that omnivory is frequent and common in zooplankton assemblages. (These authors also note long chains, looping, cannibalism, and mutual predation.)

Regardless of whether deserts are unique, desert webs are still of general importance. Deserts occupy at least one-quarter of the earth's land surface (Seely and Louw 1980; Crawford 1981) and the patterns observed in the CV and other deserts thus describe a good fraction of the terrestrial communities on this planet.

2. Should all naturally occurring trophic links be included in a community food web or should we include only "important" links? Are some links too weak or too unusual to list (May 1983b; 1986; Lawton 1989; Schoener 1989)?

I included all links in the CV web. This decision was based on four factors. The most important consideration is that it would be arbitrary and impossible for me to evaluate which links are and are not "important." Most CV consumers include 20 to more than 50 items in their diets. Which items should be included, excluded? It is probable that at least some consumers (especially in deserts) exist or are successful because they are sufficiently flexible to include a number of infrequent links that sum into an important source of energy, at least during some periods.

Second, it is not clear which links are important in terms of population dynamics. Diet and dynamics are not necessarily correlated. For example, a 1 percent representation of a rare species in the diet of a common species may produce considerable mortality to the rare species; conversely, a 100 percent representation of a common species in the diet of a rare species may scarcely affect the common species (see Polis 1981 for examples). Infrequent predation events particularly should influence the dynamics of top predators, animals that are characteristically large and (consequently) relatively rare. Further, a short but intensive predation event may not contribute much to the diet of a predator but may be central to prey dynamics (e.g., newt predation on anuran eggs; Wilbur, Morin, and Harris 1983; see Polis and McCormick 1987 for examples among desert scorpions).

Third, food webs should describe trophic interrelations within a commu-nity. Each link makes the description richer and more completely approaching the reality of the community.

Fourth, exclusion of certain links from a food web produces systematic bias against those characteristics that foster complexity.

3. Not all diet information came from studies in the Coachella Valley. How this influences the food web is uncertain. However, the overall conclusion of great complexity should not be influenced unduly by errors arising from use of these studies.

Conclusions

Desert food webs and others in the real world are much more complex than some food web theory would have us believe. This is illustrated by the food web of the Coachella Valley desert sand community. The complexity arises from the large number of species present, the frequency of omnivory, the age structure of member species, and the complexity of the arthropod and soil faunas. These characteristics occur in other desert communities and should produce equivalently complex food webs. Complexity in other communities also should be related to species diversity and the presence of omnivory, age structure, and complex arthropod and soil/benthic faunas. Evidence suggests that these features are common in communities. Further, diversity (and thus web complexity) in most nondesert habitats should be greater than that in deserts. The strong implication is that food webs from most habitats are relatively complex; at a minimum, actual community webs are much more complex than those webs cataloged by theorists.

It appears that much "food web theory" is not very descriptive or predictive of nature. The catalogs of webs used to abstract empirical generalizations were derived from grossly incomplete representations of communities in terms of both diversity and trophic connections. Consequently, theorists have constructed an oversimplified and invalid view of community structure. The inherent complexity of natural communities makes web construction by empiricists and analysis by theorists difficult.

Acknowledgments

Many people contributed ideas, data, and energy during the 9-year gestation of this paper. Sharon McCormick Carter was central to its development. Saul Frommer, Wendell Icenogle, Wilbur Mayhew, John Pinto, Ken Sculteure, Randy Ryti, Allan Muth, and Fred Andrews provided data on the Coachella Valley. The manuscript benefited greatly from suggestions by Jim Brown, Joel Cohen, Cliff Crawford, Bob Holt, John Lawton, Bruce Menge, John Moore, Chris Myers, Eric Pianka, Stuart Pimm, Ken Schoenly, Tom Schoener, Tsunemi Yamashita, and anomymous reviewers. Fieldwork was financed partially by the National Science Foundation, and the Natural Science Committee and Research Council of Vanderbilt University.

Bibliography

Andersen, D. 1987. Below ground herbivory in natural communities: a review emphasizing fossorial animals. *Quarterly Review of Biology* 62: 261–286.
Andrews, F. G., A. R. Hardy, and D. Giuliani. 1979. *The Coleopterous Fauna of Selected California Sand Dunes.* California Department of Food and Agriculture Report, 142 pp.
Askew, R. 1971. *Parasitic Insects.* New York: American Elsevier, 316 pp.

Austin, A. D. 1985. The function of spider egg sacs in relation to parasitoids and predators, with special reference to the Australian fauna. *Journal of Natural History* 19: 359–376.

Bailey, K. H., and G. A. Polis. 1987. An experimental analysis of optimal and central place foraging by the harvester ant, *Pogonomyrmex californicus*. *Oecologia* 72: 440–448.

Barrows, J. 1979. Aspects of the ecology of the desert tortoise, *Gopherus agassizi*, in Joshua Tree National Monument, Pinto Basin, Riverside County, California. *Desert Tortoise Council Symposium Proceedings* 1979: 105–131.

Bent, A. C. 1932, 1937, 1938, 1942, 1948, 1949, 1958. *Life Histories of North American Birds*. United States National Museum Bulletins 162, 167, 170, 179, 195, 196, 211.

Bond, R. 1942. Food of the Burrowing Owl in western Nevada. *Condor* 44: 183.

Bradley, R. 1983. Complex food webs and manipulative experiments in ecology. *Oikos* 41: 150–152.

Bradley, W. 1968. Food habits of the antelope ground squirrel in southern Nevada. *Journal of Mammalogy* 49: 14–21.

Bradley, W., and B. Mauer. 1973. Rodents of a creosote bush community in southern Nevada. *Southwestern Naturalist* 17: 333–344.

Brian, M. V. 1983. *Social Insects: Ecology and Behavioral Biology*. New York: Chapman and Hall, 377 pp.

Briand, F. 1983. Environmental control of food web structure. *Ecology* 64: 253–263.

Briand, F., and J. Cohen. 1984. Community food webs have invariant-scale structure. *Nature* 307: 264–267.

———. 1987. Environmental correlates of food chain length. *Science* 238: 956–960.

Brown, J. 1986. The role of vertebrates in desert ecosystems. Pp. 51–71 in W. Whitford (ed.), *Pattern and Process in Desert Ecosystems*. Albuquerque, N.Mex.: University of New Mexico Press, 139 pp.

Brown, J., O. J. Reichman, and D. Davidson. 1979. Granivory in desert ecosystems. *Annual Review of Ecology and Systematics* 10: 201–227.

Burge, B., and W. Bradley. 1976. Population density, structure and feeding habits of the desert tortoise, *Gopherus agassizi*, in a low desert study area in southern Nevada. *Desert Tortoise Council Symposium Proceedings* 1976: 51–74.

Caswell, H. 1978. Predator-mediated coexistence: a non-equilibrium model. *American Naturalist* 112: 127–154.

Cohen, J. E. 1978. *Food Webs and Niche Space*. Monographs in Population Biology, 11. Princeton, N.J.: Princeton University Press, 189 pp.

———. 1988. Food webs and community structure. In S. Levin, R. May, and J. Roughgarden (eds.), *Perspectives in Theoretical Ecology*. Princeton, N.J.: Princeton University Press.

Cohen, J. E., and F. Briand. 1984. Trophic links of community food webs. *Proceedings of the National Academy of Science* 81: 4105–4109.

Cohen, J. E., F. Briand, and C. Newman. 1986. A stochastic theory of community food webs. III. Predicted and observed lengths of food chains. *Proceedings of the Royal Society of London* (B) 228: 317–353.

Cohen, J. E., and C. Newman. 1986. A stochastic theory of community food webs. I. Models and aggregated data. *Proceedings of the Royal Society of London* (B) 224: 421–448.

Cohen, J. E., C. Newman, and F. Briand. 1985. A stochastic theory of community food webs. II. Individual webs. *Proceedings of the Royal Society of London* (B) 224: 449–461.

Coulombe, H. 1971. Behavior and population ecology of the burrowing owl, *Speotyto cunicularia*, in the Imperial Valley of California. *Condor* 73: 162–176.

Cousins, S. 1980. A trophic continuum derived from plant structure, animal size and a detritus cascade. *Journal of Theoretical Biology* 82: 607–618.

Crawford, C. S. 1979. Desert detritivores: a review of life history patterns and trophic roles. *Journal of Arid Environments* 2: 31–42.

———. 1981. *Biology of Desert Invertebrates*. New York: Springer-Verlag.

———. 1986. The role of invertebrates in desert ecosystems. Pp. 73–92 in W. Whitford (ed.), *Pattern and Process in Desert Ecosystems*. Albuquerque, N.Mex.: University of New Mexico Press, 139 pp.

Crawford, C. S., and E. Taylor. 1984. Decomposition in arid environments: role of the detritivore gut. *South African Journal of Science* 80: 170–176.

DeAngelis, D., W. M. Post, and G. Sugihara. 1983. *Current Trends in Food Web Theory: Report on a Food Web Workshop*. Oak Ridge National Laboratory Technical Memorandum 5983, Oak Ridge, Tenn.

Deligne, J., A. Quennedey, and M. Blum. 1981. The enemies and defense mechanisms of termites. Pp. 1–76 in *Social Insects*, vol 2. New York: Academic Press.

Edney, E. B., S. Haynes, and D. Gibo. 1974. Distribution and activity of the desert cockroach *Arenivaga investigata* (Polyphagidae) in relation to microclimate. *Ecology* 55: 420–427.

Edney, E. B., J. McBrayer, P. Franco, and A. Phillips. 1974. *Distribution of Soil Arthropods in Rock Valley, Nevada*. U.S. International Biological Project Desert Biome Research Memo 74–32: 53–58.

El-Kifl, A., and S. Ghabbour. 1984. Soil fauna. Pp. 91–104 in J. L. Cloudsley-Thompson (ed.), *Sahara Desert*. New York: Pergamon Press.

Elkins, N., and W. G. Whitford. 1982. The role of microarthropods and nematodes in decomposition in a semi-arid ecosystem. *Oecologia* 55: 303–310.

Evenari, M. 1981. Ecology of the Negev Desert, a critical review of our knowledge. In H. Shuval (ed.), *Developments in Arid Zone Ecology and Environmental Quality*. Philadelphia, Pa.: Balaban.

Ferguson, W. 1962. Biological characteristics of the mutillid subgenus *Photopsis* Blake and their systematic value. *University of California at Berkeley Publications in Entomology* 27: 1–82.

Franco, P., E. Edney, and J. McBrayer. 1979. The distribution and abundance of soil arthropods in the northern Mojave Desert. *Journal of Arid Environments* 2: 137–149.

Freckman, D. W., and R. Mankau. 1977. Distribution and trophic structure of nematodes in desert soils. *Ecological Bulletin* (Stockholm) 25: 511–514.

Freckman, D. W., R.. Mankau, and H. Ferris. 1975. Nematode community structure in desert soils: nematode recovery. *Journal of Nematology* 7: 343–346.

French, N., B. Maza, H. Hill, A. Aschwanden, and H. Kaaz. 1974. A population study of irradiated dune rodents. *Ecological Monographs* 44: 45–72.

Frommer, S. I. 1986. *A Hierarchic Listing of the Arthropods Known to Occur within the Deep Canyon Desert Transect*. Riverside, Calif.: Deep Canyon Publications, 133 pp.

Gallopin, G. 1972. Structural properties of food webs. Pp. 241–282 in B. Patton (ed.), *Systems Analysis and Simulations in Ecology*, vol. 2. New York: Academic Press.

Ghabbour, S., E. El-Ayouty, M. Khadr, and A. El-Tonsi. 1980. Grazing by microfauna and productivity of heterocystous nitrogen-fixing blue-green algae. *Oikos* 34: 209–218.

Ghilarov, M. 1964. Connection of insects with soil in different climatic zones. *Pedobiologia* 4: 310–315.

———. 1968. Soil stratum of terrestrial biocenoses. *Pedobiologia* 8: 82–96.

Gier, H., S. Kruckenberg, and R. Marler. 1978. Parasites and diseases of coyotes. Pp. 37–69 in M. Bekoff (ed.), *Coyotes: Biology, Behavior and Management*. New York: Academic Press, 400 pp.

Glasser, J. 1983. Variation in niche breadth and trophic position: on the disparity between expected and observed species packing. *American Naturalist* 122: 542–548.

Goldstein, D., and K. Nagy. 1985. Resource utilization by desert quail: time and energy, food and water. *Ecology* 66: 378–387.

Gordon, S. 1978. Food and foraging ecology of a desert harvester ant, *Veromessor pergandei* (Mayr). Ph.D. dissertation, University of California at Berkeley, 158 pp.

Hadley, N. 1980. Productivity of desert ecosystems. Section B in *Handbook of Nutrition*. West Palm Beach, Fla.: Chemical Rubber Company Press.

Hadley, N., and S. Szarek. 1981. Productivity of desert ecosystems. *BioScience* 1981: 747–753.

Hawkins, B., and R. Goeden. 1984. Organization of a parasitoid community associated with a complex of galls on *Atriplex* spp. in southern California. *Ecological Entomology* 9: 271–292.

Hawkins, B., and J. Lawton. 1987. Species richness of parasitoids of British phytophagous insects. *Nature* 326: 788–790.

Hawks, D. 1982. *A Checklist of the Butterflies of Deep Canyon*. Riverside, Calif.: Deep Canyon Publications, 10 pp.

Hawks, S., and R. D. Farley. 1973. Ecology and behavior of the desert burrowing cockroach *Arenivaga* sp. (Dictyoptera: Polyphagidae). *Oecologia* 11: 263–279.

Henschel, J. In press. The biology of *Leucorchestris arenicola* (Araneae: Heteropodidae), a burrowing spider of the Namib Dunes. In M. K. Seely (ed.), *Current Research on Namib Ecology—25 Years of the Desert Ecological Research Unit*. Transvaal Museum Monograph 8, Transvaal Museum, Pretoria, Republic of South Africa.

Holm, E., and C. Scholtz. 1980. Structure and pattern of the Namib Desert ecosystem at Gobabeb. *Madoqua* 12: 5–37.

Irwin, M. 1971. Ecology and biosystematics of the pherocine Therevidae (Diptera). Ph.D. dissertation, University of California, Riverside, 263 pp.

Janzen, D. 1977. Why fruits rot, seeds mold, and meat spoils. *American Naturalist* 111: 691–713.

Johnegard, P. 1988. *North American Owls: Biology and Natural History*. Washington, D.C.: Smithsonian Institution Press.

Johnson, M., and R. Hansen. 1979. Coyote food habits on the Idaho National Engineering Laboratory. *Journal of Wildlife Management* 43: 951–956.

Lawton, J. 1989. Food webs. In J. Cherrett (ed.), *Ecological Concepts*. Oxford: Blackwell Scientific.

Lawton, J., and P. Warren. 1988. Static and dynamic explanation of patterns in food webs. *Trends in Ecology and Evolution* 3: 242–245.

Levine, S. 1980. Several measures of trophic structure applicable to complex food webs. *Journal of Theoretical Biology* 83: 195–207.

Louw, G., and M. K. Seely. 1982. *Ecology of Desert Organisms*. New York: Longman, 194 pp.

Ludwig, J. A. 1977. Distributional adaptations of root-systems in desert environments. Pp. 85–91 in E. Marshall (ed.), *The Belowground Symposium: A Synthesis of Plant-Associated Processes*. Range Science Department, Science Series No. 26. Fort Collins, Colo.: Colorado State University.

McCormick, S. J., and G. A. Polis. 1986. Comparison of the diet of *Paruroctonus mesaensis* at two sites. *Proceedings of the IX International Arachnological Congress*: 167–171.

———. 1990. Prey, predators and parasites. Chapter 7 in G. A. Polis (ed.), *Biology of Scorpions*. Palo Alto, Calif.: Stanford University Press.

Macfayden, A. 1963. The contribution of the fauna to the total soil metabolism. In J. Doeksen and J. van der Drift (eds.), *Soil Organisms*. Amsterdam: North-Holland Publishing Co., 453 pp.

McKinnerney, M. 1978. Carrion communities in the northern Chihuahuan Desert. *Southwestern Naturalist* 23: 563–576.

Mann, J. 1969. Cactus feeding insects and mites. *U.S. National Museum Bulletin* 256: 1–158.

Marshall, A. 1981. *The Ecology of Ectoparasitic Insects*. New York: Academic Press, 459 pp.

May, R. 1983a. Parasitic infections as regulators of animal populations. *American Scientist* 71: 36–45.

———. 1983b. The structure of food webs. *Nature* 301: 566–568.

———. 1986. The search for patterns in the balance of nature: advances and retreats. *Ecology* 67: 1115–1126.

———. 1988. How many species are there on earth? *Science* 241: 1441–1448.

Mayhew, W. W. 1981. *Vertebrates and their Habitats on the Deep Canyon Transect*. Riverside, Calif.: Deep Canyon Publications, 32 pp.

Menge, B., and J. Sutherland. 1987. Community regulation: variation in disturbance, competition, and predation in relation to environmental stress and recruitment. *American Naturalist* 130: 730–757.

Meserve, P. 1976. Food relationships of a rodent fauna in a California coastal sage scrub community. *Journal of Mammalogy* 57: 300–319.

Miller, A. 1931. Systematic revision and natural history of the American shrikes (*Lanius*). *University of California Publications in Zoology* 38: 11–242.

Minnich, J., and V. H. Shoemaker. 1970. Diet, behavior and water turnover in the desert iguana, *Dipsosaurus dorsalis*. *American Midland Naturalist* 84: 496–509.

Mispagel, M. 1978. The ecology and bioenergetics of the acridid grasshopper, *Bootettix punctatus*, on creosotebush, *Larrea tridentata*, in the northern Mojave Desert. *Ecology* 59: 779–788.

Mitchell, J. 1979. Ecology of southeastern Arizona whiptail lizards (*Cnemidophoros*, Teiidae): population densities, resource partitioning and niche overlap. *Canadian Journal of Zoology* 57: 1487–1499.

Moore, J. C., D. Walter, and H. W. Hunt. 1988. Arthropod regulation of micro- and mesobiota in belowground detrital webs. *Annual Review of Entomology* 33: 419–439.

Morell, S. 1972. Life history of the San Joaquin kit fox. *California Fish and Game* 58: 162–174.

Noy-Meir, I. 1974. Desert ecosystems: higher trophic levels. *Annual Review of Ecology and Systematics* 5: 195–213.

———. 1981. Spatial effects in modeling of arid ecosystems. Pp. 411–432 in D. Goodall and R. Perry (eds.), *Arid-Land Ecosystems: Structure, Functioning and Management*, vol 2. Cambridge: Cambridge University Press.

Nuessly, G., and R. Goeden. 1984. Aspects of the biology and ecology of *Diguetia mohavea* Gertsch (Araneae, Diguetidae). *Journal of Arachnology* 12: 75–85.

Odum, E., and L. Biever. 1984. Resource quality, mutualism, and energy partitioning in food chains. *American Naturalist* 124: 360–376.

Ohlendorf, H. 1971. Arthropod diet of western horned owl. *Southwestern Naturalist* 16: 124–125.

Olendoroff, R. 1976. The food habits of North American Golden Eagles. *American Midland Naturalist* 95: 231–236.

Orians, G., R. Cates, M. Mares, A. Moldenke, J. Neff, D. Rhoades, M. Rosenzweig, B. Simpson, J. Schultz, and C. Tomoff. 1977. Resource utilization systems. Pp. 164–224 in

Gary Polis

G. Orians and O. Solbrig (eds.), *Convergent Evolution in Warm Deserts*. Stroudsburg, Pa.: Dowden, Hutchinson and Ross.

Otte, D., and A. Joern. 1977. On feeding patterns in desert grasshoppers and the evolution of specialized diets. *Proceedings of the National Academy of Sciences* 128: 8–126.

Paine, R. T. 1980. Food webs: linkage, interaction strength and community infrastructure. *Journal of Animal Ecology* 49: 667–685.

———. 1988. On food webs: road maps of interaction or grist for theoretical development? *Ecology* 69: 1648–1654.

Palmer, R. 1988. *Handbook of North American Birds*. Vol. 5: *Diurnal Raptors* (part 2). New Haven, Conn.: Yale University Press.

Pearse, V., J. Pearse, M. Buchsbaum, and R. Buchsbaum. 1987. *Living Invertebrates*. Palo Alto, Calif.: Blackwell Scientific Publications.

Persson, L., and B. Ebenmann. 1988. *Size Structured Populations: Ecology and Evolution*. New York: Springer-Verlag.

Pianka, E., and W. Parker. 1972. Ecology of the iguanid lizard *Callisaurus draconoides*. *Copeia* 1972: 493–508.

———. 1975. Ecology of the horned lizards: a review with special reference to *Phrynosoma platyrhinos*. *Copeia* 1975: 141–162.

Pimm, S. L. 1982. *Food Webs*. New York: Chapman and Hall, 219 pp.

Pimm, S. L., and R. Kitching. 1987. The determinants of food chain lengths. *Oikos* 50: 302–307.

Pimm, S. L., and J. Rice. 1987. The dynamics of multispecies, multi–life-stage models of aquatic food webs. *Theoretical Population Biology* 32: 303–325.

Poinar, G. 1985. Mermithid (Nematoda) parasites of spiders and harvestmen. *Journal of Arachnology* 13: 121–128.

Polis, G. A. 1979. Diet and prey phenology of the desert scorpion, *Paruroctonus mesaensis* Stahnke. *Journal of Zoology* (London) 188: 333–346.

———. 1981. The evolution and dynamics of intraspecific predation. *Annual Review of Ecology and Systematics* 12: 225–251.

———. 1984a. Age structure component of niche width and intraspecific resource partitioning: can age groups function as ecological species? *American Naturalist* 123: 541–564.

———. 1984b. Intraspecific predation and "infant killing" among invertebrates. Pp. 87–104 in G. Hausfater and S. Hrdy (eds.), *Infanticide: Competitive and Evolutionary Perspectives*. New York: Aldine Publishing.

———. 1988a. Exploitation competition and the evolution of interference, cannibalism and intraguild predation in age/size structured populations. In L. Persson and B. Ebenmann (eds.), *Size Structured Populations: Ecology and Evolution*. New York: Springer-Verlag.

———. 1988b. Trophic and behavioral responses of desert scorpions to harsh environmental periods. *Journal of Arid Environments* 14: 123–134.

Polis, G. A., and S. J. McCormick. 1986a. Patterns of resource use and age structure among species of desert scorpion. *Journal of Animal Ecology* 55: 59–73.

———. 1986b. Scorpions, spiders and solpugids: predation and competition among distantly related taxa. *Oecologia* 71: 111–116.

———. 1987. Intraguild predation and competition among desert scorpions. *Ecology* 68: 332–343.

Polis, G. A., C. A. Myers, and R. Holt. 1989. The evolution and dynamics of intraguild predation between potential competitors. *Annual Review of Ecology and Systematics* 20: 297–330.

Polis, G. A., W. D. Sissom, and S. McCormick. 1981. Predators of scorpions: field data and a review. *Journal of Arid Environments* 4: 309–327.

Powell, J., and C. Hogue. 1979. *California Insects.* Berkeley, Calif.: University of California Press, 339 pp.

Price, P. 1975. *Insect Ecology.* New York: Wiley Interscience.

Reichman, O. J. 1975. Relation of desert rodent diet to available resources. *Journal of Mammalogy* 56: 731–751.

Rich, P. 1984. Trophic-detrital interactions: vestiges of ecosystem evolution. *American Naturalist* 123: 20–29.

Rigler, F. 1975. The concept of energy flow and nutrient flow between trophic levels. In W. van Dobben and R. Lowe-McConnell (eds.), *Unifying Concepts in Ecology.* The Hague: W. Junk.

Rodin, L., and N. Bazilevich. 1964. Doklady Ahademin Nak. S.S.S.R. 157: 215–218. From E. J. Kormandy, *Concepts of Ecology,* 1969. New York: Prentice-Hall.

Ryan, M. 1968. *Mammals of Deep Canyon.* Palm Springs, Calif.: Desert Museum.

Ryti, R., and T. Case. 1986. Overdispersion of ant colonies: a test of hypotheses. *Oecologia* 69: 446–453.

―――. 1988. Field experiments on desert ants: testing for competition between colonies. *Ecology* 69: 1993–2003.

Santos, P., J. Phillips, and W. Whitford. 1981. The role of mites and nematodes in early stages of litter decomposition in the desert. *Ecology* 62: 664–669.

Santos, P., and W. Whitford. 1981. The effects of microarthropods on litter decomposition in a Chihuahuan Desert ecosystem. *Ecology* 62: 654–663.

Schoener, T. W. In press. Food webs from the small to the large: probes and hypotheses. *Ecology* 70: 1559–1589.

Schoenly, K. 1983. Arthropods associated with bovine and equine dung in an ungrazed Chihuahuan Desert ecosystem. *Annals of the Entomological Society of America* 76: 790–796.

Schoenly, K., R. Beaver, and T. Heumier. 1991 On the trophic relations of insects: A food web approach. *American Naturalist.* In press.

Schoenly, K., and W. Reid. 1983. Community structure in carrion arthropods in the Chihuahuan Desert. *Journal of Arid Environments* 6: 253–263.

Schultz, J., D. Otte, and F. Enders. 1977. *Larrea* as a habitat component for desert arthropods. Pp. 176–208 in T. Mabry, J. Hunziker, D. DiFeo (eds.), *Creosote bush: Biology and Chemistry of Larrea in New World Deserts.* U.S. International Biological Program Synthesis Series 6. Stroudsburg, Pa.: Dowden, Hutchinson and Ross.

Seastedt, T., R. Ramundo, and D. Hayes. 1988. Maximization of densities of soil animals by foliage herbivory: empirical evidence, graphical and conceptual models. *Oikos* 51: 243–248.

Seely, M. K., and G. N. Louw. 1980. First approximation of the effects of rainfall on the ecology and energetics of a Namib Desert dune ecosystem. *Journal of Arid Environments* 3: 25–54.

Silverton, J., and R. Goeden. 1979. Life history of the lacebug, *Corythucha morrilli* Osborn and Drake, on the ragweed, *Ambrosia dumosa* (Gray) Payne, in Southern California (Hemiptera-Heteroptera: Tingidae). *Pan-Pacific Entomologist* 55: 305–308.

Sivinski, J. 1985. Mating by kleptoparasitic flies (Diptera: Chloropidae) on a spider host. *Florida Entomologist* 68: 216–222.

Soholt, L. 1973. Consumption of primary production by a population of kangaroo rats (*Dipodomys merriami*) in the Mojave Desert. *Ecological Monographs* 43: 357–376.

Sprules, W., and J. Bowerman. 1988. Omnivory and food chain lengths in zooplankton food webs. *Ecology* 69: 418–426.

Stenseth, N. 1985. The structure of food webs predicted from optimal food selection models: an alternate to Pimm's stability hypothesis. *Oikos* 44: 361–364.

Sugihara, G., K. Schoenly, and A. Trombla. 1989. Scale invariance in food web properties. *Science* 245: 48–52.

Tabet, A., and J. Hall. 1984. *The Bombyliidae of Deep Canyon, Part I*. Tripoli, Libya: Al-Fateh University Publications, 63 pp.

Taylor, J. 1984. A partial food web involving predatory gastropods on a Pacific fringing reef. *Journal of Experimental Marine Biology and Ecology* 74: 273–290.

Telford, S. 1971. A comparative study of endoparasitism among some California lizard populations. *American Midland Naturalist* 83: 516–554.

Tevis, L. and I. M. Newell. 1962. Studies on the biology and seasonal cycle of the giant red velvet mite, *Dinothrombium pandorae* (Acari, Trombidiidae). *Ecology* 43: 497–505.

Venkateswarlu, B., and A. Rao. 1981. Distribution of microorganisms in stabilised and unstabilised sand dunes of the Indian desert. *Journal of Arid Environments* 4: 203–207.

Vitt, L. J., and R. Ohmart. 1977. Ecology and reproduction of lower Colorado River lizards: I. *Callisaurus draconoides* (Iguanidae). *Herpetologica* 33: 214–222.

Vollmer, A., F. Au, and S. Bamburg. 1977. Observations on the distribution of microorganisms in desert soils. *Great Basin Naturalist* 37: 81–86.

Wallwork, J. 1972. Mites and other microorganisms from the Joshua Tree National Monument, California. *Journal of Zoology* 168: 91–105.

———. 1982. *Desert Soil Fauna*. New York: Praeger Publishers, 296 pp.

Walsberg, G. 1975. Digestive adaptations of *Phainopepla nitens* associated with the eating of mistletoe berries. *Condor* 77: 169–174.

Walter, D. 1987. Trophic behavior of "mycophagous" microarthropods. *Ecology* 68: 226–228.

Wasbauer, M. S. 1973. The male brachycistidine wasps of the Nevada Test Site (Hymenoptera: Tiphiidae). *Great Basin Naturalist* 33: 109–112.

Wasbauer, M. S., and L. Kimsey. 1985. California spider wasps of the subfamily Pompilinae. *Bulletin of the California Insect Survey* 26: 1–130.

Weathers, B. 1983. *Birds of Southern California's Deep Canyon*. Berkeley, Calif.: University of California Press, 266 pp.

Welty, J. 1962. *The Life of Birds*. Philadephia, Pa.: W. B. Saunders, 546 pp.

Werner, E., and J. Gilliam. 1984. The ontogenetic niche and species interactions in size-structured populations. *Annual Review of Ecology and Systematics* 15: 393–426.

Wheeler, G. C., and J. Wheeler. 1973. *Ants of Deep Canyon*. Berkeley, Calif.: University of California Press, 192 pp.

Whitford, W. G. 1986. Decomposition and nutrient cycling in deserts. Chapter 5 in W. G. Whitford (ed.), *Pattern and Process in Desert Ecosystems*. Albuquerque, N.Mex.: University of New Mexico Press.

Whitford, W. G., Y. Steinberger, and G. Ettershank. 1982. Contributions of subterranean termites to the "economy" of Chihuahuan Desert ecosystems. *Oecologia* 55: 298–302.

Wiens, J. A. 1976. Population responses to patchy environments. *Annual Review of Ecology and Systematics* 7: 81–120.

Wiens, J. A., and J. T. Rotenberry. 1979. Diet niche relationships among North American grassland and shrubsteppe birds. *Oecologia* 42: 253–292.

———. 1981. Morphological size ratios and competition in ecological communities. *American Naturalist* 117: 592–599.

Wilbur, H., P. Morin, and R. Harris. 1973. Salamander predation and the structure of experimental communities: anuran responses. *Ecology* 64: 1423–1429.

Woods, A., and C. Herman. 1943. The occurrence of blood parasites in birds from southwestern United States. *Journal of Parasitology* 29: 187–196.

Yodzis, P. 1984. How rare is omnivory? *Ecology* 65: 321–323.

Zabriskie, J. 1981. *Plants of Deep Canyon and the Central Coachella Valley, California.* Riverside, Calif.: University of California Philip L. Boyd Deep Canyon Desert Research Center, 174 pp.

Appendixes 13.1 and 13.2 begin on page 430.

Appendix 13.1   Feeding categories of vertebrate species resident in the sand dune/sand flat
community of the Coachella Valley, Riverside County, California.

Species	Common Name	Feeding Category[a] Primary[b]	Secondary[c]
Aves			
Falconiformes			
Accipitridae			
Accipiter cooperii	Cooper's Hawk	C	A
Buteo jamaicensis	Red-tailed Hawk	C	A
Aquila chrysaetos	Golden Eagle	C	
Falconidae			
Falco mexicanus	Prairie Falcon	C	A
Falco sparverius	American Kestrel	A/C	
Galliformes			
Phasianidae			
Lophortyx gambelii	Gambel's Quail	G	A
Charadriiformes			
Charadriidae			
Charadrius vociferus	Killdeer	A	
Columbiformes			
Columbidae			
Columba livia	Rock Dove	G	A
Zenaida asiatica	White-winged Dove	G	A
Zenaida macroura	Mourning Dove	G	
Columbina passerina	Common Ground-Dove	G	
Cuculiformes			
Cuculidae			
Geococcyx californianus	Greater Roadrunner	A/C	P
Strigiformes			
Strigidae			
Otus asio	Screech-Owl	C/A	
Bubo virginianus	Great Horned Owl	C	A
Athene cunicularia	Burrowing Owl	C/A	
Tytonidae			
Tyto alba	Common Barn-Owl	C	A
Caprimulgiformes			
Caprimulgidae			
Chordeiles acutipennis	Lesser Nighthawk	A	
Phalaenoptilus nuttallii	Common Poorwill	A	

SOURCE: Species list from Mayhew (1981); dietary data from a variety of sources.
[a]Feeding categories: H = herbivore (includes fruit and nectar); A = arthropodivore
(includes insects and arachnids); G = granivore (primarily seeds; possibly some foliage);
P = plant material (leaves, seeds, fruit) eaten (secondary category only); C = carnivore
(includes vertebrate prey).
   Categories are ranked according to their importance in the diet (e.g., A/C indicates
arthropods are the most important diet item, vertebrates are second).
[b]Main food type(s). When one category is listed, at least 90% of food comes from this
category. When two categories are listed, >33% of food comes from each. When three
categories are listed, >20% comes from each.
[c]Food type(s) that form <20% of the diet or occur only during short period.
[d]Information on the diet of these taxa was obtained from general sources (e.g., field
guides) rather than specific research publications.

Species	Common Name	Feeding Category[a]	
		Primary[b]	Secondary[c]
Apodiformes			
Apodidae			
Aeronautes saxatalis	White-throated Swift	A	
Trochilidae			
Calypte anna	Anna's Hummingbird	H	A
Calypte costae	Costa's Hummingbird	H	A
Piciformes			
Picidae			
Colaptes auratus	Common Flicker	A	
Passeriformes			
Tyrannidae			
Tyrannus verticalis	Western Kingbird	A	P
Myiarchus cinerascens	Ash-throated Flycatcher	A	P
Sayornis saya	Say's Phoebe	A	
Hirundinidae			
Tachycineta thalassina	Violet-green Swallow	A	
Stelgidopteryx serripennis	Northern Rough-winged Swallow	A	
Corvidae			
Corvus corax	Common Raven	A/C	P
Paridae			
Auriparus flaviceps	Verdin	A	
Troglodytidae			
Thryomanes bewickii	Bewick's Wren	A	P
Campylorhynchus brunneicapillus	Cactus Wren	A	P/C
Salpinctes obsoletus	Rock Wren	A	
Mimidae			
Mimus polyglottos	Northern Mockingbird	H	A/C
Toxostoma redivivum[d]	Le Conte's Thrasher	A	
Turdidae			
Sialia mexicana	Western Bluebird	A	P
Sylviidae			
Polioptila caerulea	Blue-gray Gnatcatcher	A	P
Ptilogonatidae			
Phainopepla nitens	Phainopepla	H/A	
Laniidae			
Lanius ludovicianus	Loggerhead Shrike	A/C	P
Sturnidae			
Sturnus vulgaris[d]	Starling	G/A	
Vireonidae			
Vireo gilvus	Warbling Vireo	A	P
Parulidae			
Vermivora celata[d]	Orange-crowned Warbler	A	
Dendroica coronata[d]	Yellow-rumped Warbler	A	
Ploceidae			
Passer domesticus[d]	House Sparrow	G	A
Icteridae			
Sturnella neglecta	Western Meadowlark	A	P
Euphagus cyanocephalus[d]	Brewer's Blackbird	A	P

Appendix 13.1 *Continued*

Species	Common Name	Feeding Category[a] Primary[b]	Secondary[c]
Molothrus ater	Brown-headed Cowbird	H	
Icterus parisorum	Scott's Oriole	A	P
Icterus galbula	Northern Oriole	A	P
Thraupidae			
Piranga ludoviciana	Western Tanager	A	P
Fringillidae			
Pheucticus melanocephalus[d]	Black-headed Grosbeak	G	
Carpodacus mexicanus[d]	House Finch	G	
Carduelis psaltria[d]	Lesser Goldfinch	G	
Pipilo aberti[d]	Abert's Towhee	G	A
Amphispiza bilineata[d]	Black-throated Sparrow	G	A
Amphispiza belli	Sage Sparrow	A/G	
Spizella passerina[d]	Chipping Sparrow	G	A
Reptilia			
Chelonia			
Testudinidae			
Gopherus agassizii	Desert tortoise	H	
Squamata (Lacertilia)			
Gekkonidae			
Coleonyx variegatus	Banded gecko	A	
Iguanidae			
Dipsosaurus dorsalis	Desert iguana	H	A
Callisaurus draconoides	Zebra-tail lizard	A	P/C
Uma inornata	Coachella Valley fringe-toed lizard	A	P
Gambelia wislizenii	Leopard lizard	C/A	P
Uta stansburiana	Side-blotched lizard	A	
Urosaurus graciosus	Brush lizard	A	P
Phrynosoma m'calli	Flat-tailed horned lizard	A	
Phrynosoma platyrhinos	Desert horned lizard	A	P
Teiidae			
Cnemidophorus tigris	Western whiptail lizard	A	C
Squamata (Serpentes)			
Leptotyphlopidae			
Leptotyphlops humilus	Western blind snake	A	
Boidae			
Lichanura trivirigata	Rosy boa	C	
Colubridae			
Arizona elegans	Glossy snake	C	

| Species | Common Name | Feeding Category[a] | |
		Primary[b]	Secondary[c]
Chionactis occipitalis	Western shovel-nosed snake	A	
Lampropeltis getulus	King snake	C	
Masticophis flagellum	Common whipsnake	C	A
Phyllorhynchus decurtatus	Leaf-nosed snake	A/C	
Pituophis melanoleucus	Gopher snake	C	A
Rhinocheilus lecontei	Long-nosed snake	C	A
Viperidae			
Crotalus cerastes	Sidewinder	C	
Mammalia			
Chiroptera			
Vespertilionidae			
Pipistrellus hesperus	Western pipistrelle	A	
Antrozous pallidus	Pallid bat	A	
Lagomorpha			
Leporidae			
Lepus californicus	Blacktail jackrabbit	H	
Sylvilagus auduboni	Desert cottontail	H	
Rodentia			
Sciuridae			
Spermophilus tereticaudus	Roundtail ground squirrel	G/A	C
Ammospermophilus leucurus	Whitetail antelope squirrel	G/C/A	
Geomyidae			
Thomomys bottae	Valley pocket gopher	H	
Heteromyidae			
Dipodomys merriami	Merriam's kangaroo rat	G	A
Dipodomys deserti	Desert kangaroo rat	G	A
Perognathus longimembris	Little pocket mouse	G	A
Perognathus formosus	Longtail pocket mouse	G/A	A
Perognathus pencillatus[d]	Desert pocket mouse	G	
Cricetidae			
Peromyscus maniculatus	Deer mouse	G/A	
Peromyscus eremicus	Cactus mouse	G/A	
Onychomys torridus	Southern grasshopper mouse	A/G/C	
Neotoma lepida	Desert woodrat	H	A
Carnivora			
Canidae			
Canis latrans	Coyote	C	A/P
Vulpes macrotis	Kit fox	C/A/H	

Appendix 13.2   Arthropods of the Coachella Valley, Riverside County, California, categorized by diet.

Herbivores
  S.C. Acari (O. Prostigmata)[a]
    Tetranychoidae
  O. Embioptera
  O. Hemiptera
    Coreidae
    Lygaeidae (some feed on seeds)
    Miridae[a]
    Pentatomidae (2 species)[b]
    Rhopalidae
    Tingidae
  O. Homoptera
    Aphidae[a]
    Cicadellidae[a]
    Cicadidae (roots)
    Coccidae (1 species)[b]
    Delphacidae
    Fulgoroidea
    Membracidae
  O. Orthoptera
    Acrididae (6 species)[b]
    Gryllacrididae (3 species)[b]
  O. Thysanoptera
    Thripidae
  O. Coleoptera
    Anobiidae (larvae eat wood)
    Anthicidae
    Bostrichidae (wood)
    Bruchidae
    Buprestidae (larvae eat wood) (4 species)[b]
    Cerambycidae (larvae eat wood)
    Chrysomelidae
    Curculionidae
    Elateridae
    Mordellidae
    Nitidulidae
    Scarabaeidae (4 species)[b]
  O. Diptera
    Agromyzidae
    Cecidomyidae
    Drosophiloidea (also feeds on yeast)
  O. Hymenoptera
    Apidae (nectar & pollen)
    Formicidae (seeds & nectar)[a]
    Halictidae (nectar & pollen)
    Megachilidae (nectar & pollen)

[a]More than 10 species in this taxon.
[b]This number of species is the minimum number that I have identified, and, thus, is an underestimation in all cases.

O. Lepidoptera
  Arctiidae
  Coleophoridae
  Cossidae
  Geometridae (3 species)[b]
  Gracillariidae
  Hesperiidae
  Lycaenidae
  Noctuidae[a]
  Pieridae
  Pyralidae
  Sphingidae

Detritivores (with some belowground herbivory and fungivory)
  O. Isopoda
    Armadilidae (1 species)[b]
  C. Millipedia (1 species)[b]
    O. Opisthospermophora
  S.C. Acari (O. Prostigmata)[a]
    Acaroidea
    Pachygnathoidea
    Oribatei
    Trombidioidea (parasitic as juveniles)
  O. Collembola (also eats fungus)
    Entomobryidae
    Sminthuridae
    Poduridae
  O. Dermatoptera
    Forficulidae (1 species)[b]
  O. Thysanura
    Lepismatidae (also eats fungus) (2 species)[b]
  O. Orthoptera
    Blattidae (1 species)[b]
    Gryllacrididae
  O. Isoptera
    Kalotermitidae (2 species)[b]
    Rhinotermitidae (2 species)[b]
    Termitidae (6 species)[b]
  O. Coleoptera
    Alleculidae (adults eat plants)
    Anthicidae
    Dermestidae
    Lathridiidae (fungus)
    Melandryidae (wood & fungus)
    Mycetophagidae (fungus)
    Scarabaeidae
    Tenebrionidae (also eats fungus) (14 species)[b]
  O. Diptera
    Callophoridae
    Cecidomyidae (fungus)
    Ceratopogonidae (adults are parasitic)

Appendix 13.2 *Continued*

    Muscidae
    Phoridae
    Sciaridae (fungus)
    Sphaeroceridae
    Stratiomyidae
    Syrphidae
    Tipulidae (adults eat nectar)
  O. Lepidoptera
    Geometridae
  O. Siphonoptera
    Pulicidae (adults are parasitic)

Herbivorous or non-feeding adults with predaceous (Pr) or Parasitic (Pa) juveniles
  O. Coleoptera
    Cantharidae (Pr)
    Cleridae (Pr)
    Meloidae (Pr) (21 species)[b]
    Rhipiphoridae (Pa)
  O. Diptera
    Apioceridae (Pr)
    Bombyliidae (Pa, Pr)[a]
    Phoridae (Pa, Pr)
    Sarcophagidae (Pa)
    Scenopinidae (Pr)
    Syrphidae (Pr)
    Tachinidae (Pa, Pr)[a]
    Therevidae (Pr)[a]
  O. Hymenoptera
    Bethylidae (Pa)
    Braconidae (Pa, Pr)[a]
    Chalcidoidea (Pa)[a]
    Eupelmidae (Pa)
    Ichneumonidae (Pa)[a]
    Mutillidae (Pa)[a]
    Platygasteridae (Pa)
    Pompilidae (Pa) (11 species)[b]
    Pteromalidae (Pa)
    Sceleonidae (Pa)
    Scoliidae (Pa)
    Sphecidae (Pa)[a]
    Tiphiidae (Pa) (5 species)[b]
    Torymidae (Pa)
  O. Strepsiptera
    Stylopidae (Pa)

Predators
  S.C. Acari (O. Mesostigmata)
    Thrombidiidae
  S.C. Acari (O. Prostigmata)[a]
    Anystoidea
    Bdelloidea
    Coeculidae
    Cheyletidae
    Erythraeoidea
    Eupodidae
    Raphignathoidea
  O. Araneae (19 families, 39 species)[b]
  O. Scorpionida (2 families, 4 species)
  O. Solpugida (2 families, 11 species)[b]
  O. Orthoptera
    Gryllacrididae (omnivorous)
    Mantidae (2 species)[b]
  O. Hemiptera
    Nabidae
    Phymatidae
    Reduviidae (3 species)[b]
  O. Coleoptera
    Carabidae (3 species)[b]
    Coccinellidae
    Histeridae
    Melyridae
    Pselaphidae
    Silphidae
    Staphylinidae
  O. Diptera
    Asilidae (5 species)[b]
    Mydidae
  O. Hymenoptera
    Formicidae (omnivorous)
    Vespidae (omnivorous)
  O. Neuroptera
    Chrysopidae
    Coniopterygidae
    Mantispidae (1 species)[b]
    Myrmeleontidae (6 species)[b]
    Raphidiidae (1 species)[b]

Contributors

Clifford S. Crawford, Professor
University of New Mexico
Department of Biology

Diana W. Freckman, Professor
Department of Nematology
Associate Director
Dry Lands Research Institute
University of California, Riverside

Richard S. Inouye, Assistant Professor
Idaho State University
Department of Biological Sciences

William P. MacKay, Assistant Professor
University of Texas
Department of Biological Sciences

Sandra L. Mitchell, Assistant Professor
Western Wyoming College
Department of Biology

Gary A. Polis, Associate Professor
Vanderbilt University
Department of Biology

O. J. Reichman, Program Director
National Science Foundaiton
Ecology Program

Mary K. Seely, Director
Desert Ecological Research Unit of Namibia

Laurie J. Vitt, Associate Professor
University of California, Los Angeles
Department of Biology
Now at Oklahoma Museum of Natural History and
Department of Zoology, University of Oklahoma

John A. Wiens, Professor
Colorado State University
Department of Zoology

Charles S. Wisdom, Assistant Professor
University of New Mexico
Department of Biology

Bruce D. Woodward
Western Wyoming College
Department of Biology

Tsunemi Yamashita
Vanderbilt University
Department of Biology

John C. Zak, Assistant Professor
Texas Tech University
Department of Biological Sciences

Index of Genera

Subject Index

About the Editor

Gary A. Polis is associate professor of biology at Vanderbilt University. He received his bachelor's degree from Loyola Marymount University in Los Angeles in philosophy and biology and his masters and doctorate from the University of California, Riverside. His interests include several topics in evolution and ecology including the evolution and dynamics of interference competition and of age-structured populations. He is also interested in the ecology of predation, desert ecology, and the biology of arthropods, particularly scorpions and spiders. His research has been conducted in the deserts of the southwest United States, the deserts and islands of Baja California, Mexico, and the deserts of Namibia and southern Australia. He has authored over 60 scientific publications and has edited the book *Biology of Scorpions* (Stanford University Press, 1990).